Dams and earthquake

Dams and earthquake

Proceedings of a conference held at the Institution of Civil Engineers, London, on 1–2 October 1980

Thomas Telford Limited, London, 1981

Organized by the Institution of Civil Engineers in association with the British Section of the International Commission on Large Dams and the Society for Earthquake and Civil Engineering Dynamics—the British Section of the International Association for Earthquake Engineering: also sponsored by UNESCO

Organizing Committee: Mr R. G. T. Lane (Chairman); Mr D. A. Howells; Professor J. K. T. L. Nash; Mr F. F. Poskitt

Cover photograph by courtesy of Sir Alexander Gibb and Partners, Consulting Engineers

Published for the Institution of Civil Engineers by Thomas Telford Limited, PO Box 101, 26–34 Old Street, London EC1P 1JH

First published 1981

ISBN: 0 7277 0123 1

Printed in Great Britain by Burlington Press (Cambridge) Ltd, Foxton, Cambridge

CONTENTS

Opening address

MARCUS FOX, MP (Parliamentary Under Secretary of State, Department of the Environment)

The consequences of failure of a large dam are many times greater than those of any other type of structure. Civil engineers have recognized this themselves, and I am most impressed by the world-wide co-operation and exchange of information that takes place through the International Commission on Large Dams.

This Conference, which brings together both seismologists and civil engineers, has been convened to consider just one aspect of dam safety - their ability to resist earthquake. Nevertheless, the subject matter is very wide starting with the study of seismicity, moving on to the behaviour of materials and structures when subjected to earthquake, and ending with a study of the environment and risks, including induced seismicity.

A study of an atlas indicating the localities of the most severe earthquakes reveals that the UK is relatively stable, although not completely stable because occasional tremors do occur and some occasional slight damage to dams is reported. There was an earthquake near Carlisle in December last year, and there have been recent tremors in the Stoke-on-Trent area. Scotland, too, has had tremors around the villages of Comrie and Menstrie. But the majority of problems arise in other parts of the world. If British consulting engineers are to compete with engineers from other countries in obtaining commissions to design and supervise the construction of large dams in those parts of the world which are physically less stable, they must ensure that they are familiar with the problems encountered and also practised in all the modern techniques available to combat the effects of earthquake.

We all know how devastating earthquake can be. The disaster in Tangshan in north-east China, which occurred on 27 July 1976, is reported to have killed no less than 650 000 people and the damage extended to at least 150km from the epicentre.

We think of severe earthquakes occurring primarily in known seismic zones, but I am told that these are virtually impossible to predict and earthquakes have occured outside those zones in areas considered to be geologically stable.

Another phenomenon is the fact that earthquakes have been observed following the impounding of water in large reservoirs, indicating that large reservoirs can induce their own earthquakes.

The engineer is faced with many unknown factors, and the object must be to achieve greater understanding of seismic activity, to improve designs to allow for these movements and to acquire greater knowledge of the characteristics of the materials used. All these are essentials for the improvement of the earthquake resistance of dams.

Engineers must always bear in mind the devastation that a community can suffer on the failure of a large dam. Their design decisions must always be of the highest order. I trust that this Conference helps them to achieve this high standard and it gives me great pleasure, therefore, that the Institution of Civil Engineers has organized the Conference in association with the British Section of the International Commission on Large Dams and the Society for Earthquake and Civil Engineering Dynamics, the British Section of the International Association for Earthquake Engineering and UNESCO.

I am most impressed that some 140 delegates are here from over 20 countries. I welcome you on behalf of Her Majesty's Government and our people.

1 Optimum seismic input in the design of large structures

R. E. LONG, BSc, PhD, FRAS, MBCS, University of Durham

From a consideration of the nature and characteristics of seismic waves, a model for testing design for earthquake failure is proposed. The model, referred to as an integrated model, considers an earthquake as a fault slip and places this within the geology. Direct calculation of energy entering the dam is made by modelling source, geology and dam as one complete unit. The problems of implementation of this model are discussed and its potential value estimated.

INTRODUCTION

1. This paper aims to place the design of dams (and particularly earth dams) in their true seismological context. The seismologist has a growing understanding of motions in the near field of a large earthquake. The dam designer, however, generally approximates these motions to give a simple model on which to test a design. It is not apparent that such simple models necessarily test a design with all the types of motion that could cause failure. Thus this paper seeks a closer fusion of seismology and design methods.

2. The aim is not to present a complete new design method but to investigate the possibility of a method where seismological data can form an integral part of design. Normally seismic motions are assessed on one model by the seismologists, and then transferred to another model for engineering design. In the ideal situation these two models should be combined. Thus we investigate an integrated design method where earthquake source, foundation geology and dam are considered as one overall single unit.

The Seismological Problem

3. The basic reason for proposing such an approach to design is that it allows full account to be taken of the nature and characteristics of seismic motions. The need for this becomes evident by considering such motions in general terms. Motions from an earthquake are generated by relative displacements of the sides of a fault. The waves propagated from such a moving source can be considered as a set of individual wavelets. The multiplicity of wavelets received at points in the near field is well demonstrated by considering simple models. Fig. 1 shows accelerations generated at the surface by a slip propagating over a finite area of a fault (1). The particular source parameters are here chosen to show the structure of the motion The several impulses of acceleration correspond to the start and stop of the slip, each of which generate an acceleration pulse in both P and S. This illustration shows in a simple way the waves generated by a moving source. To this simple model can be added plausible variants which can account for the wide range of near field motions found in practice.

4. These waves propagate through the geology from the fault source to the dam. That the geology modifies the motions is well known. What is less well known is that the transmission function varies with the angle of incidence of the input motion onto the geology. This problem was studied in detail by Haskell (2). In engineering terms the peaks of amplification of a layer occur at frequencies which are angle of incidence dependent.

5. The effects of a layered geological foundation can be expressed either as a frequency dependent transmission function or as a series of separate wavelets emerging at the surface for a particular input. This latter approach reveals the basic mechanism. A series of multiply reflected and refracted arrivals are generated by each of the input wavelets. These reflections and refractions will be accompanied by changes of wave type, P to S and S to P. Thus a simple P input gives rise to a mixture of P and S energy at the surface of a layered geology, and similarly for an S input.

6. The input waveform therefore becomes extended in time with each input wavelet transformed into a series of (interfering) P and S wavelets which are input to the dam itself. These several P and S arrivals interfere to give a complex particle motion which is rarely rectilinear, but more ellipsoidal. Thus the dam is subjected to motion which contains virtually all possible motion in a generally random sequence.

7. This refers specifically to the motions, but the direction of propagation is not random and is predominantly directed away from the source. The source is not static but extended over the slip zone of the faulting starting at

one point on the fault and propagating outward from that point. Thus the effective angle of incidence of this energy through the geology and into the dam varies with time during the earthquake. For a large, close event the change in angle can be considerable. The effect of the layered geology changes with angle and therefore effectively with time.

8. Thus we may summarise the incident motion as a time variant multiplicity of interfering wavelets generated by a moving source and travelling into the dam via a geology which acts as a time dependent filter. This pattern of input motion is of totally different character to motions applied in a quasi-static environment where the travelling nature of the wave and the effective time dependence of the geological filter is ignored. The result is to widen the total range of input motions. An integrated model with the source modelled in the geology automatically includes a solution of the attenuation problem and the problem of assessing motion from specific events.

The inclusion of a dam

9. Into this environment we may now place a dam. A structure whose dimensions are much smaller than a wavelength would tend to move as a unit with acceleration and deceleration forces providing the major hazard. For such a case the application of a specific motion to the foundations or to the whole structure would be appropriate.

10. Such an approach is not applicable to a large dam whose dimensions are of the same order as the wavelength of the seismic energy input. For such large structures the waves are propagated into the dam so as to form a pattern of motion over the dam. This pattern is the result of the interference of waves from the multiplicity of ray paths from source through the intervening geology to the structure.

11. The surface motions resulting from a plane wave incident on an earth dam can be calculated by considering the earth dam as a simple hill in a homogeneous geology. Fig. 2, based on Bouchon (3), shows how surface motions vary with position on the surface and angle of incidence. These studies explaining the amplification of motion by topography also show in essence the problem of inputting seismic energy into an earth dam. The complexity of motions arise from a scattering of energy by the surface discontinuity formed by the dam. The implication for dam design is that motion and the implied stress distribution are assymetric and clearly very different from results obtained from applying non-travelling waves to the structures. In the real environment several such input waves, each with a different angle of approach and therefore different stress patterns within the dam, would interfere to form a continuously changing pattern of stress with time.

12. This approach leads to a number of conclusions. Firstly the type of energy incident is not restricted to any particular wave type. Secondly the response to a particular wave type cannot be considered separately from any other wave type. The word type here not only refers to P or S but also the several separate arrivals of P and S. Thirdly, since the several wave types have differing velocities and differing angles of approach into the structure, the composite motion cannot be considered as a single input time history over any one surface that might be drawn in the structure. This implies that application of a single time history, say to the base of a model, is not realistic, nor is it a realistic approximation.

The model for design

13. These observations suggest that if input motions are to be totally realistic the test model should not only include the dam and foundations, but also the fault with earthquake placed in the foundation geology. Checks for failure would then be made by allowing an appropriate slip on the fault and observing resultant stress conditions. Such a design method is clearly capable of providing a close approximation to the real problem.

14. The implication of this method however presents a number of problems. Firstly the elastic properties of the geology must be known sufficiently accurately with accurate location of faults at depth. Secondly the source parameter (area of slip, distribution of the amplitude of slip motion over the slipped region, etc.) need to be known or assessed. Thirdly the mathematical technique required to calculate the motion generated by a fault slipping in a layered geology have yet to be developed. Fourthly the integrated method would normally be implemented using finite element or different techniques, when the size of the model in comparison with the seismic wavelength would require a large number of nodes with appropriate computer problems of handling large arrays.

15. Consider these problems in order: The geology is normally investigated by a borehole. These are of limited depth and therefore give data of only the upper few metres normally restricted to the site itself. To gain accurate knowledge of deeper layers would require deeper boreholes than normal and then these would not show up lateral variation in sufficient detail. Such investigations would be required over an area much larger than the site. The solution in an earthquake environment is to measure the transmission function from fault to dam directly. Small earthquakes occurring on the fault can be recorded at site. These records can be decomposed to remove the source function when the transmission function in the time domain remains. This analysis can be done for a variety of events distributed over the fault to give the changing transmission function with position. A generalised function for a station in Iran is shown in fig. 3. This

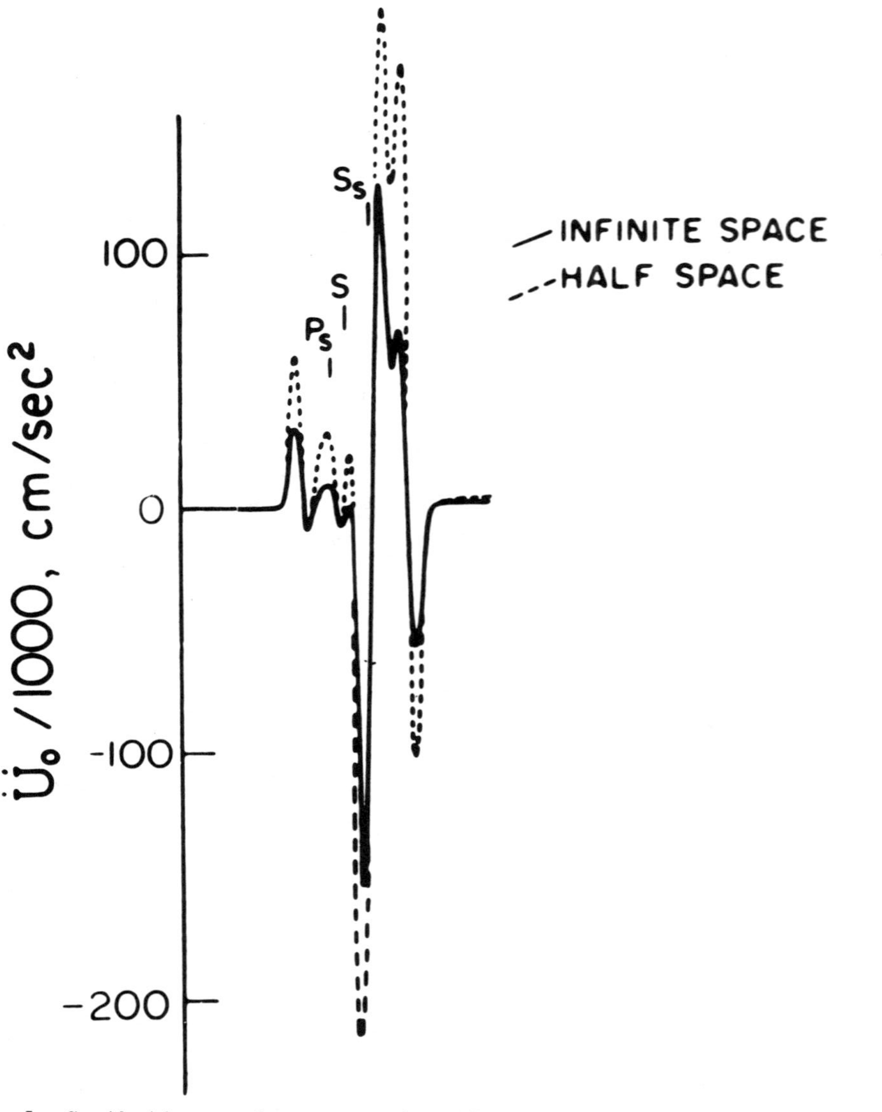

Fig. 1 Synthetic accelerograms show the complexity of motions in the near field of a simple earthquake model. After Israel and Kovack (1). P_s, S_s are the stopping phases.

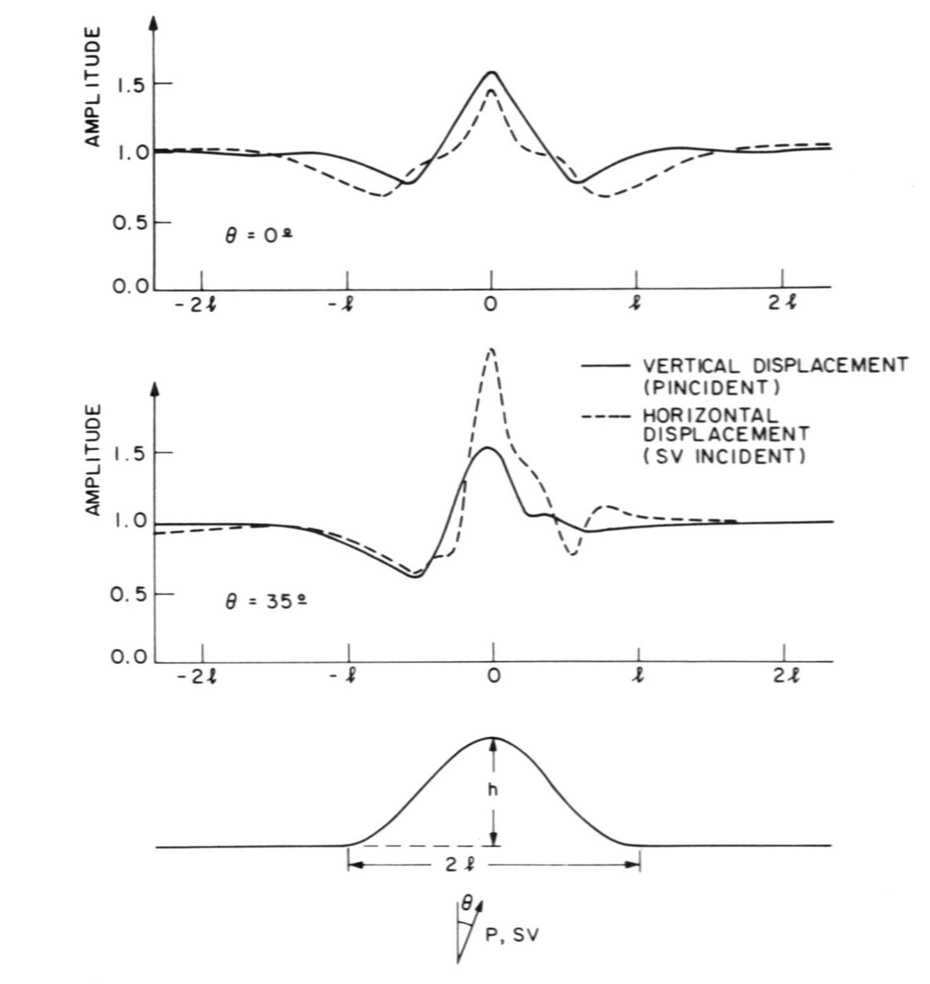

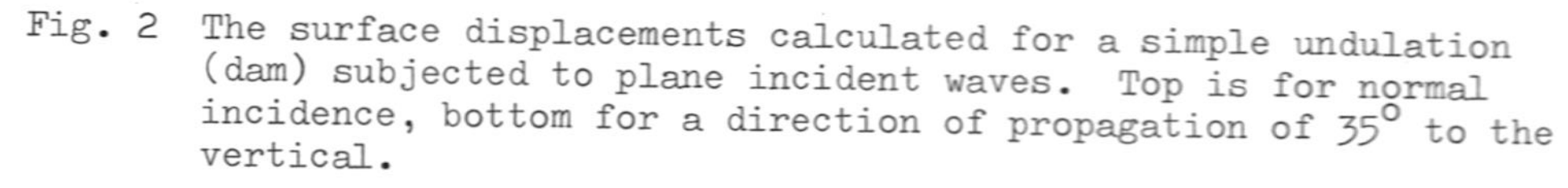

Fig. 2 The surface displacements calculated for a simple undulation (dam) subjected to plane incident waves. Top is for normal incidence, bottom for a direction of propagation of 35° to the vertical.

transmission function was obtained using homomorphic deconvolution (5) to remove the source function following an averaging of the spectra of a number of local earthquakes. The averaging process is appropriate when several events are available. It has the advantage of providing an effective impulsive source for the averaged case which has a flatter frequency spectrum than individual events and therefore gives an overall wider frequency range for the transmission function.

16. Such methods provide data for transferring motion from source to dam but do not take into account the non-linearity of the geological response to strong motion input; they can normally only be measured for small events and therefore small motions. Foundation rocks however, do not show the severe non-linearity of soils. Thus provided soils are not included the measured functions are good approximations. Alternatively a model satisfying the low amplitude transmission function can be generated and the non-linearity modelled on what should then be a firmer overall model. In essence such an approach would follow traditional seismic techniques, when measurements give a model upon which further calculation can be based. This site investigation technique can produce a model probably sufficient for the requirements of the integrated design method.

17. The position of faults at depth can be determined in an active area by location of medium size events ($M_L \geqslant 2.5$). Fig. 4 shows the results of such a survey in Iran. The several faults located can be associated with faults recognised in the surface geology. The importance of this type of data in such a tectonic region as Iran is that it places faults at depth beneath the site somewhat closer than might be considered on geological evidence alone.

18. The second problem, the range of source parameters, is a major difficulty. Whilst models have been developed for fault slip to explain observations from certain large events (6) the full range of possible parameters can at this stage only be guessed at. Some limitations can be placed from independent considerations such as a consideration of the strengths of rocks, but the range of combinations remains large. However, certain events can be modelled reasonably well to give a realistic input to the integrated model.

19. Both the source and geological section of the integrated model are derived from available data. Clearly these models may be oversimple but it is in the generation of such models that research effort needs to be applied. From the data of fig. 3 for example, the geology can be considered as a simple two-layered model. Thus the geological model becomes very simple and easy to include in the overall system. More detailed study of the transmission function could generate a more detailed model. Here the extent of suffistication needs to be assessed in terms of effectiveness.

20. In passing it is useful to realise that use of a model for the geology allows the signal ground surface to be removed for the inclusion of the dam into the integrated model.

21. We are currently investigating this overall integrated model in an attempt to identify and solve some of these problems. Sufficient data can be available and the problem basically reduces to the generation of models from the data and the mathematics for modelling the complete system.

22. The use of finite element methods allow the modelling of a fault slip within a layered geology and so this approach promises to overcome the third problem. The fourth problem remains, but is probably not so severe as might at first be considered.

Value

23. The integrated model approach is clearly a useful tool to give a full and considered picture of motions and stresses to be expected within a dam. It remains to assess its effectiveness. Is its full use really necessary? It is of course likely that a simplified approach can be adopted, but before its adoption it needs to be established that any such simplification retain those aspects of motion likely to cause failure. In this respect simple models have been tested to establish those motions likely to cause failure in earth dams. Two conclusions have been reached.

a) The well known failure by gravity sliding has been discussed by Moore and Long (7), where the worst case motion is at an angle to the sloping surface. Fig. 6 summarises the problem. The implication is that the component of motion along the direction (a) is the critical factor.

b) A second observation, the full details of which will be published elsewhere, is the formation of a tensional stress normal to the crest by vertical motions propagated into the dam, e.g. vertically incident P-waves.

24. Both these observations can be proved analytically from simple wave theory using a model whose plane waves are incident at various angles on the base of the structure.

25. In principle, observations of this type allow possible failure types to be identified with specific incident motions. Thus probability of failure can be assessed from the probability that such motions are incident at the correct amplitude into the dam. The latter probability however, still requires the use of the integrated model considered here for a realistic appraisal of the problem.

26. These conditions and their association with non-stereotyped simple motions indicates that the integrated model may yield significantly different failure probability to other design methods. Hence the work currently in progress at Durham.

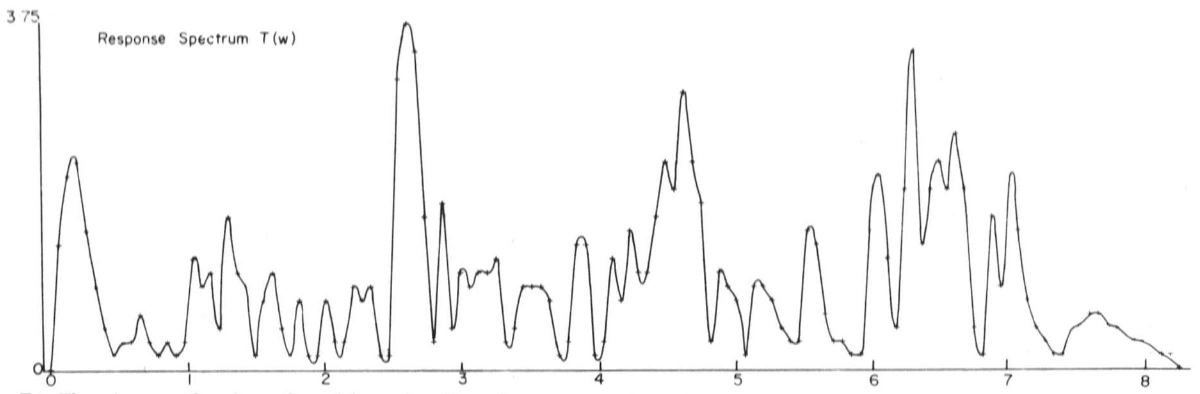

Fig. 3 The transmission function in the frequency domain calculated from site recordings. Note the major peaks characteristic of reverberation of a single layer. The smaller peaks indicate further complexity of this simple model.

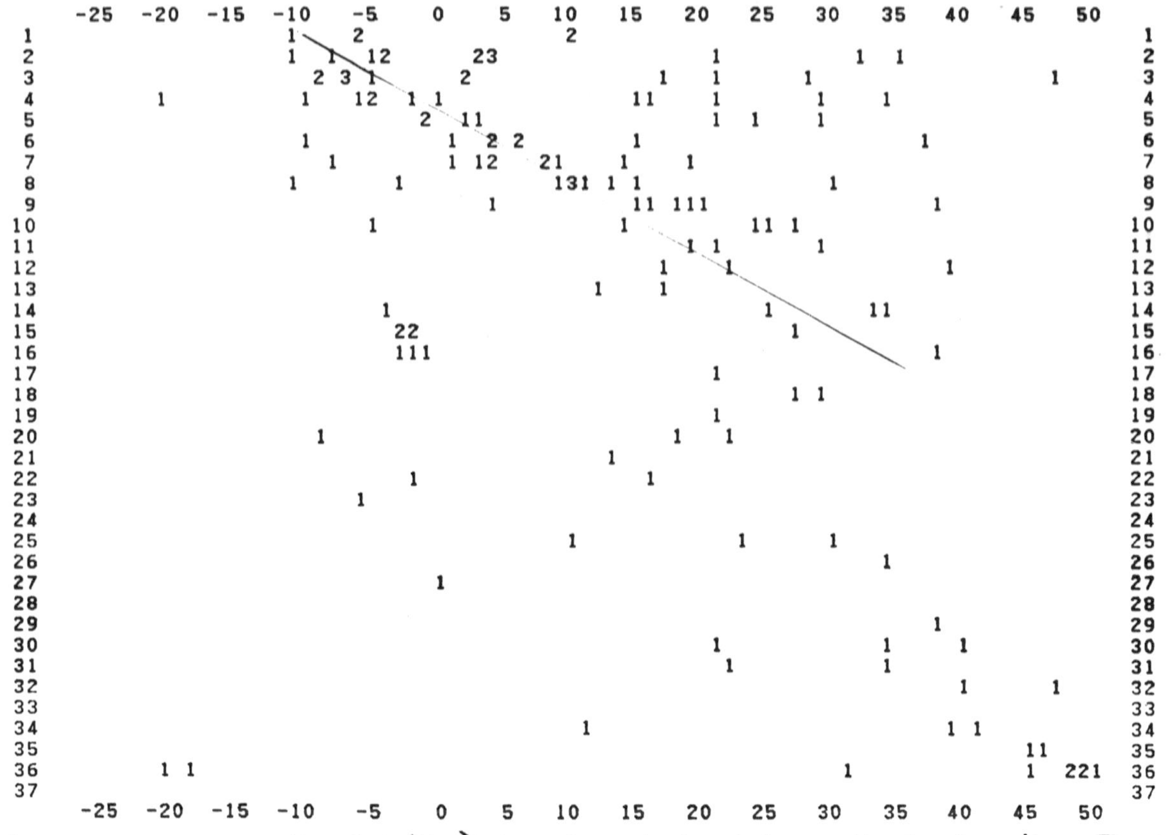

Fig. 4 An earthquake density plot ($M_L \geqslant 2.5$ of events located near the Lar dam site. The section is drawn normal to the fault strike and data integrated over about 20 km either side of the section. The axes are in units of kms and the densities per sq. km. of the section

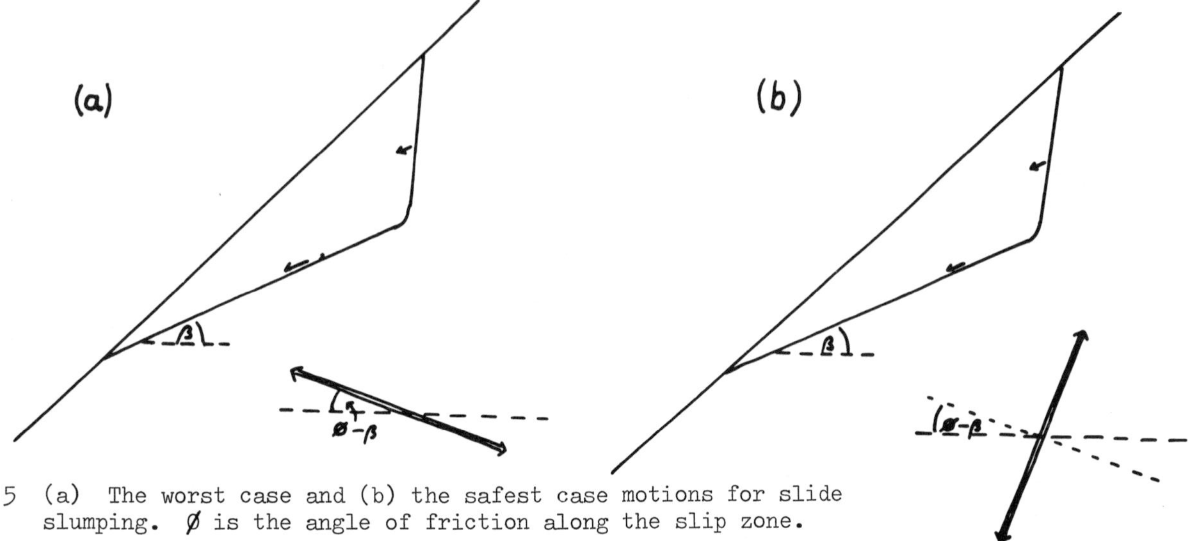

Fig. 5 (a) The worst case and (b) the safest case motions for slide slumping. ø is the angle of friction along the slip zone.

REFERENCES

1. ISRAEL, M., KOVACK, R. Near field motions from a propagating strike slip fault in an elastic half space. Bull. Seism. Soc. Am., 1977, 67, 977-994.

2. HASKELL, N.A. Crustal reflection of plane SH waves. J. Geophys. Res., 1960, 65, 4147-4150.

HASKELL, N.A. Crustal reflection of plane P and SV waves. J. Geophys. Res., 1962, 67, 4751-4769.

3. BOUCHON, M. Effect of Topography on surface motion. Bull. Seism. Soc. Am., 1973, 615-632.

5. OPPENHEIM, A.V., SCHAFER, R.W., STOCKHAM, T.G. Non linear filtering of multiplied and convolved signals. Proc. IEEE 1968, 65, 1264-1291.

ULRYCK, T.J. Applications of homomorphic deconvolution to seismology. Geophysics, 1971, 36, 650-660.

6. TRIFUNAC, M.D. A Three-dimensional dislocation model for the San Fernando, California, Earthquake of February 9th, 1971. Bull. Seism. Soc. Am., 64, 149-172.

7. MOORE, K.R., LONG, R.E. Seismic Evaluation of Dam Sites. Proc. 3rd International Congress on Engineering Geology, 1978.

2 The definition of seismic risk at dam sites

R. CHAPLOW, BSc PhD MIGeol FGS, Sir Alexander Gibb & Partners, Reading

A general methodology for defining seismic risk at dam sites is described together with a particular example of its application to the Lar Dam Site in northern Iran. Particular areas of uncertainty in carrying out such risk studies are identified and examples of the manner in which the uncertainties have been resolved are discussed.

INTRODUCTION

1. The complete definition of seismic risk at a proposed dam site requires the provision, to the dam designer, of a wide range of quantitative and qualitative information. The quantitative data include estimates of relationships between various measures of ground motion, their probabilities of exceedence together with recommendations of appropriate mean return periods to be used for design. The qualitative data include estimates of secondary and tertiary risks associated with seismic shaking. Secondary risks include consideration of such aspects as earthquake-induced landslides and soil liquefaction whilst tertiary risks are the hazards associated with seismically-induced failure of part or all of the various proposed engineering structures. Such hazards include the likely loss of life, property and facility (power generation and/or potable or irrigation water supply) resulting from partial or complete failure of structures together with loss of usage for repair of damage.

2. Experience of carrying out seismic risk studies for a variety of sites has led to the development of a general methodology for carrying out such studies. This methodology, which is considered to be generally applicable to all dam sites, is described together with an example of its recent application. Areas of uncertainty within such seismic risk studies are identified and discussed.

METHODOLOGY

3. The methodology for carrying out seismic risk studies (Fig. 1) is divided up into four main stages:

Conception

4. Immediately a dam site study commences it is necessary to define whether seismic loading of the structures must be incorporated into the design. The usual basis for this initial assessment is The Atlas of Seismic Activity (ref. 1). The absence of any record of an earthquake within 400 km of the proposed site is regarded as sufficient justification for regarding it as aseismic. Such a criterion is considered to be conservative, especially with regard to virtually any well-built embankment dam on firm foundations; such dams having been shown to be capable of safely withstanding moderate shaking with no detrimental effects (ref. 2). There is however, an increasing tendency for Project Review Panels appointed by funding agencies to require systematic seismic risk studies to be carried out even for apparently aseismic sites.

Data collection

5. The scope of this stage depends upon the status of the investigation (whether feasibility or project design), upon the scale of the proposed structures and upon a subjective assessment of likely seismic risk. Data collection commences with enquiring if an appropriate earthquake code exists. A useful list of existing codes and an assessment of their quality is provided by Dowrick (ref. 3). Lists of teleseismic events recorded around the site are obtained from either the International Seismological Centre, Newbury or the Institute of Geological Sciences, Edinburgh. Seismicity is also researched as part of the literature review carried out at an early stage of all projects utilising sources of information both in this country and in the project country. The latter sources are particularly important as local knowledge may often help in supplementing the otherwise very incomplete teleseismic records.

6. All potential dam sites are visited by an experienced member of the engineering geological section. This site reconnaissance serves to define the main geotechnical conditions and to permit the planning of site investigations. Geological mapping and the implementation of geotechnical site investigations proceed routinely as part of the preliminary works at the dam site and, in seismic areas, these are arranged to pay particular attention to such aspects as the location of active faults and to identifying those aspects of ground conditions which may be affected by, or themselves affect, seismic shaking.

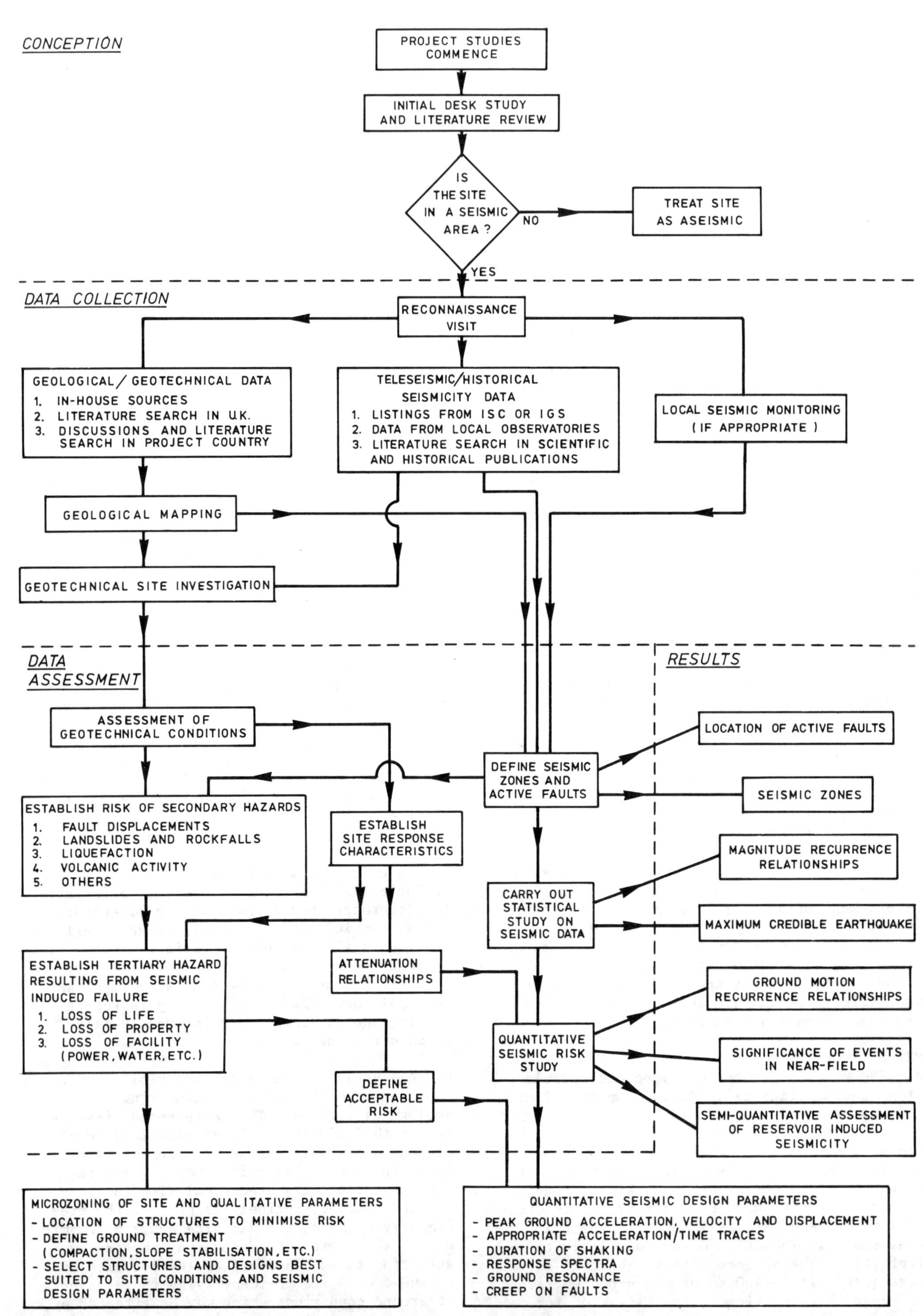

Fig.1. Methodology for defining seismic risk at dam sites.

7. The information available from teleseismic and historical records on the frequency and distribution of earthquakes around a dam site is often very incomplete. The only feasible way in which such information can be supplemented within the time scale of a dam site investigation, currently in the order of only one or two years or even less, is to install and operate a local network of seismic monitoring stations. By locating small magnitude events, active faults may be recognised and the understanding of the seismotectonics of the dam environs advanced. Such an approach is only occasionally considered justified and significant judgement is necessary to interpret the information from such local networks in relation to possible spatial and temporal changes in earthquake activity relative to the often short operating periods of the network. The provision of local monitoring facilities is of particular value in cases where a number of successive dam developments in the same area are envisaged or when reservoir induced seismicity is anticipated and it is desired to monitor such effects.

8. Information on ground response characteristics during earthquakes has been found to be very limited and often virtually non-existent at many of the dam sites studied. Recourse is generally made to using empirical attenuation relationships. A large number of these exist and can yield widely differing estimates of ground motion likely to result from a particular earthquake. Idriss (ref. 4) discusses many of the available relationships, although guidance on the selection of the appropriate relationships for use at a particular site is notably lacking in the published literature.

Data assessment

9. This third stage in defining seismic risk includes both the establishment of the seismic forces likely to act upon the site and the manner in which the site will respond to them. The assessment of both aspects is invariably based upon incomplete data and it is imperative at this stage in the study to carefully examine the sensitivity of the conclusions on seismic risk to the choice of input parameters, particularly in relation to the range of possible interpretations of geotechnical conditions, ground response characteristics and earthquake frequency and distributions. Computer based studies are used extensively in this area as they permit a rapid examination of the sensitivity of the results to a wide range of variations in input parameters.

10. Magnitude recurrence relationships based upon incomplete and inhomogeneous teleseismic records (ref. 5) and the empirical attenuation relationships are regarded as two input parameters subject to significant uncertainty and it is always considered essential to check risk assessments for a range of these parameters.

11. The uncertainties in input parameters and the need for sensitivity studies ensure that ultimately a high level of engineering judgement must be incorporated into all risk studies. Such risk studies are therefore regarded as a means of comparing risks between different sites rather than being a deterministic method of establishing absolute risk.

12. Judgement is also required in assessing the risk of secondary hazards such as soil liquefaction and landslides associated with seismic shaking. However, a considerable reservoir of knowledge of soil and rock mechanics is available to assist in such assessments of risk and of appropriate courses of action to lessen the hazard, be it by ground treatment or relocation of sensitive structures. It is, however, in the assessment of the tertiary risks that judgement is particularly required. It is a relatively straightforward process to estimate the effects, both human and financial, of the sudden failure of a dam or of part or all of its associated structures. However, to decide what is an acceptable level of risk against such a failure is a highly subjective matter. Certain structures (ref. 6) have been designed with a finite probability of partial failure during construction following a comparison of costs involved and the maximum extent of possible failure deemed acceptable from all considerations. Such a criterion is not considered appropriate for a major dam subjected to seismic shaking although it could well be appropriate for certain associated structures. The criteria adopted for dams are generally a 'no-damage' criterion for a particular seismic shaking and corresponding mean return period, and a 'no-failure' criterion for the maximum credible shaking.

Results

13. The results obtained from the risk studies are shown in Fig. 1.

APPLICATION OF METHODOLOGY TO LAR DAM, IRAN

14. The Lar Dam and Tunnels Project is located in the Alborz Mountains to the north east of Tehran, Iran and is currently nearing completion. The scheme includes a 105 m high embankment dam and associated works (ref. 7) and lies within an active seismic area (the Sangechal (1957) earthquake caused damage which extended to within approximately 15 km of the dam site).

15. The dam site is geologically complex. Mesozoic limestones containing local faulted outliers of Oligocene tuffs form the right abutment whilst on the left bank of the river the limestones are overlain by a sequence of Quaternary lacustrine and alluvial sediments. These reach a thickness of some 600 m and are separated into two parts by a horizon of trachyandesite lava flows erupted from the now dormant volcano Mount Damavand. The geotechnical investigations were carried out at the site during 1969 - 72 and included some 12 000 m of drilling, geological mapping and an extensive programme of laboratory and field testing, later supplemented by the installation of two local seismic monitoring stations.

16. The methodology shown in Fig. 1 was adopted for the definition of seismic risk and some of the more interesting and significant conclusions of the study are outlined briefly below.

Ground motion recurrence relationships

17. Teleseismic records collected from the Institute of Geological Sciences, Edinburgh and from the Geological Survey of Iran were used in conjunction with the records from the two local seismic stations installed at Lar and Lavarak to establish magnitude recurrence relationships based upon both cumulative frequency and extreme value relationships. Evidence of temporal variations in seismic activity (ref. 8) led to the production of relationships based upon both mean and peak levels of activity. Ambraseys (ref. 9) had suggested that attenuation relationships derived from western United States data underestimated ground motions in the European area (including Iran) especially in the near-field. The full analysis of seismic risk has therefore involved the use of all the various derived magnitude recurrence relationships in association with attenuation relationships from both the United States data and Ambrasey's relationships for European near-field events. Comparison of the resulting envelope of ground motion recurrence relationships with the reported occurrence of historic destructive earthquakes in northern Iran has indicated close agreement. The derived ground motion recurrence relationships were therefore regarded with a high degree of confidence.

Active faults

18. Detailed geological mapping of the Project Area revealed the presence of numerous faults which, based on geological evidence, appeared likely to be potentially active. Two particularly significant ones were recognised. The first was seen to displace the Quaternary sediments in the Lar Valley and to pass beneath the dam site. The local seismic monitoring has subsequently indicated that a number of very small seismic events could be assigned to this fault, but that the fault was not a fundamental fault in the area. These observations have had a marked influence on the design of the dam since no alternative site was possible which avoided the dam crossing the fault.

19. The second major active fault in the Project Area is the Mingun Musha Thrust which is located such that it was unavoidable that the high pressure penstock tunnel should cross it at the downstream end of the Lar-Kalan diversion tunnel. Geological mapping and the local seismic study have confirmed this fault to be a still active, fundamental structural fault in the southern Alborz. The engineering approach which has been adopted towards this area includes the adoption of a free-standing penstock with suitable expansion joints, located within a 5 m diameter, concrete lined tunnel where the penstock crosses the fault. The power station located below is also protected by an emergency butterfly valve at the base of the surge shaft. This valve is designed to close automatically in the event of rupture of the penstock.

20. The precise geological mapping of potentially active faults around Lar has been made possible by the sparse vegetation cover and excellent rock exposure. Similar geological mapping is not, however, possible at all dam sites. For example, the very thick vegetation cover at the Monasavu Dam site in Fiji has rendered geological mapping totally ineffective in locating potentially active faults. Here and elsewhere, greater reliance has needed to be placed on locating faults using other techniques, particularly by drilling and by the use of exploratory excavations.

Liquefaction risk

21. The upper portion of the Quaternary sediments on the left bank of the Lar Dam site were found to contain layers of loose sand which were subsequently shown by laboratory testing to be susceptible to liquefaction under seismic vibrations. Extensive surface and sub-surface geological studies (ref. 10) aided by the recognition of bubble structures in potentially susceptible sands (ref. 11) led to the precise definition of the extent of the problematical soils. Following a study of possible foundation improvement methods, it was decided that extensive excavation of the sediments would be required to allow the dam to be founded on lava or sediments of adequate density and strength. In the latter case the design was modified to allow vertical dropping of the dam due to foundation movement.

Rockfalls and landslides

22. Fookes and Knill (ref. 12) have described a major landslide in the Haraz valley close to Lar Dam which developed as a result of the 1957 Sangechal earthquake. Careful attention was therefore directed, during the site investigations for Lar Dam, towards identifying any unstable or potentially unstable slopes which, if subjected to seismic shaking, could move and present a hazard to the engineering structures.

23. A large landslide was recognised on the right abutment at the dam site which required complete removal to found the dam on limestone. The limestone slopes above and around the various structures on the right bank, which included the spillway, diversion tunnel and irrigation tunnel intakes, were inspected in detail and a hazard map produced based upon the assessed risk from rockfalls. This hazard map was then used as a basis for locating the various structures and designing the extensive trimming and stabilisation of the rock slopes above the dam.

Construction materials

24. The silts which occur in the Lar Valley were found to be the only feasible source of fine grained materials for the dam core. The soils did, however, have a low plasticity and were found to be wet of optimum so that a core made from them would have been liable to excessive consolidation and hence prone to

cracking, both undesirable characteristics, particularly in a seismic area. In consequence, to reduce the risk of cracking, to increase the shear strength of the core and to provide filter-compatibility between the various zones in the dam, both the core and transition zones were designed as artificial blends of the local silt and the sandy gravels, which were also available in abundance close to the dam site.

Dam cross section

25. The principal criterion in the design of the Lar Dam embankment is to ensure that a major earthquake does not cause failure, either by loss of the crest below reservoir level or by piping through cracks which may develop in the core as a result of seismic or tectonic displacement. Marwick and Germond (ref. 7) have described the dam cross-section which incorporated a number of features to increase its resistance to seismic shaking. These measures, similar to the defensive measures described by Seed (ref. 2) include the provision of ample freeboard, a downstream drainage blanket and chimney drain capable of safely passing seepage through cracks which may develop in the core. A cambered side slope of the dam is provided for greater seismic resistance at the top of the dam to match the anticipated magnification of seismic movements towards the crest. The mixed foundations beneath the dam, varying from right to left from limestone bedrock to silts and clays to lava and back to sediments, results in potential problems associated with differential settlements and differential seismic response of the foundations. The seepage control works at these various interfaces are designed to take account of the differential behaviour of the various materials. It has, however, unfortunately not been possible to monitor precisely the variations in seismic response of the various parts of the dam foundation.

Induced seismicity

26. The local seismic studies in the Lar area have indicated a concentration of minor seismic events centred close to the Latiyan reservoir which was first impounded in 1968. This concentration of activity located close to a reservoir suggests the possibility of the local seismicity being reservoir induced. The activity appears, from the local seismic study, to be associated with a steeply southward dipping fault, similar in many respects to the fault known to exist below the Lar Dam. Strong motion accelerometers are installed at Lar Dam to monitor any seismic shaking during and after impounding and it is hoped that it will be possible to reactivate the local seismic stations to provide additional monitoring facilities during impounding.

AREAS OF UNCERTAINTY

27. The methodology described above for defining seismic risk at a dam site is based upon a somewhat imprecise science. The uncertainties introduced by incomplete data and possibly inappropriate attenuation relationships, the consequent necessity of carrying out sensitivity studies and of applying significant amounts of subjective engineering judgement to the results obtained have all been briefly discussed in paragraphs 9 to 12. There are, however, a further set of uncertainties involved in defining seismic risk, related particularly to the methods adopted for the study. It is apparent that various authorities throughout the world have approaches to defining seismic risk which are different to those described in this paper. Some of these produce estimates of seismic risk similar to those obtained by the method already outlined above whereas others clearly do not produce the same results. Some of the particular differences in methodology which have been noted recently are outlined below.

Intensity versus ground motions

28. The Author's methodology for estimating ground motions at a dam site involves, first, the definition of the spatial and frequency distribution of earthquakes of differing magnitudes around the site. These distributions are then converted into recurrence relationships of ground motion, typically peak acceleration and velocity, by the application of empirical attenuation relationships suitably validated whenever possible for the particular site.

29. An alternative approach which has been adopted by others is to derive an Intensity recurrence relationship for the site under study, generally using the Modified Mercalli Intensity Scale, and then to derive ground acceleration and velocity as a function of Intensity. Ambraseys (ref. 13) has studied the correlation of intensity with ground motions for European earthquakes, has demonstrated that the correlations are weak, and recommends that the various empirical relationships he describes should not be used for design purposes. He adds that these empirical relationships are of an index nature and that they should only be considered in conjunction with the risk implicit in the use of design data associated with standard deviations of 60% to 70% of the mean.

30. However, although the correlations between intensity and ground motion appear weak, it has been frequently found, for example in Fiji, that whereas actual measurements of ground motions are almost totally non-existent, there are often numerous records of earthquake intensities either in published documents or capable of derivation from folk-lore or historical records.

31. Thus, although intermediate considerations of earthquake intensity have not been directly adopted in any of the seismic risk studies carried out by the Author, intensity records have been used as a general guide to estimating similarities of ground response between the site and areas where more systematic strong motion data are available.

Data smoothing

32. Regression analyses are used to obtain essentially smoothed relationships from partial, scattered data. The method adopted for the seismic risk studies carried out by the Author has been to subdivide the environs of the site

into a number of seismotectonic zones within which the earthquake activity can be regarded as having a particular set of characteristics in terms of recurrence and focal depths. Smoothed magnitude recurrence relationships are obtained for each zone and used as the basis for the subsequent risk assessment which requires the input of the appropriate attenuation relationships.

33. An alternative approach adopted by others is to apply the appropriate attenuation relationship to each earthquake which has been reported, to compute the anticipated ground motions at the site resulting from each event, and to then use regression analyses to obtain ground motion recurrence relationships. The reasons for the Author's preference for the former method are twofold. Firstly, the former method appears to provide a more conservative estimate of seismic risk and secondly, the adoption of a magnitude recurrence relationship as the basis of the risk study permits ready comparison of the results with the large number of corresponding published relationships from other parts of the world.

Selection of design return period

34. Having obtained from the seismic risk study an envelope of ground motion recurrence relationships for the dam site, the most difficult task is invariably the selection of appropriate ground motion mean return periods for design. There are a number of important factors considered relevant to this selection, namely:

(a) Seismic activity typically tends to occur as cycles of high activity followed by periods of relative quiescence. The length of these cycles is typically in the range of 40 to 400 years and it is probably inappropriate to consider mean return periods for design which are significantly longer than the length of these cycles.

(b) The empirical attenuation relationships are 'best-fit' lines about which the data points from which they are derived are generally reported to be lognormally distributed. The application of uncertainty factors (ref. 14) can be used to compensate for the scatter of data points about the mean line.

(c) Near-field ground motions may well be underestimated by many empirical attenuation relationships.

(d) Virtually every stage of the seismic risk study is based upon partial data and involves the input of often significant amounts of subjective engineering judgement. Such judgement tends to err on the side of conservatism. The repeated compounding of conservative judgements may yield a totally unrealistic final estimate of seismic risk.

35. The approach which has been adopted by the Author for a number of dam sites recently has been to initially recommend as a basis for design of the main dam the following ground motion mean return periods:

Design Criterion	Basis for Selecting Ground Motion
No damage	Mean return period of 100 years at 84 percentile or 500 years at 50 percentile cumulative probability level, but in any case a peak ground acceleration $\nless$ 0.1 g and generally $\leqslant$ 0.4 g.
No failure	Maximum credible shaking.

36. The above mean return periods are invariably subject to some modification as design proceeds but they are considered to form a reasonable basis for a first approximation.

CONCLUSIONS

37. The described methodology for defining seismic risk at a dam site has been applied at a number of dam sites in a variety of geological and tectonic environments. However, all dam sites are unique and the available procedures for defining seismic risk are, at best, imprecise. The engineer and geologist must therefore continue to make full use of all the available information and techniques but must always remain highly conscious of the uncertainties involved and of the absolute necessity of incorporating into any risk study a high degree of sound engineering judgement.

ACKNOWLEDGEMENTS

38. The Author acknowledges with thanks the permission, granted by Sir Alexander Gibb & Partners, to publish this paper. The local seismic study at Lar was undertaken in association with Dr R.E. Long of Durham University.

REFERENCES

1. CRAMPIN S., FYFE C.J., BICKMORE D.P. and LINTON R.H.W. Atlas of seismic activity: 1909 - 1968. Seismological Bulletin Institute of Geological Sciences, No. 5, 1976, 29.

2. SEED H.B. Considerations in the earthquake - resistant design of earth and rockfill dams. Geotechnique 29, No. 3, 1979, 215 - 263.

3. DOWRICK D.J. Earthquake resistant design. Wiley, London, 1977.

4. IDRISS I.M. Characteristics of Earthquake Ground Motions. Proceedings of the Speciality Conference on Earthquake Engineering and Soil Dynamics, ASCE. Pasadena, California, June 1978.

5. BURTON P.W. The IGS file of seismic activity and its use in hazard assessment. Seismological Bulletin Institute of Geological Sciences, No. 6, 1978, 13.

6. KNIGHT D.J. and BRICE G.J. The measurement, trial use and selection of initial design parameters for dikes on very soft clay in the Dead Sea, Jordan. Proceedings 7th European Conference on Soil Mechanics and Foundation Engineering, Vol. 3, 1979, 93 - 100.

7. MARWICK R. and GERMOND J.P. The river Lar multipurpose project in Iran. Water Power and Dam Construction, Vol. 27, 1975. Part 1 :

No. 4, April, 133 - 141. Part 2 : No. 5, May, 178 - 183.
8. LONG R.E. Seismicity investigations at dam sites. Engineering Geology, Vol. 8, 1974, 199 - 212.
9. AMBRASEYS N.N. Preliminary Analysis of European Strong-Motion Data, 1965 - 1978. Part II of the Report prepared by the EAEE Working Group on Strong Motion Studies. Bulletin of EAEE, Vol. 4, 1978, 17 - 37.
10. CHAPLOW R. The engineering geology of Lake Lar, Northern Iran. PhD thesis, University of London, 1976, 352.
11. CHAPLOW R. The significance of bubble structures in borehole samples of fine sand. Geotechnique, Vol. 24, No. 3, 1974, 333 - 344.
12. FOOKES P.G. and KNILL J.L. The application of engineering geology in the regional development of northern and central Iran. Engineering Geology, Vol. 3, 1969, 81 - 120.
13. AMBRASEYS N.N. The correlation of intensity with ground motions. Proceedings 14th Assembly European Seismological Commission. Trieste, 1975, 335 - 341.
14. DONOVAN N.C. and BORNSTEIN A.E. Uncertainties in seismic risk procedures. Journal Geotechnical Engineering Division American Society Civil Engineers, 104, GT7, July 1978, 869 - 887.

3 Substitute short duration earthquake accelerograms for nonlinear analysis

O. C. ZIENKIEWICZ, FRS, University College, Swansea, N. BIĆANIĆ, PhD, and R. FEJZO, Dipl Ing, Gradevinski Institut, Zagreb

A structure dependent short duration combisweep accelerogram is presented. It is based on the Johnson/Epstein sine sweep accelerogram and free parameters are obtained by minimizing the distance between its response spectrum and the design response spectrum -or the response spectrum of the design base motion history- at descrete points which coincide with natural frequencies of the structure. Some limited experience based on comparative analyses of the Koyna dam section is also presented.

INTRODUCTION

1. The analysis of the effects of earthquake excitation has by now become a standard requirement for dams built in known seismic areas. When only the linear response of the structure is of interest, such an analysis can be performed using either the design response spectra (graphs relating relative spectral displacements velocities and accelerations to the period -or frequency- of a simple oscilator subjected to the base acceleration history) or design base acceleration (or displacement) histories (refs. 1-3). Although requiring the determination of the natural frequencies and mode shapes of the structure, design response spectra are, after modification to allow for specific local site conditions, more frequently used.

2. In nonlinear analysis, where superposition of modal effects (the basis for the use of response spectra) is not applicable, time history analysis using a design base motion history has to be carried out (refs. 2,3). However, nonlinear analysis computational costs are usually very high and in direct proportion to the excitation duration.

3. In the final design stages a chosen or generated (refs. 1-4) design base motion history, say an accelerogram, has to be used regardless of its duration. Nevertheless, in preliminary design stages, when the structure and its computational model are being "tuned", it is extremly useful to have a substitute shorter duration accelerogram which models its parent longer duration accelerogram in such a way, that dynamic excitation of the structure by each signal results in effects of similar magnitude and location. It is obvious that in order to achieve this, the response spectra of the parent and the substitute accelerogram must be similar.

4. In this paper two recently developed substitute short duration accelerograms -sine sweep (ref. 5) and four pulse model (ref. 6)- will be described briefly, and then a structure dependent combisweep accelerogram based on the Johnson/Epstein sine sweep accelerogram will be presented in more detail.

SUBSTITUTE SHORT DURATION ACCELEROGRAMS

5. Artificial accelerograms have been generated and used (refs. 7-10) since it was recognized that the randomness of ground motion characteristics severely limits the use of actual earthquake accelerograms. Typically, the durations of these artificial and actual accelerograms are similar, and it is only recently that the desire to reduce the computational costs turned attention towards the generation of substitute short duration accelerograms.

Johnson/Epstein sine sweep accelerogram (ref. 5)

6. The basic form of the Johnson/Epstein sine sweep accelerogram is

$$\ddot{u}_g(t) = \ddot{u}^g_{max}(f)\sin[\Theta(t)] \qquad (1)$$

where $\sin[\Theta(t)]$ is the variable frequency sinusoidal signal, $f = \dot{\Theta}(t)/2\pi$ is the instantaneous forcing frequency, $\ddot{u}^g_{max}(\bar{f})$ is the maximum ground acceleration as a function of the specific forcing frequency, and $\ddot{u}_g(t)$ is the ground acceleration time history.

With a form for Θ chosen as

$$\Theta(t) = At + Bt^N, \qquad (2)$$

the relation between $\ddot{u}^g_{max}$ and $\bar{f}$ defined as

$$\begin{aligned} \ddot{u}^g_{max} &= 0.22g\bar{f} && \text{for } \bar{f} \le 1.5 \\ \ddot{u}^g_{max} &= 0.33g && \text{for } 1.5 < \bar{f} \le 3.5 \\ \ddot{u}^g_{max} &= 2.16g/\bar{f}^{3/2} && \text{for } \bar{f} > 3.5 \end{aligned} \qquad (3)$$

and by systematically varying parameters A,B and C, Johnson and Epstein were able to produce an accelerogram whose response spectra accurately represented the response spectra of the El Centro earthquake (Fig. 1). Comparative time history and response spectrum analyses of a simple 5 DOF-structure using the El Centro and the associated sine sweep accelerograms and response spectra, yielded equally satisfactory results.

Wang/Goel four pulse model accelerogram (ref. 6)

7. This substitute accelerogram is constructed by preserving the energy stored in the structure during an earthquake. To achieve this objective, the maximum vibration energy of the parent accelerogram is approximated through the Fourier Amplitude Spectrum (FAS) or the velocity response spectrum (S_v) and then, using an optimization procedure, the difference between that FAS or S_v and the FAS of the substitute accelerogram is minimized. In the process, two independent parameters related to the acceleration level and accelerogram duration are determined. In Fig. 2, a model four pulse accelerogram ES 1522 for the El Centro earthquake, its response spectrum and the response spectrum of the parent El Centro accelerogram are shown.

Structure dependent combisweep accelerogram (refs. 11,12)

8. The basic form of the combisweep accelerogram is the same as of the Johnson/Epstein sine sweep accelerogram, i.e.

$$\ddot{u}_g(t) = \ddot{u}^g_{max}(\bar{f})\sin(At + Bt^C) \tag{4}$$

where A,B and C are free parameters and the instantaneous frequency $\bar{f}$ is defined as

$$\bar{f} = \frac{1}{2\pi}\frac{d(At + Bt^C)}{dt} = \frac{A + BCt^{C-1}}{2\pi} \quad . \tag{5}$$

The modification of the Johnson/Epstein procedure is in the manner how the optimal set of parameters A,B and C is chosen. Free parameter values are systematically varied over a certain range and, since it is impossible to match the response spectra of the substitute and parent accelerogram exactly, the aim is to minimize the distance between the substitute and the target (parent) spectrum at discrete points which coincide with the natural frequencies of the structure. Thus, the associated accelerogram combines the structural characteristics with the site conditions -hence the name "combisweep"- and this implies that different structures, although built on the same site and having the same design response spectrum, will have different combisweep accelerograms.

9. The procedure is illustrated on Fig. 3 and is performed in stages (i) to (ix) as follows:

(i) Determine the desired number of natural frequencies f_i of the structure. (This is a disadvantage, but savings achieved by performing a nonlinear analysis using a significantly shorter time history are much higher than the additional cost of the eigenvalue analysis.)

(ii) Determine target spectral velocities T_v^i corresponding to natural frequencies f_i.

(iii) Select a new set of parameters A,B and C.

(iv) Substituting the cut-off frequency f_{cut} for $\bar{f}$ in (5), determine the sine sweep accelerogram duration, and if it is not within prescribed bounds, start again at (iii).

(v) Synthetize the sine sweep accelerogram using (4) and the $\ddot{u}^g_{max} - \bar{f}$ relationship defined as in Fig. 4.

(vi) For the obtained sine sweep accelerogram determine spectral velocities $S_v^i = S_v^i(A,B,C)$ corresponding to natural frequencies f_i using the Duhamel integral (refs. 1-3).

(vii) Determine the distance D = D(A,B,C) between the target and the current response spectrum as

$$D = \sum w_i (S_v^i - T_v^i)^2 \tag{6}$$

where w_i is the weighting factor for the ith natural frequency f_i of the structure. A natural choice for w_i seems to be

$$w_i = L_i = \frac{\underline{\Phi}_i^T \underline{M}\,\underline{I}}{\underline{\Phi}_i^T \underline{M}\,\underline{\Phi}_i} \tag{7}$$

i.e. weighting factors are set equal to modal participation factors (refs. 2,3), which ensures better response spectra matching in the more important lower frequency range.

(viii) If the distance D is smaller than some previously obtained least distance, memorize the current parameter set A,B and C as the optimal one.

(ix) If stages (iii) to (viii) have been performed for all permissible sets of A,B and C, synthetize the sine sweep accelerogram (as in (v)) using the optimal parameter set and evaluate its response spectrum (as in (vi)) for the whole frequency range.

EXAMPLES

10. In following examples (Figs. 5-10), the modelling of the earthquake excitation and results of the linear and nonlinear analyses of the Koyna dam section are presented. In Fig. 5, a (coarse) mesh, the first eight natural frequencies, modal participation factors and material and geometric characteristics of the section are shown. In Fig. 6, a transversal component of the Koyna 1967 earthquake accelerogram is shown. This is the parent accelerogram whose (target) spectral velocities T_v^i are determined in stage (ii) of the foregoing paragraph.

11. In Figs. 7 and 8, two Koyna combisweep accelerograms obtained as described above, are presented together with the illustration of how their response spectra match with the target response spectrum of the parent accelerogram. For the KOYNA COMBISWEEP A constant weighting factors are used, whereas for the KOYNA COMBISWEEP B weighting factors equal to modal participation factors are used. The adopted parameters for defining the $\ddot{u}^g_{max} - \bar{f}_2$ relationship (Fig. 4) are: a_{max} = 4.807 $\underline{m}$/sec^2 (ref. 13), f_1= 2.5 cps, f_2= 10.0 cps, f_{cut} = 30.0 cps and a_{fin} = 0.05 a_{max}.

12. Finaly, results of the linear analyses (explicit, no damping) using the parent and both substitute accelerograms are shown in Fig. 9, and results of the nonlinear analyses (explicit, no damping, elasto/viscoplastic degradation model for concrete - see refs. 11 and 14) using the parent and the KOYNA COMBISWEEP A substitute accelerogram are shown in Fig. 10. It is apparent that maximum displacements (linear analysis) and permanent deformations (nonlinear analysis) obtained using substitute accelerograms compare well with those obtained using the more than four times longer parent accelerogram.

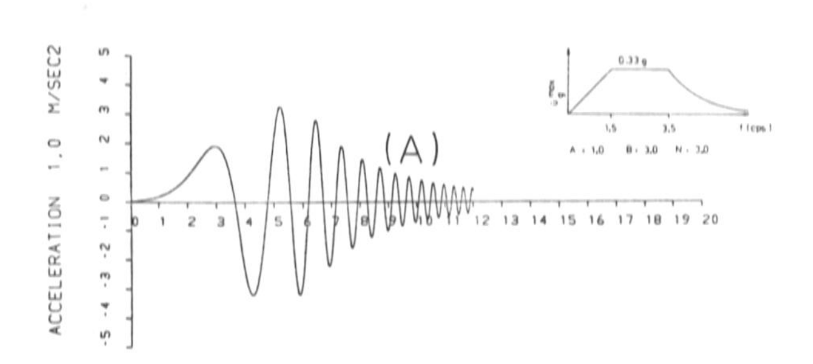

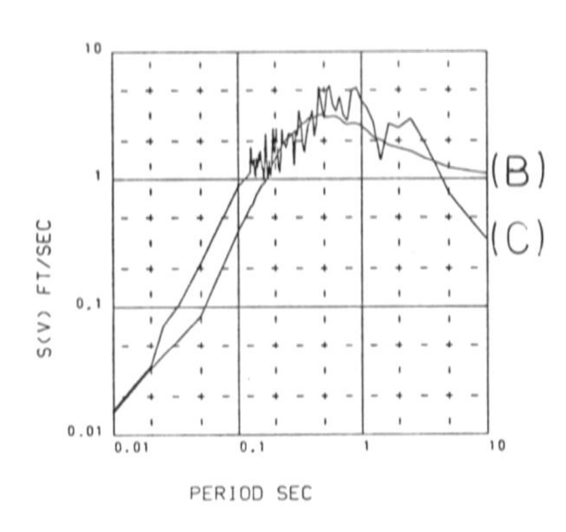

Fig. 1 Johnson/Epstein substitute sine sweep accelerogram (A), its response spectrum (B) and the response spectrum of the parent El Centro earthquake actual accelerogram (C)

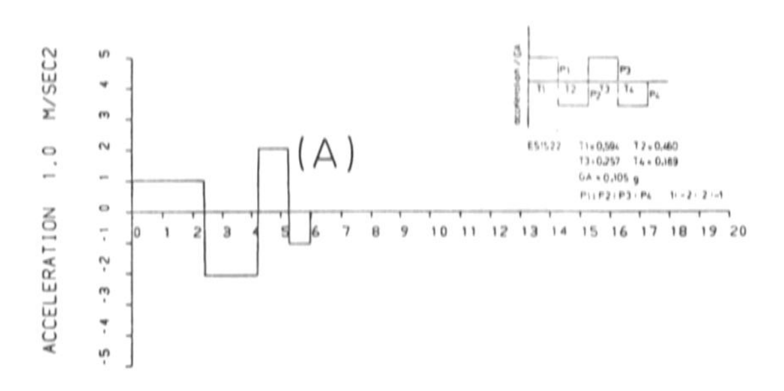

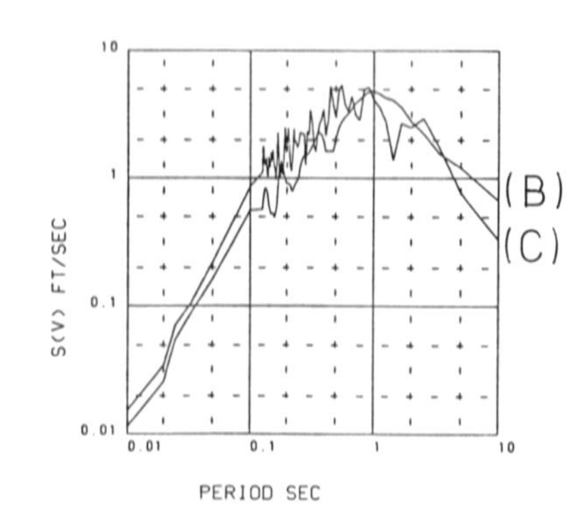

Fig. 2 Wang/Goel substitute four pulse model accelerogram (A), its response spectrum (B) and the response spectrum of the parent El Centro earthquake actual accelerogram (C)

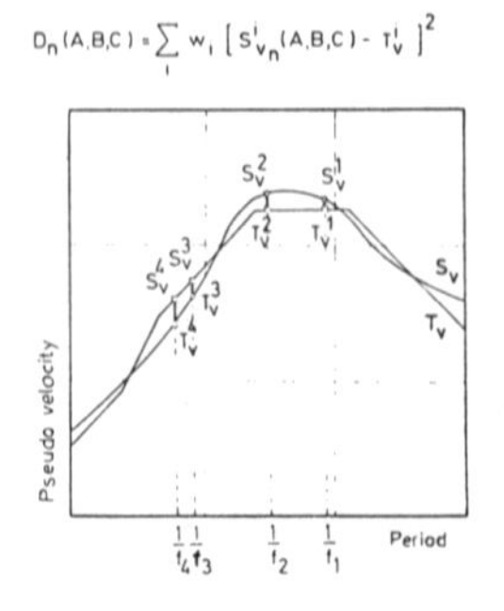

Fig. 3 Evaluation of the distance between the target (T_v) and the substitute (S_v) response spectrum

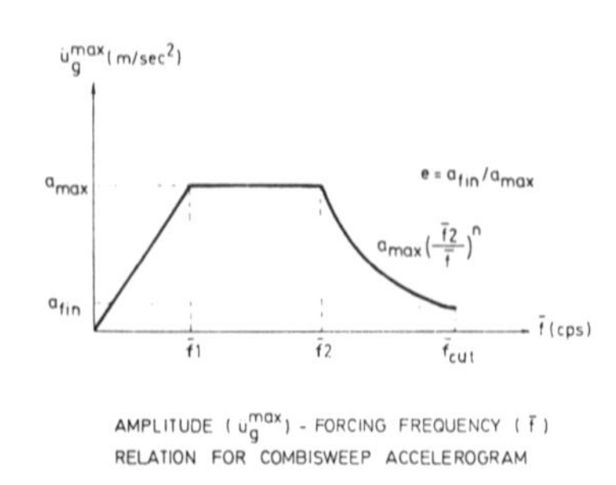

Fig. 4 Ground acceleration amplitude ($\ddot{u}^g_{max}$) vs. forcing frequency ($\bar{f}$)

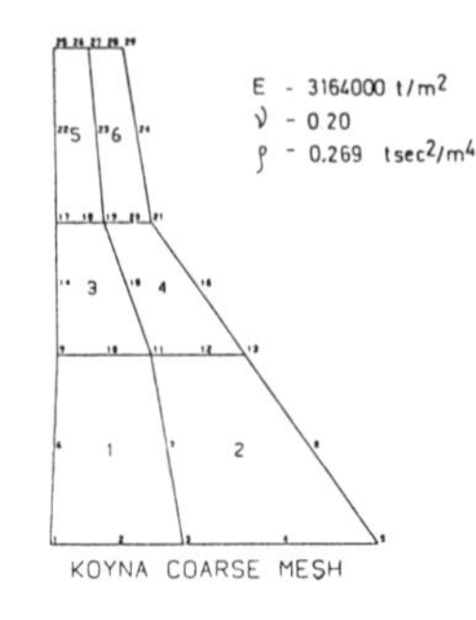

Mode	ω_i	L_i
1	19.5	30.7
2	49.6	31.4
3	71.0	21.9
4	98.2	22.7
5	156.9	13.2
6	171.1	9.6
7	207.8	8.0
8	223.4	1.2

Fig. 5 Koyna dam section mesh, material properties, natural frequencies (ω_i) and modal participation factors (L_i)

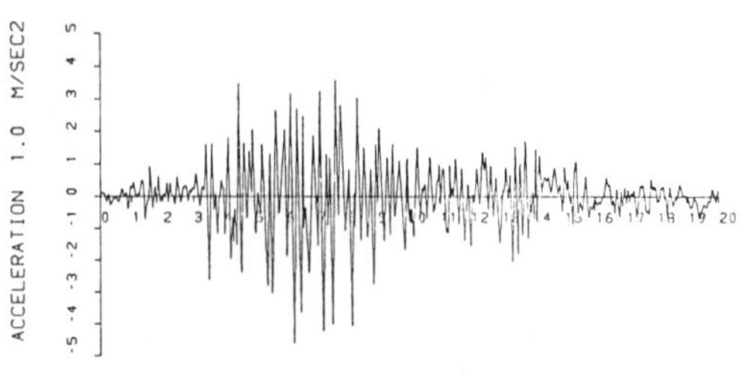

Fig. 6 Koyna 1967 earthquake actual accelerogram - transversal component

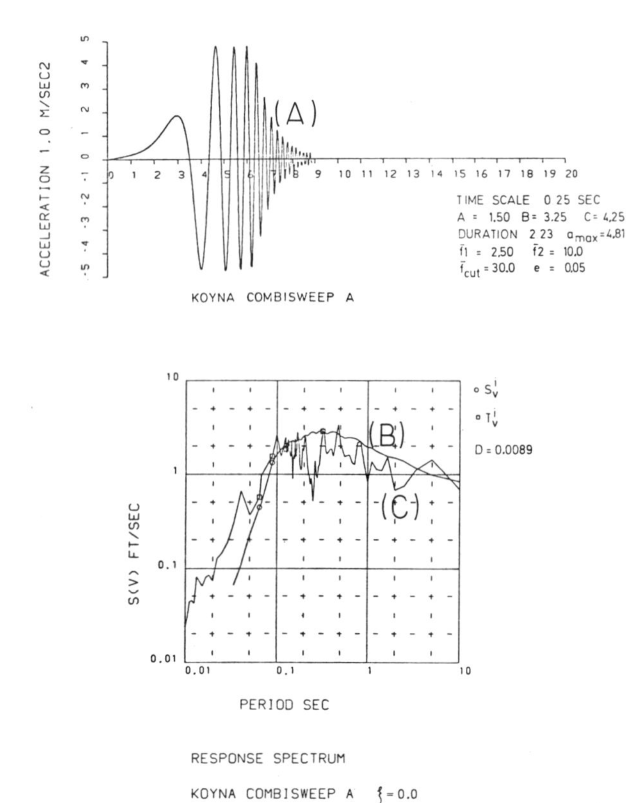

Fig. 7 KOYNA COMBISWEEP A accelerogram (A), its response spectrum (B) and the response spectrum of the parent accelerogram (C)

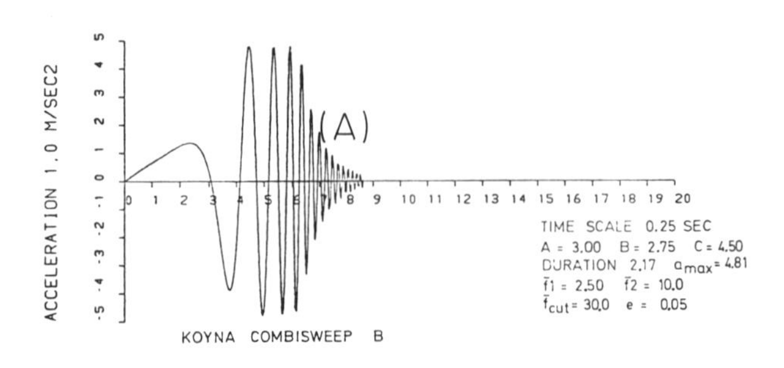

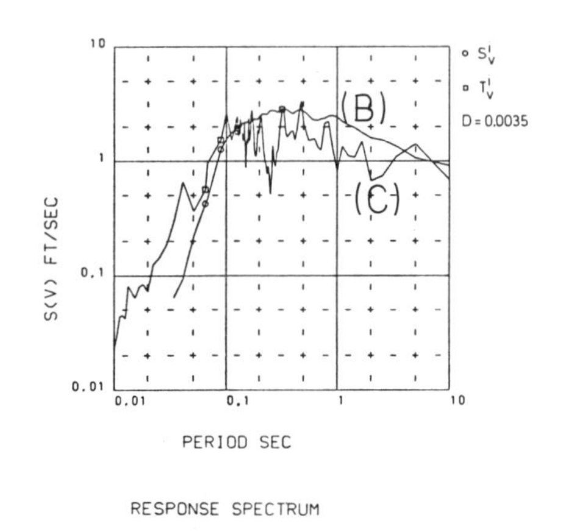

Fig. 8 KOYNA COMBISWEEP B accelerogram (A), its response spectrum (B) and the response spectrum of the parent accelerogram (C)

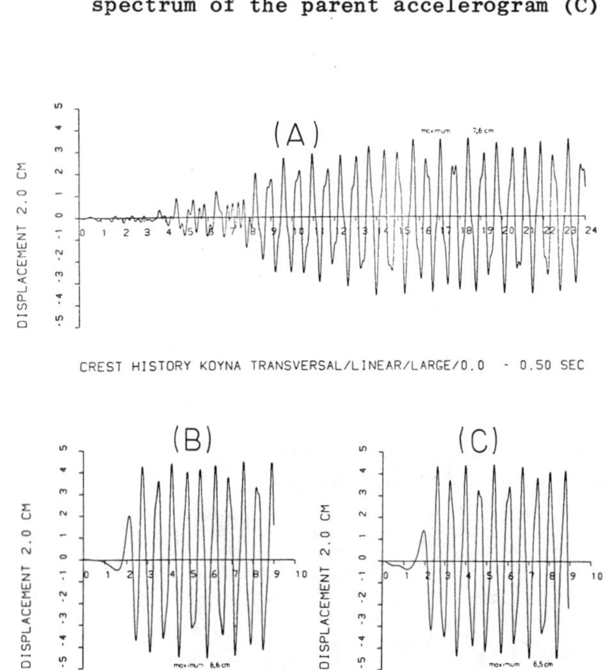

Fig. 9 Koyna dam section linear analyses results using the parent (A) and substitute accelerograms ((B) and (C))

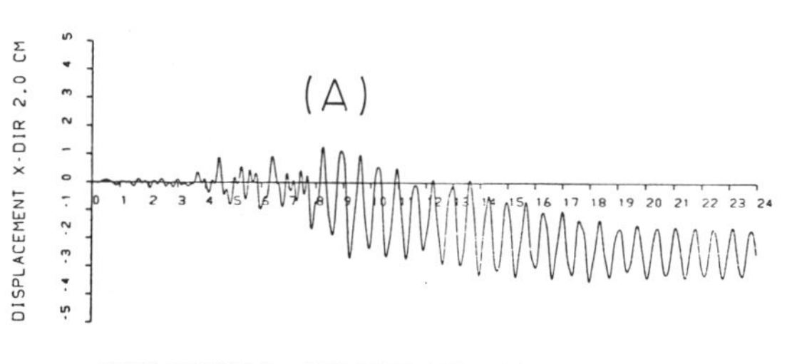

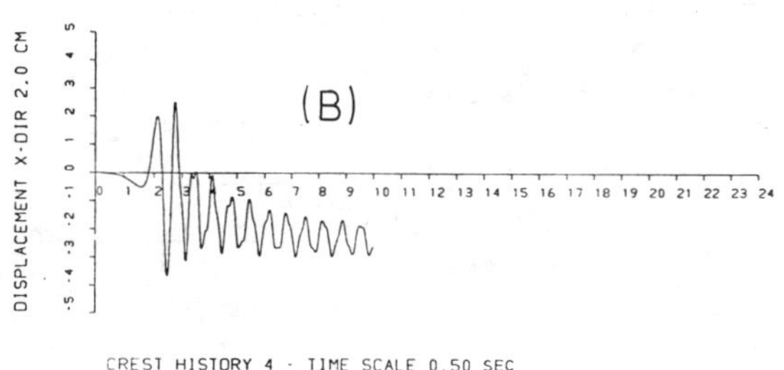

Fig. 10 Koyna dam section nonlinear analyses results using the parent (A) and the substitute (B) accelerogram

CONCLUSION

13. In this paper substitute short duration accelerograms have been considered as a useful alternative to their parent long duration design accelerograms. Two previously developed substitute accelerograms -sine sweep and four pulse model- have been reviewed and a third one -combisweep- was extensively described and its application illustrated.

14. It has been shown that :

(a) it is possible and relatively easy to obtain a good match between the response spectra of substitute and parent accelerograms, and

(b) similar maximum resonses have been obtained using substitute and parent accelerograms.

15. It is unlikely that substitute short duration accelerograms will ever fully replace the long duration ones, whose duration and selectiveness play a significant part in causing structural damage. However, bearing in mind the uncertainties of engineering modelling for nonlinear seismic analysis (the choice of stiffness and mass representation, of the mesh size, material model and of the excitation), it is felt that, at present, the use of substitute accelerograms at least in preliminary design stages, is strongly justified and warranted.

ACKNOWLEDGEMENT

The financial support granted by the Građevinski Institut (Civil Engineering Institute) and the Samoupravna interesna zajednica za naučni rad (Selfmanaged interest group for scientific work), both from Zagreb, Yugoslavia, to N. Bićanić and R. Fejzo is hereby gratefully acknowledged.

REFERENCES

1. NEWMARK N.M. and ROSENBLUETH E. Fundamentals of earthquake engineering. Prentice-Hall, Inc., Englewood Cliffs, N.J., 1971.

2. CLOUGH R.W. and PENZIEN J. Dynamics of structures. McGraw-Hill, New York, 1975.

3. DOWRICK D.J. Earthquake resistant design. John Wiley & Sons, Ltd., Chichester, 1977.

4. SHAW D.E., RIZZO P.C. and SHUKLA D.K. Proposed guidelines for synthetic accelerogram generation methods. Proceedings of the 3rd SMIRT Conference, London, 1975, K 1/4.

5. JOHNSON G.R. and EPSTEIN H.I. Short duration analytic earthquake. Proceedings A.S.C.E., 1976, 102, ST2, May, 993-1001.

6. WANG W.Y.L. and GOEL S.C. Prediction of maximum structural response by using simplified accelerograms. Proceedings of the 6th WCEE, 3, New Delhi, 1977.

7. JENNINGS P.C., HOUSNER G.W. and TSAI N.C. Simulated earthquake motions. EERL Report, California Institute of Technology, April, 1968.

8. LEVY S. and WILKINSON J.P.D. Generation of artificial time-histories, rich in all frequencies, from given response spectra. Proceedings of the 3rd SMIRT Conference, London, 1975, K 1/7.

9. CHOPRA A.K. and LOPEZ O.A. Evaluation of simulated ground motions for predicting elastic response of long period structures and inelastic response of structures. Earthquake Engineering and Structural Dynamics, 1979, 7, July, 383-402.

10. WONG H.L. and TRIFUNAC M.D. Generation of artificial strong motion accelerograms. Earthquake Engineering and Structural Dynamics, 1979, Nov., 509-527.

11. BIĆANIĆ N. Nonlinear finite element transient response of concrete structures. Ph.D. Thesis, C/Ph/50/78, University of Wales, Swansea.

12. BIĆANIĆ N. Structure dependent short duration combisweep accelerogram. Proceedings of the International Conference on Computer Applications in Civil Engineering, Roorkee, 1979, III 67-III 72.

13. KRISHNA J., CHANDRASEKARAN A.R. and SAINI S.S. Analysis of Koyna dam accelerogram of December 11, 1967. Bulletin of the Seismological Society of America, 1969, 59, Aug., 1719-1731.

14. ZIENKIEWICZ O.C., BIĆANIĆ N., HINTON E. and FEJZO R. Computational models for the transient dynamic analysis of concrete dams. Proceedings of the Conference on Design of Dams to Resist Earthquakes, London, 1980.

4 Seismic risk studies for large dam projects in Northern Iraq

M. B. TOSIC, MPhil, DIC, Energoprojekt Engineering and Consulting Company, Belgrade

Sites which today appear as seismically quiescent, in the Upper Tigris River basin, have experienced strong earthquakes in history. Because of a rather diffused pattern of seismicity and heterogeneous seismic data in this region, standard statistical methods of seismic risk analysis may be insufficient. For large dam projects, some of the historical earthquakes are almost equally important as instrumental recordings. More studies of neotectonics and of the makroseismic data of this century, in addition to greater strong-motion instrumentation, are needed in this region.

HISTORICAL SEISMICITY

A knowledge of the seismic history of a particular region is important in evaluating earthquake hazards for the region. The length of time covered by scientific seismology is negligibly short when compared with geological time scale involved in seismicity of the region. Earthquake frequency studies based on a period covering a couple of centuries are often inconclusive as to the level of seismicity that should be assigned to a region. Not seldom potentially active regions have been described as quiescent, and some of the most disastrous earthquakes of our times occurred in areas where little or no activity was observed for a period of a few centuries.

In the implementation of earthquake historical evidence into the modern studies of seismicity, the greatest contribution so far is due to Ambraseys (1961, 1962, 1968, 1970, 1971). Using old Arabic and Persian manuscripts, and other documentation pertaining to the region of the Near and Middle East, he has compiled, sorted out and made detailed studies of the earthquake historical data, demonstrating their importance in delineating the long-term seismic activity of the entire region.

Iraq has a rather well documented history of seismic activity. Alsinawi and Ghalib (1975) have prepared a chronological list, that is a catalog of the historical earthquakes that were felt in Iraq, from 1260 B.C. up to 1900 A.D. The catalog contains descriptions of 81 evants, 50 of which having source reference to Ambraseys. For many of these events, individual shocks, within a given period of time and felt over a given area, appear to have been reported collectively, and this is no doubt how the information was contained in the original documents.

This list shows that during some 360 years (847-1204 A.D.) there were 45 events, whereas in a much longer period of the next 640 years (1227-1862 A.D.) not more than 18 events were evidenced. This does not appear to be a surprising result. While the former period is characteristic for the greatest humanist and scientific achievement of the civilisation created by the Arabs, on a huge territory, the latter is marked mainly by invasions and devastations from abroad. Until the 15th century the Mongols will already have laid waste most of the Arab world, and the Ottomans, coming as the last conquerors, will keep it in a progressively decaying state for almost four hundred years. It may be possible, therefore, that the modest volume of earthquake data in this period is a result of the smaller number of written documents available to the future generation, rather than is an indication of relative seismic inactivity.

The greatest undertakings on the compilation of regional catalogs of historical earthquakes have been made by the Academia Sinica in Peking (Hsieh, 1973; Lee and Wang, 1976; Lee, Wu and Jakobsen, 1976), and by the Soviet Academy of Sciences in Moscow (Shebalin and Kondorskaya, 1977). In these catalogs, on the basis of the size of the area over which an event was felt, described effects, the duration of the event(s) and/or the number of aftershocks, and using some empirical correlations for seismic parameters, the values of magnitude and of the maximum intensity were assigned to most events (much of this information is codified in the form suitable for computer applications).

Using similar procedures, we have attempted to assign the magnitude and intensity to about 50 of the most important historical earthquakes for Iraq (about 15 excluded events, although large, reflect only marginal effects of distant earthquakes). A summary of this work is presented.

The maximum intensity could have been IX(MM) in up to about ten cases, which value is associated with the magnitude of about 6.5. Intensity X(MM) could have been reached in two or three earthquakes, but not among those in the folded series of the Upper Tigris River Basin. Because of uncertainties as to the focal depths in the region, it was some problem to correlate magnitude M with the epicentral intensity I_o(MM). Relations $M = 0.55\ I_o + 1.63$ and $M = 0.59\ I_o + 1.18$, derived for the Bekhme and Eski Mosul regions, and for the entire region in Figure 6, respectively, are valid for definite ranges of the focal depth and do not permit any significant variations in this respect. On the other hand, no reliable relations $M = F(I_o, h)$ could have been obtained because the data is insufficient. In the folded series between the Tigris and the Zagros it is often supposed that the foci, on an average, are 15 km deep. For the Caucasus region (Figure 1), we have obtained the same average focal depth h. Two relations were derived

$$(M - \frac{2}{3} I_o) = 1.65 \log (h) - 1.42 \qquad (1)$$

$$M = 0.63 I_o + 1.64 \log (h) - 1.18 \qquad (2)$$

from 878 data sets (M, I_o/MKS/, h/km/). The data were assigned weights proportional to their grades of accuracy. For h = 15 km, relations (1) and (2) give results comparable with those from the the two $M = f(I_o)$ relations in the foregoing:- for intensity V the differences in M are up to 10%, while for intensity IX they are up to 3%. Assuming that the intensities on the two scales (MM and MKS) are eqyivalent, we have used relation (2) for some recalculations in this analysis.

Earthquake historical data can be incorporated into seismic risk studies in a number of ways. The maximum magnitude in a region, M_m, assessed from the historical data, can be used for obtaining the improved magnitude-frequency relations

$$\log N(M) = a - b M \qquad (M \leq M_1)$$
$$\log N(M) = a - b M + \log (M_m - M) \qquad (M > M_1)$$

where M_1 is up to about 6.

The value of M_m is used also in the Type III distribution of extremes which, in the studies of earthquake probabilities, are more advantageous than the Type I distributions. In addition, these data can be incorporated as input in the Bayesian statistics for obtaining the improved probabilities.

INSTRUMENTAL AND MACROSEISMIC DATA (1900-1976)
Basic earthquake data- magnitude, epicenter, intensity, and focal depth determined instrumentally or from macroseismic observations, were used from different sources (ISS, USCGS, etc.). More weight has been given to the data that have been subject of more detailed analyses (Nowroozi 1976, Dewey 1976, Ambraseys 1969, 1978, Shebalin and Kondorskaya 1977, Deprem Arastirma Enstitüsü Ankara 1977).

Kaila, Rao and Narain (1974) have prepared quantitative seismicity maps of Southwest Asia. These maps show contour lines of coefficients A and b (determined by the Kaila-Narain method) of the cumulative frequency-magnitude relationship, and contour lines of the return period of earthquakes with a body-wave magnitude of six and above. For the region N 31°- N 39°, E 42°- E 49° three sets of these contours are shown in Figure 5.

Using routine computer methods, Nowroozi (1976) has relocated (epicentres and focal depths) over 600 earthquakes which occurred in and near Iran between 1920 and 1972. These data (RE in further text), in addition to other information, were used to construct a seismotectonic map of Iran. Western part of this map is also shown in Figure 5. North from latitude 32°, eleven (RE) epicentres are in Iraq; twenty additional epicentres are relatively near Iraqi border.

Ambraseys (1978) points out that (RE) epicentres are not considerably better than those originally defined by ISS, and that inaccuracies of onset times and the lack of near stations do not permit correct focal depth determinations.

For the Southern Caucasus region it is of interest to compare these data (RE) with those from the Soviet Catalog, 1977. Epicenter coordinates, in the latter, are given with possible variations of ±0.1°, ±0.2° and ±0.5° in both directions (N and E), corresponding to total variations of 0.14°, 0.28°, and 0.71°, respectively. Possible variations of the focal depth are given by its limits, while the adopted focal depth is defined as the geometric mean of these two values.

Epicenter coordinates from these sources do not appear, in general, to differ very much. Out of a total of 56 events, 32 epicentres (RE) are within the resultant limits of variation of corresponding epicenter coordinates from the Catalog. About 90% of all (RE) epicentres are within a distance of 0.71° from the epicentres in the Catalog. In fact, the two sets of epicentres differ, on an average, about 0.22° total, that is, about ±0.16° in both directions N and E.

On the other hand, focal depths (RE), h_1, compared with those from the Catalog, h_2, differ greatly, as is shown by data plot in Figure 2. The lower and upper bounds of possible variations of (h_2) are represented by bar graphs. If Q is the Briggsian logarithm of a ratio (h_1/h_2) of these focal depths, standard deviation of the Q-ratios to the mean, computed from 44 available pairs of the data, is 0.38. Focal depths h_1 are, on an average, about three times greater than the corresponding values of h_2, and are even greater than (the majority of) upper bounds of h_2. Although there have been some deep-focus earthquakes at the Caucasus, this is considered as a region of typically shallow foci.

Our objective here has not been an analysis of seismicity of the Caucasus, but a comparison of the two consistent sets of data. It can be concluded that focal depths (RE) appear to be systematically large. This is also the case with other (RE) earthquakes. Therefore, while we made use of a lot of (RE) epicentres, we did not use (RE) focal depths, for the risk studies in Iraq.

EARTHQUAKE RISK
There exist different methods for the evaluation of earthquake risk at the site of an engineering project, but these methods depend greatly on all macroseismic, instrumental and tectonic information that is available. Statistical and mathematical methods available for an assessment of the seismicity of a region and evaluations of earthquake hazard in general surpass the quality and quantity of input data.

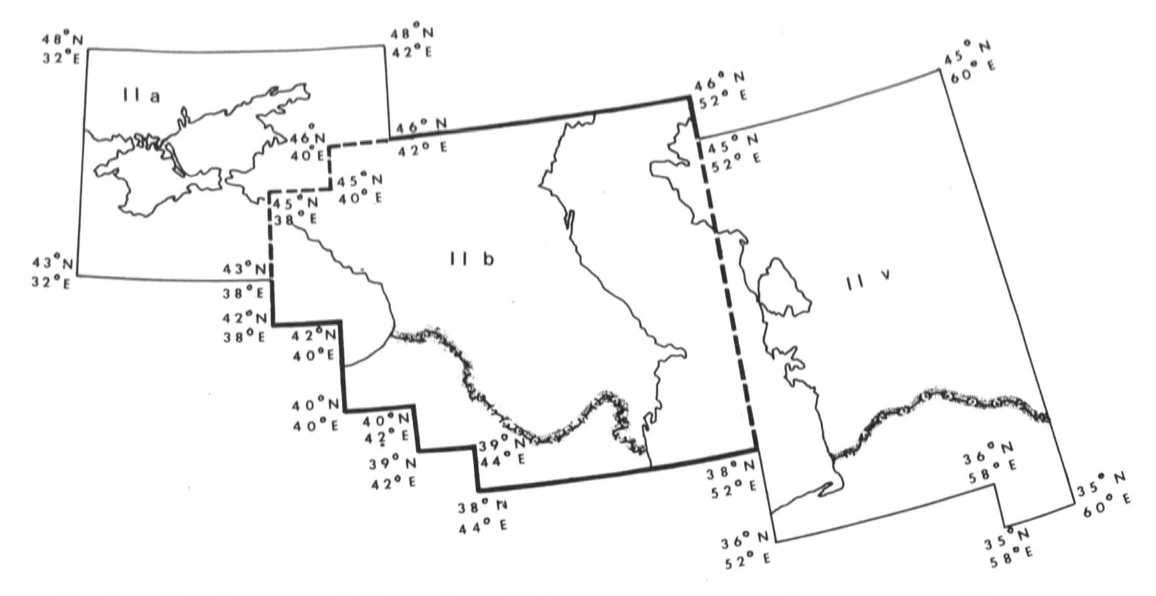

Fig.1. The Caucasus region, for which relations $M = F(I_o/MKS/, h/km/)$ were derived using data from the Soviet Catalog of Strong Earthquakes, 1977

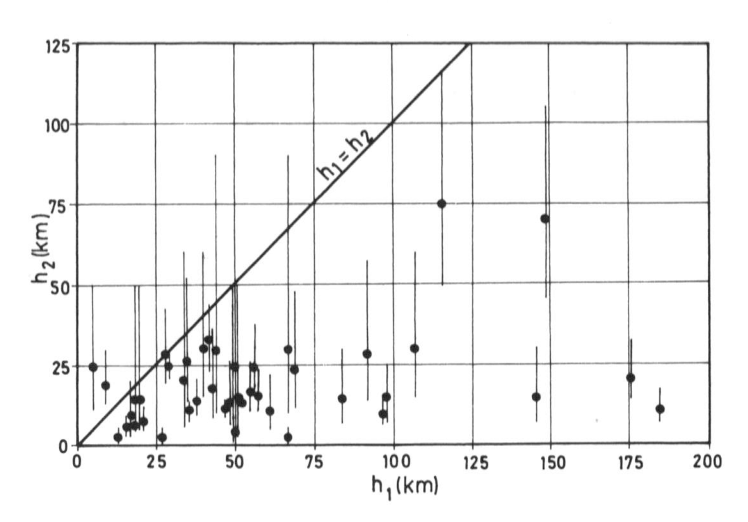

Fig.2. Comparison of focal depths for the Caucasus: (h_1) Nowroozi, 1976; (h_2) Soviet Catalog, 1977

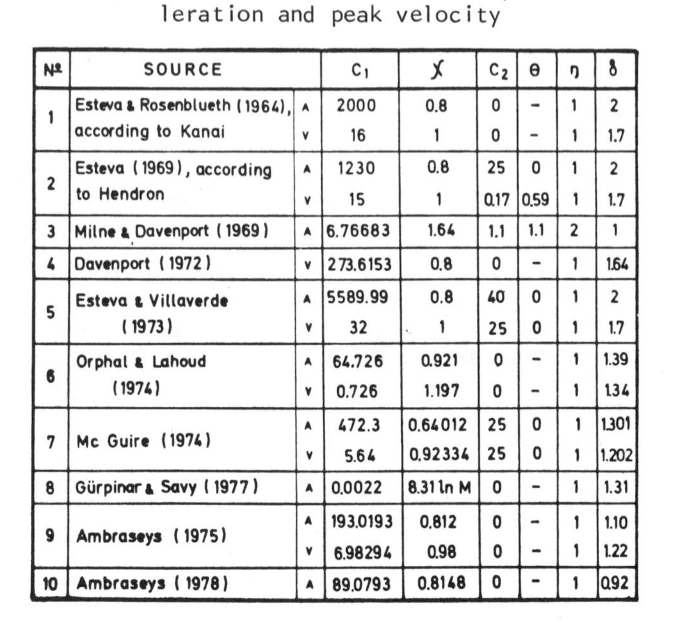

Table 1. Attenuation constants for peak acceleration and peak velocity

№	SOURCE		C_1	χ	C_2	θ	η	δ
1	Esteva & Rosenblueth (1964), according to Kanai	A	2000	0.8	0	–	1	2
		V	16	1	0	–	1	1.7
2	Esteva (1969), according to Hendron	A	1230	0.8	25	0	1	2
		V	15	1	0.17	0.59	1	1.7
3	Milne & Davenport (1969)	A	6.76683	1.64	1.1	1.1	2	1
4	Davenport (1972)	V	273.6153	0.8	0	–	1	1.64
5	Esteva & Villaverde (1973)	A	5589.99	0.8	40	0	1	2
		V	32	1	25	0	1	1.7
6	Orphal & Lahoud (1974)	A	64.726	0.921	0	–	1	1.39
		V	0.726	1.197	0	–	1	1.34
7	Mc Guire (1974)	A	472.3	0.64012	25	0	1	1.301
		V	5.64	0.92334	25	0	1	1.202
8	Gürpinar & Savy (1977)	A	0.0022	8.31 ln M	0	–	1	1.31
9	Ambraseys (1975)	A	193.0193	0.812	0	–	1	1.10
		V	6.98294	0.98	0	–	1	1.22
10	Ambraseys (1978)	A	89.0793	0.8148	0	–	1	0.92

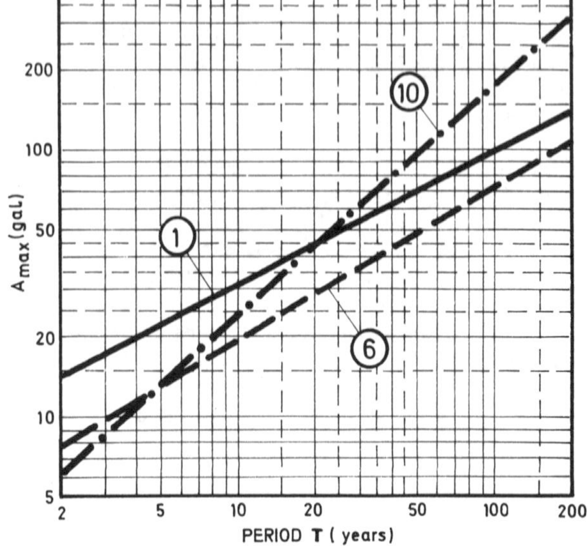

Fig.3. Peak acceleration vs. return period

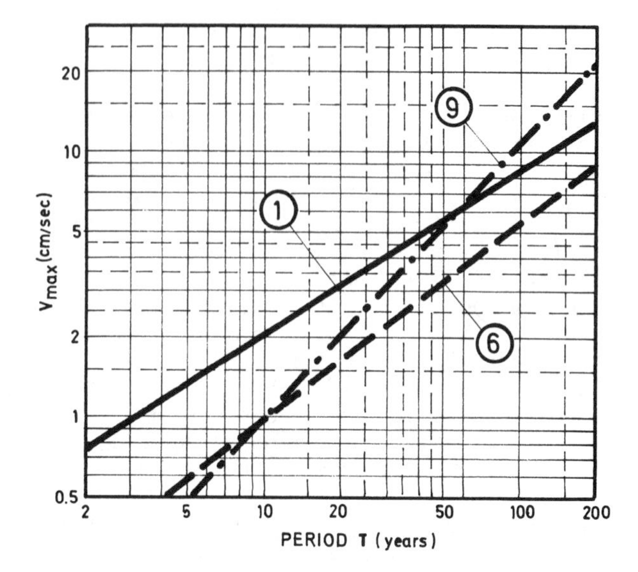

Fig.4. Peak velocity vs. return period

There are numerous attenuation laws for the peak values of strong motion parameters- ground acceleration, velocity and displacement, in terms of magnitude M and focal distance R. They are used not only for an assessment of the values of these parameters at a given site from an earthquake, but even more, for seismic risk evaluations.

If we take that the strong motion parameters are given by a general expression in the form

$$Y_{max} = C_1 e^{\chi M} (C_2 e^{\theta M} + R^{\eta})^{-\delta} \quad (3)$$

ten of these laws for peak acceleration (A), and six for peak velocity (V), can be defined by the corresponding sets of coefficients (C_1, χ, C_2, θ, η, δ) in Table 1 (with R in km, (A) and (V) are given in cm/sec^2 and cm/sec, respectively).

In relation 2(V), the velocity varies roughly as $e^M R^{-1.7}$ for large R, and is approximately constant, i.e. independent of M, when R tends to zero. The latter implies that peak velocity in focal volume (V_f) does not depend on magnitude. A physical meaning of this is that the quantity of energy E_o released per unit of volume V_f is approximately constant. This is in agreement with the basic equation of the release of energy, $E_o = c V_f$ in ergs, where $c = (1.0\text{-}1.1)10^3$ when V_f is in cu.cm.

In relation 3(A) the acceleration varies roughly as $e^{1.64M} \cdot R^{-2}$ for large R, and as $e^{0.54M}$ when R is close to zero. This shows, in a certain way that the energy is not released at a point, but in a finite volume.

All coefficients in Table 1 are constants, except χ in 8(A). Noting that only relations 2(V) 3(A) and 8(A) have somewhat different forms, all other relations can be simplified as

$$Y_{max} = C_1 e^{\chi M} (C_2 + R)^{-\delta} \quad (4)$$

On the basis of standard deviation of logarithm of the ratios of the true values of acceleration and velocity, and the computed values (A and V), these attenuation expressions seem equally reliable. However, they often give rather different results in seismic risk studies.

Relations 1 to 8 are based on strong motion near field and far field data obtained from earthquakes on the American continents, and in Japan. Ambraseys (1975, 1978) has proved that accelerations computed from some American relations are considerably smaller than those, actually recorded, on the "old" continent. Using strong motion data obtained from earthquakes in Europe, Asia Minor and the Middle East, and recorded only in the near field, he has derived relations 9 and 10. For these regions they are better than those American relations, because of smaller standard deviation of the ratios of actually recorded and computed accelerations.

However, relations 9 and 10 are valid for magnitudes of up to about six. On the other hand, seismic risk studies imply necessarily the largest possible magnitudes in a given period of time.

Since there are several magnitude scales, there must be always some doubts as to the consistency and homogeneity of magnitudes from which relations 1 through 8 have been derived and, therefore, as to the scale that is implied when any attenuation law is to be used. Earthquake magnitudes in Western United States, often listed with strong motion data, are commonly referred to the M_L scale. This is the Richter magnitude (1935), originally defined for local earthquakes in Southern California, and is based on surface waves with periods of up to one second. However, it is considered that the above relations are based on M_s magnitudes. This is a magnitude computed from the amplitude of surface waves with periods between 10 and 30 seconds, using the Moscow-Prague formula (1962).

Relations 9 and 10 have been derived using body-wave magnitudes (m_b) for the largest number of events, in addition to some data being entered in terms of M_L and M_s magnitudes. Since m_b magnitudes for these events have been found to increase appreciably with an increase of the period, they were computed from periods of up to 1.2 seconds. Both these magnitudes, and those determined by USCGS after 1963 (Mc Kenzie, 1972), for European events, appear to be relatively low.

Individual stations report M_s or m_b magnitude or both. Magnitude $M = M_s$ is commonly used for all earthquake statistics. The use of m_b or M_L magnitude, instead of M_s, is more meaningful in all studies of strong ground motion associated with the high and intermediate frequency bands.

There are conversion equations

$$M_L = 40 - (1480 - 56\,M)^{1/2} \quad (5)$$

$$m_b = 0.56\,M + 2.9 \quad (6)$$

$$\text{or} \quad m_b = 0.63\,M + 2.5 \quad (7)$$

deduced from world-wide magnitude data. Attenuation coefficients in relations 9 and 10, Table 1 were obtained using equation (6) which does not allow for smaller m_b magnitudes. Thus smaller M magnitudes were introduced into play, now probably somewhat averaged.

Attenuation coefficients as defined by Ambraseys

for 9(A) $C_1 = 2.88$ $\chi = 1.45$ $\delta = 1.10$

for 9(V) $C_1 = 1/22.91$ $\chi = 1.75$ $\delta = 1.22$

for 10(A) $C_1 = 1.31$ $\chi = 1.455$ $\delta = 0.92$

with $C_2 = 0$ in equation (4), are used directly when the original body-wave magnitudes, in a given region, are known. If this is not the case, quantitative seismicity maps or other representations which determine the magnitude-frequency relationship in terms of magnitudes m_b, can be used with thus initially defined coefficients in relation (4). Using this procedure we have obtained very satisfactory results of seismic risk studies in Montenegro, Yugoslavia.

In order to obtain best results when the relati-

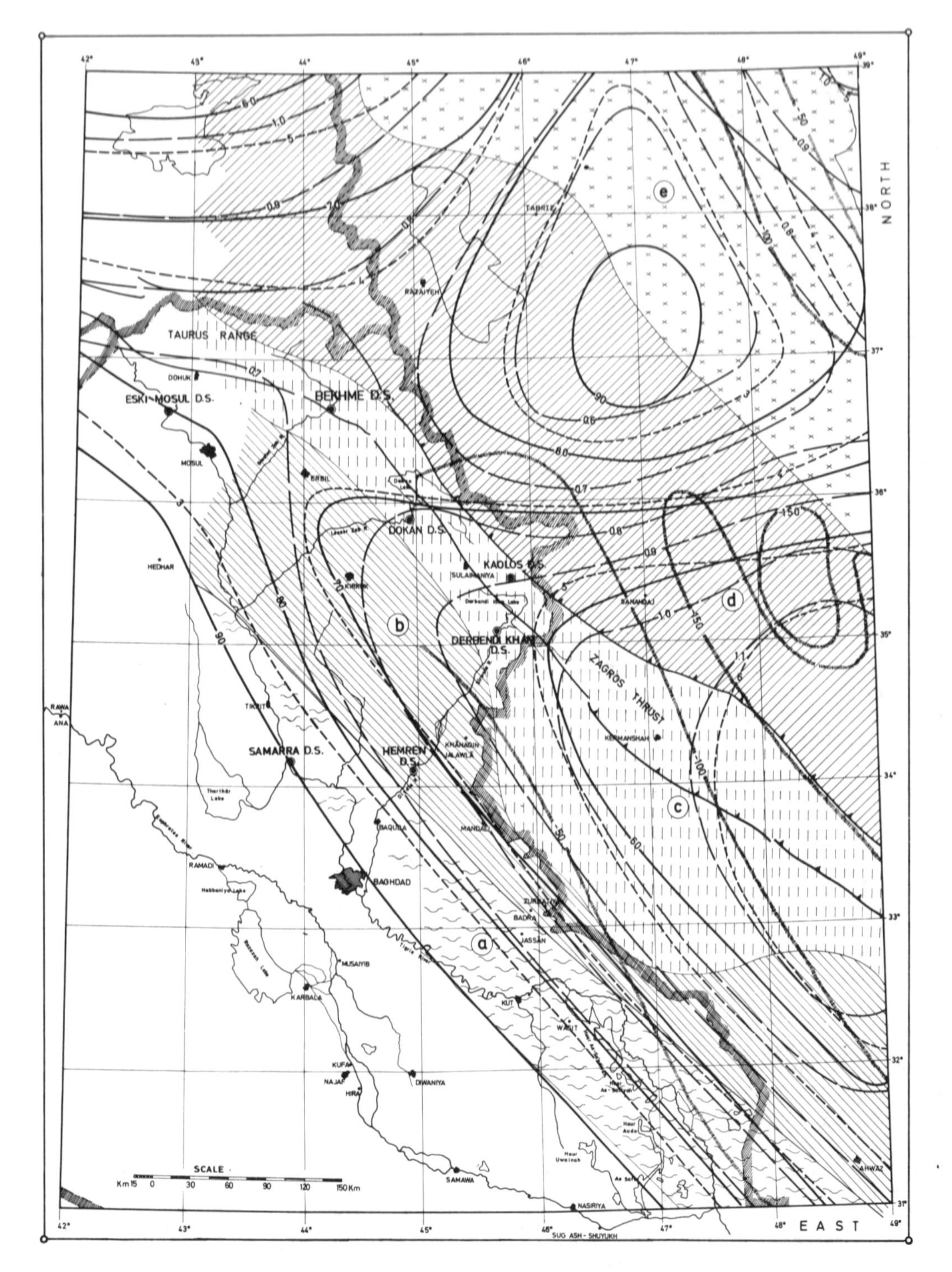

EXPLANATIONS:

RETURN PERIOD CONTOURS (Years)

b-VALUE CONTOURS (log N = A - bM)

A-VALUE CONTOURS, normalized to a 2° by 2° grid as at the equator and 14-year period of earthquake observation

AFTER K.L. KAILA, N. MADHAVI RAO, AND HARI NARAIN (1974, 1975).
(SOMEWHAT ADJUSTED)

SEISMOTECTONIC PROVINCES OF THE REGION ACCORDING TO A. A. NOWROOZI (1976)

- (a) ARVAND - SHATT - EL - ARAB FOLDED SERIES
- (b) (ZAGROS) FOOT - HILL FOLDED SERIES
- (c) HIGH ZAGROS FOLDED SERIES
- (d) REZAIYEH
- (e) MAKU - ZANDJAN

BOUGUER GRAVITY ANOMALY IN THE ZAGROS REGION, AFTER U.S.A.F. GRAVITY ANOMALY MAP OF ASIA, DURBIN ET AL, 1972. CONTOUR INTERVAL IS 50 m Gal.

Fig.5. Seismicity levels map/seismotectonic provinces

ons 9 and 10 are used in a risk study, the region should be limited to the near field, that is, to epicentral or fault-trace distances of typically several tens of kilometres and definitely less than 100 kilometres from the site.

Apart from the limitations in the use of all relations in Table 1, precautions of safety require that the risk studies be based on more attenuation laws. This implies the use of those relations in Table 1 which, for the specific range of focal distances in a region, and for the largest expected magnitudes in this region in a given period of time, will yield larger values of the seismic action.

Using the Cornell method, seismic risk studies have been made for the Hemren, Kaolos, Eski Mosul and Bekhme dam sites. This method of analysis gives peak acceleration and peak velocity in terms of average return period. Various geometric relations among each site and potential areal and lineal sources have been considered. The concept of areal sources is based on the assumption that earthquakes can occur anywhere within a given region (or in any part thereof, treated as a separate source area), and appears to be more advantageous in case of uncertainties as to the locations of epicentres. On the other hand, active seismic trends and existing dislocations, with earthquake activity rates assigned to them, can be treated as the lineal sources.

Figures 3 and 4 illustrate some of our results, obtained for the Hemren dam site. Diagram designations are consistent with the sequence numbers of attenuation laws (Table 1), with the corresponding sets of coefficients being used in obtaining these peak acceleration and peak velocity versus return period relations.

Diagrams 1 and 6 are based on earthquake frequency relationship $\log(N) = f(M)$ that we have derived from the available data (1900 - 1976), for the region considered in this analysis. Diagrams 9 and 10 are based on the relationship determined from the map in Figure 5, $\log(N) = f(m_b)$ and using the attenuation coefficients as originally defined, in terms of magnitude m_b.

The region was limited by epicentres up to a distance of R_o= 110 km from the site, and the concept of areal sources was used in this case. An average focal depth of 25 km was assumed.

It is interesting to note that, for the same con cept of areal sources we have obtained somewhat different results in other regions. For example, for the Piva dam site, in Montenegro, Yugoslavia with R_o= 75 km, peak acceleration obtained from relation 10 was greater than that from relation 1 only by a factor of about 2 for a return period of 500 years, and by a factor 1.5 for a period of 200 years. The differences in peak velocities were even smaller.

Larger differences in this case may be due to a larger area, but probably also to the locations of epicentres and to earthquake period used for obtaining these seismicity maps (Kaila et al).

Design acceleration for the Hemren site was 15 %g with a return period of 200 yrs for the dam. Larger accelerations were used on the spillway, for some elements capable to withstand large deformations without failure.

Various source configurations were analysed, and diagrams (as Figs.3 and 4) were produced for some sites. Peak acceleration for Bekhme was about 25 %g with a return period of up to 200 years.

It is known that for intensities $I_o \geq V$ (MM, MKS or MCS), in the near field, peak accelerations on "hard soil" are greater, by a factor of about two, than the corresponding accelerations on alluvium or soft rock. This has been proved from strong motion data from Europe (Ambraseys 1974), and from Western United States (Duke 1972, Donovan 1974, Trifunac and Brady 1975, Seed and Lysmer 1976), and has, therefore, the general validity. The effect of local geologic conditions on peak velocity leads to a little larger values on the hard soil, while the average maximum displacements are larger on "soft soil" by a factor of up to about two. In fact, if these relations exist between peak accelerations and peak velocities on hard and soft soils, this relation of peak displacements must also proceed from theoretical considerations.

These effects are a result of the simultaneous attenuation and amplification of seismic waves in the near field of strong earthquakes. In far field peak acceleration on soft soil should be greater than on hard soil.

If peak velocity and peak acceleration occur during the same cycle of ground motion, and if it be assumed, as a first approximation, that this motion is harmonic, the predominant period of vibration would be $T_o = 2\pi V_{max}/A_{max}$. On the basis of typical values for T_o (Seed et al, 1976) for different soils, and of peak velocity which does not differ greatly on the two types of soil, a rational determination of peak acceleration is possible for the various site conditions.

The theory of extremes has given slightly conservative results. For regions of the Zagros foot-hill folded series (b) and the High Zagros folded series (c), between longitudes E 42.5° and E 47° in Figure 5, the following Type I distributions of earthquake magnitudes can be defined:

$$\text{(b)} \quad G(M) = \exp\left[\, 398 \exp\,(-1.499\ M)\right] \qquad (8)$$

$$\text{(c)} \quad G(M) = \exp\left[2351 \exp\,(-1.786\ M)\right] \qquad (9)$$

with correlation coefficients of 0.86 and 0.95, and for areas of 58,000 and 48,000 sq.km. for zones (b) and (c), respectively.

For the entire region of the folded series (b) + (c), we have obtained

$$G(M) = \exp\left[128310 \exp\,(-2.437)\right] \qquad (10)$$

with a coefficient of correlation of 0.96.

These results are based on magnitude data (1916-1970). Distributions (8) - (10) give, on a 50% probability basis, magnitudes which exceed 7 for periods longer than about 150 years. However, as-

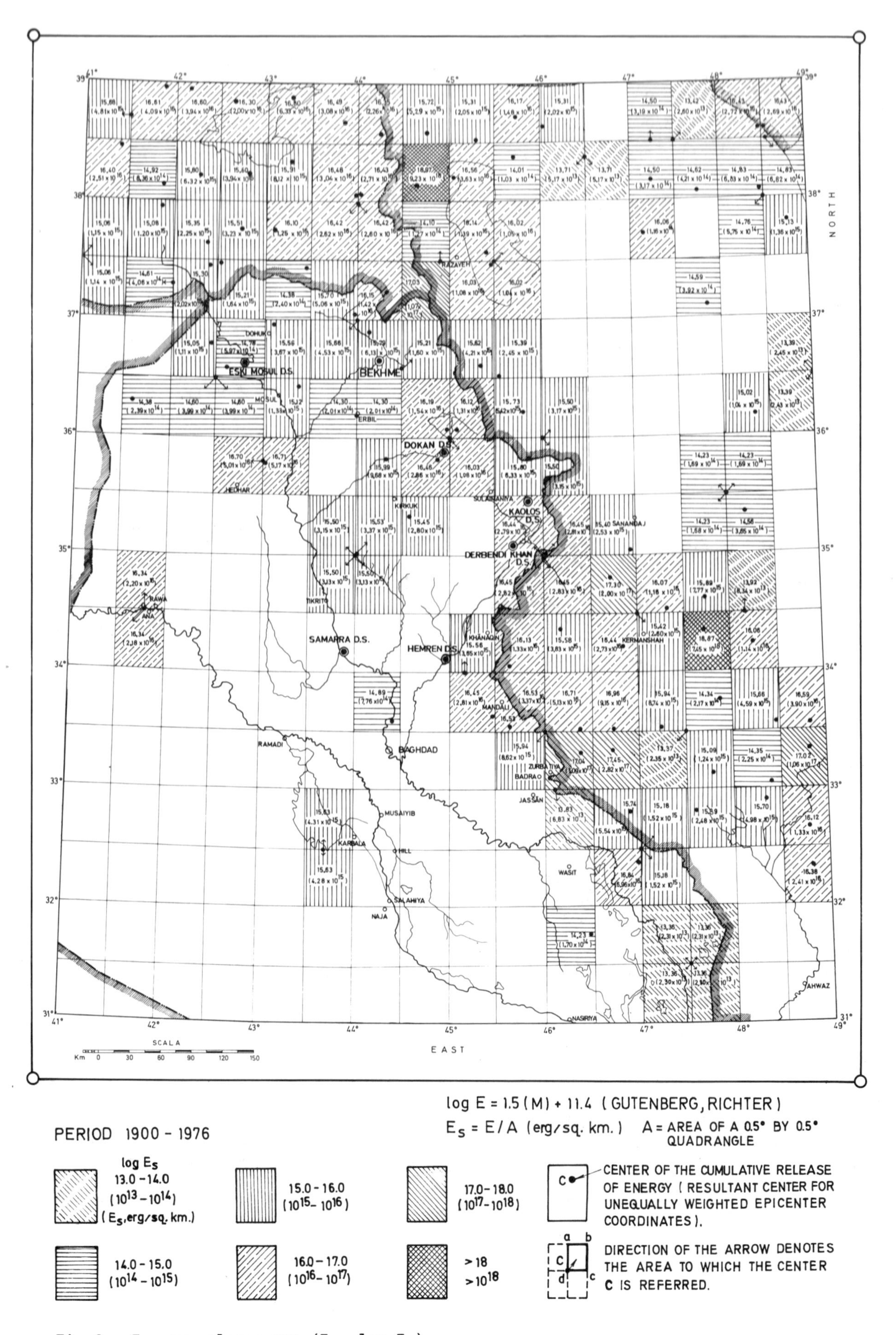

Fig.6. Energy release map (E_S, log E_S)

suming a homogeneous seismic regime over the entire regions (b) and (c), and taking that the largest credible magnitude for a period of 500 years is 7 (as confirmed by historical evidence) we have obtained design magnitudes of 6.0 (Hemren) and 6.3 (Kaolos). They are based on 50% pro babilities, and site areas of the circles of a 100 km radius.

A rough estimate shows that for the next 50 years 46,840 sq.km (i.e. 44% of the entire area of 106,000 sq.km) will be shaken with an intensity VII (MM, MKS), 6883 sq.km (6.5%) with an intensity VIII, and 520 sq.km (0.5%) with an intensi- IX. These results are again based on 50% probabilities, and an average focal depth of 25 km, and were obtained using Shebalin's attenuation constants for seismic intensities. For a focal depth of 15 km, these areas are greater, 7% for intensity VII, 20% for intensity VIII, and 80% for intensity IX. These figures may be over-estimations: with attenuation constants as determined by Chandra, McWhorter and Nowroozi (1979), we obtain smaller areas of ground shaking.

The representation in Figure 6 clearly delineates active seismic trends in the region. It was used also for an analysis of the balance of seismic energy in the region. This data can be used to compute energy flux, which requires a knowledge of the focal depth. Using areas of 2° by 2° for every site, computing energy flux, and using existing correlations between energy flux and peak velocity, attempts have been made to define "equivalent earthquakes" in terms of magnitude and peak velocity, which in a certain way should be comparative indicators of earthquake hazard for different sites.

CONCLUSIONS

Seismic risk studies reveal that, for the design of large dam projects in Northern Iraq, both historical evidence and instrumental recording of earthquakes should be considered. The largest magnitudes of up to 6.5 and peak accelerations of up to about 30 %g, should be expected for some of those sites. Because of the importance of these projects, the choice of design parameters should be based on at least a 200-year period.

REFERENCES

1. ACADEMY OF SCIENCES U.S.S.R. (editors SHEBALIN & KONDORSKAYA). New catalog of strong earth quakes on the territory of U.S.S.R. and surrounding regions, Nauka, Moscow, 1977 (in Russian).
2. ALSINAWI H. and GHALIB. H. Historical seismicity of Iraq. Bull.Seism.Soc.Am., 1975, 65, 541-547.
3. AL-TAMIMI F.S. Seismicity of Iraq. Jl.Geol Soc.Iraq, 1969, 2, 32-48.
4. AMBRASEYS N.N. Early earthquakes in North--Central Iran. Bull.Seism.Soc.Am., 1968,58, 417-426.
5. AMBRASEYS N.N. The correlation of intensity with ground motions. 14th Assem.Europ.Seis.Com. Trieste, 1974, 1, 335-341.
6. AMBRASEYS N.N. Studies in historical seismicity and tectonics. Geodinamics Today, R.Soc. London, 1975, 1, 7-16.
7. AMBRASEYS N.N. Preliminary analysis of European strong motion data (1965-1978), 1978, EAEE Bull.4, Skopje.
8. AMBRASEYS N.N. The relocation of epicentres in Iran. Geoph.J.R.Astr.Soc., 1978, 53, 117-121.
9. AMBRASEYS N.N. A reappraisal of the seismicity of the Middle East. Qt.Jl.Eng.Geol., 1978, 11, 19-32.
10. CHANDRA U., MC WHORTER J.G. and NOWROOZI N.N Attenuation of Intensities in Iran. Bull.Seism. Soc.Am., 1979, 69, 237-250.
11. CORNELL A.A. Engineering seismic risk analysis. Bull.Seism.Soc.Am., 1968, 58, 1583-1606.
12. DEPREM ARAŞTIRMA ENSTITÜSÜ BAŞKANLIĞI (T.C. Imar ve Iskan Bakanliği). 24 Kasim 1976, Kaldiran depremi raporu, 1977, Ankara.
13. GUPTA I.N. and O.W. NUTTLI. Spacial attenuation of intensities for central U.S. earthquakes. Bull.Seism.Soc.Am., 1976, 66, 743-751.
14. KAILA K.L., RAO N.M. and NARAIN H. Seismotectonic maps of Southwest Asia region, etc. Bull.Seism.Soc.Am., 1974, 64, 657-669.
15. KARNIK V. Seismicity of the European area, Parts I and II. Academia Praha, 1969/1971.
16. LOMNITZ C. Global tectonics and earthquake risk. Elsevier, Amsterdam, 1974.
17. LOMNITZ C. and ROSENBLUETH E. (editors). Seismic risk and engineering decisions. Elsevier, Amsterdam, 1976.
18. MC KENZIE D. Active tectonics of the Mediterranean. Geoph.J.R.Astr.Soc., 1972, 30, 109-185.
19. NOWROOZI A.A. Seismo-tectonics of the Persian plateau, Eastern Turkey, Caucasus, and Hin du-Kush region. Bull.Seism.Soc.Am., 1971, 61, 317-341.
20. NOWROOZI A.A. Focal mechanism of earthquakes in Persia, Turkey, West Pakistan and Afghanistan, and plate tectonics of the Middle East. Bull.Seism.Soc.Am., 1972, 62, 823-850.
21. NOWROOZI A.A. Seismotectonic provinces of Iran. Bull.Seism.Soc.Am., 1976, 66, 1249-1276.
22. ROSENBLUETH E. and SINGH S.H. Comments on Reference 14, and a reply. Bull.Seism.Soc.Am., 1975, 65, 553-556.
23. ROTHE J.R. Seismicity of the earth 1953--1965. Earth.Sci.Ser.1, 1969, UNESCO, Paris.
24. TRIFUNAC M.D. and BRADY A.G. On the correlation of seismic intensity scales with the peaks of recorded strong ground motion. Bull.Sei Soc.Am., 1975, 65, 139-162.

5 Design earthquake recurrence analysis

E. R. RIES, N. R. VAIDYA, and A. P. MICHALOPOULOS

The design earthquake for a dam can be evaluated by one of two methods: deterministic or probabilistic. For either procedure, the geology and tectonics of the site region must first be studied in order to establish seismotectonic provinces and to determine faults related to earthquake activity. In the probabilistic approach, the available historical earthquake data is then used to obtain recurrence rates for seismic events within each seismotectonic province or along any fault which might be capable of producing earthquakes in the region (300-kilometer radius area) of the proposed dam site. Based on these recurrence rates and the applicable attenuation characteristics, the contribution from the various sources is considered to obtain the recurrence interval of seismic events at a plant site.

A step-by-step methodology for the determination of design earthquake recurrence using the probabilistic technique is presented and illustrated with an example.

INTRODUCTION

1. The dynamic stresses which may be induced by earthquakes can often govern the design of a dam. This is particularly true for earth dams, where earthquake loading conditions may dictate not only the angles of the dam slopes but also may significantly influence the selection of materials, the zoning of the dam, and the construction methods (refs. 1 and 2). As a result, determination of the design earthquake (DE) is a significant aspect of dam design in potentially seismic areas of the world.

2. The seismic input is usually specified in terms of a horizontal peak ground acceleration which is associated with a design response spectrum (ref. 3), or with a recorded or synthetic acceleration time history (ref. 4). For structures whose damage or destruction could endanger public health and safety, such as nuclear power plants (ref. 5), offshore platforms (ref. 6) or other facilities (ref. 7), two levels of acceleration are often used in design.

3. The lower level acceleration corresponds to the earthquake which could reasonably be postulated to occur during the intended life of the structure. For this acceleration, the stresses should be within the elastic range of deformation as the structure should be designed to remain operable during and after the earthquake.

4. The higher level acceleration corresponds to the maximum earthquake which could reasonably be postulated to occur at the plant site. For this acceleration level, some inelastic deformation of the structure is permitted since partial damage or loss of function is acceptable as long as public safety is not endangered.

5. Regardless of the number of acceleration levels employed in design or the criteria used to establish these levels, there are two basic methods which can be applied to determine the DE for a structure. These methods, deterministic and probabilistic, are discussed in the following section.

TECHNICAL BACKGROUND

6. The deterministic procedure often used to establish a DE firstly involves a study of basic geology and tectonics of the site region with particular attention to faults and boundaries of seismotectonic provinces, i.e. of these areas which are characterized by similar seismicity and tectonic characteristics. Secondly, a review of the seismic history is made to locate the epicenter and Intensity of major earthquakes, and to relate these epicenters to faults or seismo-tectonic provinces. Based on these results, a group of conceivable DE's is postulated by selecting the most severe earthquake along each fault and in each seismotectonic province. These events are then moved along the fault or within the seismotectonic province to a point closest to the site. Using applicable attenuation curves, the site Intensity is determined from the maximum postulated events and the maximum site Intensity is chosen as the DE.

7. Although the deterministic procedure can be used to establish the DE, it does not provide the likelihood of the occurrence of such an event. Therefore, in many cases, a probabilistic analysis is preferable to quantify the risk and to permit the selection of a DE which is more representative of the overall risk embodied in the design of the particular structure, e.g., dam.

8. The probabilistic approach presented in the literature attempts to evaluate and quantify the seismic risk (Refs. 8, 9 and 10). Probabilistic concepts have often been used in the past for determining rare natural occurrences (e.g., floods, winds, etc.) for important facilities such as dams, high rise buildings, etc. By treating eartquakes as natural events the probabilistic approach estimates the rate of occurrence at the site of these events. Thus, this type of approach provides the engineer with quantitative insight to the safety aspects of facility design.

9. In this approach, as in the deterministic, the seismotectonic provinces comprising the site region, as well as capable faults, are firstly established. The site region is generally defined to be the area within a 300-kilometer radius of the site, unless very large earthquakes beyond this distance could reasonably be expected to affect the site's DE. Next, the available earthquake data are utilized to obtain the recurrence rates for events in each seismotectonic province and along each capable fault. Based on these recurrence rates and the applicable attenuation characteristics, the contributions from several seismotectonic provinces in the vicinity of the site are considered to obtain the return period of an event with a certain site Intensity. This method is described in detail in a step-by-step procedure subsequently, but first it is pertinent to discuss briefly the assumptions on which the procedure is based.

10. The probabilistic analysis rests on two basic assumptions. First, the seismic activity within any seismotectonic province or along capable faults has a uniform spatial distribution. Second, the recurrence rate for earthquakes of a given Intensity in any seismotectonic province remains the same as observed in the past. Since the projected life of a dam is very small compared to geologic time, it is reasonable to assume that earthquakes will occur in the near future with the same recurrence rate as they have in the past.

11. While the above discussions refer to earthquake Intensity, both types of analyses can be performed using instrumentally recorded magnitude data and appropriate attenuation relationships. However, since Intensity data are generally available for longer periods, it is usually preferable to perform either type of analysis with historical Intensities.

STEP-BY-STEP PROBABILISTIC PROCEDURE

12. The following steps briefly describe the general approach used in the probabilistic analysis to determine the site DE.

12.1 The boundaries of the seismotectonic provinces are determined within a 300-kilometer radius of the site as shown, for example, in Fig. 1. Several seismotectonic provinces with similar seismicity may be combined together to form equiseismic areas. Combination of the seismotec-

tonic provinces with similar seismicity may be combined together to form equiseismic areas. Combination of the seismotec-tonic provinces into larger areas may be necessary since many of the seismotectonic provinces considered may not have sufficient seismic activity to reliably calculate earthquake recurrence intervals.

12.2 Each contributory area is divided into a grid of small elements bounded either by constant latitude and longitude lines or by segments of circles for purposes of numerical integration as shown in Fig. 2. The seismic activity in each small element is assumed to originate at its geometrical center. The area within a radius of about 40 kilometers of the site may be divided into a finer grid of elements to ensure accurate numerical integration.

12.3 The recurrence rate for earthquakes in each seismotectonic province is established in the form of the following empirical relationship (refs. 11 and 12):

$$\log(N) = a_k + b_k I_o \qquad (1)$$

where N is the number of earthquakes per square kilometer per year of Intensity I_o or greater; I_o is the Intensity on the Modified Mercalli (MM) Scale; k denotes the equiseismic area; and a_k and b_k are constant characteristics of the earthquake recurrence rate associated with area k. The recurrence relation of Equation (1) is generally obtained as a least-squares fit to actual data points based on seismicity. This procedure is illustrated in Fig. 3 where the log of N has been plotted versus Intensity.

12.4 For each element, j, the seismic activity per year is computed in small Intensity increments, typically 0.2, using the recurrence relationship given in Equation (1). The Intensity scale is considered to vary continuously. Thus, the number ($N_{j'}$) of earthquakes per year in element j, with Intensities between I-0.1 and I+0.1, are obtained from the following equation:

$$N_{j'} = A_j \; 10^{(a_k+b_k(I-0.1))} - 10^{(a_k+b_k(I+0.1))} \qquad (2)$$

where A_j = area of element j.

12.5 Due to attenuation of the earthquake ground motion, the site experiences a lower Intensity than the macroseismic epicentral area. Thus, an earthquake of Intensity I_o at the epicenter would cause an Intensity I_s at the site where I_s may be estimated by using regional attenuation curves. For equiseismic areas formed by seismotectonic provinces with different attenuation characteristics, the attenuation characteristics for each province should be used for earthquakes within each individual province. A typical Intensity attenuation curve is shown in Fig. 4.

12.6 Using the attenuation relationships, the number of earthquakes from element j which will yield a site Intensity of I_s is established. The contribution to the site Intensity from all elements is then computed. This yields a seismic activity distribution for the site. From this distribution, $N(I_s)$, the expected number of earthquakes per year at the site with Intensity I_s or greater is established. The return period, denoted as $R_p(I_s)$ of an earthquake of Intensity I_s or greater, is then obtained as:

$$R_p(I_s) = 1/N(I_s) \qquad (3)$$

12.7 The acceleration corresponding to the site Intensity, I_s, may be determined using an acceleration Intensity relationship, e.g., that recommended by Trifunac and Brady (ref. 13):

$$\log a_H = 0.014 + 0.30\, I_s \qquad (4)$$

where:

a_H = Peak horizontal ground acceleration in centimeters per second squared for IV < I < X, and

I_s = Modified Mercalli Intensity at the site.

13. The above steps can best be performed with the aid of a digital computer. The following sections present and discuss the input data and the results of such analysis for a dam site.

APPLICATION OF PROBABILISTIC ANALYSIS

14. The purpose of this analysis was to provide the probability of occurrence per year of a DE for use in the design and analysis of an earth dam. The probability of occurrence of earthquakes of various Intensities was obtained by a return period computation using probabilistic methods based on the seismicity of the site region. The number and location of earthquakes that have occurred within the site

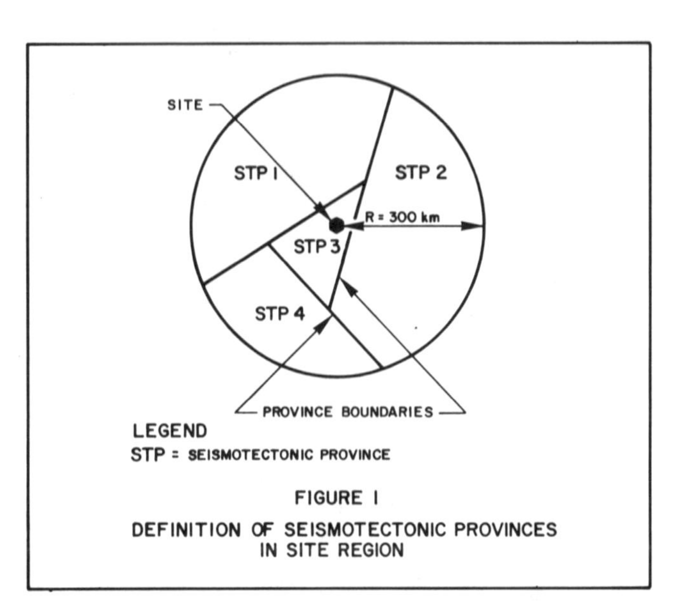

FIGURE 1

DEFINITION OF SEISMOTECTONIC PROVINCES IN SITE REGION

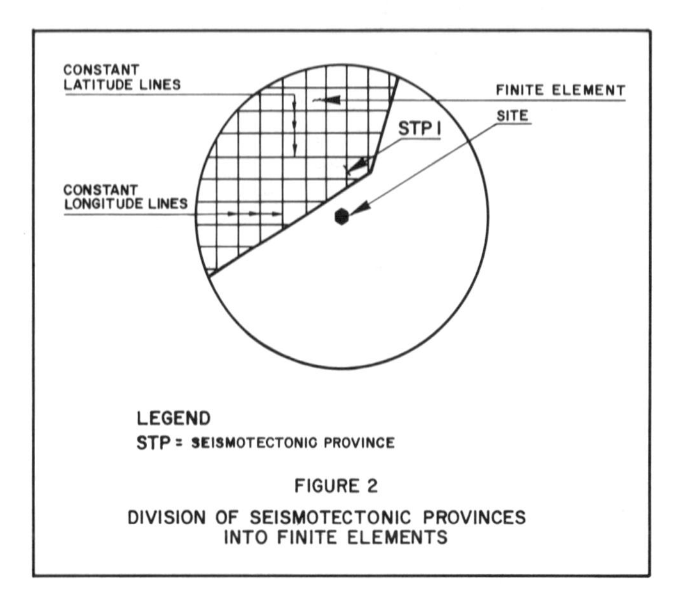

FIGURE 2

DIVISION OF SEISMOTECTONIC PROVINCES INTO FINITE ELEMENTS

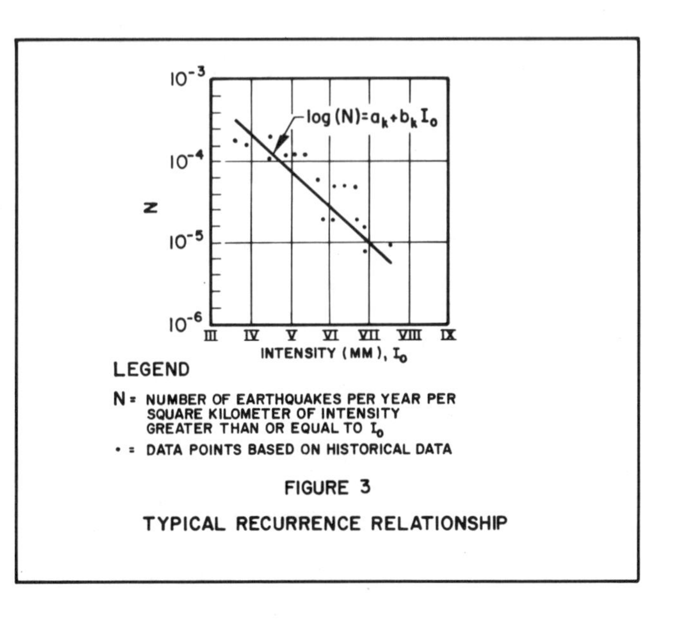

FIGURE 3

TYPICAL RECURRENCE RELATIONSHIP

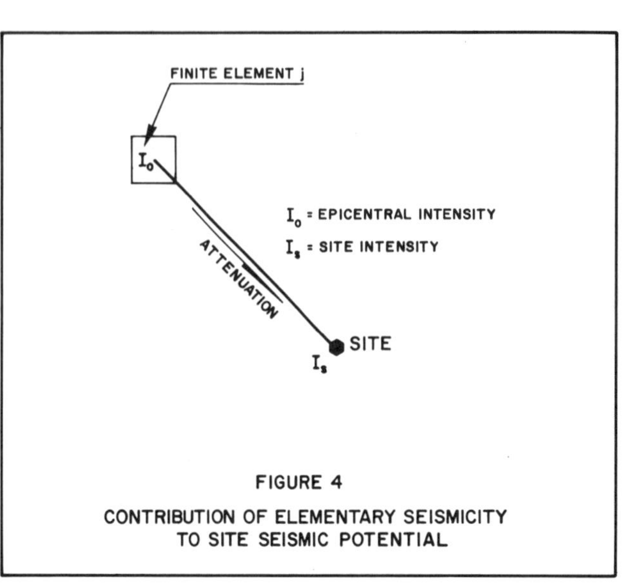

FIGURE 4

CONTRIBUTION OF ELEMENTARY SEISMICITY TO SITE SEISMIC POTENTIAL

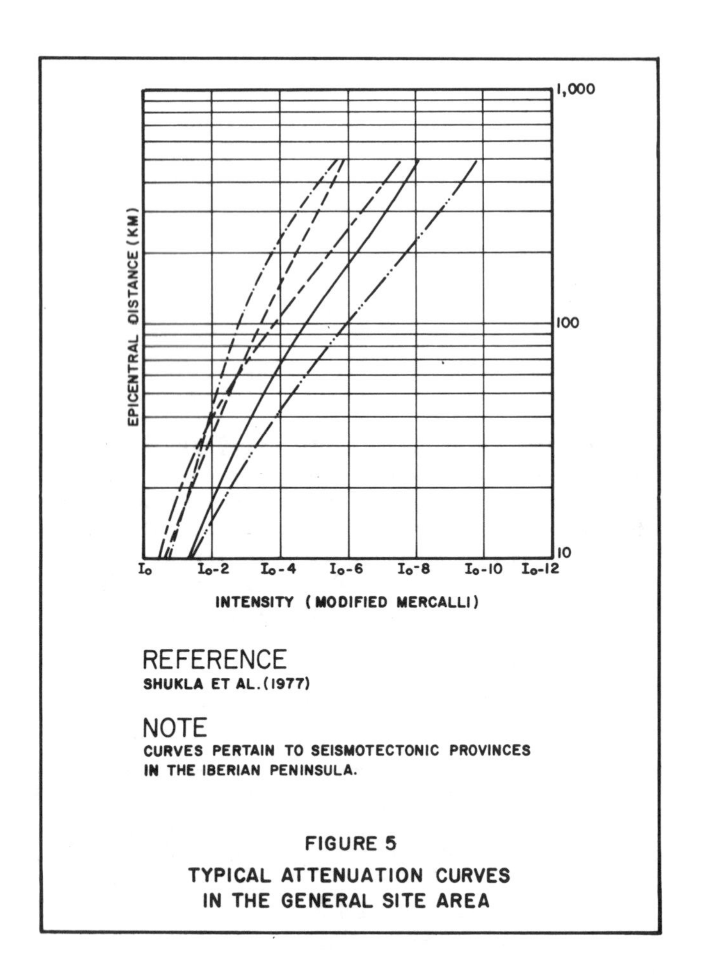

REFERENCE
SHUKLA ET AL. (1977)

NOTE
CURVES PERTAIN TO SEISMOTECTONIC PROVINCES IN THE IBERIAN PENINSULA.

FIGURE 5

TYPICAL ATTENUATION CURVES IN THE GENERAL SITE AREA

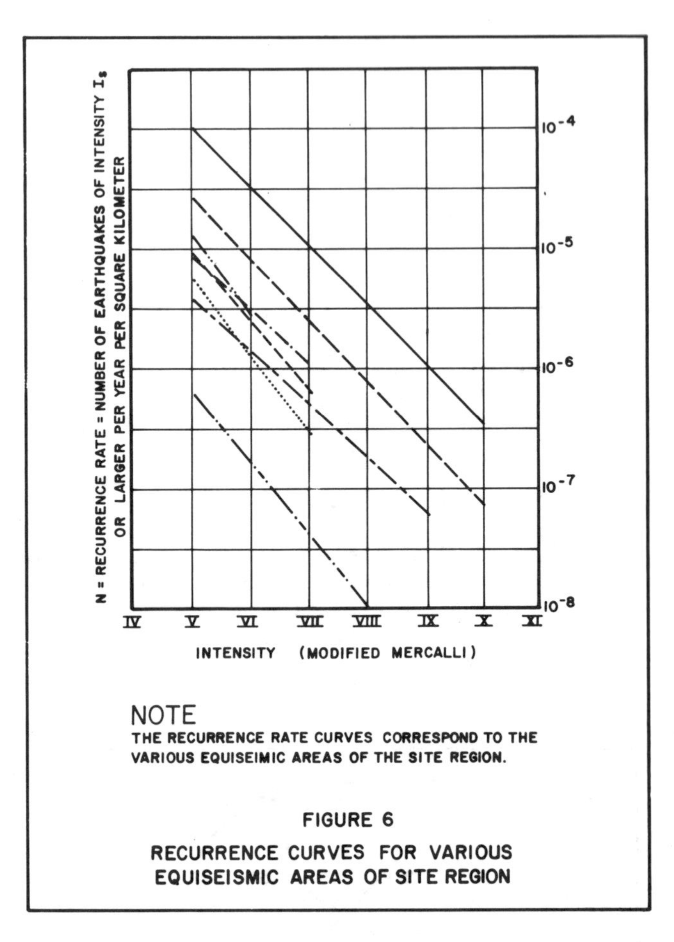

NOTE
THE RECURRENCE RATE CURVES CORRESPOND TO THE VARIOUS EQUISEIMIC AREAS OF THE SITE REGION.

FIGURE 6

RECURRENCE CURVES FOR VARIOUS EQUISEISMIC AREAS OF SITE REGION

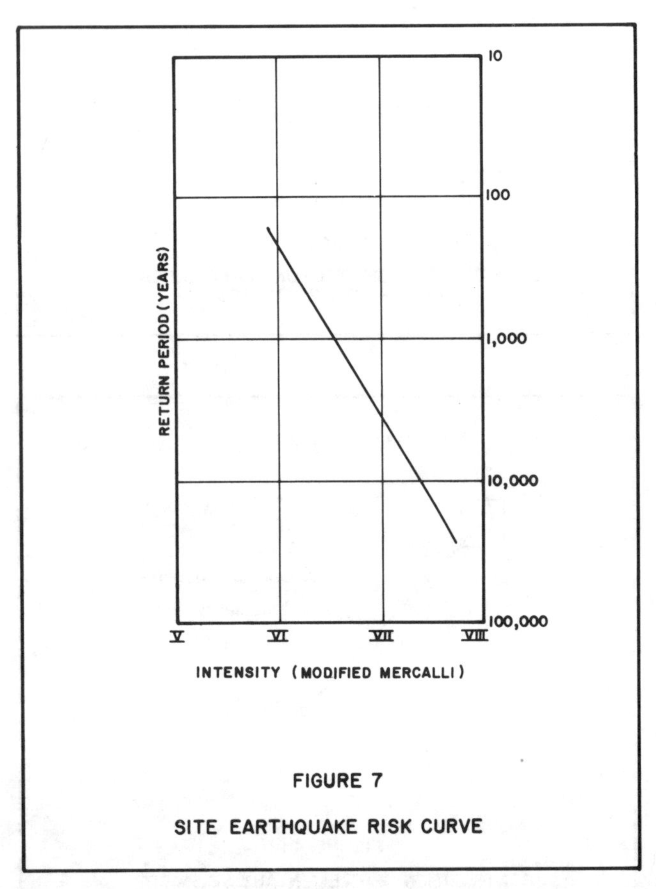

FIGURE 7

SITE EARTHQUAKE RISK CURVE

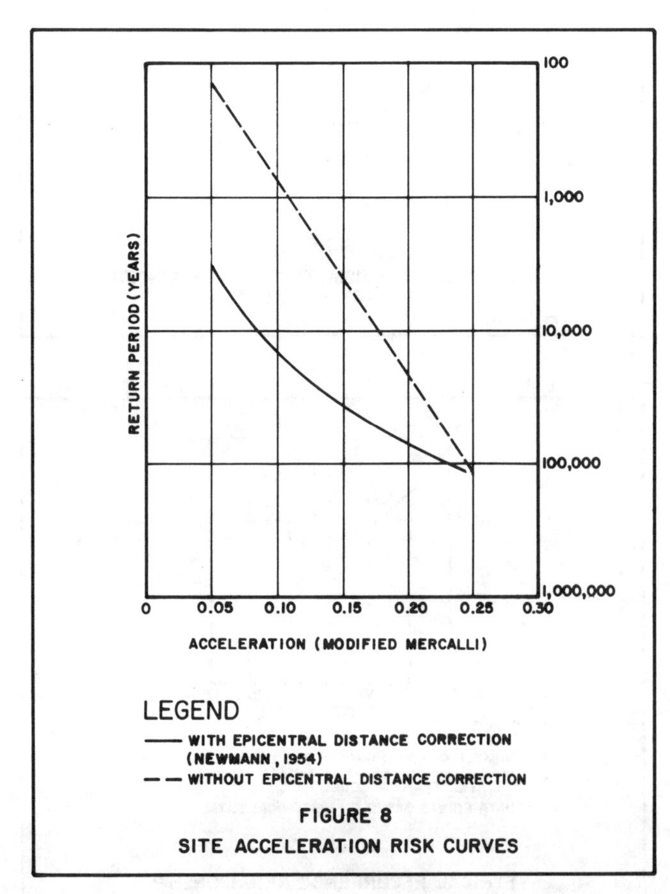

LEGEND
—— WITH EPICENTRAL DISTANCE CORRECTION (NEWMANN, 1954)
– – WITHOUT EPICENTRAL DISTANCE CORRECTION

FIGURE 8

SITE ACCELERATION RISK CURVES

region were established after an extensive search among all seismic events recorded in historic time. All earthquakes of Modified Mercalli (MM Intensity, I_{MM}, greater than or equal to V, within a 300-kilometer radius of the site, were considered as well as major offshore earthquakes which have occurred beyond this region.

15. For the purpose of this recurrence interval analysis, the number and Intensity of the earthquakes in each seismotectonic province were obtained from the earthquake list. When an earthquake was located on the border of two seismotectonic provinces, then an earthquake was attributed to each province and weighted by a factor of 0.5 for the purpose of computation. The recurrence rate relationship assumed is that of Equation (1). Coefficients a_k and b_k, which define the recurrence relationship, were obtained for each equiseismic area by choosing a least-squares fit of a straight line through a plot of number of earthquakes versus Intensity. Where insufficient data existed for a given equiseismic area, coefficient b_k was restricted between values of -0.6 and -0.4 to form a realistic basis for calculating recurrence rates (ref. 8), when the use of limited data yielded computational b_k values outside of this range, the b_k value actually used was restricted to this range.

17. Based upon recommendations by Shukla et al. (ref. 14), Howell and Shultz (ref. 15), and Esteva (ref. 10), least-squares fit curves were drawn through the observed data in the typical equiseismic areas of the site region. Typical attenuation curves for the general area of the site are shown in Fig. 5.

ANALYTICAL RESULTS

18. The distribution of earthquakes in the various seismotectonic provinces were summed to yield the number and corresponding Intensity of earthquakes in each of the equiseismic areas. A least-squares straight line fit of the form proposed by Algermissen (ref. 16) was then obtained through the points thus defined. Fig. 6 presents the recurrence rate curves for the various equiseismic areas.

19. The site risk due to all seismic sources for site Modified Mercalli Intensities greater than V is presented in Fig. 7 as a plot of Intensity versus return period. The line shown in Fig. 7 is the best fit through points with Intensities between V and VIII, since the recurrence intervals of interest are within this range.

20. The relationship between Intensity and acceleration, shown in Equation (4), proposed by Trifunac and Brady (ref. 13), was first used to compute the acceleration from the site Intensity Table 2 presents the site acceleration versus return periods. The Trifunac and Brady study, however, does not account for the effect of epicentral distance on the site Intensity. In accordance with the findings of Neumann (ref. 17) a site Intensity generated by a distant large earthquake is associated with a lower site acceleration than the same Intensity produced by a nearby earthquake. This is especially significant for sites where no nearby earthquakes have occurred, and seismic risk comes primarily from large distant events. The results presented in Table 2 do not include a distance correction although the major contributing earthquakes are distant; accordingly, the return periods are conservative. When the Neumann (ref. 17) correction for distance is included, the return periods are found to be as shown in Table 3 (Fig. 8).

TABLE 2
RETURN PERIODS FOR
SEISMIC EVENTS AT DAM SITE

ACCELERATION (G's)	RETURN PERIOD (YEARS)
0.05	130
0.10	750
0.15	3,700
0.20	28,400
0.25	116,800

TABLE 3
RETURN PERIODS FOR
SEISMIC EVENTS AT DAM SITE
WITH NEUMANN (ref. 20) CORRECTION

ACCELERATION (G's)	RETURN PERIOD (YEARS)
0.05	3,000
0.10	13,700
0.15	35,000
0.20	69,200
0.25	117,200

21. Since the probability of occurrence per year is the reciprocal of return period, the expected life of the dam can be divided by the return period to determine the probability of occurrence of the site acceleration.

SUMMARY

22. A method has been presented for the computation of earthquake recurrence periods or probabilities. Also, an example recurrence analysis based on historical earthquake data with due consideration to the tectonic and geological conditions of the site region has been described. The analysis employed probabilistic methods where the historic earthquake data was used to predict the frequency of occurrence of various Intensity earthquakes at the site.

LIST OF REFERENCES

1. Seed, H.B., 1967, Earthquake Resistant Design of Earth Dams, University of California, Berkeley, 41 pages.

2. Sherard, J.L., R.J. Woodward, S.F. Gizienski and W.A. Clevenger, 1963, Earth and Earth-Rock Dams, John Wiley and Sons, Inc., New York, 725 pages.

3. Newmark, N.M. and W.J. Hall, 1976, "Procedures and Criteria for Earthquake Resistant Design," Selected papers by Nathan M. Newmark, ASCE, New York, pp. 829-872.

4. Shaw, D.E., P.C. Rizzo and D.K. Shukla, September 1975, "Proposed Guidelines for Synthetic Accelerogram Generation Methods," Proceedings of the Third International Conference on Structural Mechanics in Reactor Technology (3rd SMiRT), London.

5. Shukla, D.K., J.F. Kissenpfennig and P.C. Rizzo, September 1975, "Safe Shutdown Earthquake Loading: Deterministic and Probabilistic Evaluations," Proceedings of the Third International Conference on Structural Mechanics in Reactor Technology, (3rd SMiRT), London.

6. Johnson, W.J., N.R. Vaidya and J.M. Musacchio, May 1980, "Seismic Evaluation of an Offshore Site in the North Aegean Sea," Proceedings of the Offshore Technology Conference, Houston.

7. Hall, W.J. and N.M. Newmark, November 1978, "Seismic Design Criteria for Pipelines and Facilities," Journal of the Technical Councils, ASCE, Vol. 104, No. PC1.

8. Algermissen, S.T. and D.M. Perkins, May 1973, "A Technique for Seismic Zoning, General Consideration and Parametes," NOAA Technical Report No. ERL 267-ESL30: Contribution to Seismic Zoning, Boulder, Colorado.

9. Cornell, C.A. and H.A. Merz, October 1975, "Seismic Risk Analysis of Boston," Journal of the Structural Division, ASCE, Vol. 101, No. ST10, pp. 2027-2043.

10. Esteva, L., 1970,"Seismic Risk and Seismic Design Decision," Seismic Design for Nuclear Power Plants, edited by R.J. Hansen, M.I.T. Press, Cambridge, Massachusetts, pp. 142-182.

11. Evernden, F.J., April 1970, "Study of Regional Seismicity and Associated Problems," Bulletin of the Seismological Society of America, Vol. 60, No. 2, pp. 393-446.

12. Chinnery, M.A. and D.A. Rogers, July/December 1973, "Earthquake Statistics in Southern New England," Earthquake Notes, Eastern Section, Seismological Society of America, Vol. XLIV, Nos. 3-4, pp. 80-103.

13. Trifunac, M.D. and A.G. Brady, February 1975, "On the Correlation of Seismic Intensity Scales with the Peaks of Recorded Strong Ground Motion," Bulletin of the Seismological Society of America, Vol. 65, No. 1, pp. 139-162.

14. Shukla, D.K., W.J. Johnson and J.F. Kissenpfennig, January 1977, "Attenuation of Modified Mercalli Intensity for Earthquakes in the Iberian Peninsula," Proceedings of the Sixth World Conference on Earthquake Engineering, New Dehli.

15. Howell, B.F., Jr. and T.R. Schultz, June 1975, "Attenuation of Modified Mercalli Intensity with Distances from the Epicenter," Bulletin of the Seismological Society of America, Vol. 65, No. 3.

16. Algermissen, S.T., 1969, "Seismic Risk Studies in the United States," Proceedings of the Fourth World Conference on Earthquake Engineering, Santiago, Chile.

17. Neumann, F., 1954, Earthquake Intensity and Related Ground Motion, University of Washington Press, Seattle, Washington.

6 Aseismic design considerations for a large arch dam

R. DUNGAR, BSc PhD, Motor-Columbus Consulting Engineers Inc, Baden

This paper describes the important aspects of the aseismic design analysis for the 226 m high, double curvature, El Cajon arch dam, to be constructed in Honduras, Central America. The seismic survey, employing both probabilistic and deterministic methods, is described. A method of computing the concrete cube strength required for both static and dynamic loading conditions is presented. A section of the paper is devoted to assessing the effects of possible construction joint opening, by using results taken from a linear elastic response calculation.

INTRODUCTION

1. The El Cajon arch dam forms the main feature of a large hydroelectric power scheme to be constructed in Honduras, Central America, and planned for completion in 1986. The double curvature dam is 226 m high and will be built in the narrow valley of the Rio Humuyo.

2. The dam is designed for conditions of both static and seismic loading, the construction site being in an area of low to medium seismic activity. Aspects of the load cases are described in this paper, and in particular the design earthquake loading is discussed.

3. The geometry is as specified for the final design investigation. Displacements and stresses were calculated for the several design load conditions, by employing a three dimensional finite element solution. Design constraints were specified, and the calculated stresses were then used to evaluate the geometry in relation to these constraints. During this evaluation period, the following questions were held in view:

a) Is the geometry, as specified, able to efficiently meet the design constraints?
b) What values of concrete strength are required for the various regions of the dam?
c) Is it necessary to adopt special measures to ensure the stability of the dam under extreme loading conditions?

4. It will be observed from the above that two basic design assessments were made, relating to the required cube strength on the one hand, and the stability of the dam on the other hand. In both cases computer techniques were developed which allowed these assessments to be made with the minimum of manual effort. During the stability assessment, the effects of opening of the vertical construction joints, which may occur during extreme dynamic loading conditions, were investigated. Consideration is also given to the hydrodynamic reservoir pressure, in relation to the hydrostatic pressure, with a view to checking for possible "cavitation effects".

Load Combinations

5. Arch dams are currently designed for a variety of different loading conditions. In the case of static loading, it is normal practice to combine the effects of self weight with those of various design temperature changes and design hydrostatic load conditions. Displacements and stresses are normally calculated by a linear elastic analysis, of the structure and associated foundation region, for each of the "basic load cases" (i.e. self weight loading, water load to a given elevation, etc.). Following this "basic calculation" a superposition exercise is used to obtain the results for each of the required "load combination" cases.

6. Load combinations may be classified according to their likelihood of occurrence during the design life of the structure, under the headings of "usual loading", "unusual loading" or "extreme loading" (ref. 1, 2). The usual load combinations, or "usual loadings" used for design purposes are chosen from those combinations of basic loads that can reasonably be expected to simultaneously occur during the design life. In all cases, self weight and silt load are normally included, together with a given temperature and hydrostatic load condition. Unusual loadings are defined from combinations that are possible, but which are unlikely to occur during the design life. Thus, for example, the load combination corresponding with the occurrence of the "probable maximum flood" (P.M.F.) is regarded as "unusual". Extreme load combinations are related to earthquakes, as will now be discussed.

Earthquake Load Combinations

7. Generally speaking, a large dam is designed

to withstand two different levels of earthquake activity. Hence, two design earthquakes, or series of earthquakes, are specified and are termed "Operating Basis Earthquake" (O.B.E.) and "Maximum Credible Earthquake" (M.C.E.).

8. An O.B.E. is an event that may reasonably be expected during the lifetime of the structure. The structure must continue to function during and after such an event and thus the loading is classed as "unusual". A full definition of this loading includes a usual static load, because such an earthquake will occur in combination with self weight, temperature, silt and hydrostatic load.

9. An M.C.E. event is defined to correspond with the most extreme earthquake action that can reasonably be expected at that site. The structure must withstand the event in such a way that repairs can be undertaken in a reasonable length of time and, more importantly, that no loss of life occurs for example due to a major flood. An M.C.E. event is classed as an "extreme" loading.

FACTOR OF SAFETY AND DESIGN CRITERIA

10. Each class of loading (usual, unusual or extreme) must satisfy certain design criteria, or what may loosely be termed "factor of safety". This may be specified as acceptable values of maximum compressive and tensile stress, in which case these stress values are normally related to either the cylinderical or cube strength of the concrete.

11. In the case of El Cajon, the design criteria are based upon the 90 day cube strength, C_u, from which a design failure envelope is defined for the concrete acting under conditions of biaxial stress. Such a design envelope is illustrated in Fig. 1, where P_{max}, P_{min} are the maximum and minimum values of the principle stress components, respectively. The factor α is defined as the ratio between the tensile strength (see later discussion) and the cube strength C_u. The biaxial strength evelope for pure tension is defined by the line $P_{max} = \alpha C_u$. The compression portion is taken from results for structural concrete (Ref. 3). Due to uncertainties concerning the accuracy of published test results (Ref. 4), the compression-tension portion is conservatively assumed as a linear relationship between P_{max} and P_{min}, as illustrated.

12. Turning now to the definition of the factor of safety, it will be observed from Fig. 1 that any condition of biaxial stress may be defined by the principle stresses or by the radial coordinates r and θ. The radial line drawn from the origin, or zero stress point, through the point $x(r, \theta)$, may be extended to meet the failure envelope, at $X(R, \theta)$. The factor of safety, F, is now defined as

$$F = \frac{R}{r} \qquad (1)$$

Strictly speaking, the factor of safety, as used in the following sections, is defined for the compression-compression range, the question of possible tensile stresses being given special consideration, as explained later.

Design Criteria for El Cajon

13. As stated above, the criteria for design varies according to the class of loading under consideration. In the case of El Cajon, the following criteria are specified:-

Usual Loading
- Compressive stresses to be within a factor of safety of 3.5
- Tension stresses to be within the tension limit of the material.

Unusual Loading
- Compressive stresses to be within a factor of safety of 2.25
- Tension stresses generally should not exceed the tensile limit, except for a small amount of joint opening in which case the apparent tensile stress is to be redistributed to a compressive resistance mechanism.

Extreme Loading
- Compressive stresses to be within a factor of safety of 1.0
- Tension stresses exceeding the tensile strength are to be redistributed to other resistance mechanisms.

Tension Stresses

14. It will be noted from the above sections that the factor F is as yet defined for compressive stresses only. The above design criteria imply that the required factor for tensile stresses is unity. The reason for this apparent reduction of safety, in comparison with that of pure compression, is that tension stresses are generally associated with bending action within an arch or cantilever, and thus any relaxation of such tensile resistance will usually be accommodated by a redistribution to a compressive resistance mechanism. However, it should be emphasized that tensile stresses should be minimized by correct shaping of the geometry, especially for usual load combinations.

15. For the compression-tension range, the factor F of Equation (1) is taken as a linear function of P_{min} and having a value equal to the nominal factor of safety (value for pure compression), when $\theta = 45^{o}$ and unity when $\theta = 135^{o}$.

Required Cube Strength

16. Turning again to Fig. 1, it will be observed that once the cube strength, C_u, has been defined for a particular location, a knowledge of the stresses, P_{max}, P_{min}, enables the factor F to be defined. Conversely, knowing the position x (r, θ), (see Fig. 1) the position X (R, θ) may be calculated for a given factor of safety and hence the required cube strength, C_u, may be obtained. The factors of

safety, for the various classes of load combination, have been specified in the above sections.

17. A knowledge of the concrete cube strength required at a given location enables the required cement content to be established. The exercise of defining the final cement content may be left until extensive material tests have been conducted under on-site conditions.

SEISMICITY OF THE EL CAJON SITE

18. The seismicity for the El Cajon site has been fully explored by conducting both a probabilistic and a deterministic seismic hazard survey. These two methods will be briefly outlined below and the method of establishing the design time acceleration input records for both O.B.E. and M.C.E. conditions is discussed. In both cases criteria relevant to the M.C.E. were first established and then the O.B.E. criteria were extrapolated.

19. The El Cajon dam site is located within the tectonic unit known as the Caribbean Plate. Generally, earthquakes occur along plate boundaries, but there are exceptions to this rule. Here, three types of sources were considered:

a) Plate boundary sources,
b) Local internal plate sources,
c) Reservoir induced sources.

Probabilistic Seismic Hazard Survey

20. The probabilistic seismic survey for the El Cajon site formed a part of the seismic survey conducted for Honduras, as reported in Ref. 5, the seismicity of the Honduras region being developed by assessing sixteen tectonic provinces. Data recorded for earthquakes occurring in each province was used to construct a recurrence relationship, or a relationship between the frequency of earthquake occurrence as a function of the Richter magnitude, for a time span of 80 years.

21. Two different seismic source models were investigated, corresponding firstly with generalized faulting of the province and secondly with seismic activity associated with known active geological faults. In addition, the effects of assuming two different attenuation relationships, which define the acceleration at a given site as a function of Richter magnitude and distance, were ascertained. A Bayesian seismic hazard model was used to obtain the expected peak ground acceleration as a function of return period.

Deterministic Seismic Hazard Survey

22. The above probabilistic seismic survey is based upon data extracted from past earthquakes, collected over a limited period of time. This indicated that the maximum credible event is expected from a near field shock of only medium magnitude, for which it is indeed difficult to obtain a realistic assessment of the expected maximum ground acceleration from the few recordings that are currently available. For this reason, it was considered more appropriate to estimate the M.C.E. event based upon deterministic methods, using known geological data.

23. Both far field and near field potentially active seismic sources were investigated. Maximum fault lengths and corresponding rupture depths were estimated and assumed stress drops used to obtain appropriate earthquake magnitudes.

24. The deterministic analysis indicated that seismic risk due to earthquake sources at plate boundaries is minimal. The El Cajon dam is located at approximately 100 km and 200 km, respectively, from two tectonic plate boundaries and accelerations at El Cajon due to the M.C.E. event, from these sources, should not exceed 0.15 g in value.

25. A microseismic survey was conducted in the site area during a three months period. Approximately 50 earthquakes of magnitude ranging from 1.50 to 2.75 were located within 20 km of the dam site. There was no evidence from this survey to suggest that any particular fault in the local Honduras Depression is active. However, it was conservatively assumed for design purposes that two normal faults, located at 14 and 22 km from the dam site, could produce the maximum credible earthquake event. A deterministic analysis of these faults indicated magnitudes of 6.5 and 6.8, respectively. The attenuation relationships of Schnabel and Seed (Ref. 6) were deemed appropriate and an acceleration value of 0.32 g obtained for the M.C.E. event. The effects of reservoir induced seismicity was considered to be contained within this acceleration value.

Design Earthquake Time Acceleration Histories

26. It is well recognized by earthquake engineers that to simply define the peak acceleration value for a given design earthquake is insufficient information. The structural response will be governed by several important additional factors such as frequency content of the earthquake, direction of energy propagation, and time duration of significant shaking.

27. In the case of El Cajon, it was decided to specify three orthogonal components of earthquake motion for design calculations. Earthquake acceleration-time histories may be specified from the records of past events or may be artificially created. It is possible to create a synthetic acceleration-time history for a given design response spectrum, but it is difficult to define an artificial three component record with a phase relationship between any two components that has the physical significance that exists in an actual situation. Thus, the Parkfield, Temblor No. 2, three component recording of June 27, 1966 was chosen

as basic data, and transformed to correspond with the El Cajon situation. The reasons for this choice includes similarity of magnitude and distance from "source" to site. A further reason is that the low stress drop event appropriate to normal faulting expected for El Cajon is best represented by the medium stress drop earthquake of a strike-slip fault, such as Parkfield, in the absence of recordings appropriate to a normal fault situation.

28. The Parkfield event is best interpreted as a multiple event and the geometry and direction of faulting, in relation to the recording station, was considered to be important. This geometric relationship was maintained by transforming the three components of the Parkfield motion into a direction appropriate to the normal fault situation of El Cajon. Thus, both large vertical and horizontal motions were obtained. Finally, the three components of motion were scaled to correspond with a peak acceleration value of 0.32 g.

29. In performing the transformation process above, it was conservatively assumed that a fault break would occur immediately upstream of the dam or in a direction corresponding with the cross canyon direction. Also, the fault break was assumed to be orientated such that maximum energy transfer would take place in the source-site direction. As previously stated, the actual faults in the vicinity of the dam are known, but at this point in time there is little or no reference in the literature to the correlation of ground motion with directionality of energy propogation, and its associated attenuation, to justify a full consideration of such effects.

Design Earthquake Response Spectrum

30. The frequency content for the design earthquake acceleration-time histories is displayed in Fig. 2 in terms of plots of pseudo spectral acceleration. Also displayed is the 84th percentile response spectrum developed from a statistical analysis of recorded accelerograms representative of M.C.E. conditions for the Auburn Dam Site, California (see Ref. 7), for which a 6 to 6.5 magnitude earthquake at 10 km is expected.

IDEALIZATION OF STRUCTURAL AND FLUID DOMAINS

31. A finite element idealization was made for the structural and fluid domain, using the EFESYS finite element system (Ref. 8). The finite element mesh used for static and dynamic computations is illustrated in Fig. 3. The dam region is idealized in terms of 42 isoparametric cubic elements, each with 20 nodes and 60 degrees of freedom, and 6 isoparametric triangle-cube elements, each with 15 nodes and 45 degrees of freedom. The foundation region is idealized using 64 isoparametric elements, each with 20 nodes. This structural idealization results in 792 deflated degrees of freedom.

32. For the dynamic computation, the fluid is idealized as an incompressible domain by the 20 node Laplace elements, illustrated in Fig. 4. Connection between the structural and fluid domain, in order to account for the fluid "add-mass" effect, is accomplished by a method described by Dungar (Ref. 9). The geometry of the fluid mesh is defined by the geometry of the structural elements at the structure fluid interface and mesh generation is accomplished by projection of this interface geometry in the upstream direction.

Static and Dynamic Modes of Deformation

33. The basic static calculation produced displacement and stress results for the "basic load cases". Hydrostatic and temperature loads were calculated in the normal way, using a "consistent load" approach. Self-weight stresses were calculated by assuming independent cantilever action.

34. The eigenvectors were calculated by the method of "subspace iteration", together with the aforementioned method of fluid-structure coupling. The resonant periods for the first ten modes range from 0.648 sec. to 0.178 sec. and the mode shapes are typical for a geometry such as El Cajon and will not be displayed here. Stresses corresponding with each eigenvector were calculated in the usual way and the resultant displacements and stresses for the basic load cases and eigenvectors were stored in a "data bank" for reuse by the EFESYS program, during post processing operations and dynamic response calculations, as described in Ref. 8.

STRESS VECTOR AND CUBE STRENGTH PLOTS

35. As explained in a previous section, the stresses for a particular load combination are analyzed for various locations on the water and air face. The EFESYS program enables stresses acting within the plane of the face to be computed from the six components of stress as calculated from a three dimensional analysis. These surface stresses are termed "local stresses".

Stress Vector Plots

36. A knowledge of the three local stress components (two direct stress and one shear stress), at various locations, enables the principle stress vectors to be plotted. Besides forming a check for the calculated cube strengths this plot enables stress details to be inspected. The areas of tensile stress are of particular interest. A typical vector plot is illustrated in Fig. 5, for the air face. A vertical projection onto the reference cylinder is here used.

Required Cube Strength: Static Loadings

37. The design criteria above include specified values of the factor of safety, F, and values of the uniaxial tensile strength factor,

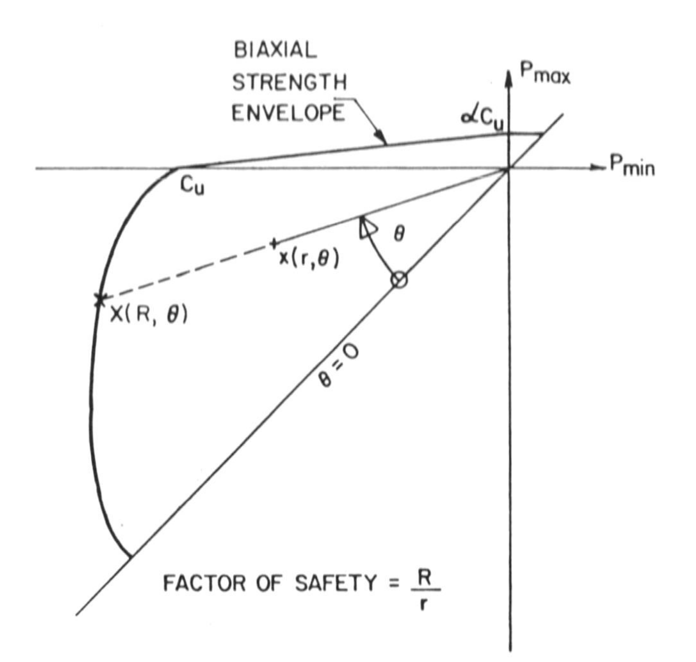

Fig. 1 Design envelope of biaxial strength

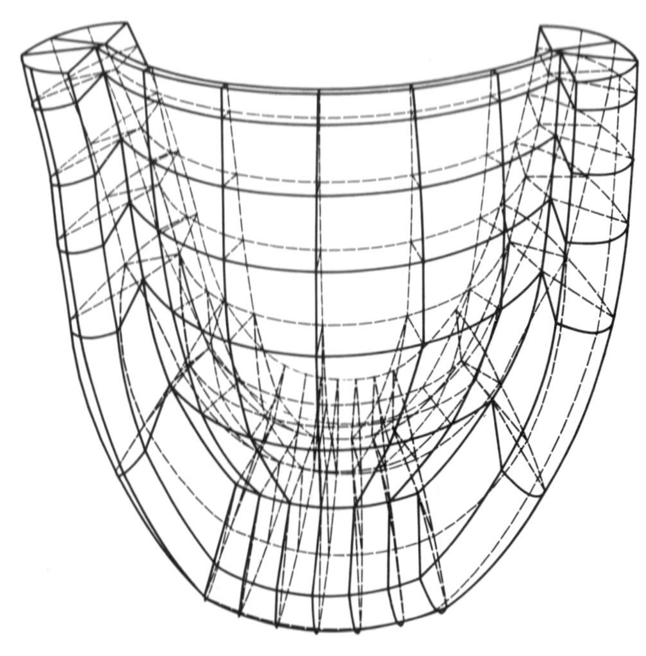

Fig. 3 Finite element mesh: dam and foundation

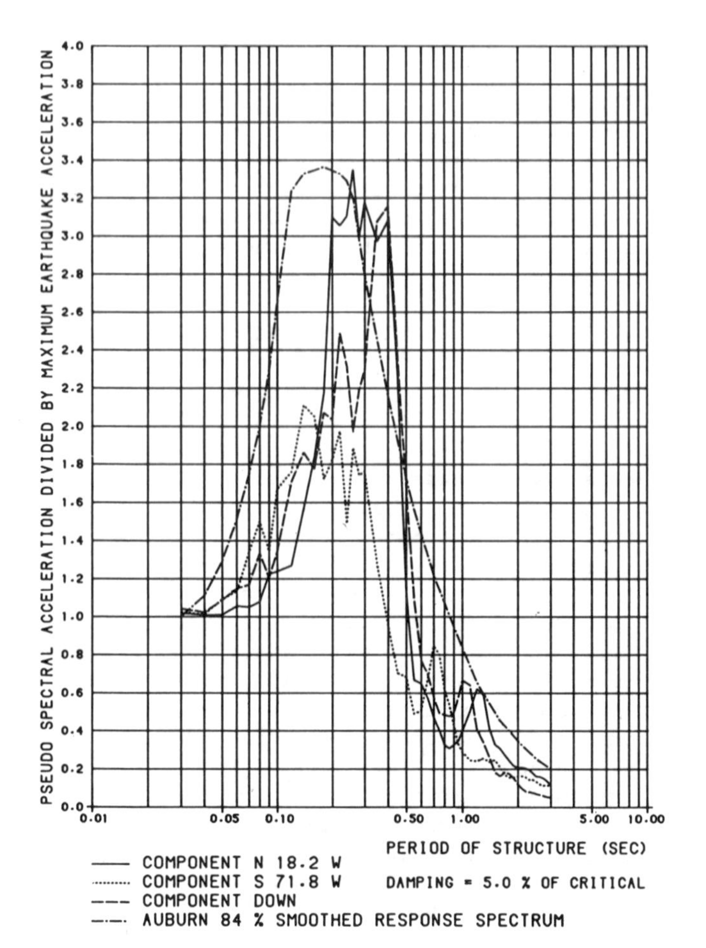

Fig. 2 Design earthquake spectral acceleration

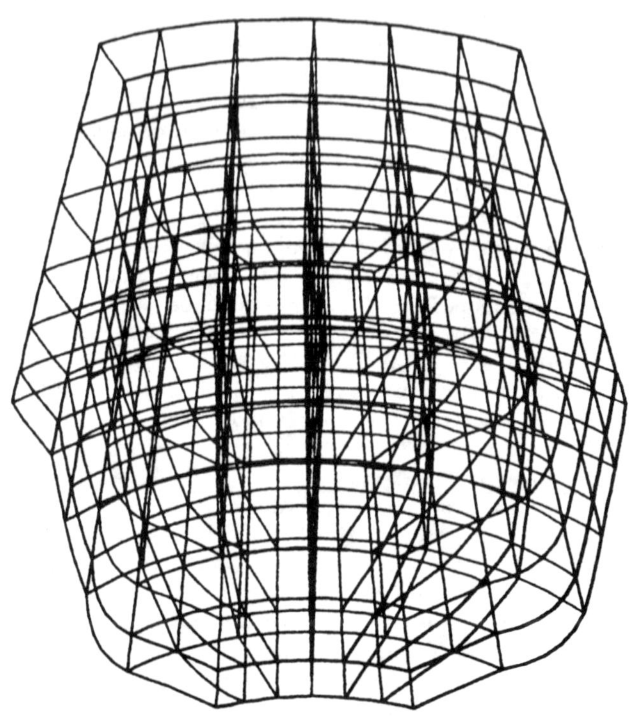

Fig. 4 Finite element mesh for fluid region

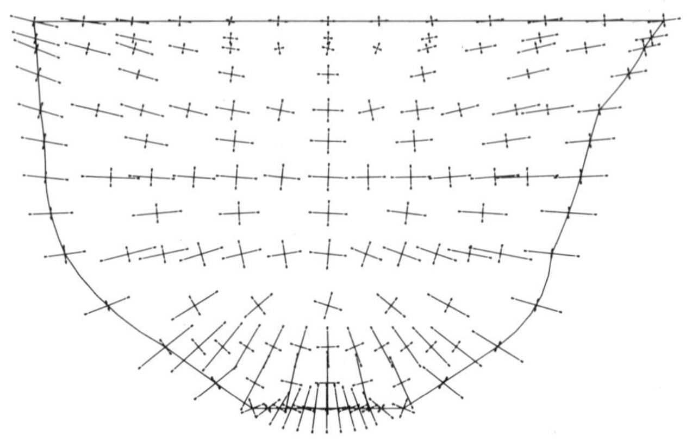

Fig. 5 Stress vector plot: usual loading

α. These values are specified for a given load combination. Knowing the local stress components the corresponding cube strength, C_u, and angle, $\emptyset$, may be computed. The EFESYS system enables contour plots of both quantities to be produced. Such plots are normally obtained for each considered static load combination. The plot of Fig. 6 corresponds with the stress vector plot of Fig. 5, ($F = 3.5$, $\alpha = 0.08$).

Required Cube Strength: Dynamic Loadings

38. In the case of load combinations with earthquake conditions the combined static and dynamic "stress picture" changes with each time step. For El Cajon, stresses were calculated for every 0.01 seconds for the first 6 seconds of earthquake motion, by employing the normal computational techniques (see Ref. 10, 11 and 12). In performing this computation, it was further assumed that any horizontal tensile surface stress would create an open joint and dissipate to compressive resistance on the opposite face. Thus, horizontal tensile stresses were set to a zero value before calculating the cube strength value.

39. It is clearly unrealistic to display the stress vectors, or required cube strengths, for each and every time increment, and thus certain information must be selected by the program. Here it was decided to select the maximum value of the required cube strength indicated at each and every location on the face and in addition the corresponding stress components were also stored. A typical contour plot of required cube strength is displayed in Fig. 7 and may be compared with the corresponding plot for a usual (static) loading combination of Fig. 6.

Envelopes of Maximum Required Cube Strengths

40. So far, we have considered the calculation, and plotting, of required cube strengths for particular load combinations (both static and dynamic). Clearly no one load combination will necessarily give the most unfavorable conditions for all points on the dam and thus the cube strength data must be sorted on a point-by-point basis. To accomplish this, the EFESYS system includes a facility for obtaining "envelopes" of cube strength by using the previously computed and stored results of the individual load combinations (static and dynamic).

STABILITY EVALUATION

41. The question of vertical construction joint opening, redistribution of stress and resulting stability evaluation will now be considered in detail. In reality, the opening of a joint, or series of joints, may occur at a large number of the time increments. It would be an exceedingly difficult task to check the stability at all such time increments and so the time steps corresponding with both the maximum and the minimum dynamic stress are chosen, for which the stress pictures were recorded and analyzed.

42. Contour plots of horizontal stress, for the water face and the air face, enable the zones of average net horizontal tension to be identified. A typical stress plot for the vertical central section, for the case of the O.B.E. earthquake acting predominantly in the upstream-downstream direction, is illustrated in Fig. 8 and the depth from the crest to the limit of average net horizontal tensile stress is indicated. The zone of net horizontal tensile stress for the complete dam is as illustrated in Fig. 9, for this and other conditions of earthquake loading.

43. The average horizontal tensile stresses of Fig. 8 may be converted to an equivalent radial load by the following equation:

Radial load/unit area of cantilever =

$$\frac{1}{R} \times \sigma_t \times T \qquad (2)$$

where R = horizontal radius of curvature,
σ_t = tensile stress at given elevation,
T = thickness of cantilever,

from which the cantilever bending stress may be obtained and combined with the corresponding stress as calculated by the computer program. The stability of each cantilever is thus checked. When necessary, account may also be taken of shear connection between adjacent cantilevers.

INTERPRETATION OF SURFACE STRESSES

44. The results of surface stress, produced by using a linear elastic finite element solution, should be viewed with care. The results of a uniaxial tension test on concrete will certainly indicate a nonlinear stress-strain relationship and by simply assuming a bending strain distribution which is a linear function of the distance from the neutral axis the stress function will be far from linear. It is appropriate, therefore, to define the tensile limit for surface stresses based upon an assumed linear stress-strain relationship, and which corresponds with the sections ultimate moment of resistance. This stress may be termed "the apparent tensile strength", A.T.S., and corresponds directly with the calculated surface stress. Hence a series of preliminary statical and dynamical beam flexure tests have been performed to estimate the A.T.S. for the El Cajon concrete.

RESERVOIR CAVITATION INVESTIGATION

45. In addition to modifications in the reservoir hydrodynamic pressure caused by compressibility effects (not considered here) the reservoir pressure cannot fall below a given value due to what may loosely be called "cavitation". The hydrodynamic pressure at the crown cantilever face is displayed in Fig. 10, in combination with the hydrostatic pressure.

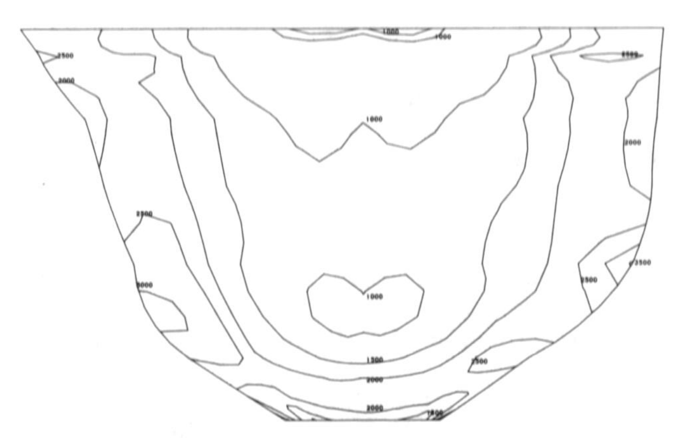

Fig. 6 Contour plot of required concrete cube strength: usual loading

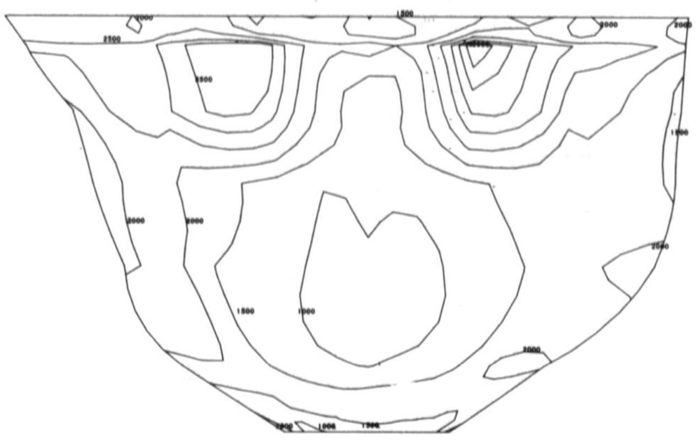

Fig. 7 Contour plot of required concrete cube strength: dynamic loading

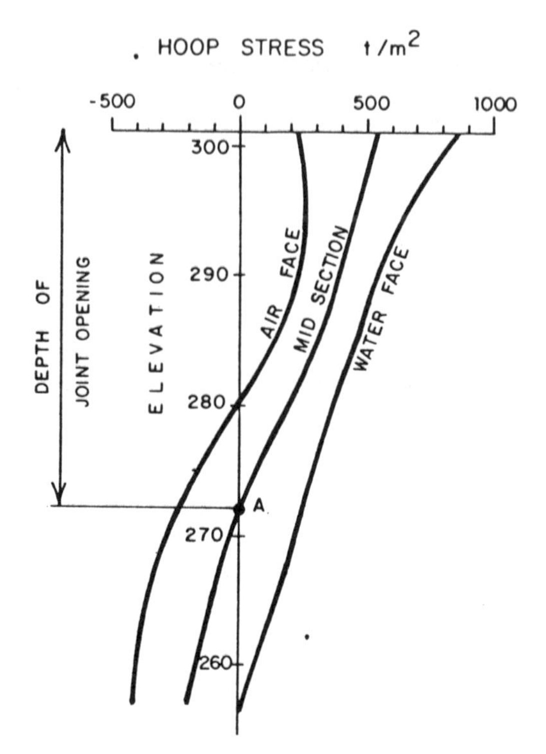

Fig. 8 Hoop stress on crown cantilever for time of maximum dynamic stress

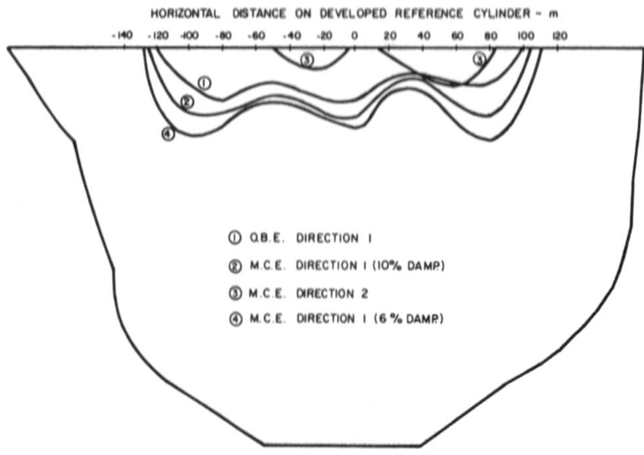

Fig. 9 Zones of tensile hoop stress for various design earthquakes

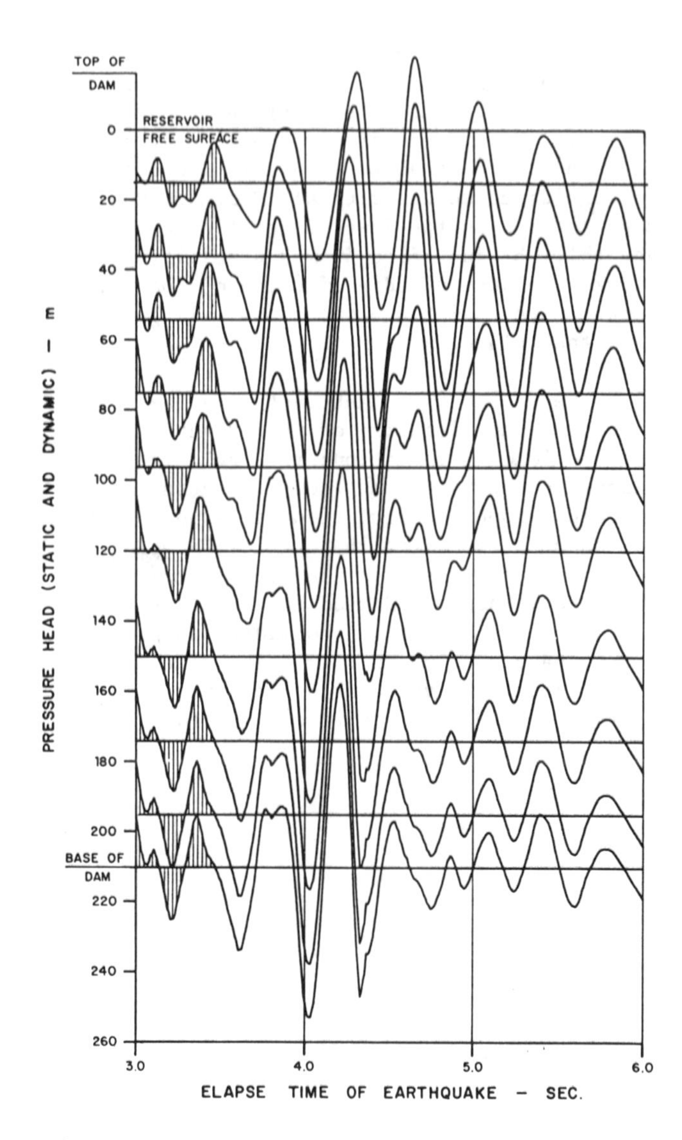

Fig. 10 Combined static and dynamic pressure head on crown cantilever, M.C.E. condition in upstream direction

These results show that only a limited portion of the reservoir experiences pressures which fall below atmospheric and other sections indicate a similar picture. Thus, it is concluded that, in this case, the effect of possible cavitation is unlikely to seriously alter the overall response of the dam.

CONCLUSIONS

46. Within the limited space available in this paper it is not possible to enter into the details of the results of the calculation for El Cajon. To date, eleven load combinations have been analyzed, including usual, unusual and extreme conditions. The results of this investigation have indicated two important and critical factors relating to the dynamic analysis of arch dams. Firstly, providing that the geometry is suitably chosen to meet the requirements for static loading, the acceptance of this geometry for earthquake load conditions will very much depend upon the allowable "apparent tensile strength" of the material. Here, the air face cantilever stress calculated from the above equilibrium solution, is considered to be critical. Cracking due to high tension in the lower foundation can generally be justified from a stability point of view because a redistribution of load to a compressive resistance mechanism will generally take place.

47. The second important factor is that the effects of nonlinearity of the structure during extreme loading conditions is a relatively unknown quantity. Thus, when conducting an elastic response calculation it is necessary to assume a value for the critical damping factor, or factors, for the various modes of vibration. Under conditions of high straining and joint opening it was considered appropriate to take a value of 10 % critical for this investigation. A value of 6 % critical was used for the O.B.E. conditions and for the earthquake acting in the cross canyon direction. However, such values may be supported only by future research in the form, for example, of large force prototype shaking or model tests.

REFERENCES

1. ARTHUR H.G. and TARBOX G.S. Bureau of Relaration's design criteria for concrete arch and gravity dams - 1975. Criteria and assumptions for numerical analysis of Dams, Swansea, 1975, 109 - 120.
2. TARBOX G.S., DREHER K.J. and CARPENTER L.R. Seismic Analysis of concrete dams, I.C.O.L.D., New Delhi, 1979, Q51, 963 - 993.
3. KUPFER H., HILSDORF H.K. and RUSCH H. Behavior of Concrete under Biaxial Stresses, Jl. A.C.I. Vol. 66, No. 8, August 1969.
4. NEWMAN J. Private Communications.
5. KIREMIDJIAN A.S. and SHAH H.C. Seismic Hazard Analysis of Honduras, Report No. 38, J.A. Blume Earthquake Engineering Center, Stanford, August 1979.
6. SCHNABEL P.B. and SEED H.B., Bulletin 198, California Mines and Geology.
7. Earthquake Evaluation Studies of the Auburn Dam area, Vol. 8, "Earthquake Ground Motions", Bureau of Reclamation, July 1977.
8. DUNGAR R. EFESYS an Engineering Finite Element System. Advances in Eng. Software, 1979, Vol. 1, No. 3. 115 - 123.
9. DUNGAR R. An efficient method of fluid structure complying in the Dynamic analysis of Structures Int. J. of Num. Methods in Eng. 1978, Vol. 13, 93 - 107.
10. DUNGAR R., SEVERN R.T., TAYLOR P. The Effect of Earthquakes on Arch Dams I.C.O.L.D., Istanbul, Sept. 1967.
11. DUNGAR R. and SEVERN R.T. Dynamic analysis of Arch Dams, Symp. on Arch Dams, London, March 1968.
12. BACK P.A.A. et al. The seismic analysis of a double curvature arch dam, Proc. Instit. Civil Eng., 1969, 43 (June) 217 - 248.

Session 1: Discussion

SEISMICITY

Mr P. Londe (Session Chairman; President of ICOLD). Earthquakes have always been known to be the most destructive and the most unpredictable of all natural calamities. However, only a few case histories of damage to dams have been reported and, at present, the assessment of the risk caused by earthquake is still of a speculative nature in relation to dams.

The damage so far observed on dams after earthquake is

(a) Concrete dams : only five dams are known to have suffered minor or moderate cracking. None have collapsed. (Rohem L.H. and Hansen K.D., Water Power, April, 1979).

(b) Embankment dams : collapses are known of 12 dams (Seed H.B., Rankine lecture, Geotechnique, 1979, Vol. 29, No. 3), (Haws E.T. and Reilly N., Paper 28). In addition 40 slope failures, mainly in sandy embankments, have been recorded.

(c) Natural slopes (soil or rock) : some have failed during earthquakes (Madison Canyon, Alaska, etc.).

The relatively low number of collapses is amazing when it is realized that there exist hundreds of thousands of dams in the world, many of them in earthquake regions. For instance, 84 000 dams have been built in China since 1949; none of them have collapsed although severe earthquakes have been experienced (Shen C. and Chen H., Paper 16). It is also interesting to recall the 1906 California earthquake, which severely damaged San Francisco, San Jose, Salinas, and Santa Rosa but did not cause any significant damage to the 33 embankment dams which felt the shocks. It seems therefore that dams have an inherent resistance to earthquake.

The engineering consequence of scarcity of case histories is that the design of dams to resist earthquakes is still of a speculative nature. 'A vital missing link seems to have been detailed observation of the field performance of dams subjected to earthquake shaking.' (Seed H.B., Paper 12.) 'Real experience with regard to seismic design' is only gained by designing earthquake resistant dams and then observing their performance under full scale earthquake loading (Seed H.B., Paper 12).

Even with such observation there remains the difficulty of assessing the actual characteristics of the ground motion to which a dam was subjected - unless thoroughly instrumented for the purpose, which is exceptional. It seems essential for further progress in the instrumentation of dams for dynamic behaviour, to a much greater extent than at present.

Ground motion is very complex. The source of the motion is a relative displacement at depth of a finite area of the sides of a fault. The source produces several impulses as it propagates along the fault. The waves produced are of different natures and each of them is refracted, reflected and transformed by the geological conditions. The result at a given site, some distance away from the focus, is a ground motion containing 'virtually all possible motion in a generally random sequence' (Long R.E. Paper 1). Like all complex data, the potential ground motion at a given site is practically impossible to forecast because of the large proportion of uncertainties in the process.

The general trend is therefore a probabilistic approach for defining earthquake risk. But serious difficulties arise as to how far extreme events of very low probability, and therefore represented by statistical samples usually too small for satisfactory analysis, are significant for the design of the dam. In addition, as pointed out by Bozovic (ICOLD Committee on Seismic Aspects of Dam Design), events usually occur in clusters after prolonged quiescent periods, making the assessment of actual return periods unreliable. The lifetime of a dam is small as compared with geological time (and therefore with return periods of strong earthquakes) but the observation period is also very short as compared with the geological time. There is some similarity with the statistical analysis of floods at a dam site, but for earthquake the uncertainty is much larger as the phenomenon is more scantily and randomly distributed in time and space.

This means that purely statistical analysis is not entirely reliable and that it should be supplemented by a thorough geological and tectonic survey of the region. This is considered as the basic step by most of the authors : survey of faulting, evidence of recent movements, nature of the rocks and soils in the vicinity of the site. The Paper by M.B. Tosic (Paper 4) is a good example of this approach.

The knowledge of these features is indispensable for a correct interpretation of the historical and even instrumental records of earthquakes. The detection of active faults is particularly meaningful.

Although the ground motions in the past were certainly very complex, only a nominal part of this complexity was recorded or observed. For instance, intensity scales give the global effect of an earthquake, but not at all the basic components of the actual motion.

N.N. Ambraseys has shown (Proceedings of 14th Assembly of European Seismological Commission, 1975) that the correlations of intensity with ground motion are weak and not reliable for design. Unfortunately the designers can usually find records of intensities and no records of actual ground motions. This is why intensities are still widely used, at least as an index for comparing sites and for preliminary evaluation of seismic risk.

The lack of data for proper knowledge of the characteristics of potential motion components at a dam site makes it imperative for the designers to make a sensitivity analysis of the parameters of the seismic risk. This is a common approach for all engineering problems where the degree of uncertainty is high (e.g. determination of design floods, stability of rock slopes, earthquake risk). This point is clearly stressed by R. Chaplow (Paper 2). This sensitivity analysis leads to defining the input characteristics in terms of the response of the dam, rather than arbitrarily from non-significant data of too small and truncated a sample of past shocks. A good example of this approach is given by T.G. Tsicnias and G.L. Hutchinson Paper 26, who show that the selection of appropriate design earthquake inputs for arch dams makes it necessary to consider vertical ground motion.

The sensitivity analysis is a fundamental tool for the designer. It reduces the detrimental effect of complexity on the reliability of simplified practical models by allowing selection of the significant parameters, and therefore the significant simplified models. Such is the manner in which all improvements have been achieved in the long history of engineering. This selection is made more and more feasible, thanks to the powerful tools we now have at hand. It is obvious that statistical, mathematical and numerical methods available at present are much better than the input data of earthquake observation. These powerful tools make it possible to proceed with a sensitivity analysis which leads to the correct design decision provided of course that experience, as defined by H.B. Seed, is associated with it.

Another fundamental factor in defining the significant input is the level of earthquake energy to be taken into account in terms of permissible or acceptable damage. The concept of two levels, as recalled by E. Ries, N.R. Vaidya and A.P. Michalopoulos (Paper 5) is common practice in nuclear plant design. The OBE (operating basis earthquake) and MCE (maximum credible earthquake) are used by R. Dungar (Paper 6) for the design of a high arch dam. There might be a difficulty however in defining an OBE (sometimes called maximum expectable earthquake) in an embankment dam, where even a low energy earthquake creates stresses outside the elastic range of deformation. The duration of the OBE is then most significant.
As a result of this complex and uncertain situation two opposite trends are present for seismic input design.

(a) It is useless and even misleading to refine, because the forecast of future ground motions is impossible. Simple models are adequate for practical design.

(b) All possible means should be developed for simulating this complex problem, by sophisticated models, now possible to handle through the more elaborate computer programs.

The first trend has been the only one for many years, when no other ways were possible. The earthquake input was a uniform constant acceleration, determined empirically from regional history, and added statically to the acceleration of gravity. Everybody agrees that such a reduction of the actual complexity of the dynamic phenomenon to a so-called 'equivalent' static condition has no theoretical justification. It should be remembered however that the overall satisfactory behaviour of a huge number of dams in the world, almost all designed in that manner, gives some *a posteriori* justification to the method.

The extreme opposite trend is represented by R.E. Long (Paper 1) who proposes a deterministic approach where the model includes the simulation of the actual slip of the fault, the transmission of the various and complex waves through the actual geology, and yields the final resulting vibratory motion within the dam body at the site. Although the finite element method offers powerful means of analysis, it may be questioned whether this integrated model is more reliable than others. In particular a large number of engineering and even geometric parameters have to be determined regarding the kinematics of the fault, the geological structure, the elastic properties of the rocks. It seems, in addition, that the mathematical treatment of such a heavy problem is beyond present possibilities, particularly if the details of the waves reaching the dam are to be obtained. In fact, this model has to solve all the problems raised by uncertainties regarding the numerical values of the parameters and it cannot remain really deterministic. The Author is conscious of all these difficulties and he rightly states that 'the extent of sophistication needs to be assessed in terms of effectiveness'.

This concern (optimization in terms of effectiveness) has led various authors to propose simplified methods for practical use. The drawback, however, is the general lack of demonstration of the validity of the model, in other words of the fact that the simplification supplies entirely significant results.

A method has been proposed by N.M. Newmark (Rankine lecture, Geotechnique, 1965, Vol. 15, No. 2) to compute the displacement of a sliding mass considered as a rigid block. The input acceleration is the pulses in excess of a yield acceleration. The maximum displacement of the mass relative to the ground is over-estimated, as the pulses in the opposite direction are not taken into account.

Another method using the model of a rigid block on a sloping surface has been proposed by S.K. Sarma and N.N. Ambraseys (Geotechnique, 1967, Vol. 17, No. 1, and 1975, Vol. 25, No. 4) for the settlement of earth dams during earthquake. It uses essentially predominant period and peak acceleration of horizontal shear waves as earthquake input. Other parameters are related to the dam structure, such as pore pressure coefficient, damping coefficient and fundamental period of the dam. Critical acceleration and angle of equivalent slip surface are deduced from a limit equilibrium static analysis.

A simplified procedure for evaluating embankment response has been proposed by Makdisi and H. Bolton Seed (ASCE, 1979, Vol. 105, GT 12) based on the shear slice theory. It enables the determination, by hand calculation, of maximum crest acceleration, natural period and shear strain, and allows the use of strain-dependent materials properties.

A very elegant procedure is proposed by N. Bicanic, R. Feizo and O.C. Zienkiewicz (Paper 3). It consists in using a short artificial accelerogram equivalent to the assumed natural accelerogram. It was already conventional to use artificial accelerograms because the randomness of actual earthquake accelerograms limits their use. This new proposal, with shorter duration accelerograms, results in lower computer costs for the same dynamic behaviour of the model. It is interesting to mention again that, for the selection of the substitute short accelerogram, the response of the dam is taken into account, that is, the significant earthquake input is not absolute data. It depends on the response of the structure. Professor Zienkiewicz and his colleagues claim that the substitute short accelerogram is fully justified only for preliminary design 'owing to uncertainties of engineering modelling for non-linear seismic analysis'. It may prove fully justified for final design as well, for the same reasons.

I would like to ask Dr Long (Paper 1) if I am right in saying that his model is a deterministic one in which every parameter is decided? In which case how can one take into account the fact that many of the parameters are uncertain?

Dr R.E. Long (Paper 1). The model is deterministic but uses sensitivity modelling; that is, where parameters can be determined, such as the position of the fault lines, the details of geological response, and so on, they may be included, but for source parameters and structural constants, for example, the sensitivity approach has to be used. It is therefore a combination. It is also probabilistic to a certain extent because you have to consider the probability of earthquakes occurring on the faults.

Mr R.G.T. Lane (Sir Alexander Gibb & Partners). It appears to me that the direction of approach of the waves is extremely important, especially with embankment dams. We do know that the vertical component of acceleration is very great. I have written a short paper to show that if there is a vertical component of the approach acceleration, then there is the possibility of longitudinal cracking in the top of the dam (The failure of earth embankments by cracking during earthquake. Proc. Instn Civ. Engrs., Part 2, 1980, Vol. 69, June, 521-527). The much more sophisticated approach that Dr Long has adopted gives the same answer. Do take into account the vertical component. Few people seem to do so.

Dr R. Chaplow (introducing Paper 2). I wish to emphasize the uncertainties in carrying out seismic risk studies at a single site and to compare these uncertainties with the range of assessed seismic risk from sites in various parts of the world.

Figure 1 shows the results of one part of the study recently carried out at the Lar Dam Site in northern Iran. Peak ground acceleration is used as an index of seismic risk and is plotted against mean return period. Various relationships have been derived from teleseismic data, historical records, strong motion records, and from local seismic monitoring. A range of relationships has been produced from the same teleseismic data by varying the form of magnitude recurrence relationship and attenuation relationships employed, although there is reasonable agreement between the relationships obtained from teleseismic data and from historical and strong motion records. The extrapolation from peak accelerations measured as part of the local seismic monitoring programme produced a totally unrealistic estimate of seismic risk.

The range of peak ground accelerations for a particular mean return period obtained from the various methods adopted for assessing seismic risk should be compared to the range indicated in Fig. 2. These relationships in Fig. 2 have been derived by applying a consistent approach to assessing seismic risk at a variety of sites having different levels of activity. Relationships have been obtained for sites in Iran, Bali, Fiji, Tanzania and South Australia and for comparison, published relationships for Iraq and Europe are also shown. The ranges of peak ground acceleration for 50 and 500 year mean return periods are show in Table 1.

These ranges are very similar and do help to emphasize the importance in seismic risk analysis of carefully considering the local

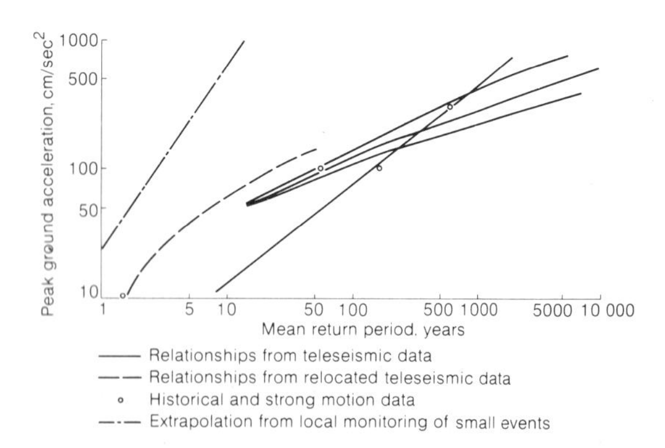

Fig.1. Summary of peak acceleration recurrence relationships for northern Iran Lar Dam site

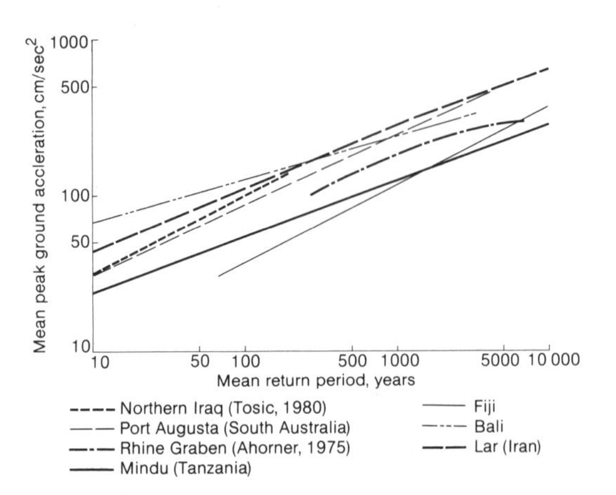

Fig.2. Peak acceleration recurrence relationships from various parts of the world

conditions at the site and the uncertainties involved in the risk assessment rather than simply taking a broad overivew of the site.

Mr P. Londe. Dr Bicanic (Paper 3) stated that as far as purely dynamic behaviour is concerned his accelerogram is equivalent to the actual one; but what of the question of liquefaction?

Dr N. Bicanic (Paper 3). I think that all time effects will be very vaguely spotted with shorter duration accelerograms. For liquefaction, where the time effect is strong, I do not see any way where shorter duration accelerograms can realistically reproduce it. I would stress that from an engineering point of view the uncertainties are so great that it is within the preliminary stages that shorter types of shocks would be more useful.

Dr E.G. Prater (Swiss Federal Institute of Technology). The question arose of the possible use of short duration accelerograms for non-linear response, such as involving liquefaction. Whereas at first sight it might appear that in this case their use is unacceptable, one feels that they should not be immediately discarded. When one looks into the method of analysis of Professor Seed and Dr Idriss it is evident that the effect of the actual time history is approximately considered by means of a strength (or fatigue) curve relating number of cycles to stress level inducing liquefaction. Thus it needs to be established if the equivalent number of cycles of given stress level produced in the earth structure is about the same for the short duration earthquake and the design accelerogram whose response spectrum is matched.

Mr M.B. Tosic (introducing Paper 4). Our interest has focused on several dam sites in northen Iraq. There are often speculations, not only here but also in some other regions, that few earthquakes were recorded over several hundred years. On the basis of a chronological list of earthquakes (Alsinawi and Ghalib, 1975 - reference 2, Paper 4) we have prepared Fig.3. It is interesting to note that two portions of the graph coincide well with the two distinctive historical periods in the development of the region. The first period of about 350 years is characterized by the tremendous humanist and scientific achievement of the civilization created by the Arabs in a huge territory. The second period of about 650 years - the period of decay- is characterized by permanent wars, and invasions from abroad. We think, therefore, that a smaller number of earthquakes from the latter period does not show relative seismic inactivity, but is merely a result of the small number of written documents available from the period.

Figure 4 is a presentation of the seismic parameters assigned to the majority of historical earthquakes from the catalogue; the integers indicate chronologically listed events, roman numerals indicate intensities, and the third (decimal) numbers are magnitudes. There is no claim that this interpretation of the historical events is definitive in any respect, and that it cannot be improved.

In assigning seismic parameters to the events from historical evidence there is an almost equal probability of underestimating or overestimating the magnitude of these events. Professor Ambraseys, in his detailed research of early Iranian earthquakes has pointed out that many earthquakes have been described collectively, which makes it difficult to locate correctly epicentral regions, and to assign dependable seismic properties to many events. Data of historical seismicity describe, most

Table 1. Peak ground accelerations

	Peak ground acceleration	
	50 year mean return period	500 year mean return period
Single site with various methods of analysis (Fig.1)	40-150 cm/s^2	180-300 cm/s^2
Range of sites with similar methods of analysis (Fig. 2)	20-110 cm/s^2	80-230 cm/s^2

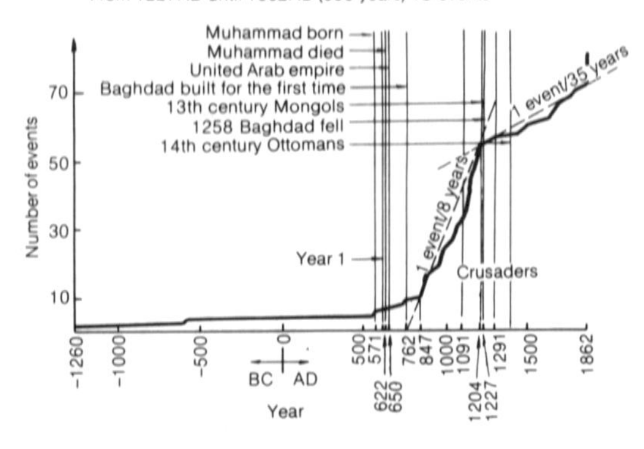

Fig.3. Cumulative number of recorded events

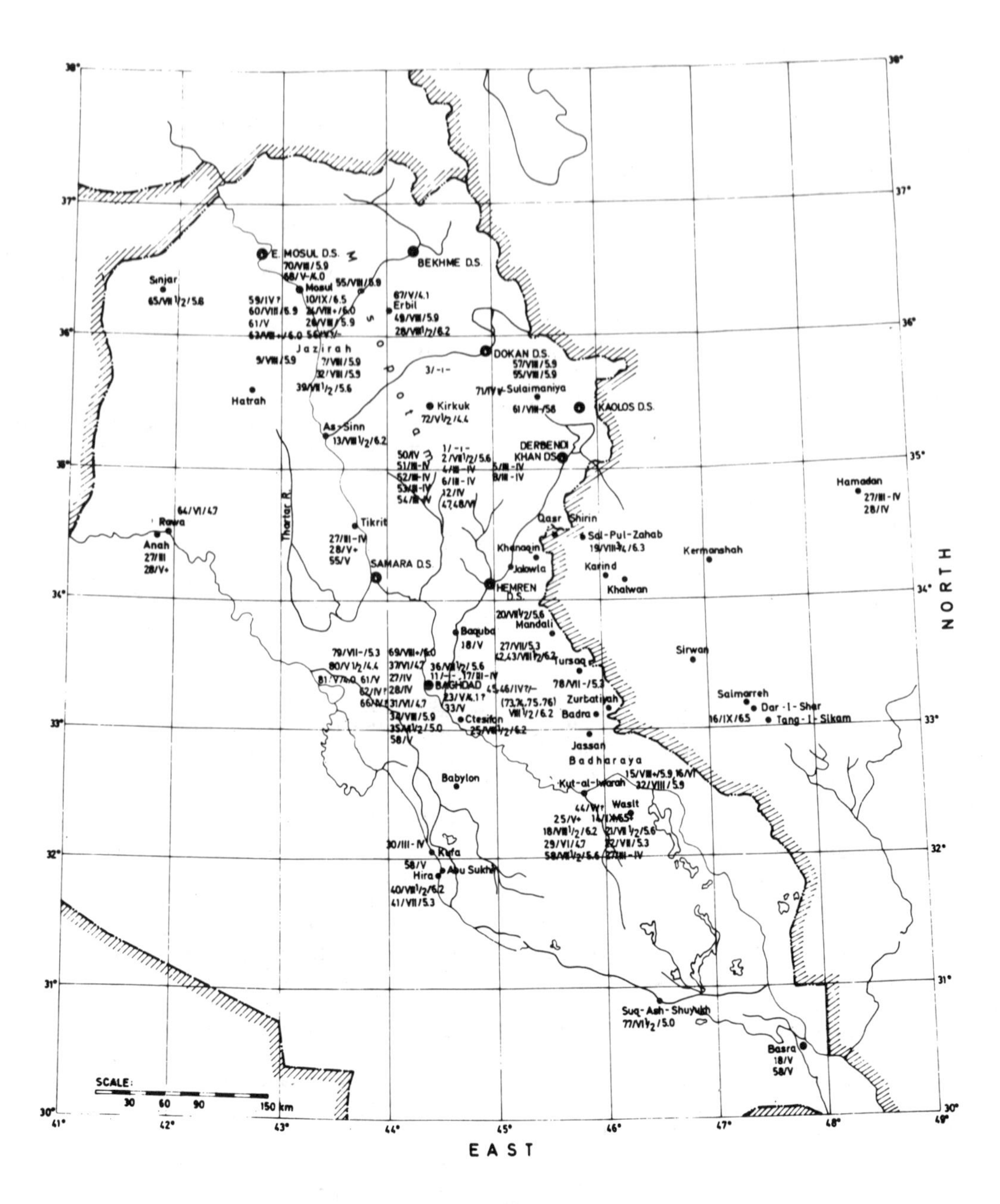

Fig.4

often but not necessarily always, the seismic history of larger inhabited places. It is therefore easier to locate local than epicentral intensities.

Dr P.L. Willmore (Institute of Geological Sciences, Edinburgh). With reference to Mr Tosic's historical analyses (Paper 4), on the much smaller scale of the UK we have increasingly begun to suspect some of the conclusions dependent on historical data and are noticing apparently significant deviations from historical patterns now that we are relying more on instrumental coverage rather than on population. The most conspicuous example is the Great Glen. It used to be an article of faith that this huge, visible, linear feature running across Scotland, with numerous historical reports of earthquakes on it, was a line of natural seismicity. However, we are now recording earthquakes up the west coast of Scotland in considerable numbers. The historical concentration around Inverness has been substantiated, but the central part of the Glen is not filling up with minor earthquakes to the expected extent. Are we looking at a fluctuation in the real seismicity, or is the historical record biased by the fact that the exceptional fertility of the Glen in relation to the surrounding territory has attracted a line of human settlement?

Mr W. Wahler (W.A. Wahler, Inc., USA). Almost all seismicity studies begin with a study of the historical record. The historical record often contains data of great value for determining the seismicity of an area or site. However, historical data when taken alone can lead to very questionable conclusions. Both the existence of a record and its absence can be equally misleading - seriously misleading. Historical data must be very carefully used.

Historical records require the presence of people and/or structures at the proper time and place, otherwise there can be no record. This does not mean that areas without a record of earthquakes have no earthquakes. It also does not mean that areas with a positive record are at the centre or even near the centre of the earthquake activity that was recorded. Therefore, both written, oral, and archeological records must be analyzed with careful consideration and correlation with the structural and lithological geology of the area and the area's plate tectonics. Without making these correlations, no sound conclusions can be drawn.

Consider also the nature of earthquake records. All existing earthquake records indicate an irregular pattern of activity both with regard to frequency and magnitude. Short records and long records both indicate irregular patterns. Most often a short history from one area will indicate a different seismicity pattern than a longer record, even for a closely related area. The critical question is always: what portion of the area's activity pattern does the record represent? Is it a period of relative quiessence or a period of concentrated activity with regard to either frequency or magnitude? Both types of periods and fluctuations exists. Because dams and reservoirs are planned to be in existence for a long time, it is prudent to consider the historical record with care - it can only indicate the minimum level of activity to be expected and really indicates nothing about the maximum level of activity that can be expected.

Let me illustrate this problem with a case history. Prior to the installation of seismic monitoring equipment in the Canadian Shield area, the region was considered by many to be non-seismic. After the area became suspect due to the recording of a few minor events, a re-examination of the records and the first scientific examination of obscure local records from the Shield area indicated that a historic 'Boston' (USA) earthquake of the eighteenth century may really have been a Canadian Shield earthquake that was felt over an area of many thousands of square miles but which was not adequately recorded and noted except in the Boston area. After the instrumental record alerted investigators to the possibility of the need to reassess the seismicity of the area, the pattern for both Boston and the Canadian Shield became subject to re-evaluation. Where previously ignored or unsuspected local records exist, such as in this case, a re-examination can correct the historical record. Where such records do not exist, the historical record cannot be corrected. So what do we have in that case?

This problem has been mitigated somewhat in more recent years since instrumental records have become available, because we have techniques by which we can now determine the location of at least some aspects of large earthquakes even when they are remote from the sensor. However, even where instrumental records are available, it is not always possible to accurately assess the earthquake potential of the area on the basis of the record because we do not necessarily possess a representative record. Assessing the seismicity of an area requires not only historical and instrumental data but also a knowledge of the geology and tectonic aspects of the site and region, and mature, sound, seismically induced judgement.

Mr P. Londe. The work of Dr Ries et al (Paper 5) and the question of operation versus designed earthquake, raises the problem of what is the definition of normal damage for a dam. It is much easier to define for a structure such as a nuclear plant or a bridge than for a dam. It is very difficult to visualize what is the damage which one could accept as normal for operation. Is it settlement? Is it a limit in cracking? It is very open, and I think the definition of acceptable damage is completely arbitrary.

Mr E.G. Aisiks (Acquasolum SA, Argentina). Several times reference has been made to flood events and their probabilistic treatment, and their similarity to earthquake events and their probabilistic treatment. I would like to point out that around 1928 the probabilistic treatment of floods became prevalent (Foster H.A. Theoretical frequency curves and their application to engineering problems. Trans. Am. Soc. Civ. Engrs, 1924, Vol. 87, pp 142-173) and many times designs were made on that basis, taking some recurrence period and using several probability distributions, such as Pearson III or Pearson I, etc. However, it soon became noticeable that nature did not exactly follow that type of law, and a more deterministic approach began to be developed, particularly in areas with scanty records (such as the USA); the unit hydrograph theory was proposed implying a deterministic approach that ended up with the hydrometeorological determination of extreme flood events. I think the most obvious period for floods is yearly because they occur seasonally, whereas we do not know what is the recurrence or the normal period for earthquake occurrences in different places. Therefore, I concur with what Dr Ries (Paper 5) says that the design for an operating basis earthquake could profit from the probabilistic treatment to educate our judgement, whereas the maximum credible earthquake (let us say the limiting event that one can foresee in a certain environment) should most certainly depend on the deterministic approach, however feeble our understanding of the cause and effect relationships.

Dr R. Dungar (introducing Paper 6). There are two new aspects of my work which I would like to include in this presentation, the first is related to the geometry definition and the second to the presentation of additional results that are not included in the original paper.

Figure 5 illustrates the computer aided design (CAD) method used to define the initial geometry of the dam. The crown cantilever is first defined by selecting five control points for both the water and air faces from which hyperbolic functions are generated, as shown. The constant thick arches are chosen as elliptical for the water face, the A and B semi-axes being defined by the distances as shown. The spline functions which control these distances are in turn specified by selected 'control points', which may be interactively defined and redefined using a graphics screen.

The dam is defined as constant thickness for the central region, illustrated in Fig. 6. The transition lines, T, and a smoothed valley profile, S, (used to provide a smooth definition of the geometry) are also interactively defined by splines. Thickening factors and angles are defined at the position S, again by interactive definition of spline functions (not shown). The resulting arch profiles may be checked, as shown in Fig. 7, in relation to the actual valley profile. Fig. 8 is a windowed plot of Fig. 7 , showing the transition from the constant thickness to the fillet region (defined as a hyperbolic function). The above method greatly reduces the required time for optimizing the dam geometry, the program being linked to stress calculation programs.

Turning to the results, Fig. 9 and Fig. 10 show stress sector plots for the MCE condition, for the water and air faces respectively, and correspond with the dynamic cube strength requirement, as explained in the Paper. It will be observed that large tensile stresses exist on the air faces resulting in relatively high required cube strength so as to prevent cracking. Fig. 11 illustrates the envelopes of required cube strength, as also discussed in the text. The zones of high strength requirement for the air faces are apparent.

Finally, Fig. 12 illustrates the dynamic air face tensile cantilever stress, as calculated for joint opening conditions, at the base of the opened cantilever. The 10% damping condition is now being used for final design calculations and the illustrated peak tensile stress will be accommodated by the dynamic flexural tensile resistance of the structure.

Mr P. Londe. The influence of the type of dam and the material on the definition of the OBE, MCE and so on, of the earthquake motion parameters is an example of the fact that the designed earthquake is partly defined by the type of structure. For instance, your OBE would be different at this particular site with an earth dam. How can you really define the operation basis for a given dam? It is arbitrary to say that there is no tension in a concrete dam because we know that a large dam is perfectly safe with some tension. It depends where it is.

Dr R. Dungar. The seismic hazard survey which was conducted for the El Cajon site was based upon tectonic and geological data only. The type of dam had no influence upon the selected seismic parameters for both OBE and MCE conditions. However, the risk that one takes in constructing a dam at a given site is a function of the type of dam and associated structural parameters taken for that design.

In the case of El Cajon, the OBE was extrapolated from the MCE, which in turn was based upon a deterministic assessment of the given geologic and tectonic data. For such a site of only medium to low seismic activity it is indeed difficult to obtain a realistic estimate of the operating basis conditions as obtained from a probabilistic assessment of historical data.

The no tension conditon referred to in the Paper was assumed only for horizontal stress components on the faces of the dam. Here the construction joints will play their part in dissipating such tensile stresses.

Mr W. Wahler. The problem of selecting an acceptable factor of safety is not one just dependent on the accuracy and/or precision of the computation (adequacy of data or technology), but is also necessarily dependent on the economics and politics of the situation.

— Crown cantilever defined as hyperbolic function by the use of control points on both faces

— Central section of arches defined as constant thickness with water face as ellipses

— Ellipse A and B axes may be adjusted interactively (using a graphics screen) by adjusting the control points

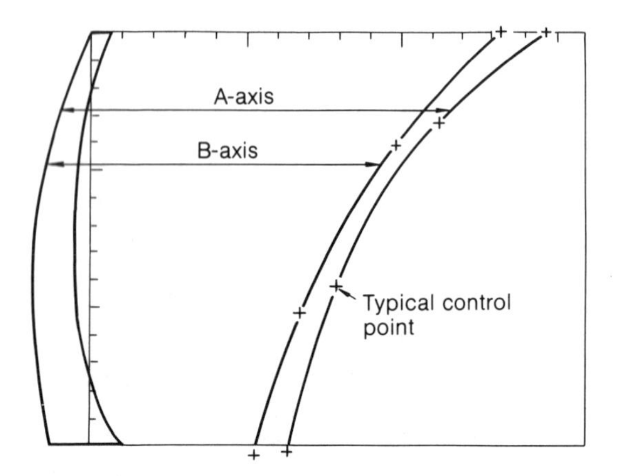

CROWN CANTILEVER SECTION CASE 41

Fig.5. Definition of dam geometry (1)

Typical arch section with both constant thickness and fillet regions

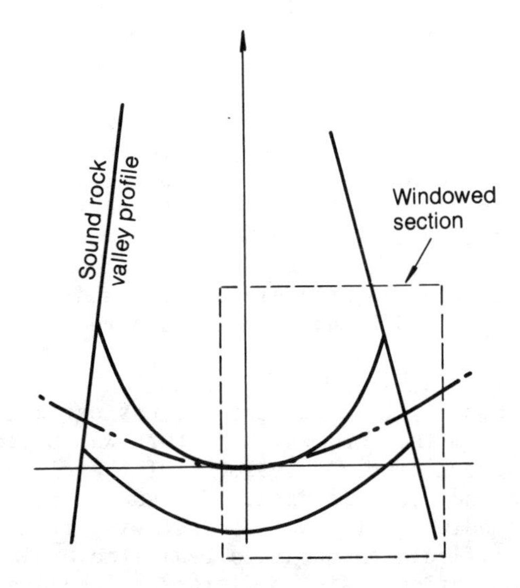

SECTION 7 ELEVATION 120.0 CASE 41

Fig.7. Definition of dam geometry (3)

Principal stress vectors for combined MCE earthquake and usual static loading conditions

WATER FACE

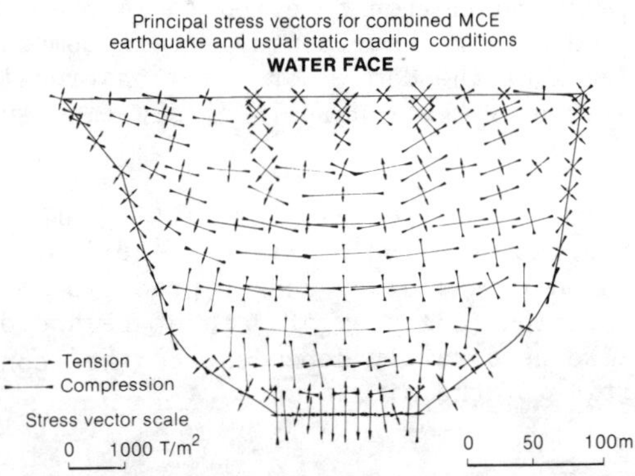

Fig.9. El Cajon arch dam - stress analysis, water face

— Arches defined as C constant thickness section and F fillet section

— Transition lines T and smoothed valley profiles S are controlled interactively by adjustment of control points

— Other thickening functions and angle functions (not shown) are used to define the fillet profiles

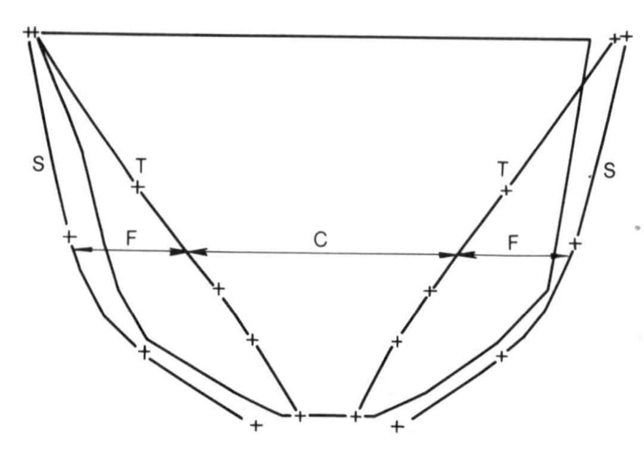

AIR FACE TRANSITIONS CASE 41

Fig.6. Definition of dam geometry (2)

Windowed section of a typical arch profile showing constant thickness section C, transition point T and fillet section F

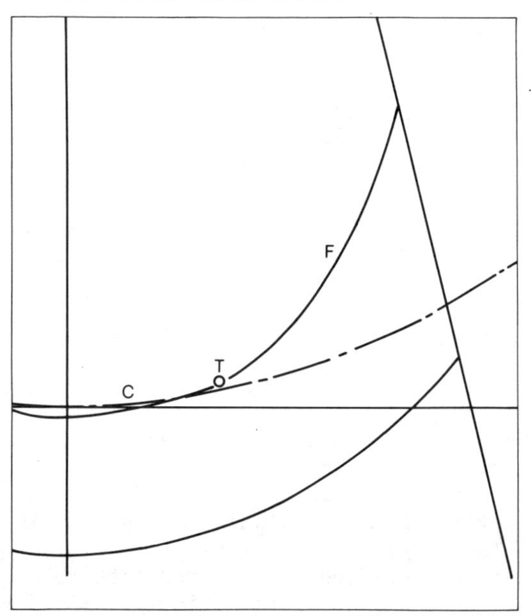

Fig.8. Definition of dam geometry (4)

Principal stress vectors for combined MCE earthquake and usual static loading conditions

AIR FACE

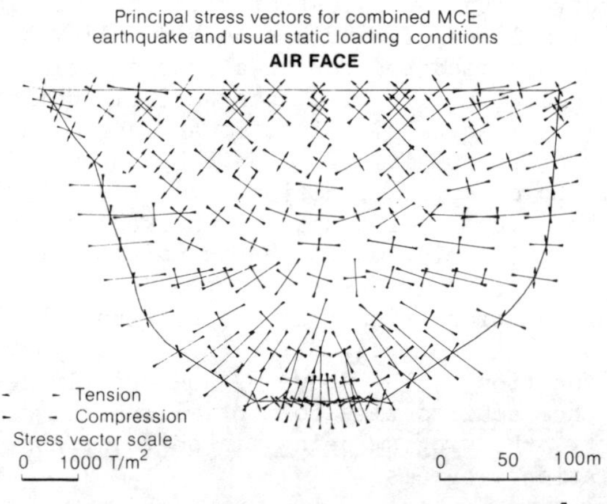

Fig.10. El Cajon arch dam - stress analysis, air face

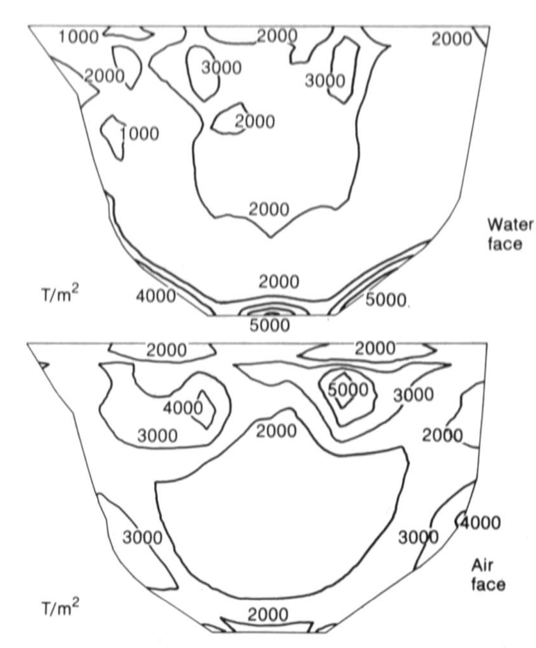

Fig.11. El Cajon arch dam - envelopes of required concrete cube strengths for usual and extreme loadings

There is some risk involved in all activities. The question that should be asked when selecting a factor of safety for design is really what risk can, or should, be tolerated. Risk abatement often costs money. When this is the case, reducing risk requires an allocation of resources. Allocating resources to one activity usually involves diversion of resources from another activity. Allocation of resources is a political and not an engineering function and when engineers violate this truism they may incur liability.

What then is the validity of a standard or acceptable factor of safety? In reality there can be no single acceptable standard factor of safety. The acceptable factor of safety for each project, or even aspect of a project is essentially and emphatically dependent on the applicability of the technology used in its design, the representativeness of the data used, the adequacy of control over its construction and operation, and the cost and the acceptability (politically) of risks involved. Thus there can be no single standard factor of safety, and it is dangerous as well as absurd to reply on anything labelled as such. A standard factor of safety can neither be absolutely safe or conservative nor universally economical. We must therefore evaluate the role of risk (political as well as economic) when selecting an acceptable factor of safety for a project.

Mr D.A. Howells (Chairman, SECED). In the design of nuclear power stations two separate levels of earthquake intensity (however this may be measured) at the site are considered. The first is the operating basis earthquake (OBE). This must cause no damage to the plant so that it can continue in normal operation after the earthquake. A more severe earthquake shaking may cause damage, but the plant must be able to shut down safely. The term used for the

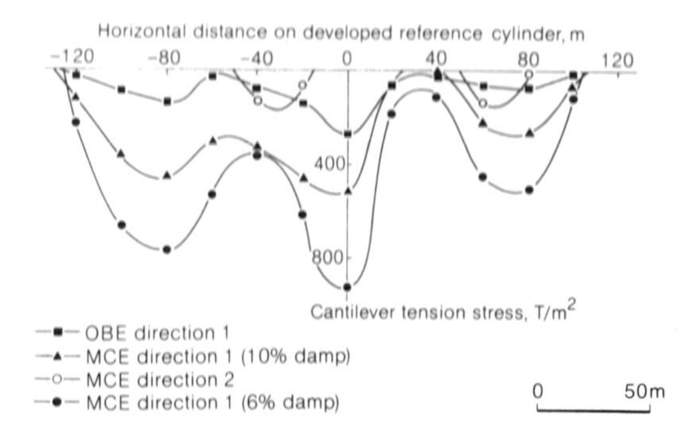

Fig.12. El Cajon arch dam - cantilever tension stress on air face: assuming joint opening and independent cantilever action

upper limit to this shaking is safe shutdown earthquake (SSE). Both these definitions are made in terms of plant behaviour. Both need to be considered because two completely different groups of plant items are involved. Both are difficult to define in terms of earthquake shaking. For the OBE, Dr Dungar's (Paper 6) definition as an 'event that may reasonably be expected to occur during the lifetime of the structure' is a step in the right direction, even though there may be much controversy about more precise definition and the choice of appropriate numbers. Remember Lomnitz's observation that earthquake coefficients in terms of fractions of g are subject to an annual inflation rate of 10%. The idea of a nuclear power plant which fails to shut down during a large earthquake is a very alarming one, so that the SSE intensity level must be quite high. The SSE in earthquake terms becomes either the maximum possible earthquake or the maximum credible earthquake. Both of these raise great difficulties in philosophical definition, let alone in numbers, but the nuclear people have to try.

This practice is extending itself into dam design. Is this necessary or helpful? In dam design we are not concerned with two different groups of plant. Nevertheless, the two different levels of shaking are separated by such wide difference of material behaviour as to merit quite different approaches.

The OBE is defined as related to the presumed lifetime of the structure. This is a very imprecise concept. But the numbers usually given for structural life are often close to those for which moderately reliable information is available about the frequency of occurrence of earthquakes, so that one can form a tolerably reliable estimate of the OBE. There exists a great amount of pragmatic judgment within the engineering profession as to the appropriate factor of safety to be applied to the assessment of environmental loads to allow for infrequent greater values of the loads. Thus the ultimate limit state may be investigated under a load factored from the OBE.

The dam designer therefore has the choice of two procedures. The first involves applying a serviceability load factor to the OBE in one stage of the calculations and a safety load factor derived from engineering consensus to the next stage. The second is to use the OBE and the serviceability for one analysis and the SSE and another appropriate load factor (close to unity) for a separate ultimate load analysis. I should be inclined to place greater confidence in an ultimate load analysis based on the OBE multiplied by a safety load factor. However, in practice the designer of a large dam will wish to calculate all possible ways and try to reconcile any discrepancies. He will also bear in mind that uncertainties affecting input data to these calculation procedures are likely to be large in relation to any discrepancies.

Dr R. Dungar. As Mr Howells has correctly pointed out, the seismic criteria used in the nuclear industry has also had an influence on the design of large dams, the concept of a safe shutdown condition being replaced by a maximum credible condition. The use of a safety load factor is implied in Paper 6, where a safety factor is used to establish the design cube strength for the different load combinations considered in the design. By these means the influence of the different types and intensities of loading may be assessed in terms of a single design result, namely the design cube strength.

Dr P. Willmore. In Paper 6 the combination of a severe earthquake with a severe flood was listed amongst the unusual circumstances. This would be true if these were independent events, so that a severe flood persisting possibly for a week in the 100 year cycle, or the 10 year cycle, or whatever one wants to take, is a low probability circumstance; and the largest earthquake which the region can sustain, occurring very close to the dam, is taken to be an independent low probability circumstance. But now let us imagine a situation where the focus of the largest earthquake which the region can sustain may be anywhere along a fault line and the stress along the fault line is gradually increasing. Then let us suppose that one has a flood at the end of a period of drought, when the water level is low, and all the political factors come into play, forcing the authority to fill the dam as rapidly as possible. Then one has an inducement situation in which the high water level would be quite likely to occur at the same time as the extreme earthquake rather close to the dam.

Dr R. Dungar. Dr Willmore has misinterpreted the definitions of the load combination for earthquake conditions. As stated in Paper 6 earthquake loading is considered together with a usual static load combination. Loading due to a flood condition is classed as an unusual event.

Professor H. Bolton Seed (University of California). I should like to consider the question of uncertainties with regard to input for design earthquakes for the design of dams and the possibility of minimizing or reducing those uncertainties through obtaining field records of earthquake ground motions. There have been two earthquakes in California in the last year or so which have thrown considerable light on the nature of ground motions produced by earthquakes.

Until about a year ago there was considerable uncertainty about the magnitude of the accelerations which might develop within a distance of 10km from the source of energy release or the causative fault of an earthquake. Most of our data base at that time was in the range of about 15km to 200km away from the causative fault. Although we could analyse this data statistically and get a mean acceleration line and mean plus one standard deviation values, and so on, it was really a matter of speculation where the attenuation curves would go at close distances to a fault, as close as 1km, or up to 10km. So there was much uncertainty.

In the 1979 Imperial Valley earthquake in Southern California, which had a magnitude of about 6.5 we were fortunate enough to obtain about 30 records within a distance of 10km, which goes a long way to filling in the uncertainty about what happens in the first 10km distance. A plot of recorded peak accelerations is shown in Fig. 13. The average peak acceleration at a distance of about 1km from the causative fault in this case was about 0.55g, which is not an incredibly high value.

This data provides a very good indication of the levels of acceleration which might be developed by a magnitude 6.5 earthquake for dams located within about 10km of such earthquakes. Since the maximum earthquake produced by reservoir induced seismicity seems to be of the order of 6.5 or less, this kind of data is useful in telling us the design motions we might anticipate from earthquakes of this type.

An interesting feature of the Imperial Valley earthquake was the extremely high vertical accelerations recorded at some of the recording

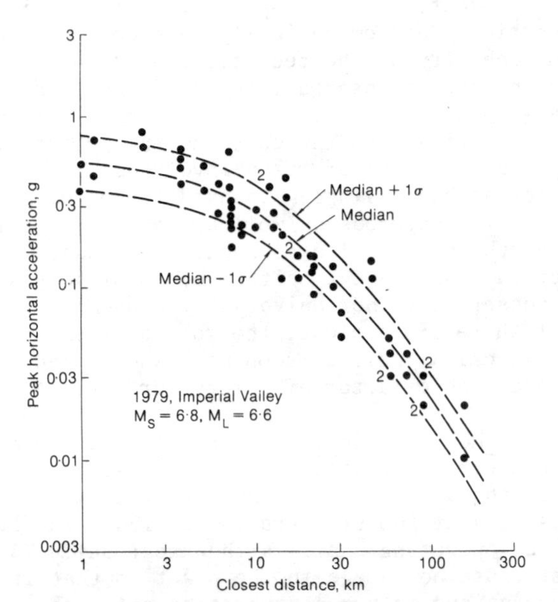

Fig.13. Regression analysis of peak accelerations recorded during Imperial Valley earthquake, 15 October 1979

stations. The highest of all was 1.6g recorded at a station near the end of the fault, but another station recorded a peak vertical acceleration of about 1g. This is much higher than we normally think about for the design of dams and it may make us wonder if perhaps we should design for higher vertical acceleration than we have done in the past.

However, it is important to note that there was a significant time differential of several seconds between the occurence of the peak vertical and the peak horizontal accelerations; consequently it appears that it may not be necessary to consider peak vertical motions occurring at the same time as peak horizontal motions. We can probably separate them for design purposes, find the effects of each, and then combine them in evaluating dam performance.

Another uncertainty which has existed since 1971 has been the significance of the peak horizontal acceleration of 1.25g recorded at a station on a ridge of rock above the abutment of Pacoima Dam. The rock below the instrument station was badly cracked, and there has been much debate about the significance of this 1.25g horizontal acceleration recorded in close proximity to the dam. Because of this record, some dams have been designed for corresponding high accelerations. Other people have argued that the high acceleration recorded at this instrument station was very much influenced by the topography of the area where the station was located. As a result many engineers have considered that a more appropriate value for the peak acceleration developed near the base of the dam would be in the range of 0.6 to 0.8g, with an average of about 0.7g. This was a matter of considerable debate until a very recent earthquake series which occurred in another part of California. In the Pleasant Valley reservoir earthquake series of 1980 we had five magnitude 6 earthquakes, affecting a very small localized area, in the course of two days. The earthquakes occurred on the Hilton Creek fault, only about four miles from the Long Valley dam, which is one of the dams which has been selected for detailed instrumentation in California; as a result we have a lot of records for that dam in this earthquake series.

At the crest of Long Valley dam the peak acceleration was 0.49g. On rock, near the toe of the dam, the maximum acceleration was about 0.24g. However on rock above the crest of the dam, a maximum acceleration of 0.99g was recorded. This indicates that very high accelerations may be recorded on rock up above the abutments of dams, but down near the toe the motions are likely to be considerably lower.

These kinds of records do help to clear up some of the uncertainties with regard to the nature of ground motions recorded in close proximity to faults and to dams and some of the uncertainties regarding the significance of the Pacoima Dam record for use in the design of dams in the future.

7 Materials behaviour under earthquake loading

P. BERTACCHI, Chairman Committee on Dam Materials (ICOLD),
Director CRIS (ENEL), Milan

The Committee on Materials for Dams of ICOLD was asked by the Committee on Seismic Aspects of Dam Design (S.A. of D.D.), of which Mr. Lane acts as Chairman, to collect information and to prepare a document on material behaviour when subjected to repeated high loadings, as from earthquakes.
The said paper will have to be included in the final report of the Committee on S.A. of D.D., together with other reports dealing with design, computation and environmental aspects related to seismic risks.
In fact, it has been ascertained that few information is available on materials subjected to earthquakes and so it seemed particularly useful to search thoroughly knowledges on this subject.
Therefore, the Committee on Materials sent a circular letter (March 1978) to all National Committees of ICOLD requesting information concerning the following topics:

a. Laws which regulate resistance, deformation and creep of concrete in connection with rate of loading in uniaxial and pluriaxial conditions. Damping coefficient. Shock effects between two masses of concrete.

b. Behaviour of concrete and foundation rock when subjected to low cycle fatigue, with alternate frequency not higher than those of earthquakes (15 cycles/s); the behaviour at very high loads, repeated after recovery of the deformation caused by the previous load, is very interesting.

c. Behaviour of embankment materials and foundation soils, when subjected to alternating exceptional short lived loads, with reference to material density and to saturation degree; phenomena of liquefaction and settling.

d. Behaviour at repeated loads of other materials for dams (bituminous materials, water stops, etc.).

A wide documentation was sent by Japan and USA, references by Great Britain and further information by Italy, New Zealand and Czechoslovakia, besides what dealt with in a report already presented by USSR to ICOLD in 1977.
Therefore, it became necessary to complete this information with any other easily found in the literature, paying particular attention to the two central topics.
So at last it was decided to prepare two separate documents as follows:

A. Embankment Materials Behaviour under Earthquake, prepared by Mr. Ghionna and Mr. Battaglio on behalf of Mr. Bertacchi (Italy), already delivered to the Committee on S.A. of D.D.

B. Concrete Behaviour under Earthquake, being prepared by USCOLD on behalf of Mr. Veltrop; its delivery is foreseen within October 1980.

Both documents are a review of the state of the art of knowledges in the relevant fields.
The Committee on Materials for Dams intends to develop and extend thoroughly investigations and researches also after presentation of the final report by the Committee on S.A. of D.D. Further information will concern laws which regulate resistance, deformation and creep in connection with rate of loading in uniaxial and pluriaxial conditions, the previous evaluation of risks of liquefaction phenomena, as well as the behaviour of other materials for dams under repeated loads.

In the following text, some notices on the topics dealt with in the two above said reports will be presented: the report A) is here condensed, whereas of the report B) some flashes will be given, necessarily incomplete, on the main subjects developed.

A. Embankment Materials Behaviour under Earthquake

After a brief discussion of the importance of investigations on the behaviour of earth and rockfill dams under seismic actions, keeping into account the various types of failure which may occur during an earthquake, the report examines the influence of soil properties which must be kept into account during design.
It is therefore necessary to know the state of stresses and dynamic soil behaviour, the dynamic soil properties, the liquefaction potential.

For typical elements of a dam and its foundations, an earthquake induces a dynamic stress distribution superimposed to the static stress. It is necessary to distinguish between sufficiently impervious saturated soils and highly pervious materials such as gravels and rockfills.
In the first case one has to assume that the earthquake stresses are applied under undrained conditions, that is with no drainage or any significant dissipation of excess pore water pressure, in the second one significant dissipation of excess pore water pressure may occur even during the relatively short duration of earthquake shaking.

Under undrained conditions two different behaviours may be discerned; the one is inherent to loose to medium dense sand which are susceptible to a rapid increase of the pore pressure in large zones of the embankment with reduction in the shear strength and potentially large movements leading to almost complete failure, the other is inherent to very compacted cohesive soils and very dense sands for which the build up of pore pressure is much lower and consequently the post-earthquake behaviour is a limited permanent deformation of the embankment.

The main mechanical properties that are to be kept into account are: dynamic stress-strain relationship; dynamic shear strength. These properties depend on the type of soil, its initial relative density or degree of compaction, its degree of saturation and the rate of loading. Furthermore for a cohesionless soils, under certain dynamic conditions, liquefaction may occur.

With the term liquefaction it is intended a complete loss of shear strength which can occur when loose cohesionless soil is subjected to shear stress, either monotonic or cyclic. The term liquefaction has also been used to denote a partial loss of shear strength due to build up of pore water pressure. When liquefaction is induced on saturated cohesionless samples of low to medium density by cyclic shear stresses, the sudden collapse of the soil structure is preceded by a gradual build up of pore water pressure with negligible shear strain.

When dense saturated cohesionless soil is subjected to cyclic shear stress at a level somewhat lower than the static strength, the pore water pressure may build up gradually until it reaches the applied confining pressure, or to a condition of initial liquefaction.
As soon as the shear strain exceeds a certain limit, however, the soil becomes "dilatative" causing a drop in the pore water pressure, with a consequent recovery of the effective stress.

With respect to sand, gravelly soils exhibit considerably greater stability because their high permeability would either preclude a full development of pore water pressure or reduce the duration of fully liquefied condition during the period of earthquake shaking.

The factors which influence liquefaction of cohesionless soils are primarily: the coefficient of earth pressure at rest, K_o; the strain history; the age of the soil deposit; the soil type, gradation and structure, the relative density.
Some of these factors are studied at laboratory, by using reconstituted sand specimens: generally, the methods which have been proposed to evaluate the liquefaction potential are classified as follows:

1. Empirical criteria of liquefaction potential based on field observation during earthquakes;

2. Comparison of computed shear stresses in the field with liquefaction resistance determined in the laboratory;

3. Prediction of liquefaction in the field by analyses based on mechanical models of soil elements.

As a consequence of the above, the most important soil properties and characteristics which are needed in soil dynamics and earthquake engineering are: dynamic moduli; damping ratio; liquefaction parameters; shearing strength in terms of strain rate effects.
Some of these are best measured or studied in the field, others in the laboratory and some can be measured both in laboratory and in situ.

The most important laboratory techniques are: cyclic triaxial tests; cyclic simple shear tests; cyclic torsion tests; shake table tests; resonant column test.
Laboratory techniques such as cyclic triaxial tests and cyclic shear tests are employed when it is required the knowledge of soil behaviour at high strain amplitudes. Field seismic measurements can establish values of the shear wave velocity "V_s" within a rock or soil mass.
Field techniques usually develop shearing strains in the field of 10^{-6} or less, therefore these tests develop low amplitude strains.
The low amplitude shear modulus is calculated from:

$$G_o = \rho V_s^2$$

ρ is the mass density of the soil.

The most important field techniques are: cross-hole; down-hole and up-hole; resonant footing; surface waves techniques; seismic refraction techniques.

Both laboratory and field techniques are necessary for satisfactory solution of many dynamic soils problems; however, combining results from both laboratory and field has often been disappointing or impossible, because of the large differences between them.

Generally, laboratory methods provide good correlations when undisturbed samples are used: in dealing with liquefaction of cohesionless soils, owing to the difficulties to obtain undisturbed samples, there are some limitations in evaluating the liquefaction potential using laboratory results. One of the most important uncertainty is the knowledge of the relative density of the sand deposit.

In order to provide a good evaluation of this characteristic, a technique has been developed to estimate the relative density by means of the results of quasi-static cone penetration tests performed in a large "calibration chamber", in which the sand sample was pluvially deposited and subjected to one dimensional compression.
Using this approach, it is possible to correlate the relative density of sand deposits with the point resistance of the penetrometer.

B. Concrete Behaviour under Earthquake

It is particularly difficult to take into account separately the behaviour of the material itself and that of the whole structure.
The researches, which we are acquainted with, aimed at examining only some aspects of concrete behaviour under repeated high loads, so that we think this review is still incomplete.

A first aspect concerns relationship between some parameters and the rate of loading.
It seems generally ascertained that compressive, tensile and consequently flexural strength are influenced by the rate of loading; impulsive loadings increase strength as well as moduli of elasticity, that can act as a moderate reason of safety.
It is stated that dynamic strengthening of materials is associated with their reserve of plasticity. However, creep behaviour with respect to rate of loading is very doubtful.

Instead, we think a negative aspect can be found in the low cycle fatigue of material when subjected to high loads. In fact, it has been noted that concrete, and to some extent rock too, can support well a very great number of slow cycles at low stress values; at higher stress values, i.e. when the first phenomena of micro-fissures occur, concrete supports a number of cycles more and more limited; in a border-line event, after few cycles at very high load levels, but lower than statical strength, it can be possible to get failure also keeping constant the load.

All these phenomena have to be still examined carefully and compared each other; particular studies have to be devoted to phenomena of recovery of deformation after very high load, residual strength and self healing, if any.

As regards damping, it is not presently possible to measure it in quantity on the material itself. In general coefficients of viscous damping have been determined experimentally by dynamic tests on model or on prototype structures, low level excitation being provided by mechanical shakers or impact loading.
Viscous damping is defined in such cases as a percentage of critical damping (the lowest damping for which no free vibration can occur). Values of 2 to 7 percent of critical damping are considered reasonable.

8 Liquefaction potential of Wash sands

T. D. RUXTON, MA, Binnie & Partners, London

An enclosing embankment, mainly of sand, built on the sandy foreshore of the Wash, a large bay on the east coast of England, would be a feasible way to form a freshwater reservoir. No special provision would be needed to make the embankment impervious if it was built to heights between 12 and 24m, a range which is also economic.

To assess the possibility of liquefaction of embankments and foundations, samples of the Wash sand were prepared at various densities and subjected to load-controlled undrained triaxial tests. Particle size and grading was examined and a scanning electron microscrope was used to examine particle shape and the microtexture of features on the surface of a number of sand grains taken from the samples. Similar tests were made for comparison on sand samples from Ottawa (Illinois, USA), Valgrinda (Norway), Zeeland and Beisbosch (Netherlands), Leighton Buzzard (England) and the Dee estuary (England/Wales border). The sands from Ottawa and Leighton Buzzard had been found to liquefy in earlier laboratory tests and the sand samples from Valginda and Zeeland were from sands that have liquefied to cause coastal flow-slides. The sand from the Beisbosch has been used to build enclosing embankments about 6m high for freshwater reservoirs. From these comparisons a tentative scale of liquefaction potential was developed.

A blasting test was carried out in the foreshore sands to simulate the triggering effect of vibrations due to earthquakes or the passing of heavy construction vehicles.

INTRODUCTION

1. A feasibility study was undertaken from 1971 to 1975 into building freshwater reservoirs on the intertidal foreshores of the Wash, a large bay on the east coast of England. Similar detailed studies have been carried out for the west coast sites, the Dee estuary, 1966 to 1971 and Morecambe bay 1967 to 1972 and again from 1976 to 1978. Preliminary studies have also been carried out for several other coastal sites (ref. 1).

2. Although intertidal sites are unfavourable, because of exposure to the sea, they have been considered because opposition to building further reservoirs inland in England and Wales has grown stronger in recent years. The opposition comes mainly from farmers and farming organisations who see existing farms made uneconomic due to drowning valuable valley bottom land leaving only sloping and hill-top land. The adverse economic effect is equally important in predominantly stock raising or arable farming areas.

DESCRIPTION OF RESERVOIRS

3. The reservoirs considered would be within enclosing sand embankments from 12 to 24m high founded on the sandy foreshore (Fig. 1). The embankment crest level would be 14m above Ordnance Datum which is about 0.2m below mean tide level. These embankments would be built by dredger placing and no special provision is proposed for making them impervious. The surface would be protected against wave attack by rock riprap over two or more sand and gravel filter layers. Above the level of wave attack and on its seaward side, the outer slope of the enclosing bank would be covered with top soil and grassed (Fig. 2).

4. The reservoirs would be sited on foreshores between the rivers discharging into the bay so that land drainage and fluvial flows could discharge freely to sea and the navigation to the ports could continue as at present.

5. Water to fill the reservoirs would be abstracted from rivers at intakes sited many kilometres inland and conveyed in tunnels driven through Kimmeridge Clay strata about 80m below the reservoir floor. Tunnels, in the same strata, would also be built to convey the water away from the reservoirs to purification works and on to supply. Shafts would be built to convey the water from the tunnels to the reservoir and vice versa.

POSSIBILITY OF LIQUEFACTION

6. Liquefaction of the sands forming the reservoir embankments and their foundations has been considered even though coastal flow-slides, such as have been reported in the Dutch province of Zeeland (ref. 2), are not known to have occurred in the Wash or along nearby coasts. Also earthquakes and tremors that might trigger flow slides, have occurred only rarely in England.

7. Dutch experience suggests that the regular accretion of sediments may have prevented coastal flow-slides, a situation that could be altered locally if reservoirs are built and river outfall channels cut new meanders and erode material adjacent to deposits of loosely packed sand. The existence of such loosely packed layers was expected as evidence from other similar coastal sites suggests that conditions near the shore may encourage deposition of loose layers. Also parts of the foreshore are said to be quick after large tides and evidence of sand blows, due to the release of water when a mass of sand settles after liquefying, has been found in the foreshore (ref. 3). Site investigation of the foreshore confirmed the presence of sand layers offering less than 10 kgf/cm^2 resistance to penetration by a Begemann cone at depths down to 15 and 20m. The full depth of sediments on the foreshores varies from 10 to more than 25m.

8. In an introduction to a report on liquefaction Castro (ref. 4) briefly describes typical liquefaction failures. He first mentions the Zeeland failures, which are thought to have been triggered by erosion by tidal currents and surface forces induced by abnormally high tides, and then mentions a variety of other failures. These include some massive flow failures of natural slopes triggered by small causes and failure of sand embankments placed at very high void ratio.

9. Other failures, reported in 1968, are six submarine slides in Central Norway and a slide of a foreshore during placing of dredged sand at Helsinki (ref. 5).

SCOPE OF INVESTIGATION

10. Detailed site investigations have been made of the full depth of foreshore sediments in the area proposed for reservoirs (ref. 6). Samples were recovered during these investigations, many of which were subjected to laboratory tests. Static cone penetrometer testing was used extensively. Also two trial embankments were built on the foreshore. The first of these, built in 1972, was a straight 50m long (at crest) section of bank built as two separated halves, the halves being raised alternately in steps of about a metre. During the early stages of construction the dredged sandfill remained quick and many sand blows occurred in sand that had settled out of the deposited sand-water mixture.

11. As these site investigations were put in hand an investigation, by Dr's. P.L. Bransby and G.W.E. Milligan, of the liquefaction potential of Dee estuary sand, using load-controlled undrained triaxial tests, was nearing completion in the Engineering Department of Cambridge University. This investigation included comparative tests on samples of sand from Ottawa (Illinois), known under the trade name of Banding sand, Valgrinda near Trondheim and Zeeland and also of the fraction of Leighton Buzzard sand between BS sieves Nos. 100 and 170, the fraction thought to be most prone to liquefaction on the basis of an earlier investigation. They agreed to make two tests on a sand sample from the Wash (ref. 7). The Department afterwards agreed to further tests being made on further samples of Wash sand with equipment already set up. These tests were made by Mr. M.G. Ratnayake, a post-graduate student from Aston University (ref. 8).

12. Also in the Cambridge Engineering Department at the time were Dr's. N.K. Tovey and D.W. Krinsley who were using the scanning electron microscope to study the surface of soil particles and had recently commented on the possible importance of surface microtexture of soil particles on the angle of shearing resistance, especially at low stress levels (ref. 9). Dr. Tovey agreed to examine sand grains sampled from the Wash, Dee estuary, Ottawa, Valgrinda, Zeeland and Beisbosch and to develop a tentative scale of liquefaction potential based on particle size grading, shape, surface microtexture and density of packing (refs. 10, 11 and 12).

13. A blasting test was carried out on the foreshore to investigate liquefaction potential of the sediments and whether compaction, if it were needed, could be achieved by the use of explosives as suggested by Florin (ref. 13) and Rao (ref. 14).

Load-controlled triaxial tests

14. Load-controlled undrained triaxial tests were performed with the objects of attempting to produce liquefaction failure such as occurs in a flow slide and provide some indication of the likelihood of liquefaction.

15. The work done was similar to that of Castro (ref. 4) when he tested three different sands. He found that Banding sand, from Ottawa, gave liquefaction at relative densities similar to those at which sands occur in the field, while the other two sands only liquefied when the initial relative density was close to zero. Liquefaction was shown to be a result of a sudden large increase in pore pressure causing low effective stresses rather than being due to an exceptionally low value of internal friction angle.

16. A single strain-controlled test was performed but no further tests of this type were made because earlier tests by Bjerrum et al (ref. 15) on Valgrinda sand, which is known to form flow slides in its natural state, had failed to produce liquefaction, even though very loose samples of the sand produced exceptionally low values of internal friction angle.

17. Five types of Wash sands were tested, four were from in situ samples from the foreshore and the fifth was from sand placed in the first trial bank after dredging. The sands used for comparative tests were from two in situ samples from the foreshore of the Dee estuary, single in situ samples from Zeeland and Valgrinda, a single sample after dredging from Beisbosch, a single sample of Banding Sand from Ottawa

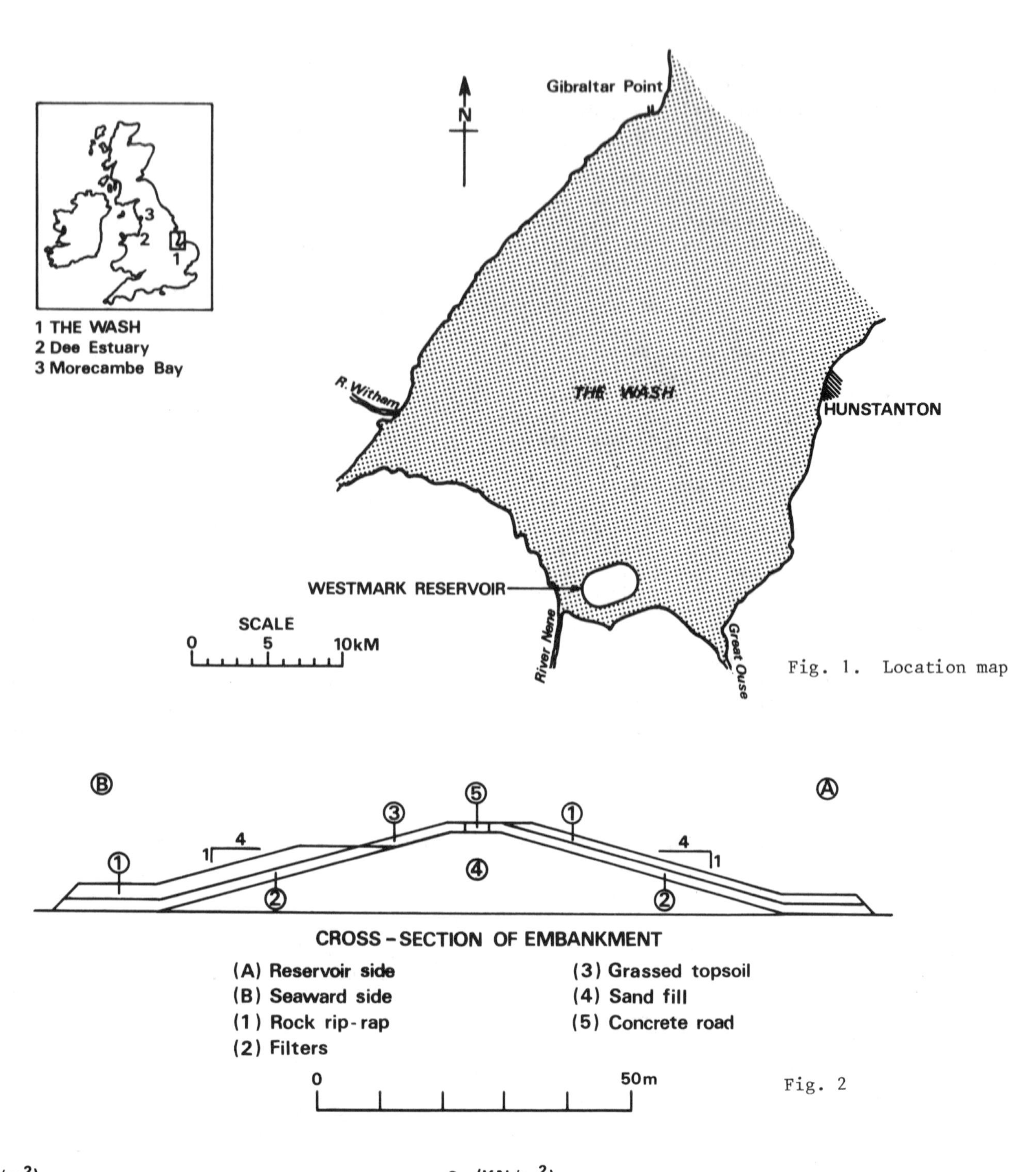

Fig. 1. Location map

Fig. 2

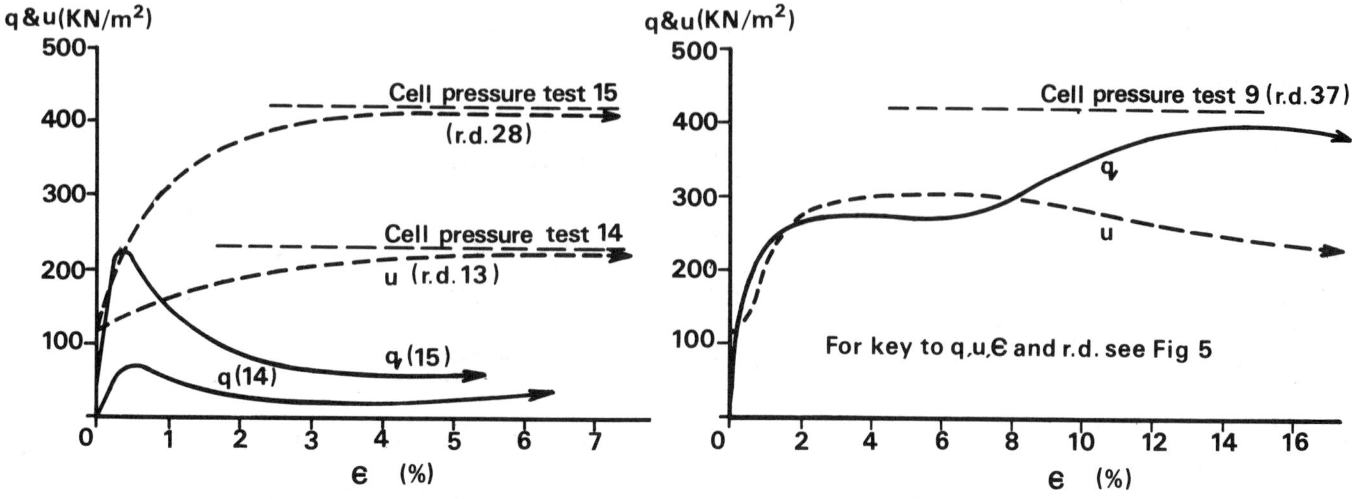

Fig. 3. Full liquefaction: banding sand

Fig. 4. Partial liquefaction: Valgrinda sand

(Illinois) and a single selected fractional sample from Leighton Buzzard sand. In all 47 load-controlled tests were made using Wash sands and 28 on the other sands (Table 1).

18. The load-controlled triaxial apparatus was designed to record automatically the rapid changes that occur at liquefaction. Electric transducers linked to an ultra-violet light recorder were used to record axial load, pore pressure and axial deformation. The recorder paper was run at 80 mm/s which was satisfactory even for the most rapid observed change from 1 to 25% axial strain in 0.18s during one failure described below. Further details and details of sample preparation and test procedure are given in ref. 7. Considerable effort was expended attempting to achieve test samples with high voids ratios. Three methods of deposition were used, quick deposition under water through a wide-necked funnel, deposition through a narrow-necked funnel and placing under water with a spoon. All methods could produce equally high voids ratios. The first method was used because it appeared to produce more uniform samples.

19. In addition to the triaxial tests, tests were performed to determine the maximum and minimum voids ratio for each sand in water. The results of these tests and the range of initial void ratios of the triaxial test samples are given in Table 1.

20. Three different types of behaviour were observed in the triaxial tests as follows:-

(a) full liquefaction the pore pressure in the sample increased to a value close to the cell pressure and thus the effective stress became small. The deviator stress, having reached a peak at low strains, decreased rapidly and the sample suddenly collapsed completely under a comparatively small load.

(b) partial liquefaction - a sudden rapid deformation of the sample early in the test, accompanied by a rapidly rising pore pressure and falling deviator stress, halted before complete collapse of the sample. Further deformation resulted in a dilative response.

(c) dilative response the sample showed the normal' behaviour for a loose sand in which a relatively small additional pore pressure developed initially. As the strain increased the pore pressure decreased, the

Table 1

Tests made, voids ratios, specific gravities and minimum relative densitites

type of sand	number of tests	Voids ratios max.	Voids ratios min.	Voids ratios initial value in test	specific gravity	minimum relative density (%)
Wash						
foreshore 1971	2 (1)	1 125	0.768	0.903 to 0.922	2.65	57
foreshore No 1 (2)	4			0.822 to 0.913		
foreshore No 3 (3)	5			0.810 to 0.936		
foreshore No 6 (4)	5			0.816 to 0.910		
after dredging	31	1.08	0.60	0.639 to 0.920		33
Banding (Ottawa)	2 (1)	0.850	0.546	0.766 to 0.810	2.65	13
Valgrinda	4 +3 (1)	1.085	0.806	0.741 to 0.830 0.914 to 0.982	2.74	91 37
Zeeland	1 +3 (1)	0 850	0.630	0.709 0 766 to 0.794	2.65	64 26
Leighton Buzzard	2 (1)	1.060	0.726	0.948 to 0.957	2.65	39
Dee estuary	13 (1)	0.850	0.614	0.651 to 0.825	2.65	12

(1) Bransby & Milligan, others tested by Ratnayake

(2) 0.5m below foreshore surface

(3) 0.5 to 1.5m below foreshore surface

(4) 2.0 to 2.5m below foreshore surface

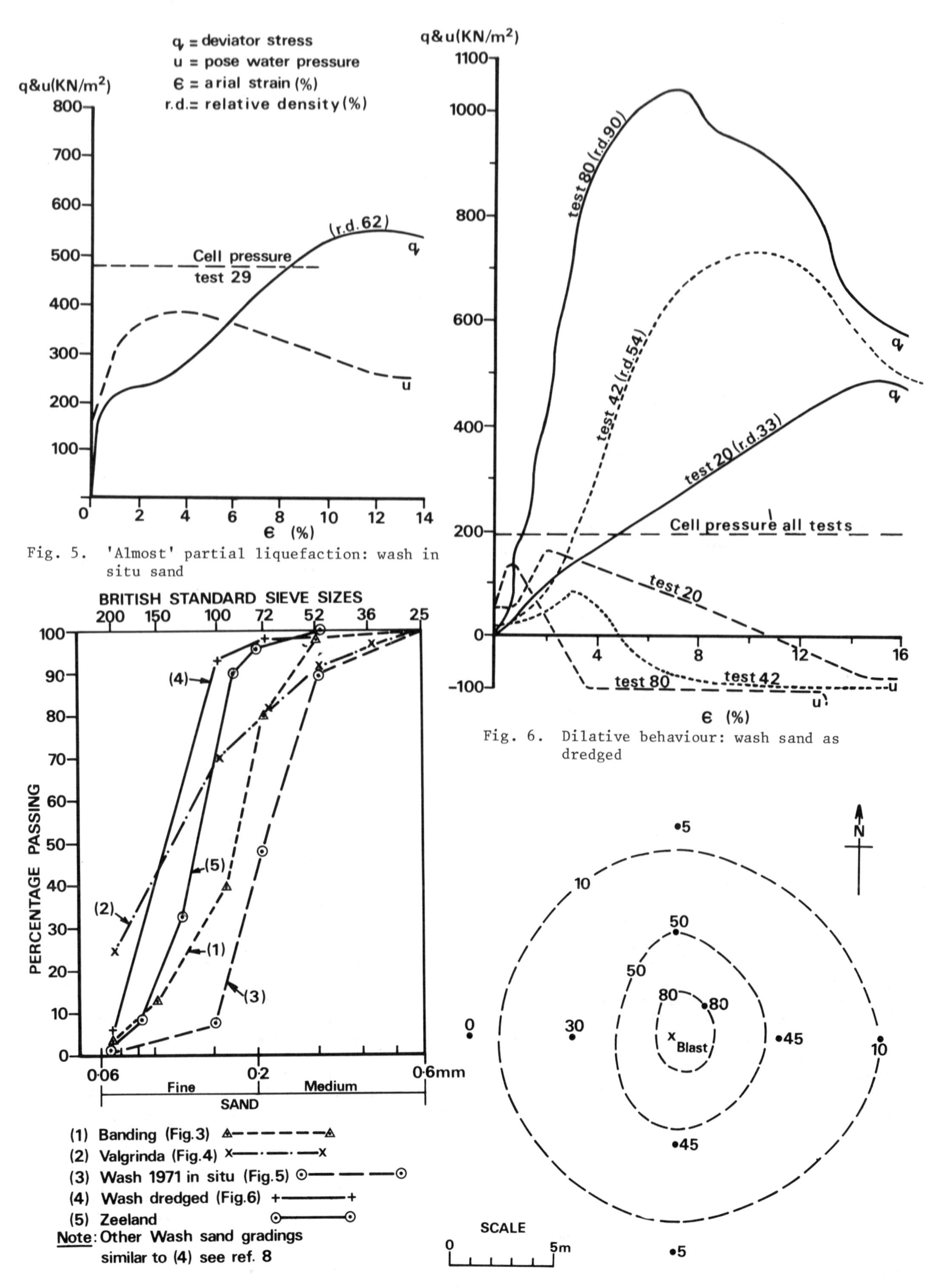

Fig. 5. 'Almost' partial liquefaction: wash in situ sand

Fig. 6. Dilative behaviour: wash sand as dredged

Fig. 7. Particle size distribution

Fig. 8. Blasting test (mm settlement)

reduction continuing until failure occurred at relatively high strain and load.

21. Full liquefaction was observed only in Banding sand (Fig. 3) though partial liquefaction was observed in both tests of Leighton Buzzard sand and in the loosest sample of Valgrinda sand (Fig. 4). This sample had a relative density of 37% and mobilised a friction angle over 30° under test. Full liquefaction would probably occur in samples at very low relative densities.

22. Four tests of in situ foreshore sands from the Wash showed a considerable pore pressure build up in the early stages which almost put the tests in the partial liquefaction category, before normal dilative behaviour took over. In one of the tests the pore pressure rose to a substantial proportion of the cell pressure and had still not returned to its starting value when the sample failed (Fig. 5). In view of the high relative density of this sample (62%) it seems likely that liquefaction would be obtained in samples of similar in situ sand at low relative density. All the tests of samples of dredged sand from the Wash showed normal dilative behaviour (Fig. 6).

23. Pore pressure dissipation was slower in the tests using in situ sands than in tests using dredged sand. Also the in situ sand had finer grading and contained some silt.

24. Tests were undertaken on samples prepared at various densitites. Results for Wash sand after dredging are shown in Fig. 6 from which it can be seen that pore pressure dissipated more rapidly from the sample with higher densities. These samples also had greater strength.

25. The tests of Zeeland sand showed an increase in pore pressure initially followed by a decrease in pore pressure at higher strains. The minimum relative density of 26% is quite low and it is doubtful whether liquefaction would be obtained even at very low relative densities in the laboratory test. Comparison with the work of Geuze (ref. 16) and Koppejan (ref. 2) is difficult due to the considerable differences between the values obtained for maximum and minimum voids ratios by Geuze and those reported in Table 1. The observed values of internal friction angle at failure were between 32.2° and 36.6°, considerably lower than those obtained by Geuze. It appears, therefore, that the sample of sand tested was not very similar to the sand tested by Geuze.

26. Tests on Dee estuary sands showed normal dilative behaviour.

27. These results suggested that the order of decreasing susceptibility to liquefaction of the sands tested is: Banding, Leighton Buzzard, Valgrinda, Wash (loose in situ samples), Zeeland, Wash (dense in situ and dredged samples) and Dee estuary sand.

28. Also the test results suggest that all types of sand from the Wash would have little susceptibility to liquefaction except when in a very loose state.

PARTICLE SIZE AND GRADING

29. Particle size distribution of the various sand samples were carried out in accordance with BS 1377:1967 Test 7A (wet sieving). Results are shown in Fig. 7. Dr. Tovey also carried out size analysis by direct measurement of sand grain dimensions for some of the very small samples (about 50 grains), the grains of which were examined under the scanning electron microscope at low magnification.

EXAMINATION OF SHAPE AND TEXTURE

30. Dr. Tovey, who undertook the examination of sand grains, considers that the following parameters are among those which affect the liquefaction potential of sand:

(i) Particle size and grading. A sand of low permeability, i.e. one containing a high proportion of silt, is less able to dissipate pore pressures than a coarse sand, and is thus more liable to liquefaction.

(ii) Particle shape. The more rounded the grains the lower their resistance to shearing and the greater the liquefaction potential.

(iii) Density of packing. A loose sand is more able to move into a denser state than a dense sand and, despite the greater permeability, more liable to liquefaction.

(iv) Microstructure of surface features. A particle with many irregular surface protrusions will tend to resist both shearing and liquefaction. If high effective stress levels are encountered, such a surface might be smoothed off.

(v) Electro-chemical effects. Attractive and repulsive forces between particles depend on the surface characteristics of the particles and the properties of the pore fluid.

31. Information about the first and third factors were gained from normal site investigation. The scanning electron microscope at Cambridge was used at low magnification to examine both size and shape of a selected sample of sand grains so adding information about the second factor. It was also used at high magnification to examine the surface texture of the grains. Electro-chemical factors are difficult to assess without sophisticated tests and were not considered.

32. The consideration of microtexture is important only at low effective stresses, such as occur during liquefaction, since at high effective stresses these minor surface features would probably be broken and abraded.

33. The selection of sand grains used by Dr. Tovey was made so that the particle size distribution was as near as practicable the same as the distribution found in the larger samples from which they were taken.

34. Examinations of the surface texture took account of a wide range of features from which the history of the sand as well as its likely susceptibility to liquefaction could be deduced. These features include: presence of concoidal fractures, cleavage surfaces, plates, adhering debris, deposition due to chemical precipitation, crystal growth, chemical etching and smoothness. Direct evidence of glaciation was suggested by the shape of some grains of all sands except Banding sand which is an ancient deposit of the Ordovician age with well rounded grains. Beach action or fluvial action seems to be absent on all samples but there is evidence in the Banding sand of weak chemical attack, probably due to diagenetic or post-depositional modification over long periods of geological time.

35. The particles of Wash sand examined showed wide variations. Some were well rounded, with smooth surfaces, some were angular and some appeared to have been recently shattered. The median size was about 0.22mm. The particles had evidently come from a number of sources and Tovey suggested that action in a very vigorous mechanical environment' caused the shattering.

36. Of special interest was the considerable solution precipitation on all types of sand examined. This was particularly noticeable in the samples of Valgrinda and Zeeland sand and in some types of sand from the Wash. On some Wash sand grains the precipitated deposits had grown with well defined angles which Dr. Tovey suggested probably arose from crystal growth in conditions where the rate of precipitation is slightly above some threshold value. The implications of this are that sands used in a subaqueous environment could undergo further surface-texture modification by further solution precipitation leading to generally smoother microtexture and possibly also to cementing action. The rates at which these two processes occur cannot be assessed but if the former predominates then the sand could become more susceptible to liquefaction with time.

37. After describing the examination of particle shape and microtexture in detail Dr. Tovey suggested the following probable order of susceptibility to liquefaction based mainly on these two factors but also with some knowledge of particle size, grading and density.

38. Banding sand has grains of moderate size which generally are well rounded and have smooth microtexture and they would therefore appear to be highly susceptible to liquefaction.

39. Valgrinda sand has particles that are more angular but nevertheless the corners are rounded and the surfaces are smoothed by a solution precipitation layer. Furthermore the sand includes a large range of particle sizes down to about 1 micron diameter. This sand would therefore appear to be fairly susceptible to liquefaction.

40. Zeeland sand has similar characteristics to those of the Valgrinda sand but the formation of mechanical plates or surface protrusions is slightly more pronounced and hence the Zeeland sand would seem to be rather less susceptible to liquefaction than the Valgrinda sand.

41. Wash sands are generally more angular than Zeeland or Valgrinda sand and the Beisbosch sand, mentioned below, but there are great differences between the individual Wash samples. As mentioned above these differences were found in particle shape and surface texture, which were both found to vary with depth at one site. Differences were also found in particle size and grading but these were less pronounced. A high degree of solution precipitation on some grains leads to a rougher texture and greater resistance to liquefaction though this could be partly offset in artificially placed deposits if placed in a looser condition because of the greater roughness.

42. Beisbosch sand has slightly larger particle sizes and particles are more rounded than Wash sand but the effects of these factors are partly offset by the rougher surface texture of some Wash sands. Beisbosch sand has some small particles, up to 10 micron size, adhering to its surface, which if firmly fixed would increase the resistance of this sand towards liquefaction.

43. Finally, the Dee estuary sand contains large particles, a number of which are very angular and some of which include extensive plate formation. Although solution precipitation is again apparent, some of the fine grains are very rounded but, on subjective assessment, the susceptibility to liquefaction appears less than that of Zeeland sand.

44. Thus the order of decreasing susceptibility is probably Banding sand, Valgrinda sand, Zeeland sand, Wash sand, Beisbosch sand and the Dee sand although some of the Wash sand would probably be less susceptible than Beisbosch sand and might be more appropriately ranked between Beisbosch and Dee sand instead of in the order given (ref. 11). This order is similar to that deduced from the load-controlled triaxial tests.

BLASTING TEST

46. A blasting test was carried out at one of the borehole sites to investigate liquefaction potential of the sediments and whether compaction could be achieved by the use of explosives.

A single 2.3 kg (5 lb) charge of gelignite was buried 4.5 m deep and detonated during a low tide period. Settlements of the foreshore were measured and a sounding was made using a static cone penetrometer. The test was then repeated. Settlements of about 40mm were recorded at 5m from the hole and 10mm at 10m distance (Fig. 8). The soundings revealed that resistance to penetration was reduced by the blasts at some depths and increased in others, though the changes are small and probably not significant.

47. No further blasting tests were made because the heterogeneity of the foreshore would have made interpretation of results very difficult if not impossible. Also this method was relatively expensive.

CONCLUSION

48. The second trial bank was subsequently built to the full proposed height of reservoir embankments. The area of foreshore forming its foundation included several areas of loose sand. These areas settled as the bank was built. Final settlements were from 100 to 400mm. The dredged sand-fill was found to have a high relative density where it first settled after discharge from the dredge pipe. Much of the sand was then spread by bulldozer. This movement decreased the density but subsequent bulldozer movements, travelling to and fro over the fill as successive layers of sand were spread, restored the density to a high value well above the limit at which the sand might be susceptible to liquefaction.

THANKS AND ACKNOWLEDGEMENTS

Thanks are due to Mr. Kåre Senneset, research engineer of the Geotechnical Laboratory, Technical University of Trondhiem for taking considerable trouble to find a sample of Valgrinda sand having a grain size distribution similar to the sand described in ref. 15. Also for the help of the late Dr. L. Bjerrum and Dr. N. Janbu who arranged this for us.

Thanks are also due to Mr. W.J. Heijnen Soil Mechanics Laboratory, Delft for providing samples from borings made in the vicinity of a large flow-slide in the Province of Zeeland, Netherlands. The slide occurred some years ago at the Noord-Beveland bank of the Ooster Schelde.

Professor A. Casagrande is thanked for providing a sample of Banding sand from Ottawa, Illinois and Professor P. Knoppert for providing a sample of sand from the Petrusplaat ringdijk, Beisbosch, Netherlands.

All other samples were provided from investigations carried out for Mr. O. Gibb, Director of the former Central Water Planning Unit of the Department of the Environment.

The author is grateful to the Department of the Environment and to Binnie & Partners for agreeing to the publication of this paper. The views expressed are those of the author.

REFERENCES

1. TAYLOR L.E. The concept of freshwater storage in estuaries. Estuarine and coastal land reclamation and water storage. Edited by KNIGHTS B. and PHILLIPS A.J. Saxon House, Farnborough, 1979, 135-151.
2. KOPPEJAN A.W., VAN WAMELEN B.M. and WEINBERG L.J.H. Coastal flow-slides in the Dutch province of Zeeland. Proceedings of the 2nd International Conference on Soil Mechanics and Foundation Engineering 1948, Vol V 89-96.
3. EVANS G. Intertidal flat sediments and their environments of deposition in the Wash. Quarterly Journal of the Geological Society of London 1965, 121, July 209-245
4. CASTRO G. Liquefaction of sands. Harvard University Soil Mechanics Series No. 81, 1969.
5. ANDERSON A. and BJERRUM L. Slides in subaqueous slopes in loose sand and silt. Norwegian Geotechnical Institute, 1968.
6. BINNIE & PARTNERS The Wash water storage scheme; feasibility study: volume 3; Geotechnical investigations, 1976.
7. BRANSBY P.L. and MILLIGAN G.W.E. A report on some laboratory tests to study the liquefaction potential of Wash sand. Cambridge University Engineering Department, 1972.
8. RATNAYAKE M.G. The potential to liquefaction of Wash sand. University of Aston in Birmingham, 1972.
9. TOVEY N.K. and KRINSLEY D.H. Discussion on: Effect of particle characteristics on soil strength, by KOERNER R.M. Journal of the Soil Mechanics and Foundations Division, ASCE, Vol 97 No SM4, Proc. Paper 7393 April 1971, 691-3.
10. TOVEY N.K. Liquefaction: A study of sand grains from Ottawa, Norway, Holland, the Dee Valley and the Wash using a scanning electron microscope. Cambridge University Engineering Department, June 1972
11. TOVEY N.K. An examination of sand grains from a sample from the Petrusplaat reservoir ringdijk Beisbosch, in the scanning electron microscope. School of Environmental Sciences, University of East Anglia, Norwich, December 1974
12. TOVEY N.K. Contact prints of scanning electron micrographs of sand grains from a sample from the Beisbosch ringdijk and the Wash. School of Environmental Sciences, University of East Anglia, Norwich. December 1974.
13. FLORIN V.A. and IVANOV P.L. Liquefaction of saturated sandy soils. Proceedings of the 5th International Conference on Soil Mechanics and Foundation Engineering 1961, 107-111.
14. RAO P.S. Compaction of non-cohesive soils by explosions, Indian Journal of Power & River Valley Development, 1972, 29-33 and 28.
15. BJERRUM L., KRINGSTAD S. and KUMMENEJE O. The shear strength of a fine sand. Norwegian Geotechnical Institute, Publication No. 45, 1961.
16. GEUZE, E.C.W.A. Critical density of some Dutch sands. Proceedings of the 2nd International Conference on Soil Mechanics and Foundation Engineering 1948 Vol III, 125-130.

9

Dynamic behaviour of rockfill dams

M. NOSE, and Dr Eng K. BABA, EPDC, Tokyo

It will be discussed on dynamic behavior of rockfill dams during earthquakes, and will give briefly the results of earthquake observations at an actual rockfill dam, vibration failure tests of slopes of large size crushed stones performed on a large shaking table and dynamic material tests by dynamic triaxial material testing apparatus.

An example of a more rational and practical dynamic analysis method based on those results will be described and then an evaluation will be given on the results of analysis and factor of safety.

I. Introduction

A number of irrigation earth dams have been constructed since olden times in Japan, some of them which now exist to serve their original purpose dating back 1,000 years. This fact is eloquent of the wisdom our predecessors gained by their experience in frequent strong earthquakes which must have attacked these dams and practiced in their design and construction of aseismic structure. On the other hand, large filltype dams to be used for multi purpose including power generation, water supply and flood control have a relatively short history of not more 30 years in Japan, with the result that our knowledge of their construction material and the behavior of their embankment during earthquake, which is indispensable for their final design and construction, can not be considered leaving nothing to be desired. In particular, the large rockfill dams with a height of more than 100 m which are mostly chosen zone type seem to require us to make further research in order to clarify their complicated dynamic behaviors during earthquake.

Electric Power Development Co., Ltd. (EPDC) this writer et al. are now serving took the part of a pioneer in the field of construction of rockfill dams in Japan. They have been executing construction of dams on their own responsibility worthy of their owner, committing themselves to all stages of work covering site investigation, design, supervision of construction and maintenance. Their primary concern throughout these stages has been to take special note of the circumstances particular to Japan which is subject to frequent strong earthquakes. With their years of experience and accumulation of research efforts, EPDC has been increasingly confident of their ability to construct dams of even more earthquake-proof structure. In the following, a brief of the results of the studies is introduced, while introducing the dynamic behavior of a rockfill dam during earthquake and one of its stability analytical methods.

II. Earthquake Observation on Dams

To have the behavior of a dam and its surrounding rock during earthquake is one of the most fundamental and important factors that must be taken into consideration in our design of dams of aseismic structure. The earthquake observation equipment generally used in Japan where a seismic observation network of considerable minuteness has been consolidated is required to permit (1) automatic recording of waves necessary for earthquake engineering, (2) simultaneous observation of earthquake motion felt on several points which dam, its attached structures and its surrounding rock and (3) extensive observation of earthquake motion of both strong and weak, providing not only ease of adjustment and sufficient stability and reliability of performance, but high water-tightness and ruggedness. The nearly developed electromagnetic seismographs now in wide use in Japan satisfy all of the requirements just mentioned. The accelerograms introduced this paper are all what were recorded by the electromagnetic seismographs of the PTK-130 type manufactured by Katsujima Mfg., Co. of Japan.

Fig.-1 shows the arrangement of seismographs installed on the KUZURYU Dam which was completed in 1964. A rockfill dam of the inclined impervious core type, 128 m high, 355 m in crest length and 6.3×10^6 m^3 in volume and reservoir storage capacity of 353×10^6 m^3 for 220 MW power generation. With its site located no more than dozens of kilometers away from an active tectonic zone in central Japan, this dam has been subject to earthquakes with a magnitude of between four and seven and with an epicentral distance of 40-50 km in general, at the rate of about once every several years.

Figs.-2 through -4 show the accelerograms recorded in 1969 when an earthquake with magnitude 6.6 and an epicentral distance of 40 km hit the dam. These accelerograms, which seem to indicate the behavior of a dam during earthquake that is of much interest, are introduced in the following. The No. 1 through No. 14 accelerograms are those recorded at four points on the dam (recording of vertical direction was not available at EL 449 m) and on the rock of the left bank, by the use of three-component seismographs. These records indicate the following. First, the No. 1 through No. 3 for the dam crest each show a fairly predominant peak, 2.4 Hz for the dam axis direction, 2.0 Hz for the river flow direction and 3.7 Hz for the vertical direction. As for the impervious core (No. 7 through No. 14), the seismograms records at higher elevations each showed a predominant peak, while those records at lower elevation showed several peaks similar to those recorded on the rock of both banks. The No. 4 through No. 6 recorded on the rock of the left bank showed a predominant peak at nearly 1.0 Hz, also showing another predominant peak at almost 5 Hz. These results clearly show that the input at the foundation rock, though it contains wide-ranging vibration components to some extent, led to the response of the dam which is characterized by a remarkably predominant vibration period. This fact seems to indicate that the response of the dam predominantly shows a natural vibration period, irrespective of short period components contained in the input. Also, it was found that the predominant period of the dam tend to become shorter in relation to the river flow direction (rectangular to the dam axis), dam axis direction and the vertical direction in that order. In addition, this predominant period was found representing the 1st order, the 2nd order or higher order period. As for the foundation rock, the examination of the Nos. 4 and 5 recorded at the same elevation as the dam crest in comparison with Nos. 13 and 14 recorded at the same elevation as the dam bottom clearly shows that the frequency components in the case of the latter recording, compared to those in the case of the former recording, developed a considerably complex power spectrum between 1 Hz and 10 Hz. The absolute values of acceleration in relation to the river flow direction in particular at a higher elevation and that at a lower elevation, the former coming to some two times the latter. However, the acceleration in relation to the dam axis direction was found to be almost the same both for the former and the latter. Moreover, the No. 1 recorded at the dam crest center in relation to the dam axis direction showed the acceleration to be some five times that at a lower elevation, while similar No. 5 and No. 14 recorded in relation to the river flow direction showed the acceleration to be some 2.5 times that at a lower elevation. The No. 5 and No. 2 recorded at a higher elevation and at the dam crest, respectively, showed much the same acceleration. Figs. 5, 6 and 7 show the relationships in terms of acceleration between the foundation rock and the dam crest, together with other recorded acceleration on this dam. These figures seem to suggest that the acceleration on the foundation rock exceeding the level of 100 gal leads to the response acceleration of the dam crest going below the ratio of 1 : 1. This indicates the possibility of the damping of the dam including the radiation damping reaching unexpectedly high values.

Fig.-8 shows a typical example of deformation of a dam when subjected to excess acceleration. The MIBORO Dam as shown in the figure represents the first rockfill dam constructed via modern technique in Japan, which stands 131 m high, 405 m in crest length, and 8 × 106 m^3 in volume. This dam, eight months after began impounding reservoir, was attacked by an earthquake which was M = 7.2, epicentral distance 20 km. As shown in the figure, however, it was confirmed that no damage was caused to the dam except that it suffered a settlement of 3 cm and a displacement of 5 cm in the downstream direction. There was no deformation witnessed on its embankment body. The acceleration on the rock is estimated to have been almost 250 gal (no seismographs was installed at that time).

III. Experimental Study

To conduct experimental studies on the stability of a rockfill dam during earthquake, it is necessary to make a simulated scale model of the small size material used for prototype dam and place it on a shaking table, so that it may conduct vibration tests to make a qualitative study of the failure mechanism of a dam. This, therefore, precludes a similitude between model and prototype in the strictest sense. However, if the vibration model test is conducted taking due account of the size effects of model material, the tests seem to prove significant in their own way.

Fig.-9 shows the relation between acceleration and frequency that results in a failure in the slope made of crushed stone of the 10-20 mm grain size. As shown in the figure, the failure acceleration has almost nothing to do with the frequency till the level of 15 Hz or so is reached, but a considerably high acceleration is required at a higher frequency to cause a failure to the slope. Fig.-10 shows the acceleration of failure in relation to its slope gradient, indicating that a gradient between 1 : 2.0 and 1 : 2.5 causes a change in the curve, and that the slope stability will not increase in proportion to the gradient which will be made more gentle than 1 : 2.5.

Fig.-12 and -13 show the results of vibration failure tests conducted on a simulated model dam with an impervious core made of crushed stone of the 10 mm grain size and soil which was placed on a shaking table (Fig.-11) also with a port of reservoir filled with water. The tests were conducted with constant frequency set at 15 Hz and with the acceleration being increased gradually so that it may clarify how the slope develops a failure to let water overtopped the dam. For this purpose

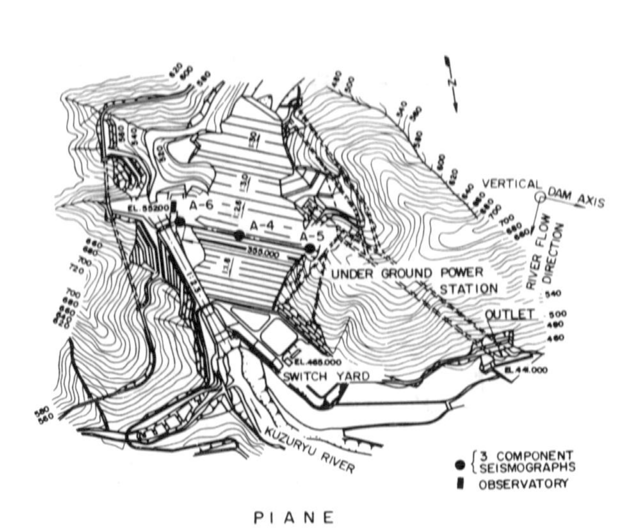

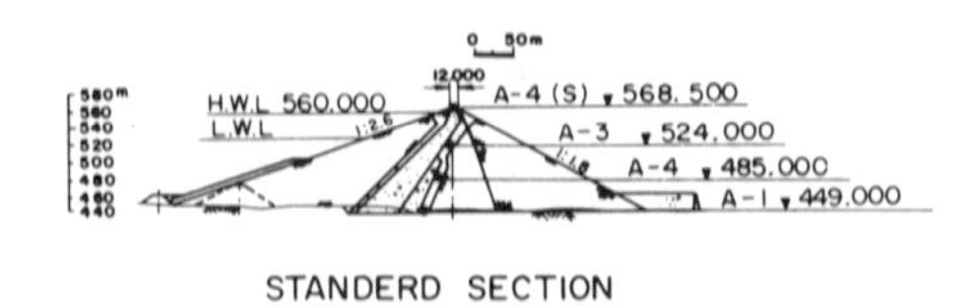

Fig.1. Arrangement of Seismographs on KUZURYU Dam

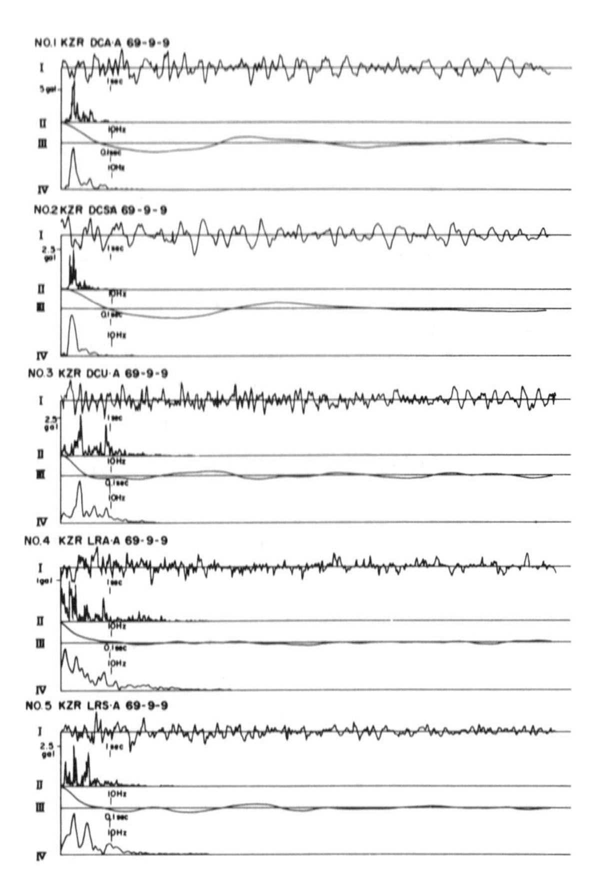

Fig.2. Earthquake Record on KUZURYU DAM

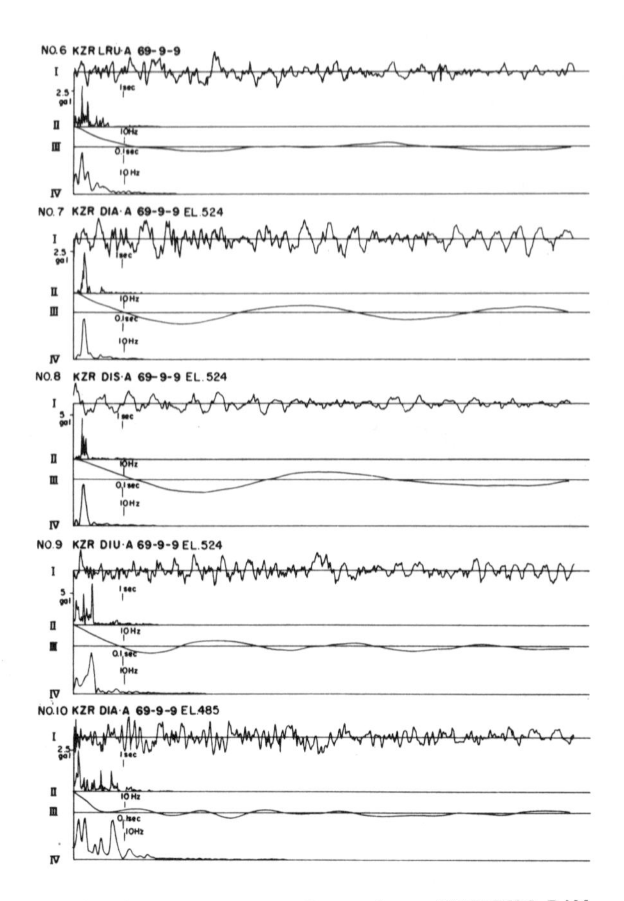

Fig.3. Earthquake Record on KUZURYU DAM

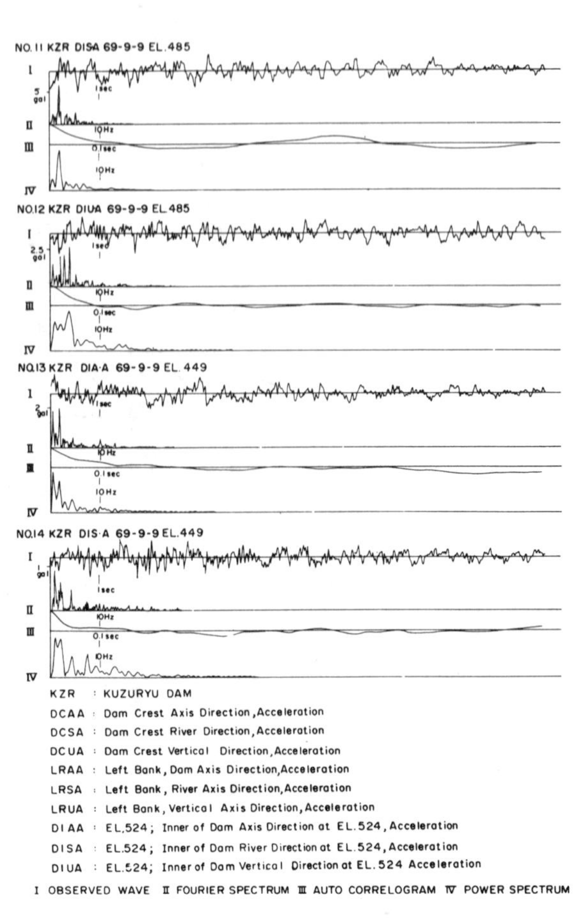

Fig.4. Earthquake Record on KUZURYU DAM

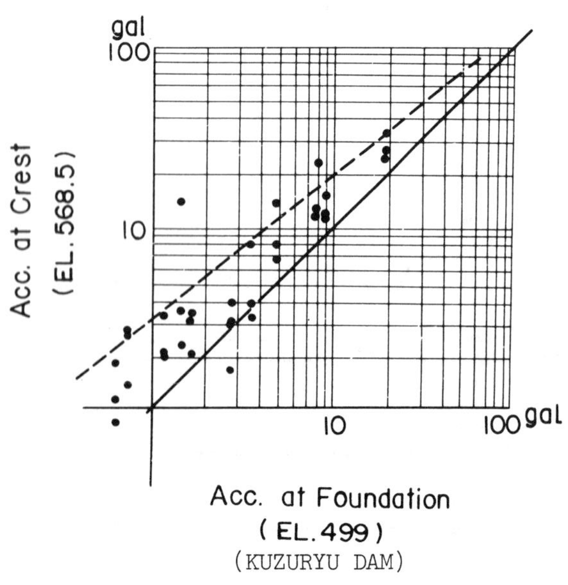

(KUZURYU DAM)
Fig.5. River Flow Direction

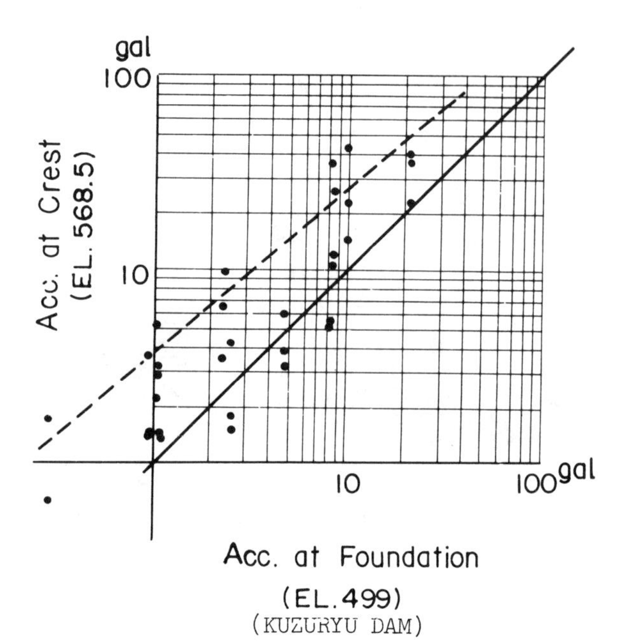

(KUZURYU DAM)
Fig.6. Dam Axis Direction

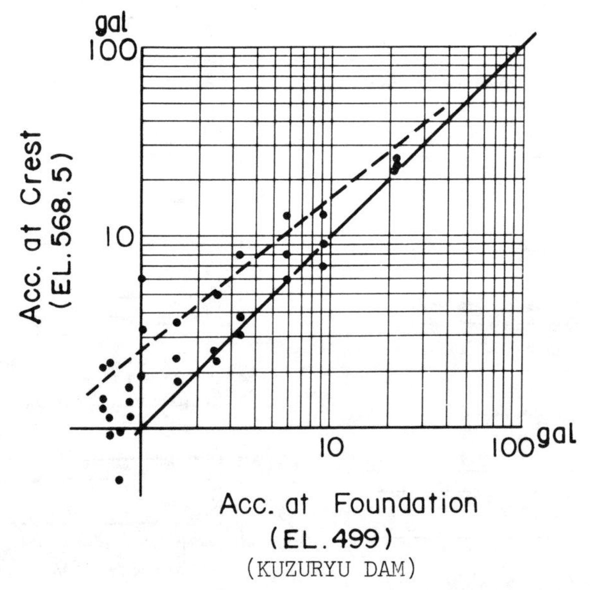

(KUZURYU DAM)
Fig.7. Vertical Direction

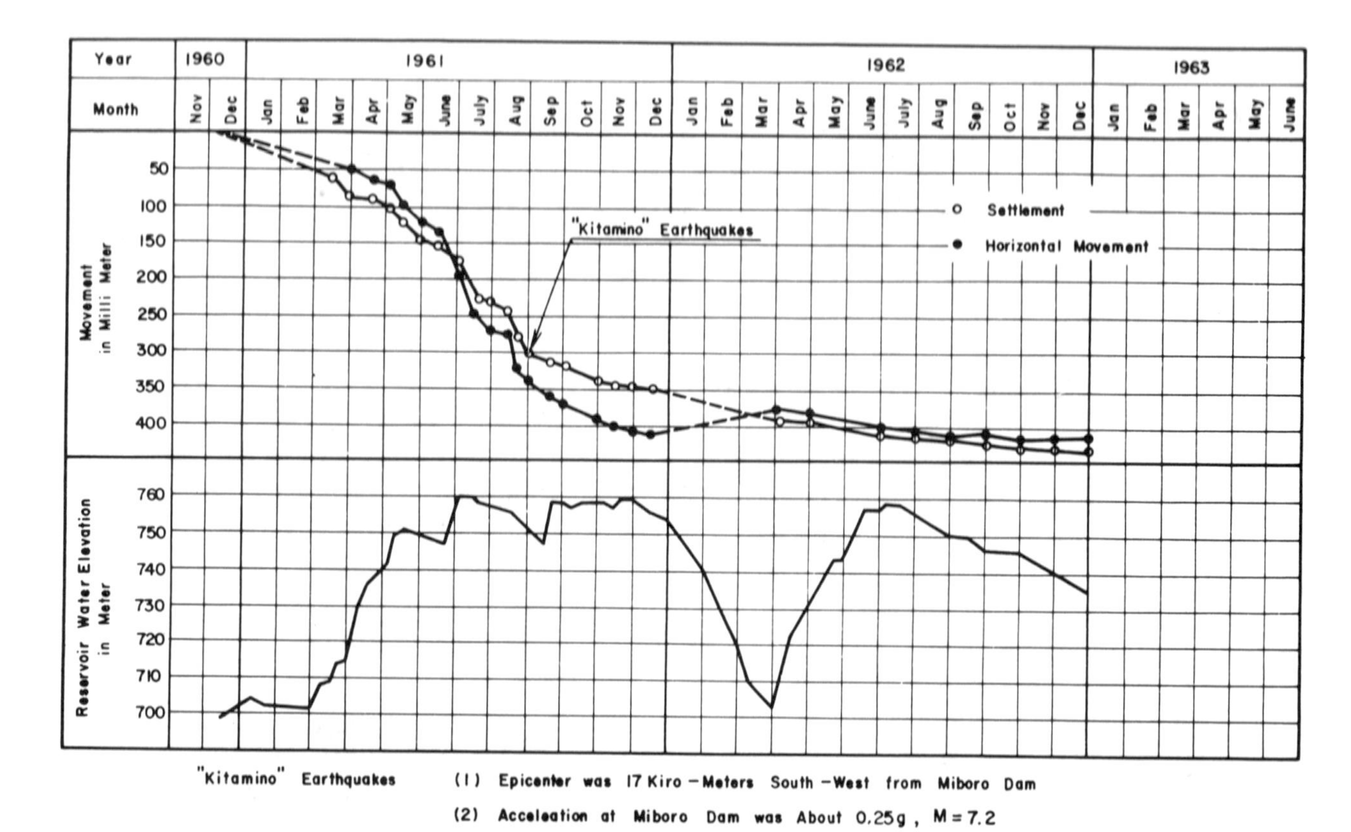

Fig.8. Water Level and Movement Curve of MIBORO Dam

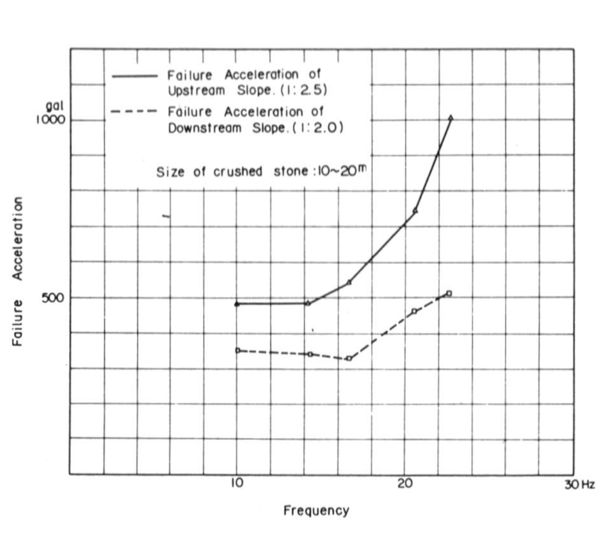

Fig.9. Failure Acceleration - Frequency Curve

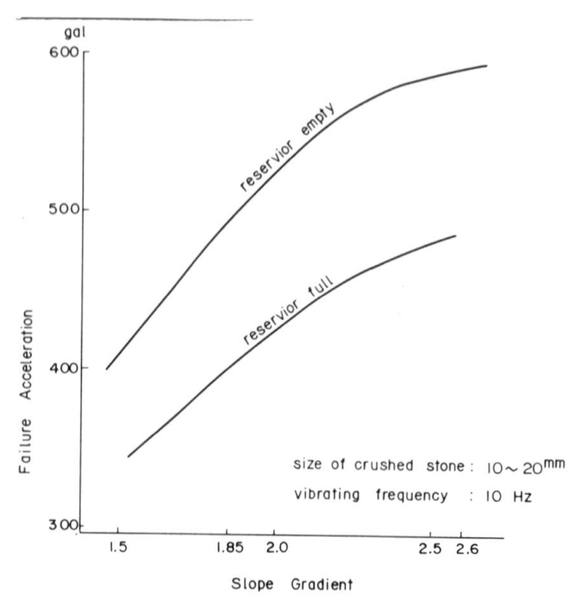

Fig.10. Failure Acceleration - Slope Gradient Curve

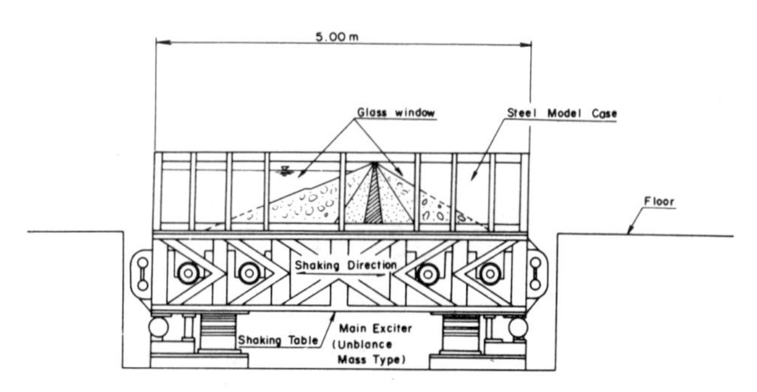

Fig.11. Vibration Failure Test of Rockfill Dam Model

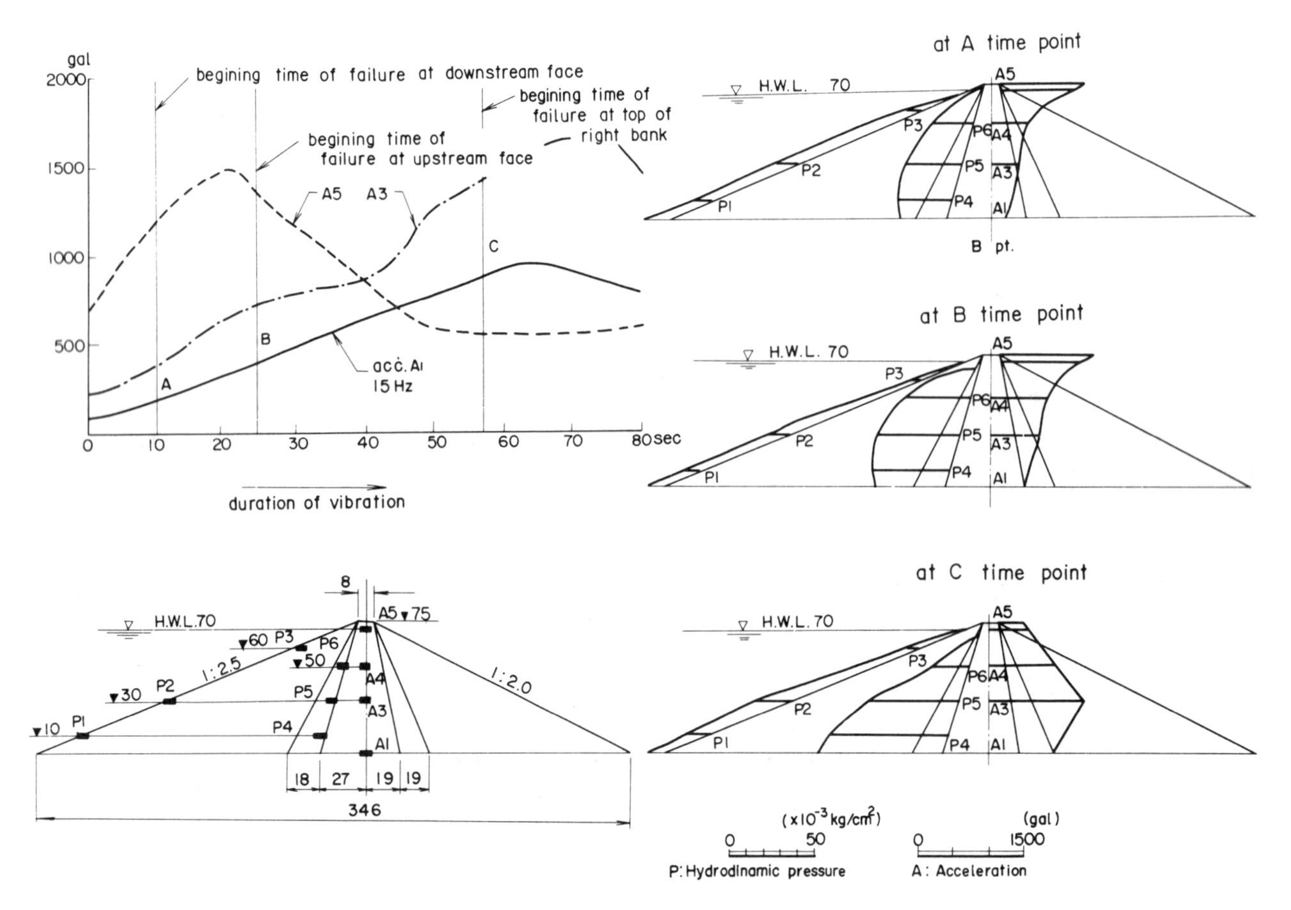

Fig.12. Failure Test of Rockfill Dam Model

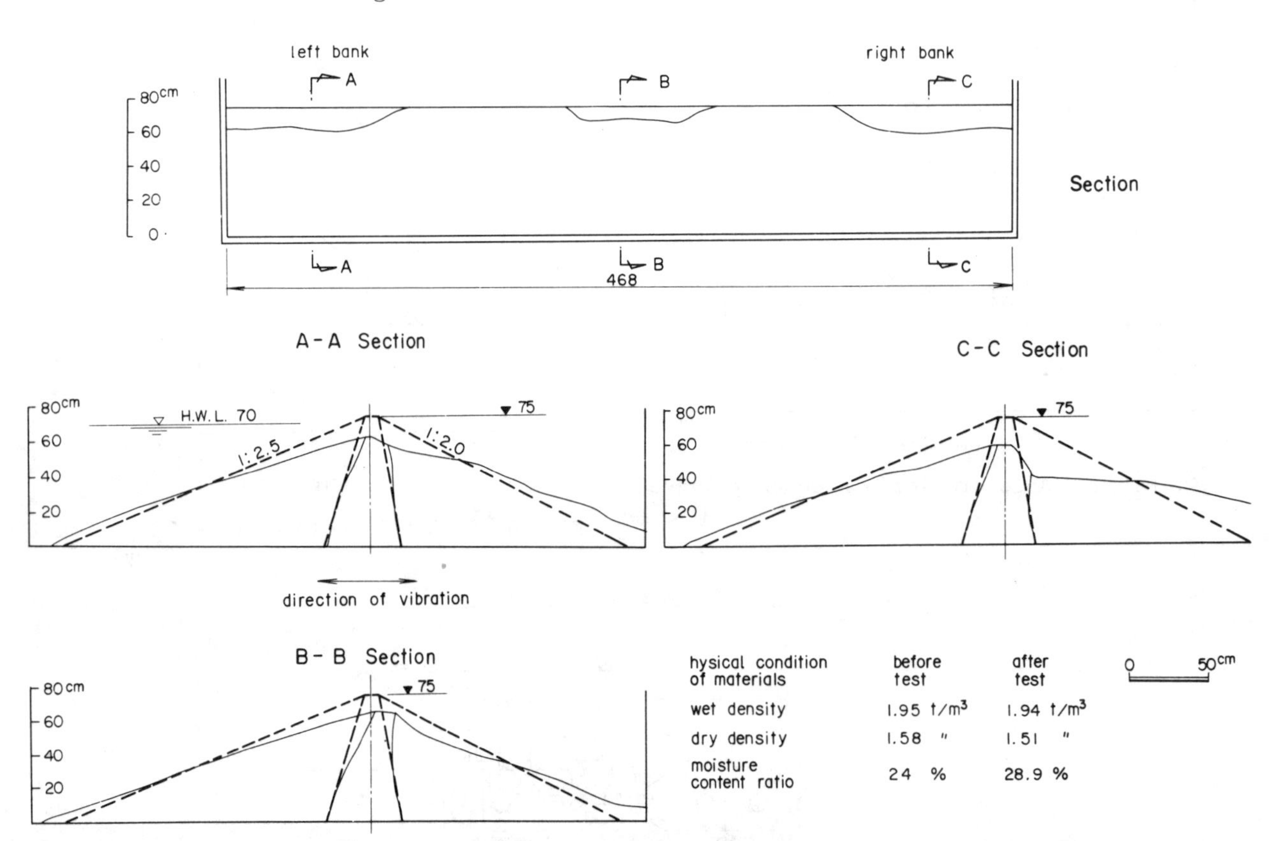

hysical condition of materials	before test	after test
wet density	1.95 t/m³	1.94 t/m³
dry density	1.58 "	1.51 "
moisture content ratio	24 %	28.9 %

0 50cm

Fig.13. Profile of Model After Failure

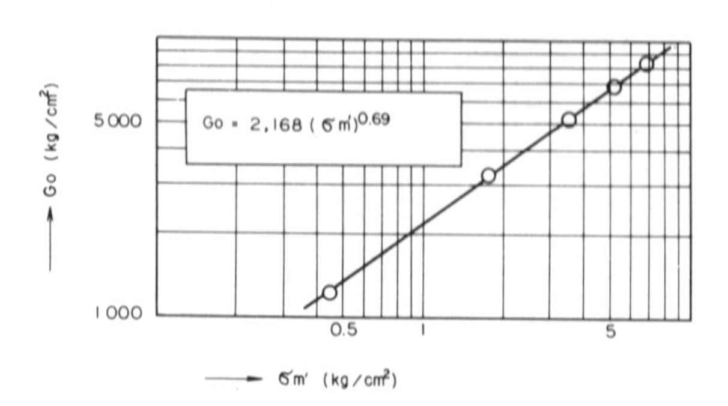

Fig.14. Go ∿ σm' Relation for Core Material

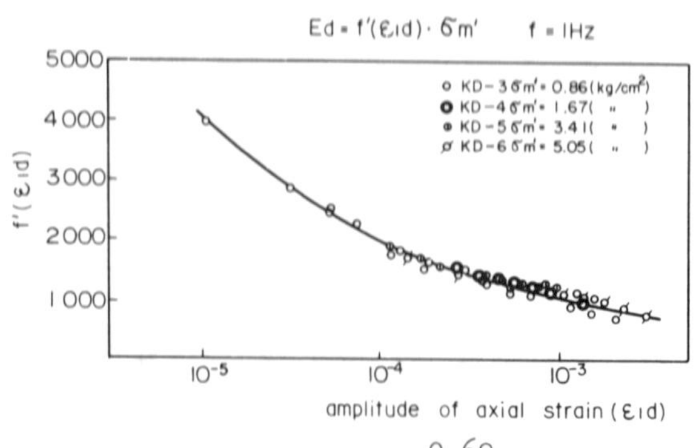

Fig.15. $Ed/(\sigma m')^{0.69}$ ∿ εid
(Ogata, Watanabe, Miura. 1978)

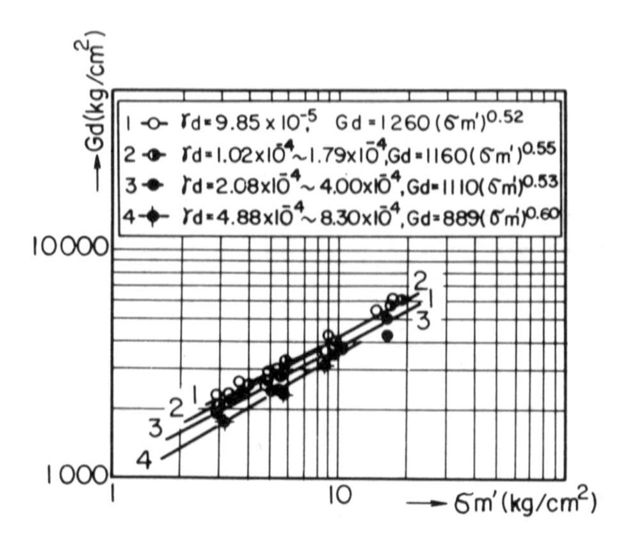

Fig.16. Gd ∿ σm' Relation

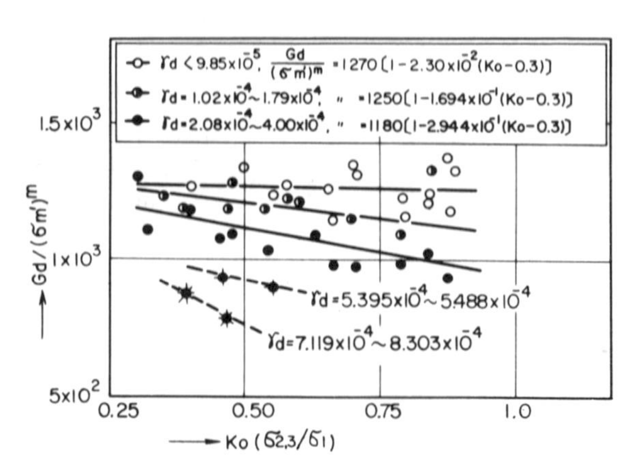

Fig.17. $Gd/(\sigma m')^m$ ∿ Ko Relation

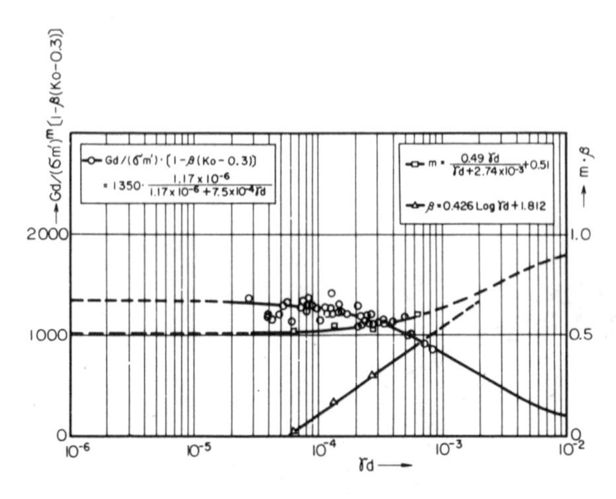

Fig.18. Gd ∿ γ Relation

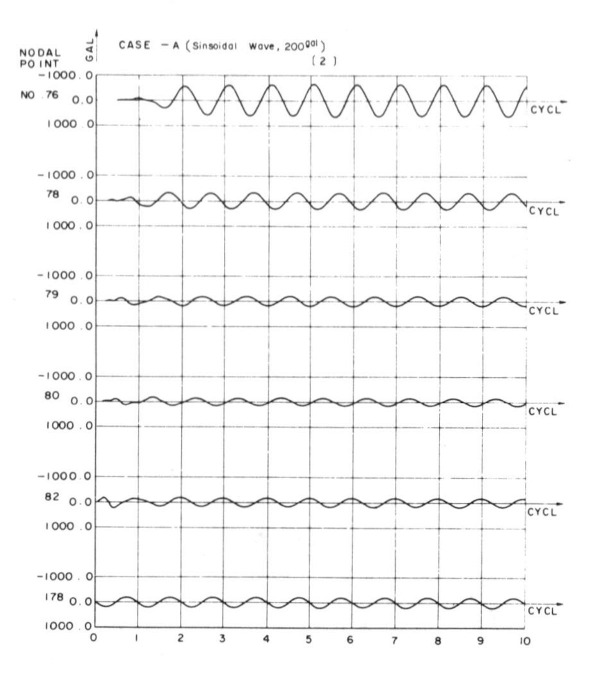

Fig.19. Horizontal Acceleration during Earthquake

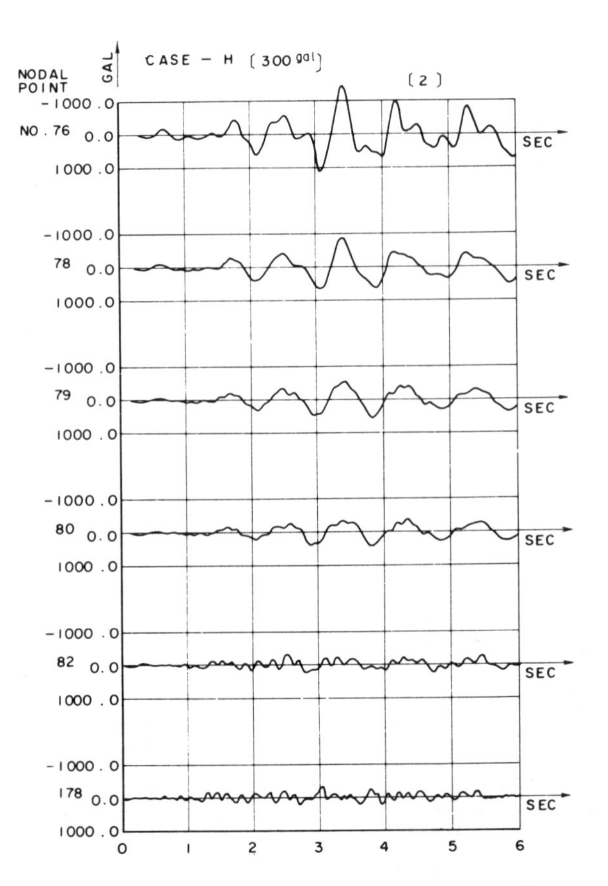

Fig.20. Horizontal Acceleration during Earthquake

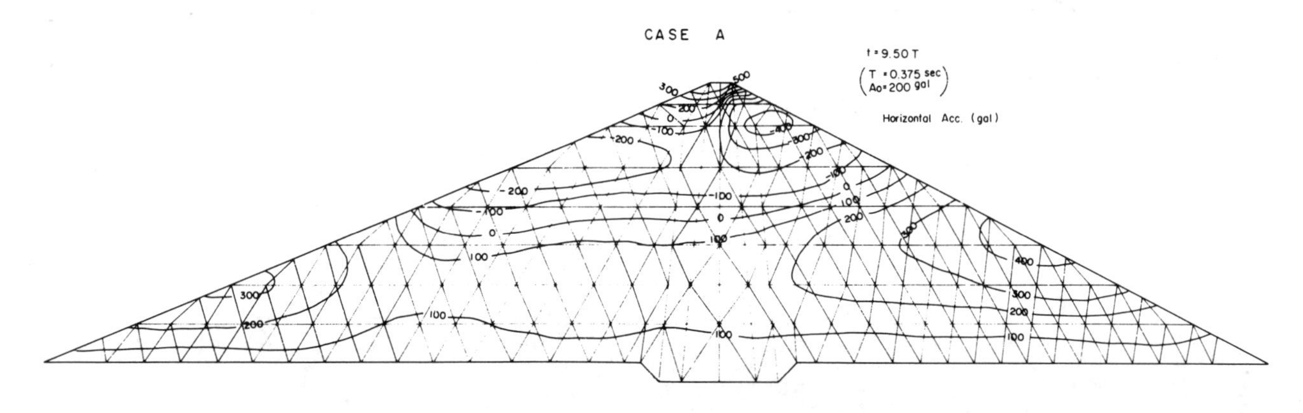

Fig.21. Distribution of Horizontal Acceleration (CASE A, t = 9.50 T)

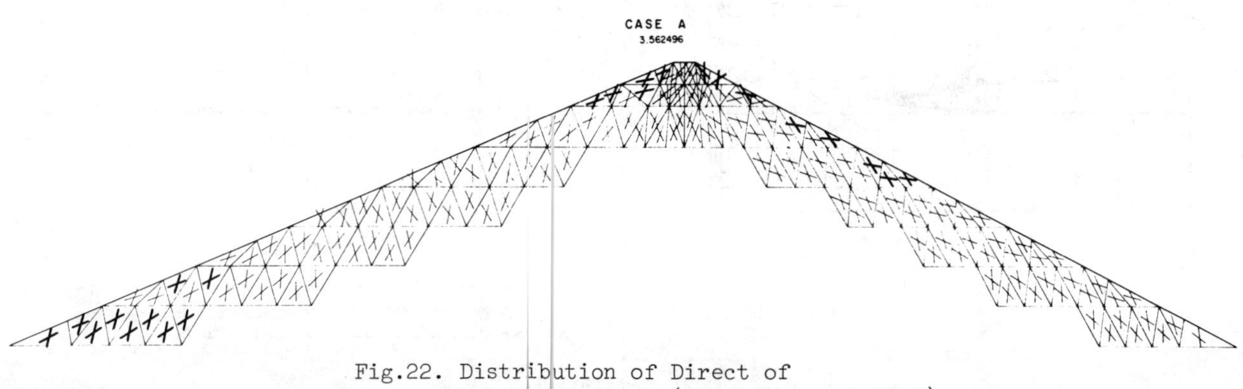

Fig.22. Distribution of Direct of Mobilized Plane (CASE A, t = 9.50 T) Bold Arrows are Less Than 1.0 of Partial Safety Factor

accelerometers, hydrodynamic pressure meters and pore pressure meters were molded in the model dam. The failure process began with the collapse of the top of the downstream side of the steeper slope, followed by the stone slipping down from the top of the upstream slope, when the top of the core began to show a sudden decrease in acceleration due probably to the crack developed on the top of the core. With the dam crest showing further decrease in acceleration to fall below the acceleration of the shaking table, the top of the core began to collapse in the downstream direction leading to the water leaking therefrom. Because of the effects of the side walls made in conducting the test, the collapse of the crest provides evidence that the model was vibrated in the 2nd order mode. As for the inside condition of the model dam, the accelerometers (A1-A5) showed their corresponding values to the mode. The hydrodynamic pressure meters (P1-P3) indicated almost constant at relatively small values. The pore pressure meters (P4-P6) indicated fairly large values in a distribution pattern of much interest. The results of these tests can not be applied directly to the prototype, but seem to serve as one pattern of special interest which represents the failure process of a real dam during earthquake.

Furthermore, the same vibration tests as mentioned above which were conducted on a model dam made of large-size crushed stone of the mean grain size of 60 cm mixed with crushed stone of the size one-third as much at the ratio of 3 : 1 in weight showed that the small-size stone developed subsidence to form a high density layer, down which the remaining large-size stone began to slip as if it were sliding down of slide. This phenomenon seems to suggest that the use of small-size stone simply for the purpose of increasing the density of embankment structure is not always recommendable.

IV. Numerical Analysis

On the basis of the results so far obtained, it will be now supposed to formulate each step necessary for numerical analysis. Here, some of the items that seem to have important effects on the outcome of numerical analysis are discussed as follow.

The first item concerns how the physical property value should be expressed. As far as core material is concerned, measurement was taken by means of a geophysical exploration method of the distribution of S-wave velocity and P-wave velocity to determine a velocity distribution model so that the results of earthquake observation may be compared with the above model. Then, considering the fact that the ratio of stress within an embankment under construction to principal stress can be considered constant, and that some 70% of the load on the core can be considered, the average effective principal stress $\sigma m'$, the shearing elasticity Go and Poisson's ratio ν are expressed as follows.

$$\sigma m' = 0.3(1+\nu)\rho_Z$$
$$Go = \rho V_S{}^2 = \rho(140Z^{0.34})^2 \qquad \text{.......... (1)}$$
$$\nu = 0.45 - 0.006Z^{0.60}$$

$$Go = 2{,}168(\sigma m')^{0.69} \qquad \text{.................... (2)}$$

Here, Z : Depth below surface

Fig.-14 shows the relations as expressed in the equations above. The equation (2) above is for micro-strain amplitude. On the other hand, on the basis of the results of dynamic triaxial material tests, there is proportional relation between $\sigma m'$ and the dynamic secant modulus (Fig.-15). Therefore, after determining, Poisson's ratio taking into account the void ratio reliance of Go's, the dynamic shearing modulus G for core material is expressed as follows.

$$G = 395\frac{(2.97-e)^2}{1+e}\cdot\frac{1}{1+\dfrac{\gamma}{1.33\times10^{-4}}}(\sigma m')^{0.69} \qquad (3)$$

Similarly, the results of dynamic triaxial material tests conducted on rock material provide the relations as shown in Figs.-16, -17 and -18. Therefore, the following equation is obtained.

$$G = 440\frac{(2.97-e)^2}{1+e}\cdot\frac{1.56\times10^{-3}}{1.56\times10^{-3}+\gamma}(\sigma m')^{0.55} \qquad (4)$$

As for the damping factor h, the following equation is formulated, taking into consideration the results of dynamic triaxial tests and various seismograms, and assuming the radiation damping factor to be 5 per cent on the basis of various seismograms, multireflection and exciter test results.

$$h = 0.23\frac{\gamma}{\gamma+1.33\times10^{-4}} + 0.05 \qquad \text{........ (5)}$$

However, the equations above are required to be corrected on the basis of the results to be obtained by conducting further material tests, vibration tests, etc. and by making further earthquake observations, so that they may hold good for higher levels of strain.

Figs.-19 and -20 show on example of numerical analysis. Case A (Fig.-19) shows a numerical analysis made in order to obtain a stationary response solution, assuming the sinusoidal wave to be an input, while Case H (Fig.-20) assumes the real seismic wave to be an input. Fig.-21 shows the acceleration distribution for the embankment body, and Fig.-22 shows the elements leading to the local safety factor falling below 1, in order to evaluate the results of numerical analysis made in Case A. To make a final evaluation, we obtained a sheering stress in the mobilized plane with regard to elements with a safety factor of less than 1, so that, when the plane formed through connection of these direction of sheering stress in the elements goes out the embankment body, it maybe obtained a macro-slip safety factor regarding the plane thus formed as a slip plane. As shown in Fig.-22, stability of the slope is discussed by the representative circles obtained through the groups of slip circles at the moment for one eighth of one cycle after the condition of the stational response. Regarding Case A, even if the acceleration amplitude became considerably large with the short period (for example 600 gal with a period of 0.125 sec.), the resultant strain was found small making it difficult

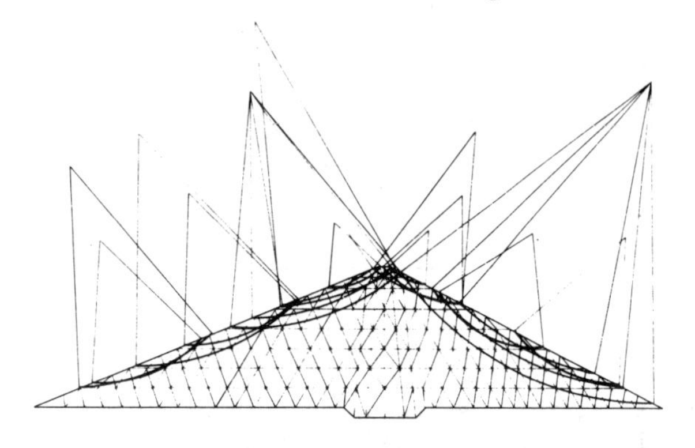

Fig.23. Supposed Slip Civeles for Safety Evaluation

for slip instability to occur. Regarding Case H, the short period components, through selective resonance, showed a tendency to disappear at the crest and its vicinity. In either case, the damping factor is affected the most, which fact seems to require further study in the future.

V. Conclusion

The dynamic behavior of rockfill dams during earthquake has so far discussed, on the basis of the results of earthquake observations and experimental studies, also introducing an example of a numerical analysis made of these results. The non-linearity shown by a dam which is suffered by a strong earthquake still leaves much to be examined further with respect to dam material and dam structure. In either case, a further inquiry into the high levels of strain seems to be required. For example, a further inquiry needs to be made into such fundamental aspects as high levels of strain develops when it collapses in a model and in specimen of material tests, and stone particles in conducting slope failure tests by means of a shaking table. At the same time, earthquake observations must be made of the embankment body of a real dam in order to clarify its real behavior during earthquake, especially to work out measures for strong earthquakes. As for the vibration tests to be conducted on a real dam by means of exciters, it seems rather practical to conduct vibration measurements by means of blasting if possible to be made in the vicinity of a dam. This is because the vibration force generated by exciters can not be easily transmitted to the embankment body.

All of the studies it was shown that a vibration phenomenon is conspicuous at the crest section, a part representing one third or more of height of dam. As a matter of fact, the crest part is the most liable to vibration of all the parts of a dam, thus naturally attracting serious attention in designing dams. Notwithstanding, it may be feared that no serious structural consideration has so far been paid to this part of a dam.

The fact that a number of structurally weak points were remarked on the crest section on the downstream side seems to suggest the need for a reexamination of future design and execution of dam construction. The research on this aspect will be of much interest to us. For example, the destruction of the crest as remarked as a failure pattern in the experimental studies earlier mentioned, and its special acceleration distribution as resulting from the numerical analysis will be worthy of serious attention. This seems to require to pay at least the same serious design consideration to this part of a dam as that to the upstream side. At the same, it seems that it should be made a further inquiry into the weak points or the reliability of other parts as remarked as a result of clarification of their dynamic behavior during earthquake, so that it maybe clarified not only the overall safety of a dam, but also its local or partial safety. This is what earthquake engineering is primarily concerned with.

10 Friction parameters for the design of rock slopes to withstand earthquake loading

S. R. HENCHER, BSc PhD DIC FGS, Geotechnical Control Office, Hong Kong

Experiments to investigate the friction between rock surfaces under transient loading are described and data presented for four rock types. The method used involved a vibratory test rig and the back analysis of sliding blocks of rock. Friction values obtained in this way are compared to data from inclined plane sliding and direct shear tests. It was found that :

1) the displacement of a block due to vibration loading was greater than would have been predicted using peak friction angles obtained from static tests;

2) the apparent shear strength of the surfaces at the commencement of sliding was higher than the equivalent static strength and was dependent on the rate of loading;

3) once sliding had begun, frictional resistance reduced considerably and was several degrees lower than the equivalent static angle of friction.

The implications of these findings for slope stability analysis are considered and a finite displacement method recommended for design against earthquakes.

INTRODUCTION

1. This paper describes experiments to investigate frictional resistance developed between rock surfaces when loaded dynamically rather than statically. The results are relevant to the design of rock slopes to withstand earthquake loading.

The occurrence of rock slope failure due to vibration loading

2. Rock slope failures resulting from vibration loading commonly occur due to earthquakes and to a lesser extent blasting. As far as the author is aware there are no published back analyses of such failures that have provided values for shear strength at failure. It has, however, been demonstrated that displacements in a progressively failing slope may be related to vibration magnitude (ref. 1, 2).

The design of rock slopes to withstand earthquake loading

3. The first task for design in a seismic region is to determine the magnitude and duration of forces that might affect the slope within its lifetime. Existing earthquake records and the way in which such earthquakes are related to the tectonics of the region are statistically studied and the ground motion characteristics to be included in design determined. Earthquake ground motions are however extremely complex and would be most difficult to incorporate in design. Simplifications are necessary. The most usual approach is to consider earthquake loading as a static horizontal driving force. This gross simplification has been justified because of the questionable quality of other input data in most designs (ref. 3). An alternative method is to calculate the actual displacements likely to be caused in the slope due to the design earthquake (ref. 4, 5, 6). The advantage of this method is that finite displacements are determined for forces that would imply total failure by using a pseudo-static force in a limit equilibrium method. The finite displacement method reflects the important transient nature of the inertial loading. It has been shown that simplifications for ease of calculation can be made by representing complex ground motions as simple acceleration pulses (ref. 6).

Following such an analysis, decisions can be made on the basis of "permissable displacements" either in terms of some controlling factor of shear strength such as roughness wavelength (ref. 7) or in terms of the post earthquake condition of the slope. The importance of considering the post earthquake condition is emphasised by the many reports of failures that occur days or even months following an earthquake in slopes that were previously stable (ref. 8, 9).

Shear strength developed to resist transient loads

4. Whatever method is used to design a slope to withstand seismic loading, a representative shear strength must be adopted. The experiments described here were designed to investigate whether the frictional resistance developed between rock surfaces under a transient load is the same as that developed where

forces are applied more slowly as in a direct shear test. A number of observations suggest that this may not be the case :

(i) When a load is applied rapidly there may be a time lag between measured stress and strain (ref. 10). According to the adhesional theory of friction, frictional resistance is proportional to the true area of contact which varies directly with normal load. If the normal load is reduced rapidly, the area of contact may adjust more slowly resulting in a higher measured strength than if the load were reduced at a lower strain rate.

(ii) Frictional resistance is low for surfaces that have only been in contact for a short period prior to shearing (ref. 11, 12). Scholz and Engelder (ref. 12) were able to relate this time dependence to the change in real area of contact with time. In the case of a sliding block, the period of contact between any two points on the surfaces is only momentary; hence frictional resistance may be low during sliding.

(iii) Compressive strengths of many materials are dependent upon the rate of loading, higher strengths being achieved for faster rates. The shear strength of rough rock joints, particularly at high stresses, involves shearing through intact rock. One would therefore expect shear strength under rapid loading to be higher. The tests reported here were carried out on flat surfaces but loading rate dependency may be important for natural rock joints.

ROCK DESCRIPTIONS AND SAMPLE PREPARATION

5. Tests were carried out using four different rock types. Rocks were chosen on account of their homogeneity and small grain size relative to the size of the sliding surfaces. Surfaces were saw cut and then ground using a 220 grade diamond wheel.

The rock types are described briefly below :

Darleydale sandstone

6. This rock is a poorly sorted subarkose consisting mainly of angular quartz and feld-spar grains of average size 0.25mm. Surfaces were scratched using minerals defining Moh's scale of hardness and examined by microscope. The main factor controlling scratching strength of the surface is the bonding of grains by pressure solution. Pores are partially infilled with iron oxides and clay minerals, possibly providing some additional strength. Many grains are fractured and are therefore weak compared to the strong intergranular bonding. Surface wear results from the fracture and rounding of embedded grains. Whole grains are not generally plucked from the surface.

Permian sandstone

7. This rock is a mature sandstone consisting of mainly quartz grains of average size 0.25mm, showing high sphericity. The rock is friable with weak bonding between grains which are generally unfractured. When the surface is scratched whole grains are removed from the surface rather than individually modified. Surface wear occurs by the progressive loosening of grains as the free surface changes. As a result this rock does not show as marked a reduction in strength with dis-placement nor with the removal of loose rock flour as other rock types.

Portland limestone

8. This rock is a poorly washed oobiosparite consisting largely of spherical ooliths of approximately 0.25mm diameter. The ooliths consist of clay size particles of calcium carbonate often surrounding a central nucleus. The ooliths are coalescent with secondary cementation by sparry calcite. When the surface is scratched with minerals harder than 2 on Moh's scale, asperities are smoothed and grooves cut with shiny surfaces. Fine white flour is produced.

Delabole slate

9. This rock is a fine grained metamorphic rock with a poorly developed slatey cleavage. The rock is well indurated and surface wear takes place by smoothing of the surfaces with the accumulation of very fine rock flour. Specimens were cut and ground parallel to the cleavage.

INCLINED PLANE SLIDING

10. To investigate frictional resistance under transient loading a simple test was devised using the model of a block on an inclined plane. In order that the results from the vibration tests could be compared to values obtained from more conventional 'static' tests such as direct shear, for similar ground surfaces, it was first necessary to study the frictional resistance developed by a block on an inclined plane where sliding was induced by the weight of the block alone. Variations in frictional resistance with displacement were measured for each rock type so that the effects of rate of loading in vibration tests could be distinguished. The static tests were carried out by placing a weighted thin slice of rock on a large milled surface of rock and then gradually tilting the surface by means of a screw device until sliding occurred. Each run involved approxi-mately 160mm differential displacement and runs were repeated for up to 11.0m total displace-ment. Angles of sliding from inclined plane sliding tests are referred to as ϕ_{st} (ϕ_{static}).

The main conclusions from these tests are summarised as follows :

(i) The inclined plane sliding test gives repeatable results for ground surfaces of rock.

(ii) Peak angles of sliding from inclined plane tests agree well with angles of friction obtained from direct shear test data for similar ground surfaces.

(iii) Residual angles of sliding of nearly a third of peak values may be obtained by repeated sliding.

(iv) Much lower residual angles of sliding are reached for surfaces where rock flour is removed between runs rather than allowed to accumulate.

(v) Different residual angles of sliding are obtained for surfaces under similar conditions if the earlier stages of the test are conducted in a different manner.

(vi) The area of wear is controlled by the asymmetric stress distribution over the area of contact (ref. 13).

VIBRATION TESTING

Experimental method

11. A general view of the test apparatus is given in fig. 1. The test rig consisted of a tilting table to the top of which could be fixed large, milled slabs of rock, (see fig.2). This assembly was bolted onto a base plate that could slide horizontally on low friction bearings. The base plate was fixed to end plates by means of extension springs that could be adjusted to accelerate the table by 0 to 1g horizontally at frequencies of 1 to 5 Hz, where g is the acceleration due to gravity. When a locking device was released, the tilting table moved backwards and forwards in damped harmonic motion. The base acceleration was measured by an accelerometer and plotted against time by a chart recorder. The sliding block consisted of a block of steel with a thin slice of ground rock fixed to the underside. The calculations of frictional resistance relied upon measuring incremental displacements of the block accurately over a period during which the applied forces were known. To do this each test was filmed and each frame of the film examined under a microscope. Matching of the film record with the acceleration record was carried out graphically using a visual display terminal rather than on the basis of individual film frames.

Further details of the experimental method, checks on accuracy and preparation of specimens are given in ref. 14.

Calculation of frictional resistance for incremental displacements

12. Where a block sits on a slope inclined at β^{o} which is less than the angle of friction, ϕ^{o}, the block is stable. If a horizontal acceleration greater than a critical acceleration is applied to the block, directed out of the slope, displacement will occur. This is illustrated diagrammatically in fig. 3. The horizontal axes represent time.

In diagram (i) the base acceleration is given.

In diagram (ii) the inertial acceleration acting upon the block due to the base acceleration is given. The inertial acceleration is equal to the base acceleration but has a phase difference of 180°.

In diagram (iii) the shearing component of the inertial acceleration is added to the down slope gravitational acceleration acting on the block.

In diagram (iv) the magnitude of the maximum retarding acceleration due to frictional resistance is given.

In diagram (v) the maximum retarding acceleration is superimposed on the shearing acceleration.

In diagram (vi) the velocity of the block resulting from the accelerations in diagram (v) is given and in diagram (vii) the cumulative displacement is shown.

From time t_o until t_1 the shearing acceleration exceeds the maximum retarding acceleration. The block begins to move at t_o and reaches maximum velocity at t_1.

The block begins to slow down from time t_1 as the frictional resistance exceeds the shearing force.

At time t_2, when area A = area B, (velocity = 0), the block will stop sliding.

The block will then sit stable on the slope until time t_3 when again the shearing force exceeds the frictional resistance.

In this way the displacement of a block may be calculated for a known magnitude, frequency and duration of base acceleration providing the frictional resistance is known. Conversely, if a block is seen to travel a certain distance in a time during which the applied forces are known, the frictional resistance acting for that time can be determined. Such an approach was followed in this study to investigate the effect of vibration on friction.

13. Fig. 4 illustrates the method used to calculate frictional resistance from the test data.

At times t_1, t_3 and t_5 the displacements of the block d_1, d_2 and d_3 relative to the basal surface were measured from consecutive frames of the film. Average velocities (v_1 and v_2) were calculated for intermediate times t_2 and t_4 so that :

$$v_1 = \frac{d_2 - d_1}{\Delta t} \quad \text{and} \quad v_2 = \frac{d_3 - d_2}{\Delta t}$$

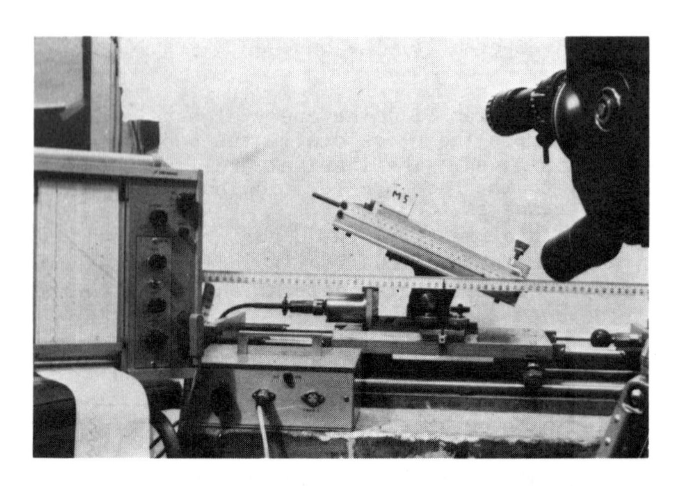

Fig. 1 General view of test apparatus

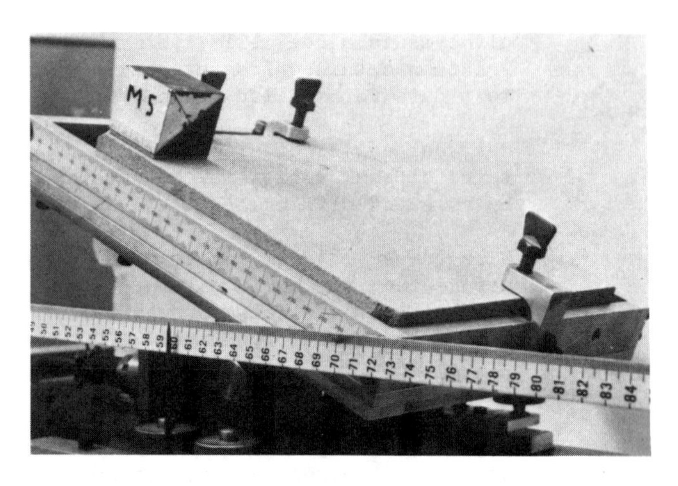

Fig. 2 Inclined plane and block

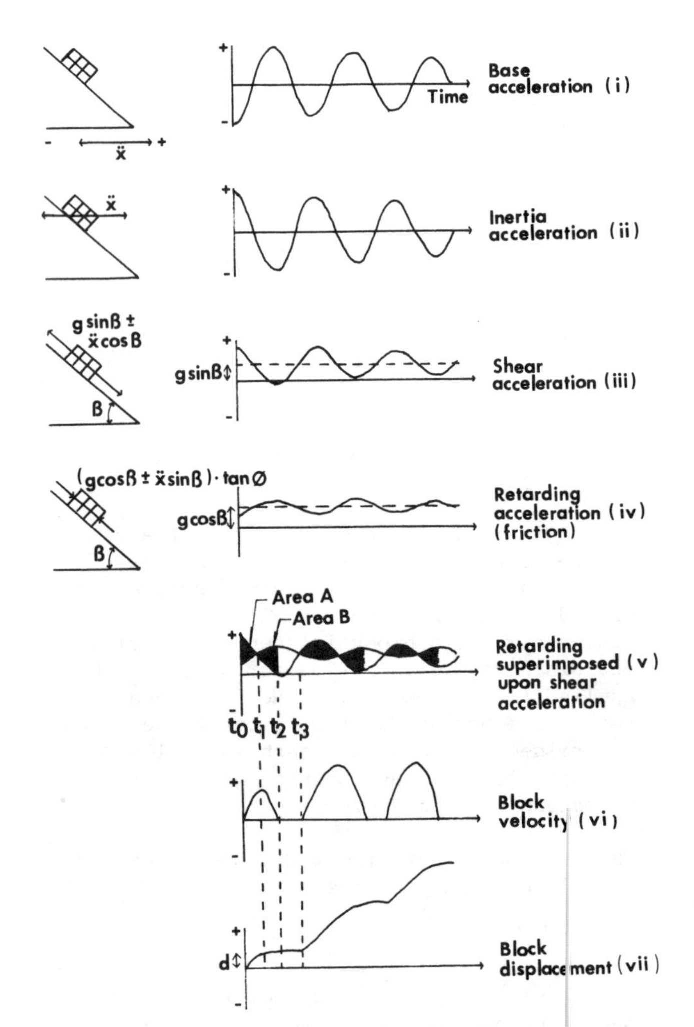

Fig. 3 Behaviour of a block on an inclined plane subjected to horizontal vibrations

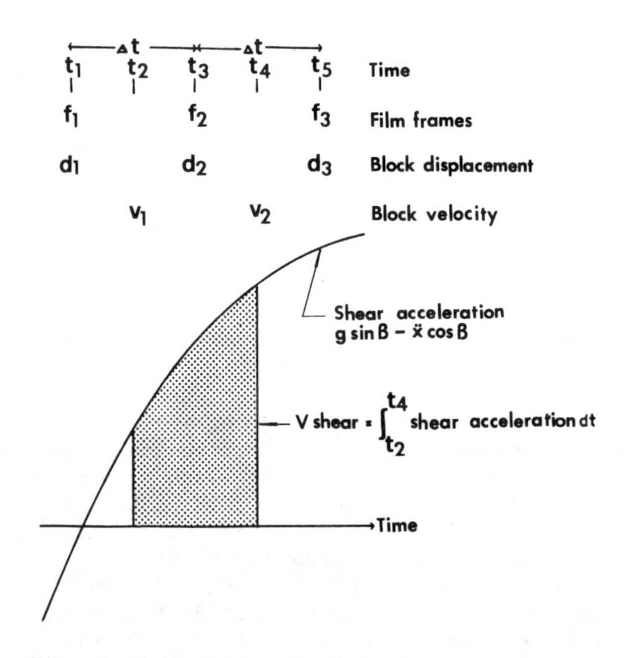

Fig. 4 Method to calculate frictional resistance for incremental displacements

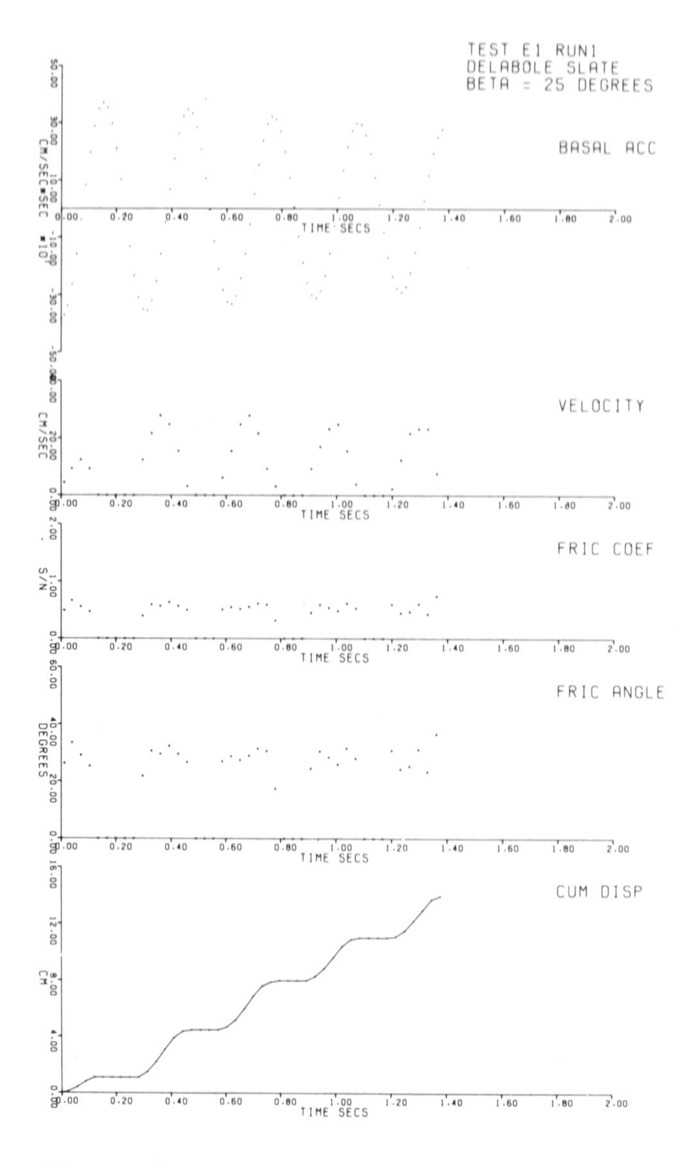

Fig. 5 Example of vibration test results

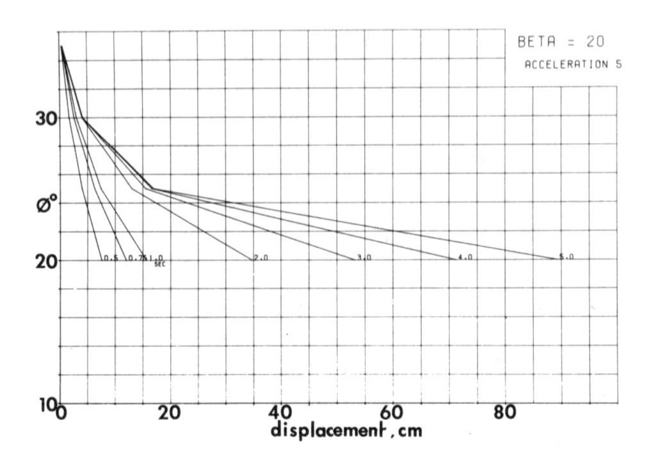

Fig. 6 Example of graph used to determine $\emptyset_{av}$

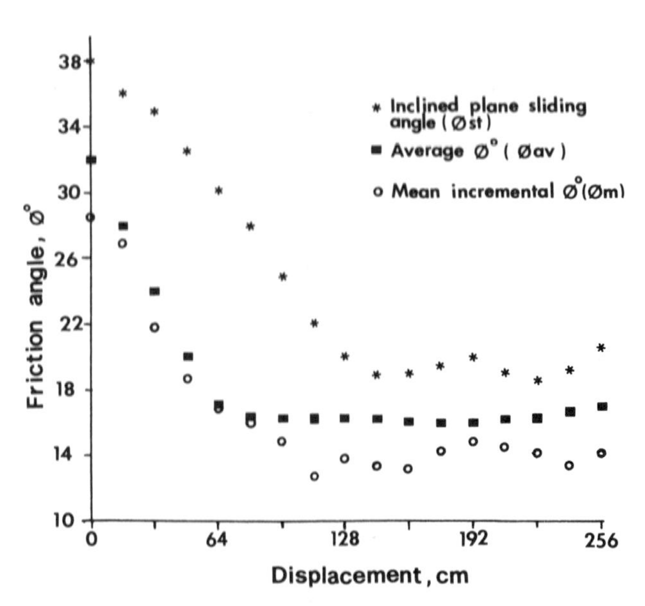

Fig. 7 Friction angles from vibration tests compared to friction angles from inclined plane sliding tests

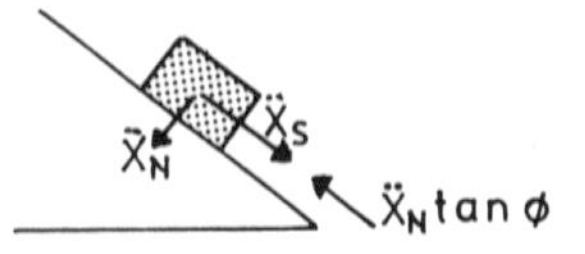

$\ddot{X}_S$ = shear load acceleration

$\ddot{X}_N$ = normal load acceleration

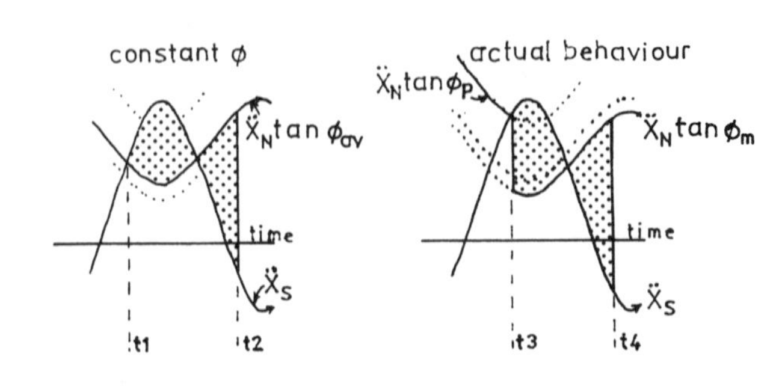

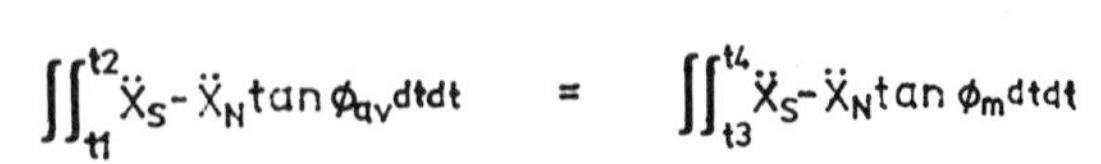

= measured displacement

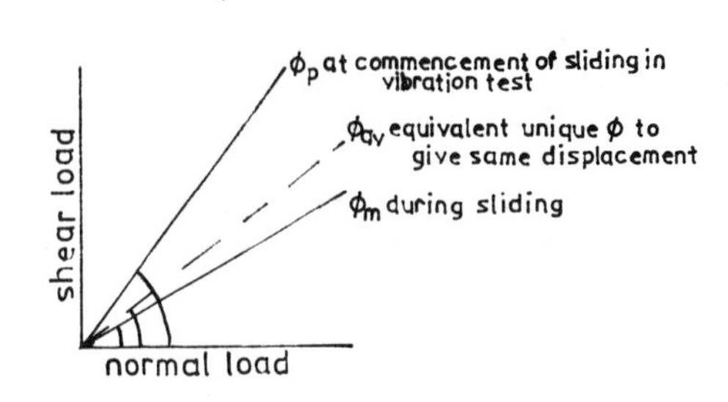

Fig. 8 Diagram to explain the difference between $\emptyset_m$, $\emptyset_{av}$ and $\emptyset_p$

where Δt is the time interval between film frames. Between times t_2 and t_4 the block was accelerated by a known shear load. By integrating the shear load acceleration against time for that period, the velocity (v_{shear}) ideally gained by the block can be found (disregarding friction), so that :

$$\text{for } \emptyset = 0^{o}, \quad v_1 + v_{shear} = v_2$$

However friction slows the movement of the block so that :

$$v_1 + v_{shear} > v_2$$

and $v_1 + v_{shear} - v_2$ equals the integral of the deacceleration due to frictional resistance against time.

Now assuming a totally frictional material, where the frictional resistance is directly proportional to load applied normally between the surfaces, then :

$S = \mu N$

where S = frictional resistance
N = normal load
μ = coefficient of friction.

Referring to fig. 3,

The applied shear load acceleration is $g \sin\beta + \ddot{x} \cos\beta$

the normal load acceleration is $g \cos\beta - \ddot{x} \sin\beta$

and the deacceleration due to frictional resistance is $\mu(g \cos\beta - \ddot{x} \sin\beta)$

where $\ddot{x}$ is the inertial acceleration.

$$\text{Then } v_1 - v_2 + \int_{t_2}^{t_4} (g \sin\beta + \ddot{x} \cos\beta)\, dt$$

$$= \int_{t_2}^{t_4} \mu(g \cos\beta - \ddot{x} \sin\beta)\, dt \qquad (1)$$

A typical graphical output from a program used to solve this equation for μ from test data is given in fig. 5.

<u>Calculation of frictional resistance for larger displacements</u>

14. Average values of $\emptyset$, where $\mu = \tan \emptyset$, were determined for larger displacements of the block i.e. for individual cycles of motion and for the complete length of sliding by the following method. Theoretical displacements were calculated for different sections of each acceleration record using a range of possible values for $\emptyset$. These calculated displacements were then compared to actual measured displacements to arrive at an average value for $\emptyset$ ($\emptyset_{av}$).

A typical output for a single slope angle is given in fig. 6. Referring to this figure, if the block was seen to slide 15 cm in the first 2.0 sec. of the test run, the operative angle of friction would have been 24.5°. If in the first 0.5 sec. the block only slid 3 cm then the operative angle of friction would have been 26°. The average angles of friction ($\emptyset_{av}$) obtained in this way were compared to values calculated from incremental displacements.

RESULTS

15. An example of the results obtained, based on data from individual film frames, is shown in graphical form in fig. 5. The results generally show scatter mainly due to reading errors in displacement measurements but possibly in part due to real variations in $\emptyset$ during sliding. The errors are important for incremental values of $\emptyset$ but tend to cancel out for complete runs. Results were treated statistically to give mean values from the incremental angles of friction ($\emptyset m$) and plotted as histograms and cumulative frequency curves of incremental friction values.

16. One test on Portland limestone consisted of 17 runs with the angle of slope being progressively reduced from 30° to 10° . Rock flour was removed for the first 14 runs and accumulated for the final 3. Values for $\emptyset m$ (mean incremental values for $\emptyset$), $\emptyset_{av}$ (average operative values of $\emptyset$ calculated from total displacement of the block), and $\emptyset_{st}$ (angles of sliding from inclined plane sliding tests), are plotted against displacement in fig. 7. Table 1 contains data from this test and from tests on three other rock types.

DISCUSSION

17. Fig. 7 shows that the reduction in friction with displacement in vibration tests followed the pattern observed in inclined plane sliding tests. This was also shown by tests on other rock types. The measured angle of friction was lower for the vibration tests and displacements calculated on the basis of inclined plane sliding test results would have clearly underestimated the true displacements.

18. Fig. 7 and table 1 also reveal that $\emptyset m$ was generally less than $\emptyset_{av}$. This means that although the frictional resistance during sliding was calculated at one value ($\emptyset m$), the total displacement was less than would be expected on the basis of $\emptyset m$ and indicated a higher average friction angle ($\emptyset_{av}$). This may be explained where the block commenced sliding at a later time and at a higher shear load : normal load ratio than would have been the case if strength was controlled before sliding by $\emptyset m$. Once sliding, the friction angle reduced to $\emptyset m$. This concept is illustrated in fig. 8. The times at which the block commenced sliding at the start of each displacement cycle were found by extrapolation of displacement increment data and the shear load : normal load ratios at those times calculated. Friction angles ($\emptyset_p$) when the block began sliding in vibration tests were found to be very high and are given in table 1 together with comparable angles of sliding ($\emptyset_{st}$) from inclined plane sliding tests. The peak angles of friction ($\emptyset_p$) occurred at the beginning of each cycle of movement and

therefore several values of $\emptyset_p$ were obtained for each test run. It was found that the value of $\emptyset_p$ generally decreased during each test run and it is this range of values that are given in table 1.

Table 1. Friction angles from vibration tests compared to inclined plane sliding test results

Rock Type	Test No.	Run No.	Dispt. (cm)	$\emptyset^o_m$	$\emptyset^o_{av}$	$\emptyset^o_{st}$	$\emptyset^o_p$ (Range)	Comments
Portland limestone	1	1	0	28.7	32.0	38.0	51/48	Rock flour removed
		2	16	27.0	27.0	35.5	51/47	
		3	32	21.9	23.0	34.5	46	
		4	48	18.8	19.0	32.5	41/40	
		5	64	16.8	17.0	30.0	33/32	
		6	80	16.0	16.3	28.0	32/28	
		7	96	14.9	16.3	25.0	31/28	
		8	112	12.8	16.3	22.0	30/27	
		9	128	14.0	16.3	20.0	31/29	
		10	144	13.5	16.3	19.0	32/28	
		11	160	13.2	16.0	19.0	28	
		12	176	14.4	16.0	19.5	30/28	
		13	192	15.0	16.0	20.0	29/28	
		14	208	14.4	16.3	19.0	29	
		15	224	14.2	16.3	18.5	30/27	Rock flour accumulated
		16	240	13.5	16.8	19.0	30/27	
		17	256	14.0	17.0	20.5	30/28	
Darleydale sandstone	1	1	0	27.9	29.0	33.0	44/41	
		2	16	25.9	27.3	28.0	43/42	
	2	1	0	26.7	29.5	33.0	38/30	
		2	16	27.7	28.0	28.0	38/32	
		3	32	27.6	27.8	27.5	38/29	
	3	1	0	28.0	30.0	33.0	39/33	
		2	16	28.1	28.8	28.0	39/33	
		3	32	27.6	28.0	27.5	39/33	
	4	1	0	25.0	28.8	33.0	40/34	
		2	16	24.3	27.5	28.0	44/37	
Permian sandstone	1	1	0	23.9	27.0	32.5	44/39	
		2	16	24.1	26.5	28.5	44/43	
	2	1	0	24.6	26.8	32.5	40/33	
		2	16	22.5	25.5	28.5	40/33	
Delabole slate	1	1	0	28.2	29.3	29.8	43/40	
		2	16	26.3	27.5	28.0	44/42	
	2	1	0	27.4	29.0	29.8	38/32	
		2	16	27.1	28.5	28.0	38/32	

Variation in peak friction angle with rate of loading

19. In table 2, peak friction angles ($\emptyset_p$) are given for the beginning of each displacement cycle in two tests on Permian sandstone. Table 2 also gives the angle of inclination of slope (β^o), the horizontal acceleration acting on the block at time of failure ($\ddot{x}_c$) and the peak accelerations before and after commencement of sliding, ($\ddot{x}_1$ and $\ddot{x}_2$). It is clear that $\emptyset_p$ varied both with angle of slope and within each run. The decrease in $\emptyset_p$ during each run is considered not to be due to wear of the surfaces as the results from test 2 run 2 were almost identical to those for run 1 for the same surfaces. It has been shown, (ref. 14), that the rate of change of the ratio of shear load to normal load decreases both with angle of slope and with time where the horizontal basal acceleration is damped harmonic, and a clear relationship was found between this rate of change and $\emptyset_p$ (fig. 9). Peak friction angles at the beginning of each displacement cycle were higher for higher rates of change. At lower rates, $\emptyset_p$ approached the values obtained from inclined plane sliding tests ($\emptyset_{st} = 32^o$ for Permian sandstone). $\emptyset_p$ was never found to be lower than the equivalent $\emptyset_{st}$ from inclined plane sliding. The implication from these results is that the critical acceleration for the initiation of sliding will be higher for high frequency vibrations than for low frequency vibrations.

Table 2. Peak friction angles ($\emptyset_p$) from Permian sandstone tests.

Test No.	Run No.	Dispt. cycle	$\emptyset^o_p$	β^o	$\ddot{x}_c$	$\ddot{x}_1$	$\ddot{x}_2$
					cm sec^{-2}		
1	1	1	43.9	25	336	-370	359
		2	43.1		321	-348	336
		3	39.2		249	-324	313
1	2	1	43.5	25	328	-374	359
		2	40.7		275	-351	336
		3	42.6		311	-325	315
2	1	1	40.0	20	357	-371	357
		2	38.7		331	-351	331
		3	36.9		298	-325	310
		4	35.7		275	-305	290
		5	35.4		270	-282	270
		6	33.7		240	-263	249
		7	33.3		232	-244	232
2	2	1	40.0	20	359	-371	359
		2	38.8		334	-351	334
		3	36.8		296	-321	309
		4	35.6		273	-301	290
		5	33.2		230	-282	268

Values for $\emptyset$ during sliding ($\emptyset_m$)

20. Once sliding commenced, the frictional resistance between the rock surfaces was no longer a function of the rate of loading, being fairly constant for complete runs at different inclinations. Variations in $\emptyset_m$ for repeated runs can be attributed to decreasing frictional resistance with displacement. In most tests, mean values of 'sliding' friction were less than the corresponding 'static' values for all rock types.

21. The high peak friction angles at commencement of sliding and low values during sliding can be explained by the mechanisms outlined in para. 4. The high values may reflect a delay in the reduction of area of contact with rapidly decreasing normal load and increasing shear load. The low values result from the poor bonds that can be formed as contacts are repeatedly made and broken.

CONCLUSIONS FROM VIBRATION TESTS

22. (i) The displacement of a block due to vibration loading is greater than would be predicted using friction angles obtained from 'static' tests.

(ii) The friction angle for the initiation

of sliding is higher than the equivalent 'static' friction angle and is higher for higher rates of loading.

(iii) The friction angle during sliding is lower than the 'static' friction angle.

IMPLICATIONS FOR DESIGN OF ROCK SLCPES TO WITHSTAND EARTHQUAKES

23. It is suggested that a rock slope design should be checked for seismic loading in two stages :

(i) A limit equilibrium analysis should be carried out using a pseudo-static force incorporating the peak ground acceleration expected at the site. A peak friction angle obtained from direct shear tests corrected for dilation should be used in this analysis. The use of a peak value will tend to result in conservative design as the actual frictional resistance may be much higher where the applied vibration is of high frequency.

(ii) If the limit equilibrium analysis shows that sliding will occur, then an estimate of the resulting displacement may be made by using Sarma's method (ref. 6). Peak acceleration and predominant period of ground motion are required for this calculation. Total displacements in the vibration tests would have been slightly underestimated for two rock types if residual angles of friction had been used for calculation. On the basis of the tests conducted it seems likely that the use of a friction angle 5° lower than the residual value obtained from direct shear tests will give an upper bound for the displacement. The implications of this displacement in terms of shear strength and redistribution of loads should be taken into account in deciding whether the design is acceptable or not.

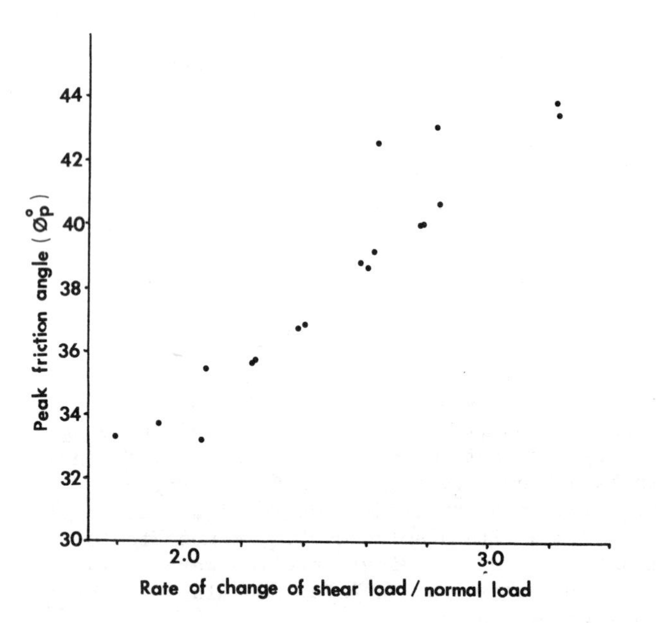

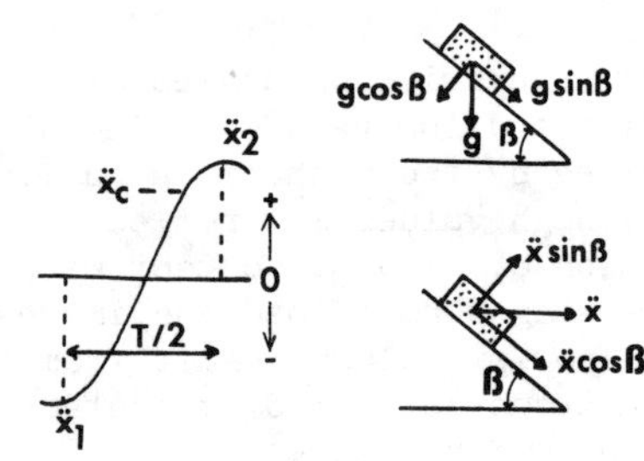

$\ddot{x}_c$ = inertia acceleration at time of failure

$$\phi_p^\circ = \tan^{-1}\frac{g\sin\beta + \ddot{x}_c\cos\beta}{g\cos\beta - \ddot{x}_c\cos\beta}$$

$$\text{Rate of change of S/N ratio} = 2\left(\frac{g\sin\beta+\ddot{x}_2\cos\beta}{g\cos\beta-\ddot{x}_2\sin\beta} - \frac{g\sin\beta+\ddot{x}_1\cos\beta}{g\cos\beta-\ddot{x}_1\cos\beta}\right)/T$$

Fig. 9 Relationship between ϕ_p and rate of change of shearload : normal load ratio

24. It must be emphasised that these recommendations are based on a few experimental results using small samples with idealised surfaces. Until more information is available on the behaviour of rock slopes in earthquakes however, these recommendations seem justified.

ACKNOWLEDGEMENTS

This research was supported financially by the Natural Environment Research Council. I would like to thank Mr. M.H. de Freitas of the Engineering Geology Division, Imperial College, London for his constant encouragement and guidance, Mr. E. Rodgers for constructing the test rig and Miss M. McKinlay and Dr. S.K.Sarma for their helpful advice. The paper was read critically by Mr. J. Bryant and Mr. A. Clover.

REFERENCES

1. ORIARD I.L. Blasting effects and their control in open pit mining. Proceedings of International Conference on Stability in Open Pit Mining, 1971, 197-222.

2. KO K.C. and McCARTER M.K. Dynamic behaviour of pit slopes in response to blasting and precipitation. Applications of Rock Mechanics. Proceedings of 15th Symposium on Rock Mechanics, South Dakota, 1973.

3. HOEK E. and BRAY J.W. Rock Slope Engineering. Institution of Mining and Metallurgy, London, 2nd Edition, 1977.

4. NEWMARK N.M. Effects of earthquakes on dams and embankments. Geotechnique, 1965, 15, No. 2, 139-160.

5. SEED H.B. A method for earthquake resistant design of earth dams. Jnl. Soil Mech. Fdn. Div. ASCE. 1966, 92, SM1, 13-40.

6. SARMA S.K. Seismic stability of earth dams and embankments. Geotechnique, 1975, 25, No. 4, 743-761.

7. HENDRON A.J., CORDING E.J. and AIYER A.K. Analytical and graphical methods for the analysis of slopes in rock masses. U.S. Army Engineers Waterways Experiment Station, NCG Technical Report No. 36.

8. U.S. GEOLOGICAL SURVEY. The San Fernando, California earthquake of February 9, 1971. U.S. Geol. Survey Prof. Paper 733, 1971.

9. CRANDELL D.R. and FAHNSTOCK R.K. Rockfalls and avalanches from Little Tahoma peak on Mount Rainier, Washington. U.S. Geol. Survey Bulletin 1221-A, 1963.

10. ROBERTSON E.C. Viscoelasticity of rocks. State of Stress in the Earth's Crust. W.R. Judd, Ed., Elsevier, New York, 1964, 181-234.

11. DIETERICH J.H. Time dependant friction in rocks. J. Geoph. Research, 1972, Vol. 77, No. 20, 3690-3697.

12. SCHOLTZ C.H. and ENGELDER J.T. The role of asperity indentation and ploughing in rock friction. Int. J. Rock Mech. Min. Sci. & Geomech. Abstr. 1976, Vol. 13, 149-154.

13. HENCHER S.R. A simple sliding apparatus for the measurement of rock joint friction, discussion. Geotechnique, 1976, No. 4, 641-644.

14. HENCHER S.R. The effect of vibration on the friction between planar rock surfaces, Ph.D. Thesis, University of London (Imperial College), 1977.

11 Peculiarities of the seismic-resistant analysis of earth dams with pervious gravelly shells

A. J. L. BOLOGNESI, Universidad de Buenos Aires, and Bolognesi-Moretto Ingenieros Consultores Buenos Aires

The seismic resistant analysis of zoned earth dams on good foundations with relatively thin impervious core supported essentially by well compacted pervious gravelly shells which do not build up large dynamic pore pressures is considered as well as the design features resulting from such analysis and requirements.

INTRODUCTION
Both on dynamic analytic procedures and on the knowledge of the dynamic properties of materials, great progress has been made during the last fifteen years.

Since strong earthquakes are required to verify and adjust the tests and analytical procedures already available and enough of these events have not occurred yet, present methods are only an aid to judgment but "the overall results can be extremely useful in guiding the engineer in the final assesment of seismic stability".

It follows a secuence of publications which have defined the earthquake resistant design of earth and rock fill dams.

Junio, 1965
N. M. Newmark
EFFECTS OF EARTHQUAKES ON DAMS AND EMBANKMENTS. (ref.1)
Mayo, 1966
H. Bolton Seed and Geoffrey R. Martin
THE SEISMIC COEFFICIENT IN EARTH DAM DESIGN. (ref.2)
Setiembre, 1967
N. N. Ambraseys and S. K. Sarma
THE RESPONSE OF EARTH DAMS TO STRONG EARTHQUAKES. (ref.3)
Junio, 1973
H. Bolton Seed, et al.
ANALYSIS OF THE SLIDES IN THE SAN FERNANDO DAMS DURING THE EARTHQUAKE OF FEBRUARY 9, 1971. (ref.4)
Noviembre, 1975
Arthur Casagrande
LIQUEFACTION AND CYCLIC DEFORMATION OF SANDS. A CRITICAL REVIEW. (ref.5)
Julio, 1978
Faiz I. Makdisi and H. Bolton Seed
SIMPLIFIED PROCEDURE FOR ESTIMATING DAM AND EMBANKMENT EARTHQUAKE-INDUCED DEFORMATIONS. (ref.6)
September, 1979
H. Bolton Seed
CONSIDERATIONS IN THE EARTHQUAKE-RESISTANT DESIGN OF EARTH AND ROCK FILL DAMS. (ref.7)
1971
NISEE (National Information Service for Earthquake Engineering). (ref.8)

These publications and their references are an excellent guide to the evolution of ideas and developments of laboratory and analytical tools and to the present knowledge on the subject.

The general use of advanced programs has been made possible since the creation of the National Information Service for Earthquake Engineering, University of California, Berkeley.

LIMITATION OF THE SUBJECT
The design of earthquake-resistant embankment dams and its appurtenant structures is so complicated that only a very restricted aspect of it can be considered in the space available. Inside the frame provided by Seed's Rankine Lecture, this paper is limited to the "types of soil and conditions which do not build up large pore pressures or cause significant strength loss due to earthquake shaking and associate displacements". Furthermore, it deals essentially with the determination of a balanced cross-section.

To precise the use of concepts and terms Seed's Rankine Lecture will be used as a reference. Unless stated otherwise, expressions between quotation marks refer to that lecture.

REQUIREMENTS FOR THE UPSTREAM SHELL
Until recently there was a consensus that in a well-graded, well-compacted pervious gravelly shell the full drained strength was available even when submerged and subjected to a strong earthquake shaking.

This is now valid only for the essentially dry downstream shell.

Based on the results of laboratory cyclic loading tests and, so far, scant field evidence, the possibility that excess pore water pressures may develop in the upstream submerged shell of pervious gravelly shells has been put forward in California (ref.7-9-10). If sustained by further field information, it will have an important influence on the design of upstream shells. This paper assumes a practically fully drained upstream shell, which may need the introduction of one or more of the following requirements:
a) "If the permeability of the shell material for a dam becomes sufficiently high, say of the order of 1 cm/s, then it may be impossible for an earthquake to cause any build-up of pore pressures in the embankment since the pore pressures can dissipate by drainage as rapidly as the earthquake can generate them by shaking".

If specified, a gravel that fulfills such conditions can be produced at no excessive cost by appropiate methods.

b) Since rock fills have higher permeabilities than gravelly shells (of the order of 10^2 cm/s) all the material produced by the rock excavation of the appurtenant structures must be placed in the upstream shell. Also, if the cost diference is not decisive a sizable part of the upstream shell must be preferable rock fill.

c) Dynamic pore pressures can also be controlled by chimney rock drains, (permeability coefficient of the order of 10^2 cm/s) installed in ordinary pervious gravelly shells (permeability of the order of 10^{-1} cm/s), with appropiate outlets.

Analytical procedures for evaluating the simultaneous generation and dissipation of pore pressures in cohesionless soils during earthquakes have been proposed by Seed et al (ref.11). An approximate solution can be obtained by the Terzaghi theory of consolidation. The volume of water to be drained can be computed by the method proposed by Lee and Albaisa (ref.12), assuming 100% pore pressure development. Nevertheless, the chimney drains must be designed taking into consideration the possible distorsions that can be created by earthquakes.

Because of stratifications caused by construction methods, the relation between the horizontal and vertical coefficients of permeability is frequently of the order of 10.

ANALYSIS PROCEDURES

Essentially, there are two procedures involving dynamic analysis of embankments:
a) The SEED-LEE-IDRISS analysis procedure, applicable to any type of soil and condition, is the one that has been more extensively used when large pore pressures build up due to the earthquake shaking. The application of this method is outside the purpose of this paper.

b) The NEWMARK analysis procedure, applicable to the "types of soil and conditions which do not build up large pore pressures or cause significant strength loss due to earthquake shaking and associate displacements". This method applies to the type of dam considered in this paper. The method has been improved by Seed and Martin (ref.2), Ambraseys and Sarma (ref.3) and Makdisi and Seed (ref.6). The essential steps of this analysis procedure, in accordance with the last and more refined version, which is the one by MAKDISI and SEED, are (for the procedures a) or b), the steps between quotation marks are alike):
a) "Determine the cross-section of the dam to be used for analysis".
b) "Determine with the cooperation of geologists and seismologists, the maximun time history of base excitation to which the dam and its foundations migth be subjected".
c) Determine the cyclic yield strength of the materials to be used in the dam.
d) "Determine the dynamic properties of the soils comprising the dam, such as shear modulus, damping characteristics, bulk modulus, Poisson's ratio, which determine its response to dynamic excitation. Since the material characteristics are non-linear, it is also necessary to determine how the properties vary with strain".
For a given potential sliding mass:
e) Determine the yield acceleration,i.e. the acceleration at which the potential sliding surface would develop a factor of safety of unity.
f) Determine the earthquake induced accelerations in the embankment using dynamic response analysis.(Accelerations and stresses from finite element response analysis).
g) Evaluate the magnitude of the displacements along the direction of failure surface by double integration of the effective acceleration on the sliding mass in excess of the yield acceleration as a function of time.

Makdisi and Seed described in detail the application of this analysis to a number of cases and based on the results, proposed a simplified procedure (ref.6)

had been altered.

17. The specimens used for the first series were all laboratory compacted samples prepared to the density of the embankment. Both static and dynamic properties were determined from these samples and the results were reflected in the initial design. The second series of tests was carried out primarily to justify the modified cross-section and to better define the dynamic properties of the core and the shell by sampling the compacted embankment soil undisturbed. Some laboratory compacted samples were also used in the second series as needed.

18. For the static analysis of the final cross-section, most of the static test results from the first series of the tests were retained and used with some supplementary test results from the second series. The static soil parameters determined from these tests are discussed in the next section. The dynamic properties of the core and the shell were, on the other hand, completely re-evaluated for the final design section. The results of the dynamic property tests shown hereafter are values from the second series, and the dynamic response analysis was carried out using these values.

19. The undisturbed block samples were obtained by two sampling processes: one by the more conventional method of placing a wooden box frame over a cube of soil, and the other by inserting a thin-walled tube into a soil column trimmed to a diameter slightly in excess of the tube diameter (Mori et al., 1979 and Mercuson et al., 1979). In both methods of undisturbed sampling, the top 30 cm of the compacted embankment fill was removed prior to sampling.

20. The block samples of 30 cm cubes were trimmed in the laboratory into cylindrical specimens 6 cm in diameter. However, the trimming of the block samples proved to be very difficult because of the presence of uncrushed sandstone and mudstone aggregates in the sample, and, in the end, this sampling procedure was abandaned.

21. Sampling of the soil with a thin-walled tube of 7.3 cm in diameter was also difficult and time-consuming, but satisfactory samples were recovered, and these samples were used to determine the strain-dependent shear modulus and damping properties of the soils. These specimens were set in a resonant column-cyclic triaxial test unit, an apparatus having features of both a resonant column and of a cyclic triaxial in one system so that one sample can be subjected to a wide strain range. The results of the tests are shown in Fig.9 a and b for strain-dependent shear modulus, and in Fig. 10 a and b for strain-dependent damping ratio.

22. The resonant column test results and cyclic triaxial test values overlapped around a 10^{-3} shear strain range, and the transition from the resonant to triaxial system for shear modulus occurred rather smoothly for all three mean confining pressures of 1, 2, and 3 kg/cm^2, while the effect of confining pressure were difficult to distinguish for the damping curves and a single average curve was drawn to represent all the confining pressures in the following dynamic analysis.

STATIC ANALYSIS

23. A static stress analysis carried out to investigate the stress and strain distribution within the dam and to determine the static stability. The static stress state inside the embankment is important also for the dynamic response analysis because the effective confining pressure affects the strain-dependent shear modulus, as shown in Fig.9.

24. The static analysis was carried out in two stages: the first stage was the determination of the stress and strain changes due to the successive construction of the embankment. A finite element computer program, "ISBILD" (Ozawa, et al., 1973), was used to simulate the phased construction of the embankment. The program takes the non-linear behaviour of soil into account by conducting elastic analysis based on the placement of each layer and changing the soil parameters thereafter according to the new stress state computed. The soil parameters used for this analysis are given in Table 1, and the finite element mech is shown in Fig.11.

25. After the stress state was computed, the second stage of the calculation was made by determining the steady state seepage force. In order to calculate the stresses due to the steady state seepage force, saturated-undsaturated seepage analysis was conducted using the finite element technique (Neuman, 1973 and Komata, 1973). Unlike the conventional seepage analysis which incorporates water flow only through the saturated portion of the dam, the saturated-unsaturated analysis procedure computes the moisture migration taking place in the partially saturated zone as well as the water flow through the saturated zone of the embankment.

26. Fig.11 shows the equi-potential lines determened from the seepage analysis. The figure shows that the phreatic surface reaches only half way into the core. However, with the soil suction, water will reach the drain filter and will be eventually pumped out at the relief well. For the following dynamic analysis, both the core and the lower core were assumed saturated.

27. With the seepage force determined, the stress state of each element with the reservoir full could be calculated. Fig.12 shows the contours of the deviator stresses, $\sigma_1'-\sigma_3'$, at this stage. It should be noted that the deviator stresses are smaller in the upstream slope than in the downstream, due primarily to the reservoir water pushing the dam in the downstream direction. Thus, the factor of safety of each element shown in Fig.13 is higher in the upstream side.

DYNAMIC RESPONSE ANALYSIS

28. From the result of the static analysis, the mean effective confining pressure of each element was determined, and, based on this value, a maximum shear modulus of each element was assigned by referring to Fig.9. For the dynamic analysis, a finite element computer program, "LUSH" (Lysmer, et al, 1974), was used. The program calculates

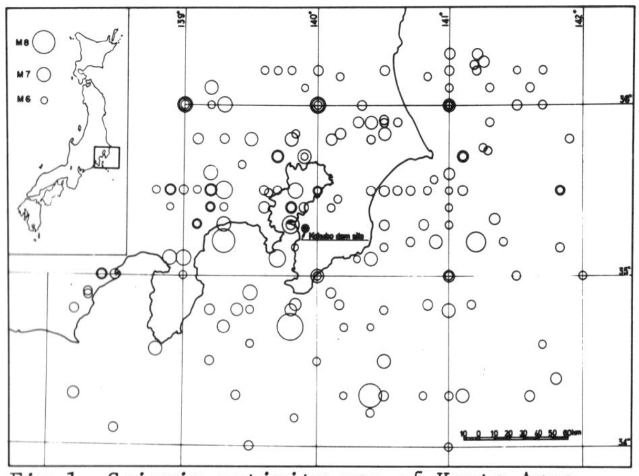

Fig.1 Seismic activity map of Kanto Area

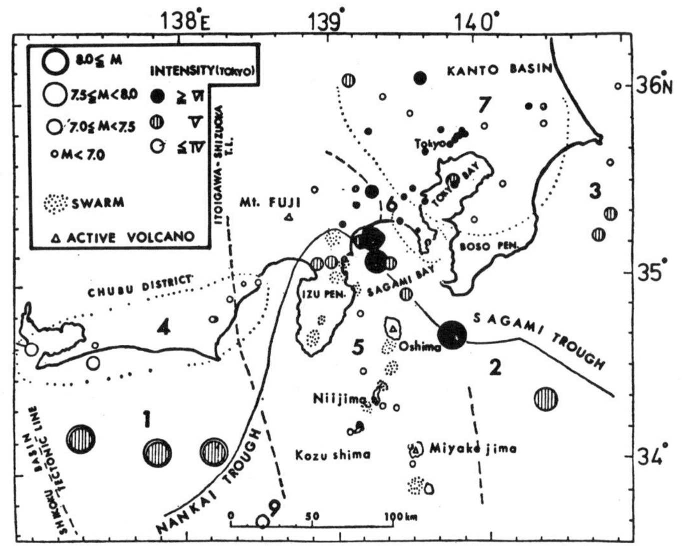

Fig.2 Active troughs and groupes (after Kakimi et al., 1974)

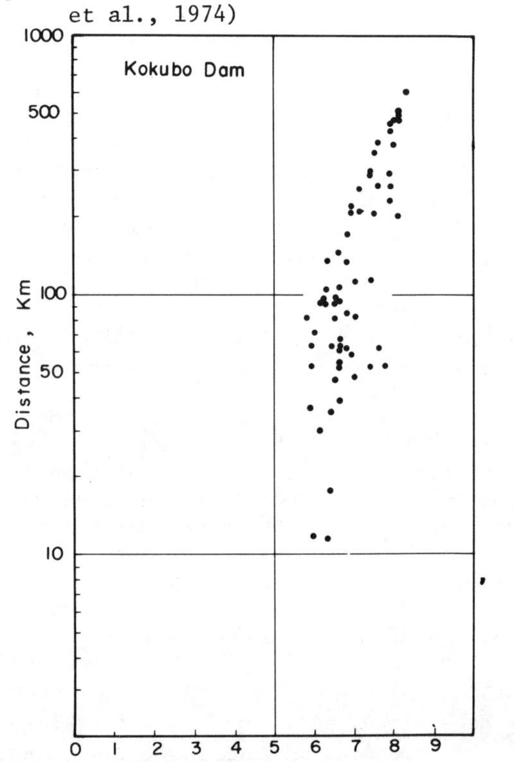

Fig.3 Earthquake magnitude and their epicentral distances

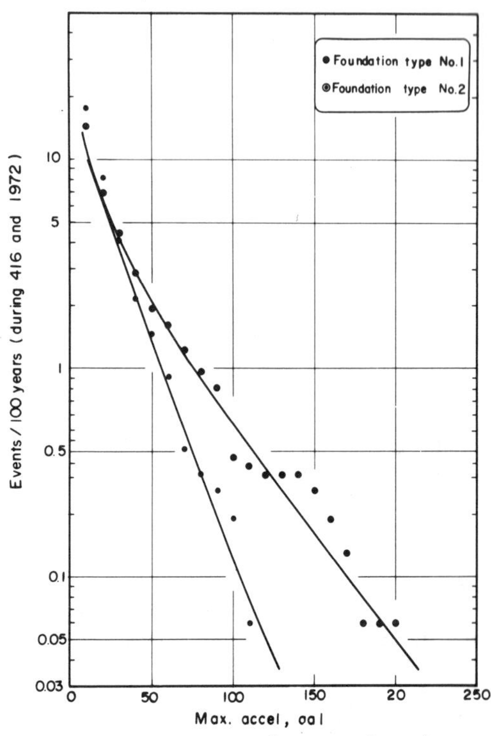

Fig.4 Recurrence curves for earthquakes recorded from 416 to 1972

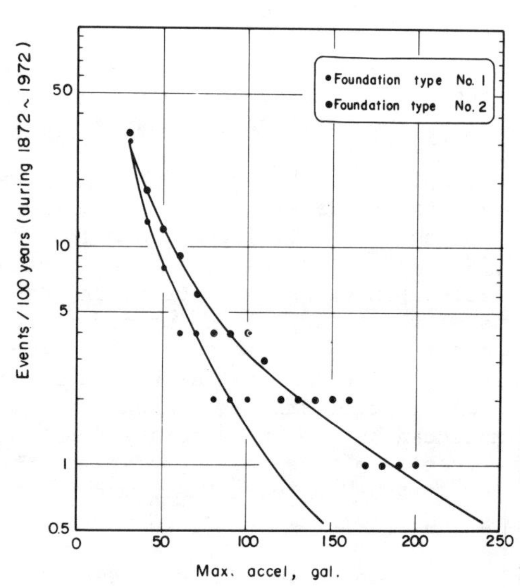

Fig.5 Recurrence curves for earthquakes recorded form 1972 to 1972

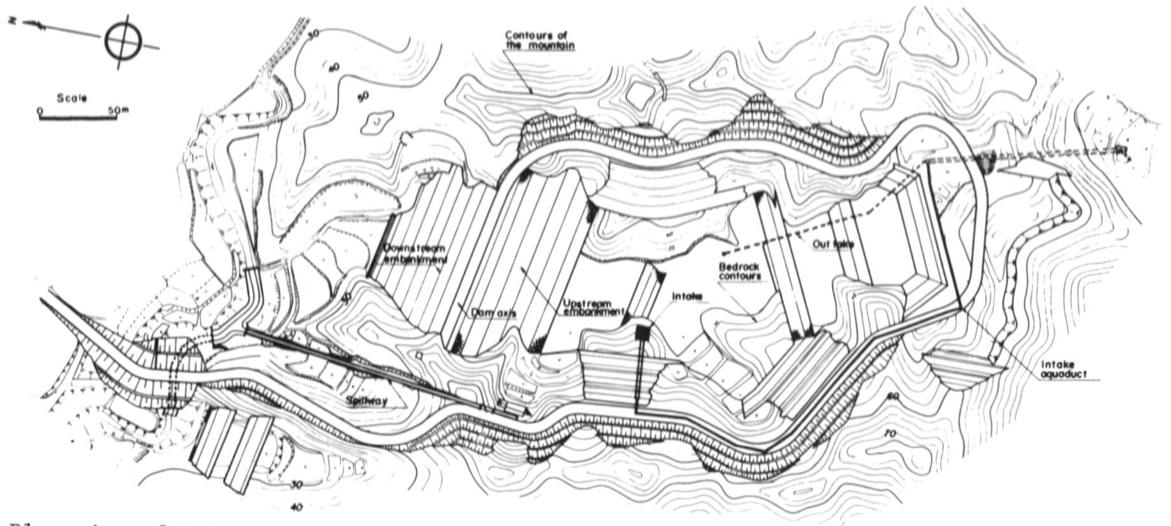

Fig.6 Plan view of Kokubo Dam

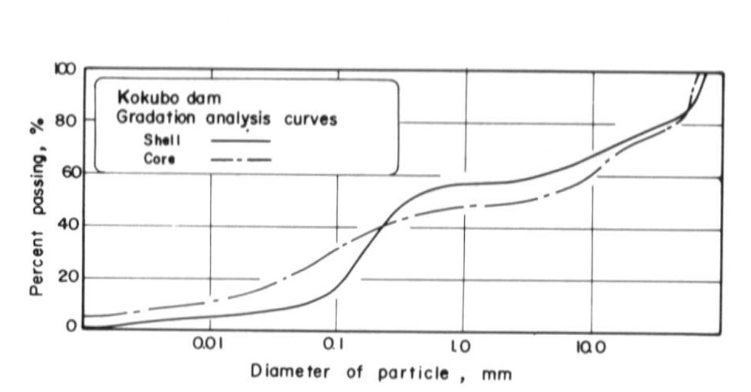

Fig.7 Average gradation curves of the shell and the core

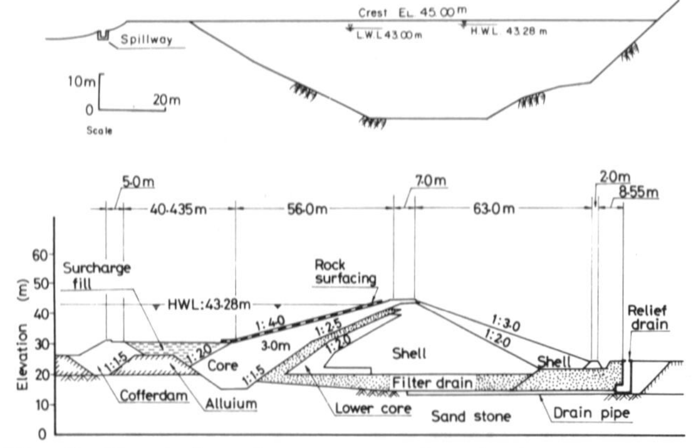

Fig.8 Frontal and cross-sectional views

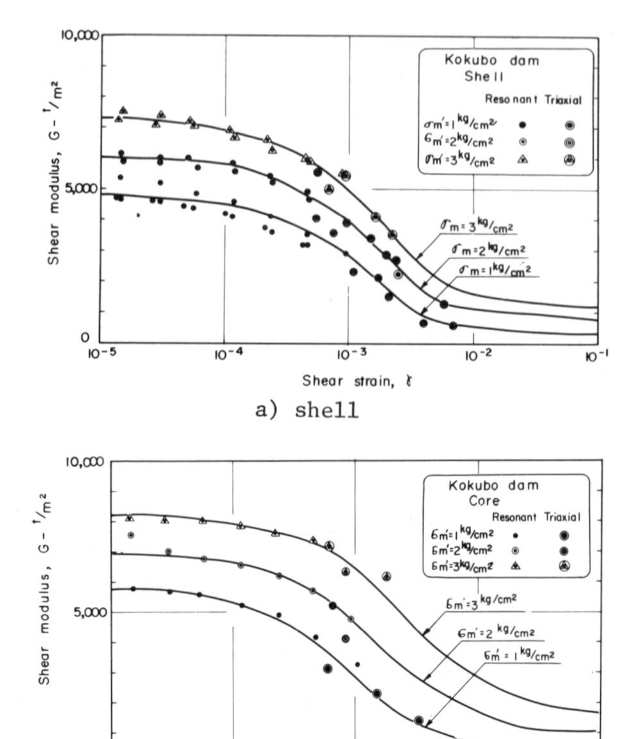

a) shell

b) core

Fig.9 Strain-dependent shear modulus

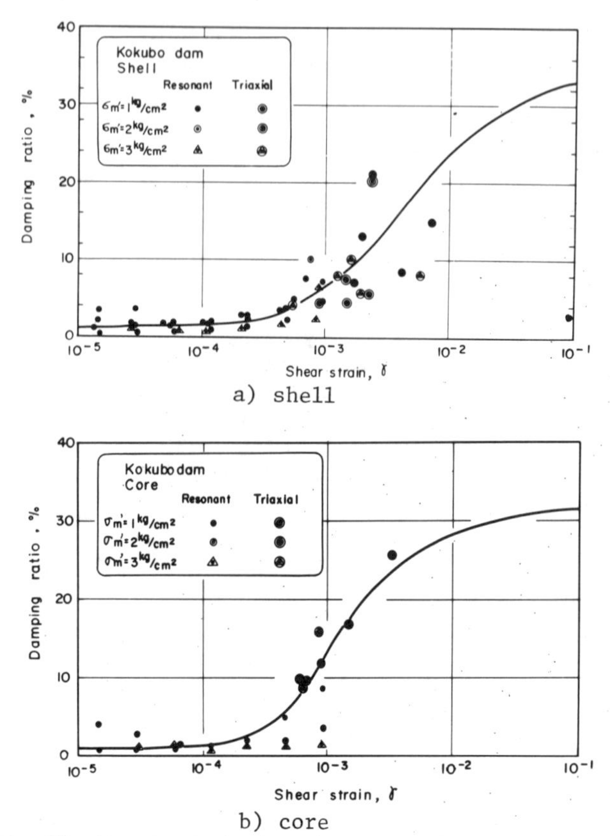

a) shell

b) core

Fig.10 Strain-dependent damping ratio

the dynamic behaviour of soils by using the method of complex response and simulating the non-linear behaviour of soils by an equivalent linear model. The same mesh used for the static analysis was also used for the dynamic analysis. For the input motion, the north-south component of the Taft record was input at the base as an upward-propagating shear wave with maximum accelerations of 150 gal and 250 gal and a predominant period of 0.45 sec. The maximum accelerations computed at the crest of the dam were 522 gal and 760 gal, respectively.

29. For the purpose of studying the dynamic behaviour of the dam, time variations of three stress components, σ_x, σ_y, and τ_{xy} were also computed for 10 elements: five in the core and five in the shell.

STRAIN POTENTIAL AND DEFORMABILITY

30. From the dynamic response analysis discussed above, three stress components, σ_x, σ_y, and τ_{xy} were calculated. Because the input motion was a horizontal shear motion, vertical stress component did not show appreciable changes, while the lateral stress, σ_x, and the shear stress, τ_{xy}, showed large variation. Therefore, in order to truly simulate the field stress condition, all three stress components should be variably applied to each specimen. The soil should be consolidated under the in-situ static stress state shown in Fig.14, then dynamic stresses should be applied. In reality, this type of testing is not very practical, so an attempt was made to run a laboratory triaxial test using a more conventional apparatus to take the three stress components into consideration.

31. First the samples were consolidated in a triaxial cell under major and minor principal stresses determined from initial static stress condition as shown in Fig.14. Then, three dynamic stresses at each time increment were added to the corresponding three static stresses and a new Mohr circle was constructed. From this Mohr circle, the major and minor principal stresses were calculated. At this stage, the three stress components which existed at the start were now represented by two stresses. In order to perform a cyclic triaxial test, the lateral confining pressure must be kept constant. This condition was easily met by using saturated core samples as the variation in the ambient pressure does not change effective stresses of saturated soil. Another word, the deviator stress calculated from the major and minor principle stresses were applied in the axial direction with lateral stress constant.

32. As mentioned in the static analysis, the static deviator stresses and shear stress in the core zone were rather small, which means the static Mohr circle had a very small diameter. Thus, any additional stresses tended to cause the Mohr circle to expand, requiring the application of compressive stresses. Fig.15 shows the results of a triaxial loading test of one of the core specimens.

33. With saturated soils the changing allaround pressure could be ignored, but with unsaturated soils this is not possible. In order to study the effect of very rapid variation in lateral pressure, a simple test referred to as the C-UU test was conducted. First, the samples were consolidated at confining pressures of 1 and 3 kg/cm^2 and tested rapidly without drainage. Then, with new set of samples consolidated under the above pressures, the confining pressures were either very rapidly increased or decreased and, again the samples were rapidly loaded to failure. The result shown in Fig.16 makes clear that the fact that lateral pressure changes have a profound effect on sample strength. Therefore, for an unsaturated sample consolidated at a given lateral pressure, σ_{3s}, and subjected to a new stress state, σ_{1ds} - σ_{3ds}, as shown Fig.17, a computation is made so that the Mohr circle expressed by σ_{1ds} and σ_{3ds} will have the same effect when the confining pressure is at σ_{3s}. This could be achieved by determining the ratio of the σ_{1ds} - σ_{3ds} circle to the failure circle σ_{1dfs} - σ_{3ds} and multiplying this ratio by the failure curve σ_{1sf} - σ_{3s} to obtain the new Mohr circle of σ_{1s} - σ_{3s}. The Mohr failure envelope is the curve determined in the C-UU test discussed above.

34. The maximum and residual axial strains developed during the tests on specimens of both the core and the shell are presented in Table 2 and shown in Fig.18 a and b in the direction of major principal stress for maximum input accelerations of 150 gal and 250 gal.

35. It is clear from both the table and the figure that the core and the shell underwent very small deformation, except for the shell elements designated by I and J. These two elements are, as seen in Fig.18, from the downstream slope of the dam, and they would exhibit large deformation due to the large tensile stress applied during an earthquake--(in the laboratory, this tensile stress was simulated by very large axial stress using the procedure described for unsaturated material).

36. These axial strain values are the strain potential of each element and do not express the deformability of the dam because the deformability of an element is strongly affected by the strain potential of adjacent elements. Therefore, the very large strain exhibited by elements I and J do not necessarily suggest slip failure of the downstream embankment, though cracks may develop and minor sliding may occure under strong shaking.

37. In order to compare the result of this study with the conventional pseudo-static analysis, a slope stability analysis was carried out of the core. It is clear from Fig.19 that the core is very strong; even with the horizontal seismic coefficient, k_h, equal to 0.5, a factor of safety of 2 is preserved. The seismic coefficient of 0.5 approximately corresponds to the case of 250 gal input motion with a maximum crest acceleration of 760 gal and an average acceleration of the core along the upstream face of the dam of about 500 gal.

38. With the maximum dynamic deviator stress reaching about 40 to 60 percent of the static

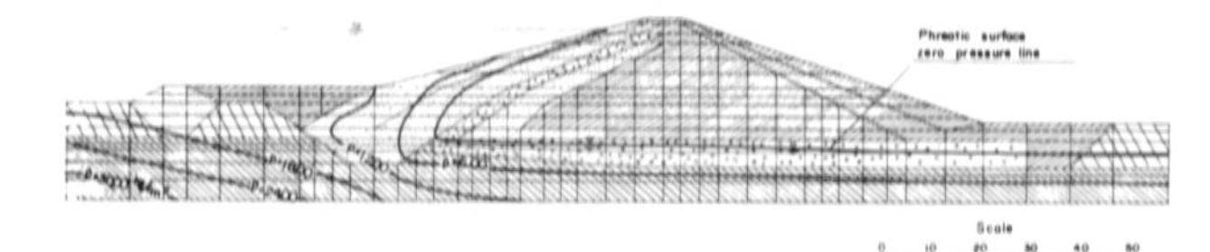

Fig.11 Finite element mesh and equi-potential lines

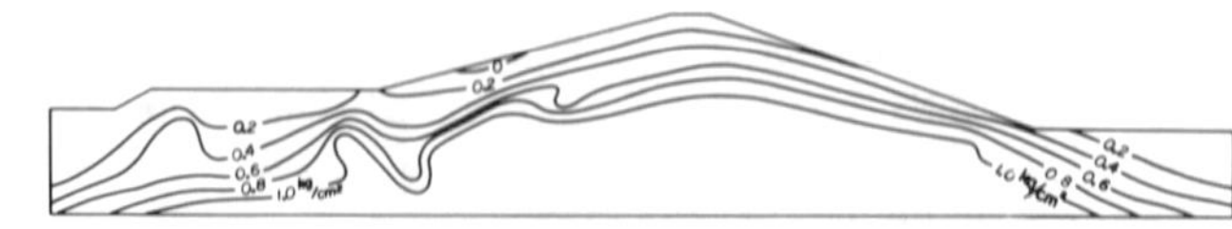

Fig.12 Static deviator contours

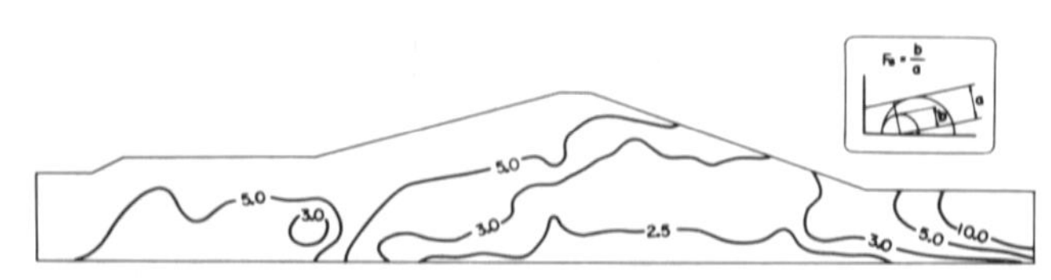

Fig.13 Static factor of safety

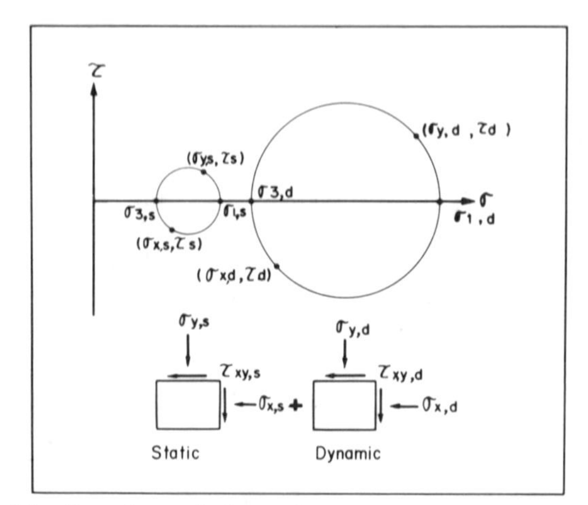

Fig.14 Static and dynamic stress states and Mohr circles

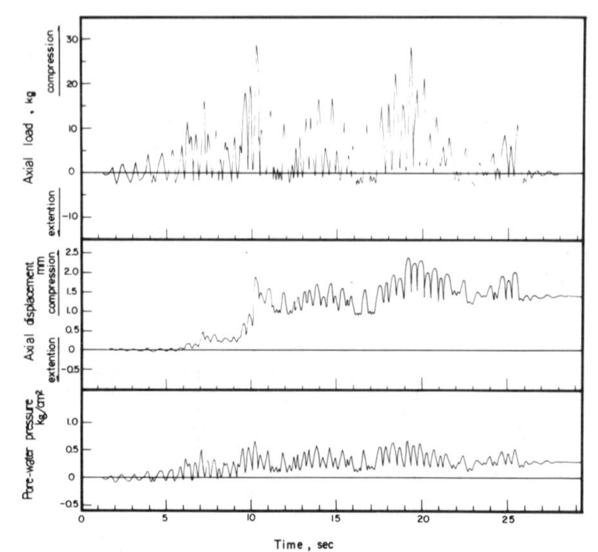

Fig.15 Dynamic loading of the core sample

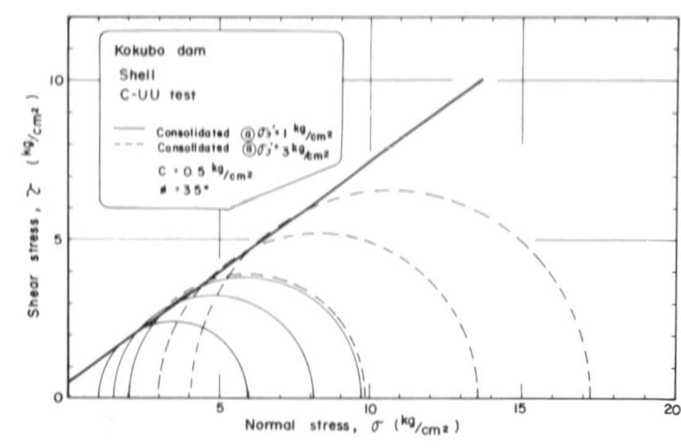

Fig.16 C-UU test of the shell material

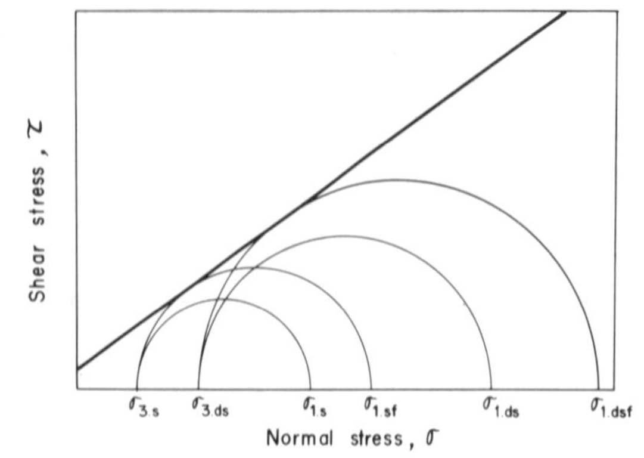

Fig.17 Schematic illustration of stress state transfer

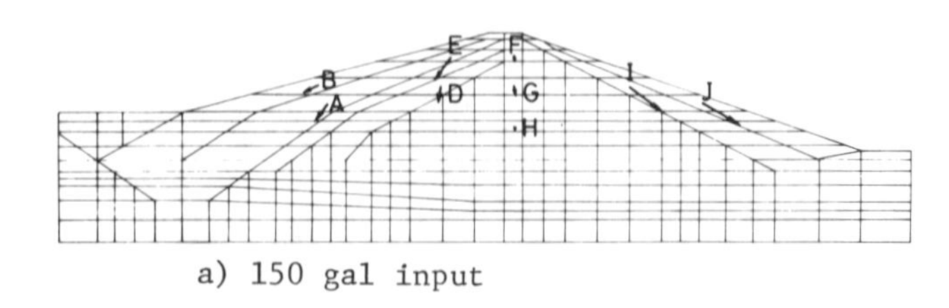

a) 150 gal input

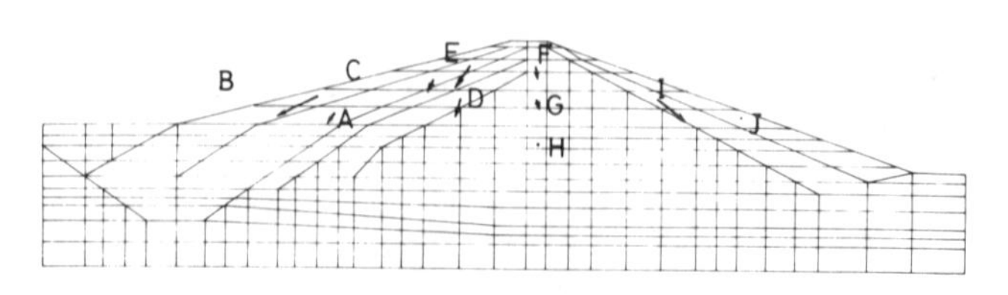

b) 250 gal input

Fig.18 Strain potentials

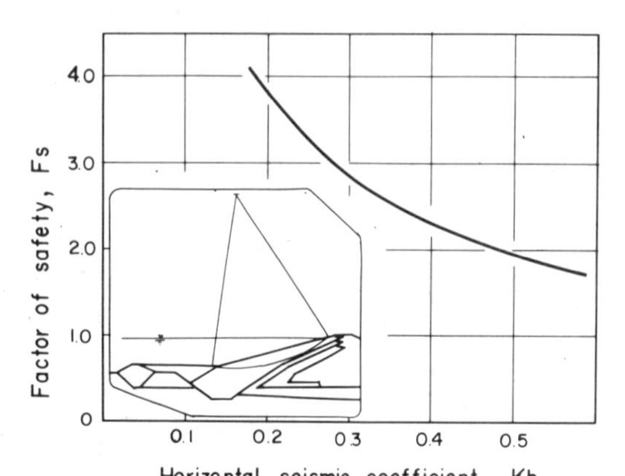

Fig.19 Pseudo-static slope analysis of the core

strength, in general, and the pseudo-static analysis indicating a factor of safety of about 2, only slight deformation of the samples was expected, although the deformation observed seems to be rather small. But, overall, the dam seems to resist dynamic loads well.

CONCLUSION

39. In this study, starting with a general outline of the dam, a complete analysis was carried out with primary emphasis being placed on the dynamic behaviour of the dam. The dynamic numerical analysis showed that lateral stress is as important as shear stresses, and an attempt was made to simulate the effect of three dynamically varying stress components in a dynamic triaxial apparatus. The dam has a very gentle downstream slope of 1:3.5 and an upstream slope of 1:4, and both the core and shell materials, in spite of their rather light compacted weight, possessed high static strength and tended to resist large deformation or strain potential except for two elements of the downstream slope which primarily failed due to tensile stress in the lateral direction.

40. In conclusion, the Kokubo Dam is not likely to undergo large deformation under sever seismic loads. The average strain potentials calculated over the entire dam were in the order of 0.1 % and 1.5 % for the 150 gal and 250 gal input motions, respectively. And these values seem to be in general agreement with pseudo-static analysis.

ACKNOWLEDGEMENT

The authers wish to thank Mr. H. Komada at the Central Research Institute of Electric Power Industries for his generous permission to use the saturated-unsaturated seepage analysis computer program.

REFERENCES

Aseismic Design Code. Ministory of Construction, Dept. of Public Works, 1977.

Japanese Meteorological Agency. Science table, Maruzen, 1980.

Kakimi, T. and Suzuki, I. Seismic activities in Kanto District and tectonic movements, Maruzen, Tokyo, 1974, (in Japanese).

Komada H. The analysis of unsteady seepage flow in saturated-unsaturated porous media, Cent. Res. Inst. of Elec. Power Ind., Report No.377015, 1978.

Lysmer, J. et al. LUSH: A computer program for complex response analysis of soil-structure systems, Earth. Engin. Res. Cent., Univ. of Calif., Berkeley, EERC 74-7, 1974.

Mercuson, W. F. and Franklin, A. G. State of the art of undisturbed sampling of cohesionless soils. Proc. of the Inter. Symp. of Soil Samp., State of the Art on Current Practice of Soil Sampling, Singapore, 1979, 55-56.

Mori, K. and Ishihara, K. Undisturbed block sampling of Niigata sand. Proc. of the Sixth Asian Reg. Conf. on Soil Mech. and Found. Engin., Singapore, 1979, 39-42.

Neuman, S. P. Saturated unsteady seepage by finite elements, Proc. of Ame. Soc. of Civil Engin., Vol.99, No.HY12, Dec. 1973.

Ozawa, Y. and Duncan, J. M. ISBILD: A computer program for analysis of static stresses and movements in embankments, Dept. of Civil Engin., Univ. of Calif., Berkeley, Rep. No.73-4, 1973.

Seed, H. B. and Idriss, I. M. Rock motion accelerations for high magnitude earthquakes, Earth. Engin. Res. Cent., Univ. of Calif., Berkeley, EERC 72-2, 1969.

CONVERSION TABLE

To convert	To	Multiply by
kg/cm^2	kN/m^2	98
t/m^2	kN/m^2	9.8
g/cm^3	kN/m^3	9.8
t/m^3	kN/m^3	9.8

Table 2 Residual and maximum axial strain in percent

		Axial strain in percent			
		150 gal input		200 gal input	
Element	Zone	Residual	Maximum	Residual	Maximum
A	Core	0.15	0.36	0.30	0.83
B	Core	0.10	0.25	1.40	2.40
C	Core	-	-	0.36	0.91
D	Core	0.08	0.44	0.48	1.32
E	Core	0.20	0.47	0.83	1.32
F	Shell	0.03	0.12	0.40	0.92
G	Shell	0.07	0.15	0.27	0.45
H	Shell	0.01	0.11	0.05	0.10
I	Shell	0.30	0.11	1.06	1.72
J	Shell	0.30	0.45	10.	> 10.

Table 1 Values of Stress-Strain Parameters for Static Analysis of Kokubo Dam

Parameter	Symbol	Core		Shell	Filter	Foundation	Alluvial clay
		Unsaturated	Saturated				
Unit weight (kg/cm^3)	γ	1.838	1.873	1.827	2.101	2.269	1.717
Cohesion (kg/cm^2)	c	0.80	1.00	0.38	0.50	0.38	0.25
Friction angle (degree)	ϕ	37.2	32.0	39.5	40.0	40.67	31.5
Modulus number	κ	985.6	250.0	707.0	1746.0	833.0	308.8
Modulus exponent	η	0.211	0.467	0.367	0.325	0.345	0.222
Failure ratio	R_f	0.788	0.850	0.883	0.683	0.730	0.902
Poisson's ratio parameters	G	0.423	0.260	0.451	0.485	0.407	0.361
	F	0.254	0.260	0.471	0.249	0.230	0.143
	d	9.123	5.132	5.851	11.500	14.395	2.733

14 The response of concrete to short term loading

J. W. DOUGILL, King's College, London

The response of concrete to short term loading is reviewed with particular emphasis on the mechanisms involved in energy dissipation and damping. The review leads to a view of the essential components to be included in constitutive relations for concrete and rock-like materials. A number of approaches to developing such relations are discussed.

INTRODUCTION

1. The object of this paper is to review the behaviour of concrete and similar materials with a view to establishing the factors that are likely to be important to the behaviour of dams under earthquake conditions and so need to be included in descriptions of material behaviour used in analysis. In doing this, attention will be focussed on non-linear aspects of material behaviour and the various mechanisms by which energy may be dissipated and so contribute to damping. This approach separates the discussion from applications in which a simplified view of structural behaviour is adopted, either by restrictions to linear behaviour or by substituting an acceleration spectrum for a fully sequential prescription of ground motion. Essentially then, we are looking for the kind of information that is required in the design of large or particularly important dams which might be designed on the basis of one or more suitably scaled long ground motion records. Lane (1), in his report to the 13th Congress on Large Dams, has noted that the design process for such dams has much in common with a research project in the need to capitalise on the most recent information and techniques available. In the context of analysis and material behaviour, this is most evident in improvements in numerical techniques for analysis and in the development of appropriate constitutive models for material behaviour.

2. It is becoming more generally recognised that our capacity to perform extensive high speed computation with model materials has outstripped our knowledge of material behaviour itself and the range of fully validated constitutive models for use in analysis. If a proper balance is to be achieved between the quality of the input and the available computational capacity, further work will need to be done both in observing material behaviour experimentally and in devising suitable models. The present paper is intended to provide a framework for these studies. The main discussion is concerned with concrete with a few remarks being included on rock and masonry.

Concrete in uniaxial compression: Monotonic loading and Discontinuity

3. It seems useful to start with the behaviour of concrete in uniaxial or nominally uniaxial compression. As with all tests on concrete the results are affected by the mix proportions and materials used, the age of the concrete and the curing conditions, but are also often influenced by the testing techniques adopted including the shape and size of specimen.

4. Figure 1 gives results from McHenry and Schideler (2) showing how the strength of concrete is affected by the rate of loading. There are significant differences between results from cubes and cylinders but the state of stress is different in these types of samples with the cube being affected more by restraint from the testing machine platens. This restraint tends to inhibit internal cracking and leads to strengths 15-20% higher than those obtained on cylinders. To avoid these effects, some workers have used longer samples. However, much of this work has been done under quasi-static conditions with rates of loading slower than 1 MN/m^2sec. and so at rates one or two orders of magnitude less than would be relevant in the design of a concrete dam. In spite of this, such tests do provide valuable insight into the mechanisms involved as energy is dissipated during loading.

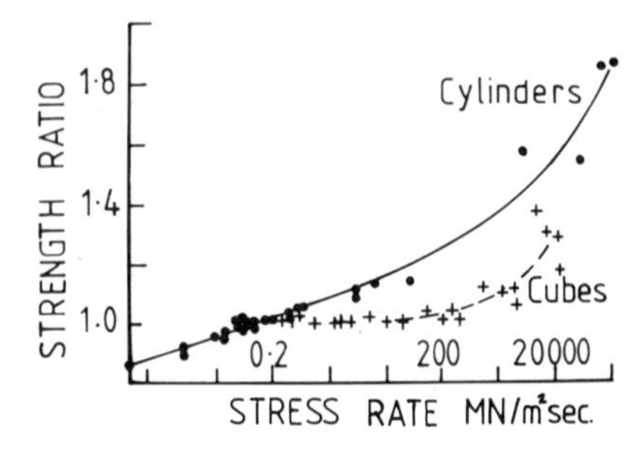

Fig.1. The effect of loading rate on compressive strength of concrete (2).

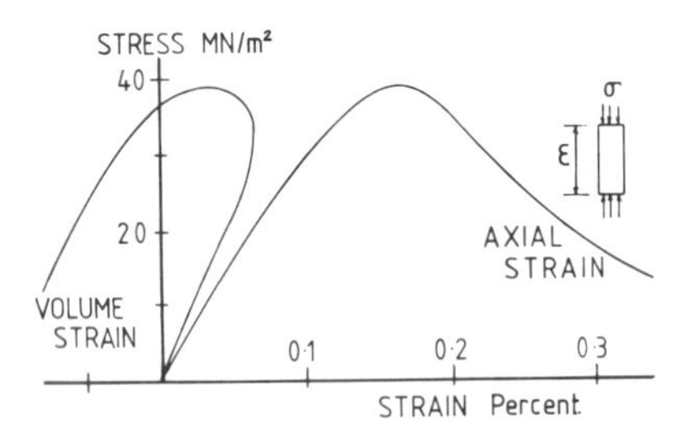

Fig.2. Typical stress strain curves for concrete loaded in uniaxial compression from (11).

5. Typically, the stress strain curve for concrete under short term loading in uniaxial compression has the form shown in Figure 2. Before loading, the concrete already contains some flaws or micro cracks (3) due to internal shrinkage during curing. Some of these may close, during the initial stages of loading leading to a slight increase in stiffness as described by Brace (4) for rocks. More importantly, additional micro-cracks develop during loading leading to the emission of acoustic energy and an increase in the strain measured at right angles to the direction of loading over and above that expected on the basis of elastic behaviour and Poisson's ratio. With continued loading, the micro-cracks increase in size causing reduced stiffness and load capacity. Thus, if the axial strain is increased, the induced stress reaches a maximum value before decreasing with further increases in strain and damage.

6. Besides leading to a reduction in load capacity at a given strain level, cracking is accompanied by an increase in voids which counters the effect of bulk compression of the solid material. Because of this, a stage is reached where the volume ceases to decrease with increasing axial contraction, this being followed by an increase in volume with further cracking. The stress level at which this change in sign of the rate of change of volumetric strain occurs is somewhat below the peak stress and fairly easy to determine without ambiguity. Maybe because of this, the stress level so determined has achieved a status of its own with different workers taking it to mark the "onset of unstable crack propagation" (5), the fatigue limit or the "true" strength of concrete (6) in the sense that it is supposed that concrete can sustain indefinitely any constant stress below this level. However, it is an open question whether there is sufficient experimental evidence to support these views and particularly the idea that the limit provides an upper bound to the useful strength of concrete in compression.

7. Attempts have also been made to identify a limiting stress level below which cracking is insignificant. This limit has come to be known as "discontinuity", the term being originally coined by K. Newman in 1964 (7) but developing its present meaning more akin to the "onset of stable crack propagation" (8, 6), somewhat later. The discontinuity level is taken to correspond to the point of departure from linearity of the stress strain curve rather in the way that the initial yield stress is defined for a hardening metal. In these terms, discontinuity is found to occur at a stress of 50-60% of the peak stress in uniaxial compression.

8. The idea of a discontinuity level involves a simplification as concrete is not truly elastic even under low stress. The principal value of the concept is in design following its adoption as a lower bound for strength under short and long term loading and also slow cycle fatigue. In analysing concrete dams, a similar limit might well be appropriate in those situations where a linear elastic analysis with nodal or structural damping might be used in contrast to an analysis based on a fuller appreciation of material behaviour.

Repeated Loading

9. More information is obtained from a compression test if a series of load or deformation cycles are used and in which the maximum strain is progressively increased with each cycle. Such tests have been undertaken on a variety of materials including concrete (9-12), rock and cement paste with essentially similar results of the form shown in Figure 3.

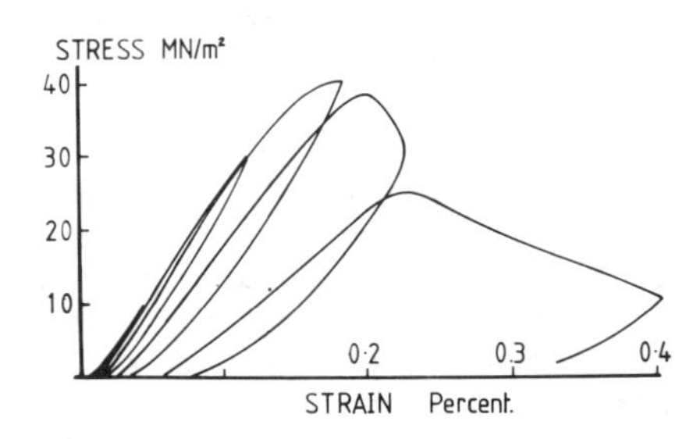

Fig.3. Results from a cyclic loading test on concrete (14).

10. Before going into detail, two points should be noted. First, the envelope to the stress-strain curve obtained using repeated cycles of deformation corresponds quite closely to the curve that can be expected if monotonically increasing deformation is used (10). Second, having established a loop in stress-strain space by a loading-unloading-reloading sequence; if the cycle is then repeated, there is only a marginal change in the shape and size of the loop during subsequent cycles. Clearly the first loading over a given strain range has a significantly different effect to that of subsequent loadings.

11. Returning to Figure 3, the most significant feature is the evidence of loss in stiffness with increase in deformation. This is most easily seen by noting how the initial (i.e. at low stress) slope of the reloading

branch of the stress strain curve changes with an increase in the range of deformation cycle. The results from both concrete and rock are similar and shown in Figure 5. This shows a continuous decrease in stiffness with applied strain, a loss in stiffness being apparent at very small strains. This is at odds with the concept of a discontinuity limit or at least with the idea of using such a limit as a basis for a description of material behaviour rather than as a design aid.

12. The idea that the first passage over a particular strain range has a significantly different effect than subsequent cycles is supported by acoustic emission data. In tests involving repeated loading, Rusch (13) and subsequently Spooner and Dougill (14) noted that detectable noise was produced only when the highest previously attained stress was exceeded, with the exception that noise became more general when the applied stress was within about 20% of the peak value.

13. The occurrence of both acoustic emission and loss of stiffness during first loading is considered to be due to the initiation and stable propagation of micro-cracks within the concrete structure. In addition the results suggest that repeated loading over a given strain range does not lead to increased micro-cracking except at high stresses. The results also point to the existance of a further mechanism of energy dissipation, presumably involving friction across the surfaces of the cracks either within the mortar matrix or at the aggregate cement paste interface. Such a mechanism would lead to energy dissipation during repeated cycles of loading as described by Jaeger and Cook (15).

Physical Measures of Degradation and Damping

14. In an attempt to quantify the energy dissipated by the two different mechanisms, Spooner and Dougill (14) compared the behaviour of concrete with that of an idealised material which,
(a) stores an amount of strain energy W_ε during loading to a given strain, this same amount of energy being recovered on unloading,
(b) dissipates an amount of energy W_γ during the first loading to a strain ε with no further energy being dissipated by this mechanism during subsequent unloading or reloading over the same strain range, and
(c) dissipates an amount of energy W_δ during each and every loading or unloading over the strain range 0 to ε.

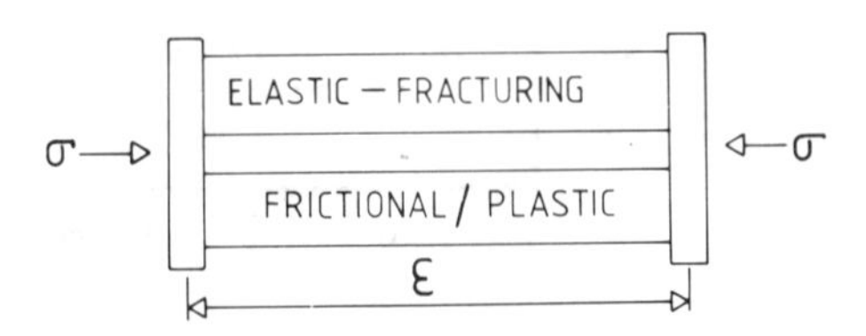

Fig.4. Rheological model giving rise to energy dissipation components W_γ and W_δ.

15. The mechanism giving rise to W_γ can be considered to be linked with the propagation of micro-cracks and degradation of stiffness. Again, if a physical base is required, W_δ can be thought of as being due to friction. In these terms W_γ is a convenient measure of damage whilst the quantity W_δ provides a measure of damping. It will be evident, also, that the prescription adopted for the ideal material leads to a view of material behaviour based on a model in which the two modes of energy dissipation occur in separate elements linked in parallel as shown in Figure 4. The model can be generalised and clearly suggests the kind of formulation to be preferred in devising constitutive equations for use in analysis.

16. The quantities W_γ and W_δ can be determined from the results of cyclic loading tests and values for one concrete are given in Figures 6 and 7. The amount of damage at low strains is small, but the energy dissipated in frictional damping may be appreciable with repeated load cycles. At higher strains, the energy dissipated in damage, through micro-cracking in the first cycle of loading, is significant and would need to be considered in assessing material behaviour during only a few cycles of loading.

Damping

17. A commonly used measure of damping is the relative damping capacity S. This is the ratio of the energy dissipated in a cycle of loading to the peak strain energy in the same cycle. Thus, $S = 2W_\delta/W_\varepsilon$ and can be determined from results of the sort shown in Figures 6 and 7. Some results from Spooner (11), shown in Figure 8 show that damping increases with stress level. This is to be expected as additional micro-cracks are formed with increasing strain and these lead to energy dissipation by friction between the crack surfaces during subsequent strain cycles. Accordingly, we would expect both the damping capacity and the energy component W_δ to be dependent on the energy dissipated in damage W_γ.

18. Although the trends derived from Figures 6 and 7 are as expected, the values for damping capacity obtained appear to be too high when compared with those obtained by other methods. This could be because the tests were undertaken at a relatively low strain rate but, more probably the W_δ component is augmented by energy dissipated at the interfaces between the sample and the machine platens. However, Ashbee et al (17) have described in detail how this can be avoided and the type of instrumentation required to obtain reliable measurements of damping capacity from uniaxial tests using cyclic loading at rates around 0.1 Hz. Following their studies they note that the logarithmic decrement δ for concrete is in the range 0.03 to 0.15. Particular results from Ashbee et al (17) for concrete in compression, Jordan (18) for concrete in tension and from tests on chimneys (19) suggest that, for low frequency cycling, the logarithmic decrement is at the lower end of this range, the respective values from these

sources being 0.037, 0.044 and 0.03.

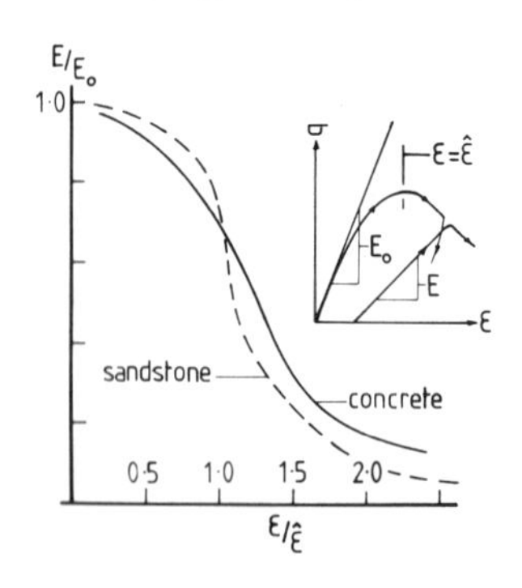

Fig.5. Effect of deformation history on the reloading modulus. Data from Refs. 10 and 14.

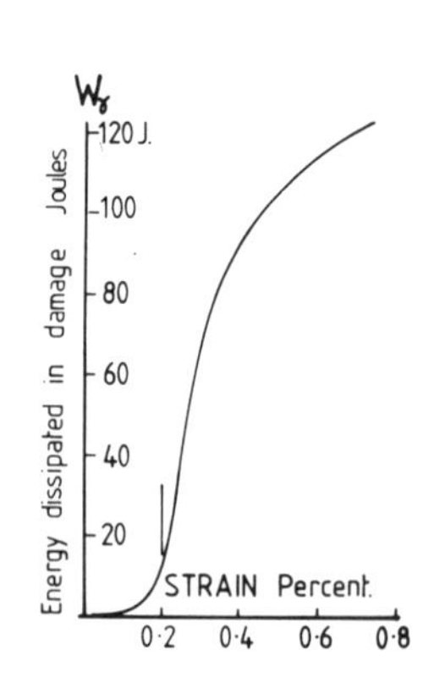

Fig.6. A typical result for W_γ the energy dissipated in damage (14).

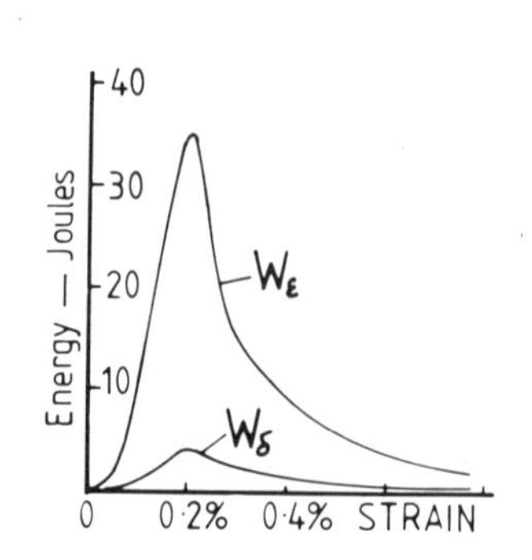

Fig.7. A typical result for W_δ the energy dissipated in damping (14).

19. At high frequencies, the resonant frequencies of specimens vibrated by an electro magnetic driver can be measured. Such tests give the dynamic properties of the material at low stress levels. Table I is a summary of results obtained by Jones (20) on a single concrete from measurements of resonant frequency in flexural, longitudinal and torsional modes over a frequency range of 400-10,000 Hz. These results suggest that damping is proportionately the same for each mode and that damping is effectively independent of frequency at these rates of excitation.

Table 1. Damping properties of concrete: data from Jones (20).

Concrete Condition	Equivalent Logarithmic Decrement
7 days - moist cured	0.029 - 0.038
25 days - moist cured	0.021 - 0.026
176 days - air dried after initial moist cure	0.013 - 0.018

Values of the Logarithmic Decrement are found from $\delta = \pi/Q$ where Q is the band width measure given by Jones.

20. The effects of age and drying shown in Table I have also been observed by other workers. The results shown in Figure 9 from Cole (21) are typical in showing a reduction of 50% in the logarithmic decrement over the period from 1 month to 30 months. Similar reductions are caused by drying and results of this sort suggest a view of damping at low stress as being due to damping in the stiff gel structure together with an additional contribution due to enforced movement of moisture in the voids and capillary pores. Presumably this last mechanism has a viscous element so that we should expect damping to be dependent on frequency within a band in which the cyclic disturbance is not so rapid as to make the viscous effect negligible or so slow that these effects relax entirely within a single cycle. Interestingly, Cole and Spooner (22) have found that the logarithmic decrement of hardened cement paste increases as the frequency is reduced below 2.5 Hz although it is constant at higher frequencies (Figure 10). These data appear to bridge the gap between dynamic tests and results from slow cycle or quasi-static tests. Also it seems entirely probable that the values of most relevance to large concrete dams subject to vibration are likely to occur in this range of greatest uncertainty.

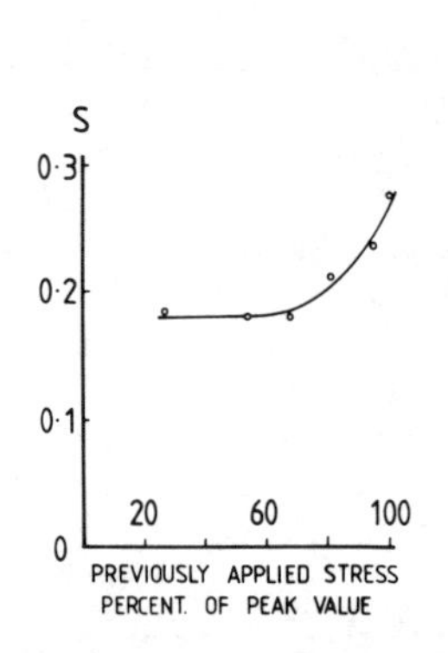

Fig.8. The influence of stress level on damping (11).

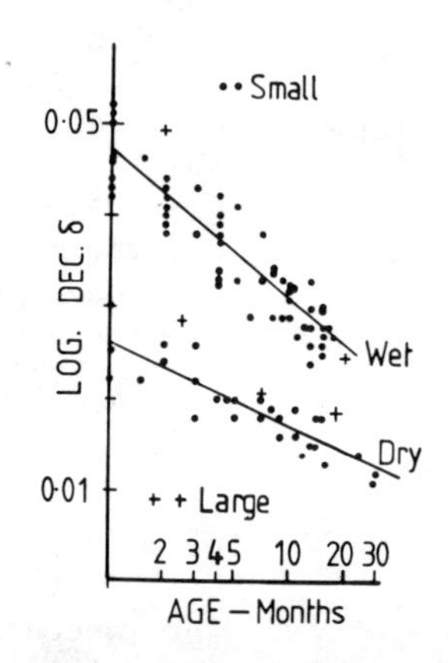

Fig.9. Effect of age and moisture conditions on damping capacity (21)

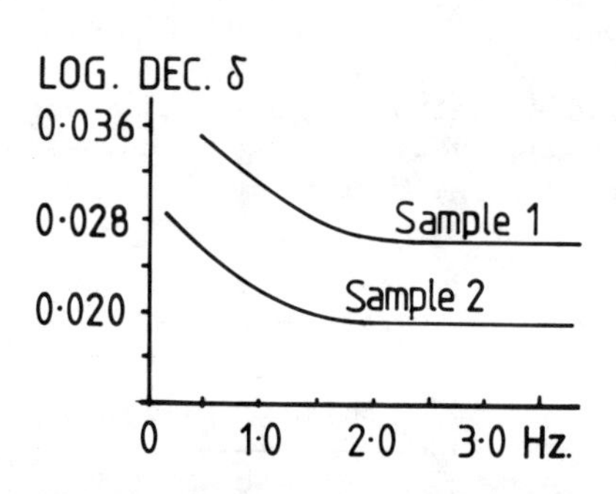

Fig.10. The influence of frequency on damping (22).

Rate of Loading

21. Reference has already been made to results reported by McHenry and Shideler (2). These trends have been confirmed in later studies (23, 24) with both the range of linear behaviour and the peak stress in compression increasing with an increase in the rate of loading. These effects are apparently not common to all states of stress and conditions of concrete. For instance, Saucier and Carpenter (25) have shown that the tensile strength of a dry mature concrete is unaffected by the rate of loading up to around 70 MN/m^2sec. On the other hand, the loading rate had an effect on samples that had been soaked in water for a month before testing. With these, the rapid strength was almost identical to that obtained with dry samples with the strength obtained at slower rates being 17% lower. This difference between the strengths of wet and dry samples at normal rates of loading is quite usual and was also found between the static compressive strength of the dry and soaked specimens.

22. Besides the moisture conditions, the concrete composition and type of aggregate are also influential. With a stiff limestone aggregate, Sparkes and Menzies (23) found the compressive strength was increased by only 4% for a tenfold increase in loading rate whereas the increase was around 16% when a lightweight aggregate was used.

23. As yet, there are not sufficient data available to be certain of the effect of high rates of loading particularly on strength in tension and shear. It follows that in dealing with the serviceability requirements of earthquake resistant structures that it is inadvisable to adopt an overall increase in permissible stress under dynamic conditions as is sometimes done with blast resistant structures.

Combined Stresses

24. Early work on strength under the effect of combined stresses was motivated by the need to understand the behaviour of columns with hooped reinforcement and so was concerned with concrete in triaxial compression (26). Later studies focussed on the problem of shear in reinforced concrete members and dealt with behaviour in combined compression and tension (27, 28). With the 1960's and the need for information to support the design of more complex structures, such as prestressed concrete pressure vessels, a more general view of failure was developed. Some of these later studies have gained from co-ordination within a collaborative programme (29) which led to improvements in testing technique and broad agreement on many aspects of behaviour particularly under compressive states of stress. Because of this, it is convenient to give an overall view of results rather than to go into each contribution.

25. Values of peak stress or strength obtained for proportional loading and biaxial states of stress are shown in Figure 11 as multiples of the uniaxial compression strength. In this form the results represent behaviour of a wide range of concretes under compressive states of stress although there are differences between concretes of differing strengths when one or more stresses is tensile.

26. In biaxial compression, the form of the relationship between each of the applied stresses and its corresponding strain is similar to that already shown for uniaxial compression. The volumetric strain exhibits the characteristic turnover but at different strain levels as the tendency to increase in volume is inhibited by the mean normal compressive stress. This influence on cracking extends to post-peak behaviour with the material being more unstable the more tensile the value of the mean normal stress. Post peak behaviour and softening are known to occur in uniaxial tension (31). However, the softening curve is far steeper than in compression and difficult to observe experimentally. It follows that behaviour in biaxial and triaxial tension may be supposed to approach that of an ideal brittle material even more closely.

27. For the proportional loading paths employed, it is possible to represent results up to the peak stress by relations between the octahedral shear stress τ_0, the mean normal stress σ_0 and the corresponding strains γ_0 and ε_0 (32, 33). This leads to stress strain relations in terms of varying secant values of shear and bulk moduli as shown in Figure 12. Unfortunately, no data are available on the effects of unloading, or slow cyclic loading, to complement the stress strain relations with information on damping and so allow an analysis of dynamic effects in the manner used by Seed et al (34) for earth dams.

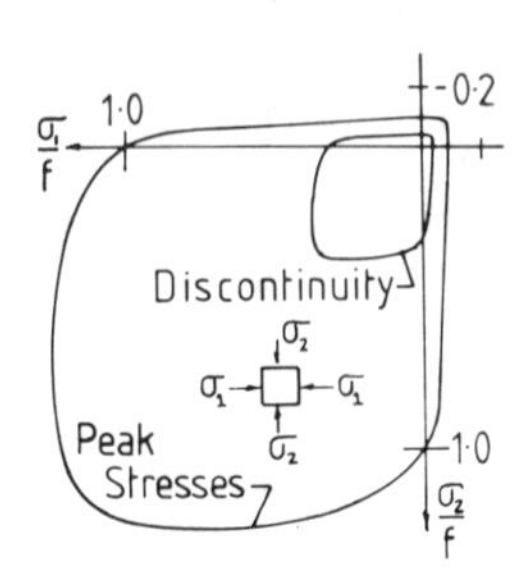

Fig.11. Peak stress and discontinuity envelopes for a concrete tested with proportional biaxial loading (30).

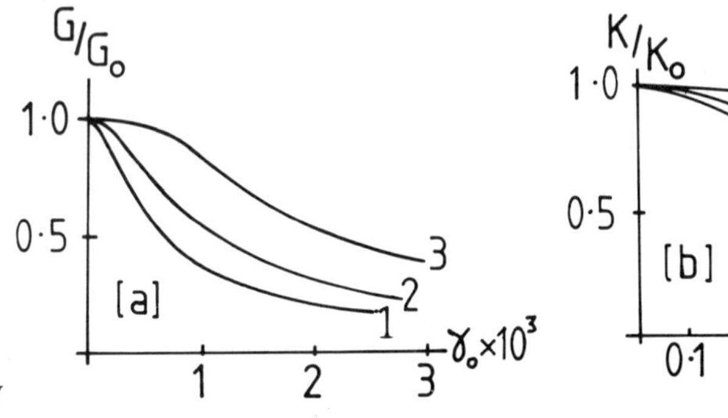

Fig.12. Secant moduli for three concretes (32). (a) Shear modulus, (b) Bulk modulus.

28. Extensive results are now available on the behaviour of concrete in triaxial compression (35, 36). Again there has been success in describing behaviour in terms of shear and bulk moduli with Kotsovos and Newman (36) adding terms to account for dilation accompanying the octahedral shear stress. Newman's tests included load paths in which the volumetric and deviatoric stresses were applied sequentially. In these, the same relation obtained between τ_0 and γ_0 irrespective of the sign of the octahedral shear stress superimposed on the hydrostatic compression. This tends to confirm that the influence of the third (cubic) invariant of stress is negligible and so supports the use of expressions involving only the first and second invariants. It is also of interest that data from triaxial tests involving sequential loading have been used successfully to predict strains observed with proportional loading in biaxial compression. This suggests a degree of path independence under compressive stress that could encourage the development of total rather than incremental theories in descriptions of material behaviour. Whether this is justified or not remains an open question until rather more critical tests of path-independence are undertaken.

Constitutive relations - general view.

29. Although apparently simple, there are a number of disadvantages attached to the use of stress-strain relationships based on isotropic behaviour and stress dependent moduli G and K. The assumption of path-independence effectively limits the analysis to a form of non-linear elasticity which is inappropriate if unloading occurs. Also, unless a load path is precisely reversed, the theory provides no discrimination to determine whether unloading occurs or not. This could be an important limitation with cyclic or vibrational loading.

30. Comment is necessary on the orthotropic "non-linear elastic" models occasionally used to describe degradation under different combinations of principal stress. Besides the difficulty of recognising unloading, these models have the further limitation that the orthotropy must be maintained if the model is to be applicable. Such models are therefore restricted to situations in which the directions of principal stress at each point in the material do not change during loading. Accordingly, these models have little general interest or useful application.

31. In order to achieve greater generality, a number of workers have proceeded on the basis of the theory of plasticity. A usual idealisation is to take concrete to be an isotropic-hardening material with the initial yield surface coincident with the state of discontinuity or, alternatively, with the shape of the yield surface being determined by the strength of the material found under proportional loading. There are a number of questions here. Most important could be whether the yield surface has been properly identified and whether an associated flow rule is appropriate. Clearly though, an isotropic hardening model with elastic behaviour for load paths enclosed by the yield surface cannot give a satisfactory account of behaviour under repeated loading. Some relief from this difficulty is achieved by postulating that load paths within the yield surface are accompanied by hysteretic behaviour (37) but the theory is complicated and one is uncertain of the generality achieved. Basically, if plasticity models are to be used at all to describe behaviour involving cyclic loading. it would seem that they should include some form of kinematic hardening in the development of the flow rule. In addition, there seems to be a need to lean more heavily on physical evidence of the phenomena involved when assessing the form of theory likely to be most suitable.

32. The review of material behaviour suggested three main strands that need to be woven into the texture of a theory describing the mechanical behaviour of concrete. First, there is degradation of stiffness due to cracking at a dimensional level less than that characterising the structure of the material. Second is a description of energy dissipation under repeated loading, possibly by friction but sensibly independent of rate effects. And third, there is the need to include a viscous term to provide time dependence and more particularly damping under moderate rates of loading at small amplitudes. Recognition of these effects motivated the development of a theory for degrading materials (38), led to coupled theories involving plasticity and degradation (39, 40) and has influenced the choice of parameters in extensions of Valanis' endochronic theory (41) to concrete (42).

Theoretical description of degradation

33. A view of degradation is provided by considering a material in which the only form of energy dissipation is through loss in stiffness. A network formed of linear elastic brittle members of varying strengths and stiffness would behave in this way (38, 43). In a sample of such a material, stress and strain can be related at any stage by the generalised form of Hooke's Law

$$\sigma_{ij} = S_{ijkm} \varepsilon_{km} \quad (1)$$

34. During loading, progressive damage occurs to cause changes in the stiffness S_{ijkm} so that

$$\dot{\sigma}_{ij} = \dot{S}_{ijkm} \varepsilon_{km} + S_{ijkm} \dot{\varepsilon}_{km} \quad (2)$$

where the superior dot indicates differentiation with respect to some convenient monotonically increasing parameter such as time. The form of equation 2 suggests separating the stress rate $\dot{\sigma}_{ij}$ into two components, an elastic component $\dot{\sigma}_{ij}'$ and a fracturing component $\dot{\sigma}_{ij}''$ to account for the effect of change in stiffness. Clearly,

$$\dot{\sigma}_{ij}' = S_{ijkm} \dot{\varepsilon}_{km} \text{ and } \dot{\sigma}_{ij}'' = \dot{S}_{ijkm} \varepsilon_{km} \quad (3)$$

35. These definitions of stress rate allow a theory to be constructed by analogy with the theory of hardening plasticity (44), for a material that has been called the progressively fracturing solid (45, 38). A loading function F is introduced to distinguish between loading,

which is accompanied by degradation, and linear elastic unloading and reloading. F is taken to be a function of the strains ε_{ij} and parameters H_K which change with degradation. These parameters might well be linked with the W_γ component of energy found from cyclic loading tests described earlier. Now, during progressive fracture

$$F(\varepsilon_{ij}, H_K) = 0, \tag{4}$$

from which it follows that the fracturing component of the stress rate tensor can be written

$$\dot{\sigma}_{ij}'' = \alpha_{ij} \frac{\partial F}{\partial \varepsilon_{km}} \dot{\varepsilon}_{km} \text{ for } \frac{\partial F}{\partial \varepsilon_{km}} \dot{\varepsilon}_{km} > 0 \tag{5}$$

where α_{ij} is a symmetric tensor still to be determined. At this stage, two approaches are possible. The first leads to an associated flow rule. For this, the α_{ij} are taken to be independent of the strain rates $\dot{\varepsilon}_{km}$ which together with Il'iushin's postulate of plasticity (46) leads to the flow rule

$$\dot{\sigma}_{ij}'' = -K \frac{\partial F}{\partial \varepsilon_{ij}} \frac{\partial F}{\partial \varepsilon_{km}} \dot{\varepsilon}_{km} \tag{6}$$

with K being found from the loading function.

36. In order to complete the theory, the rates of change of the moduli $\dot{S}_{ijkm}$ need to be determined. To do this, it is assumed that the change in stiffness accompanying an increment of deformation $\delta\varepsilon_{ij}$ is independent of the precise loading path connecting the strains ε_{ij} and $\varepsilon_{ij}+\delta\varepsilon_{ij}$. This leads to severe restrictions on the form of loading function, which apparently must be linear in the ε_{ij}, but provides a complete theory in the sense that the loading function leads to both an associated flow rule and an account of degradation. This theory has been investigated in detail (47) using a range of independent linear loading functions in a similar manner to that proposed by Sanders (48) in plasticity.

37. The restrictions on the form of the loading surface and the need to use a large number of functions makes the theory somewhat cumbersome and suggests it may be worth developing a non-associated flow rule or at least relaxing some of the conditions determining the theory.

38. As an example, consider a material with the loading function

$$F = F(\varepsilon_{ij}, \sigma_{ij}'') \text{ where } \sigma_{ij}'' = \int_0^t \dot{\sigma}_{ij}'' dt \tag{7}$$

so that during progressive fracture

$$\frac{\partial F}{\partial \varepsilon_{ij}} \dot{\varepsilon}_{ij} + \frac{\partial F}{\partial \sigma_{ij}''} \dot{\sigma}_{ij}'' = 0. \tag{8}$$

Equations 3 and 5 are both satisfied if the change in moduli are written

$$\dot{S}_{ijkm} = \frac{\alpha_{ij}\varepsilon_{km}}{\varepsilon_{rs}\varepsilon_{rs}} \frac{\partial F}{\partial \varepsilon_{pq}} \dot{\varepsilon}_{pq} + R_{ijkm} \tag{9}$$

with $R_{ijkm}\,\varepsilon_{km} = 0$ and $\dot{S}_{ijkm} = \dot{S}_{kmij}$

We shall take R_{ijkm} to be identically zero and note that the $\dot{S}_{ijkm}$ must be symmetric so that $\alpha_{ij} = \lambda\varepsilon_{ij}$. The scalar λ can be found from equation (8) and the flow rule (5) so giving the following expressions for $\dot{\sigma}_{ij}''$ and the change in stiffness:

$$\dot{\sigma}_{ij}'' = - \frac{\varepsilon_{ij}}{\frac{\partial F}{\partial \sigma_{pq}''} \varepsilon_{pq}} \frac{\partial F}{\partial \varepsilon_{km}} \dot{\varepsilon}_{km}, \quad \dot{S}_{ijkm} = \frac{\dot{\sigma}_{ij}'' \varepsilon_{km}}{\varepsilon_{pq}\varepsilon_{pq}} \tag{10}$$

39. Equations 10 are not the only possible outcomes of equations 3 and 5 and the symmetry condition in 9. There is room to use experimental evidence to influence the choice of the R_{ijkm}. This flexibility follows from dropping the condition for path independence in the small. The consequences of this have yet to be fully explored.

Energy dissipation under repeated loading

40. The concept of kinematic hardening brought into plasticity theory by Prager (49) enables us to describe some of the main effects of cyclic loading. In essence, a hardening rule is adopted that forces the yield surface to move in stress space as plastic deformation occurs, so providing a restricted range of elastic behaviour on unloading and reloading. Different forms of hardening rule have been devised (50, 51) and there are many possibilities still open for exploration. Clearly, it is also possible to combine the effects of isotropic and kinematic hardening with a suitable choice of loading function (52).

41. Most work in plasticity has been undertaken using loading functions to define a yield surface in stress space and to determine the plastic component of deformation ε_{ij}^P. Examples of flow rules derived in this way are readily available (53). However, in looking to combine degradation and cyclic plasticity in a single model there is much to be gained by working in terms of strains throughout. As with degradation theory, we divide the stress rate tensor into two components, one elastic $\dot{\sigma}_{ij}^E$ as before and the other plastic $\dot{\sigma}_{ij}^P$. The theory then follows in the conventional manner but with an exchange of static and kinematic variables.

41. As an example, consider the effect of adopting a loading function f of the form

$$f(\zeta_{ij}, \sigma_{ij}^P), \text{ with } \zeta_{ij} = \varepsilon_{ij} - \alpha_{ij}. \tag{11}$$

For a stable hardening material for which continuity and normality both obtain it follows that

$$\dot{\sigma}_{ij}^P = M \frac{\partial f}{\partial \zeta_{ij}} \frac{\partial f}{\partial \zeta_{km}} \dot{\varepsilon}_{km}. \tag{12}$$

The value of M is found from the condition that f=0 during plastic deformation. In using this, the style of kinematic hardening is set in the treatment of the $\dot{\alpha}_{ij}$. In this, we follow Prager (49) and write

$$\dot{\alpha}_{ij} = Q\,\dot{\sigma}_{ij}^P$$

where Q is a scalar and so find the flow rule:-

$$\dot{\sigma}_{ij}^P = \frac{1}{\frac{\partial f}{\partial \zeta_{pq}}\left[Q\frac{\partial f}{\partial \zeta_{pq}} - \frac{\partial f}{\partial \sigma_{pq}^P}\right]} \frac{\partial f}{\partial \zeta_{ij}} \frac{\partial f}{\partial \zeta_{km}} \dot{\varepsilon}_{km} \tag{13}$$

We note that if $\frac{\partial f}{\partial \sigma_{pq}}P$ is zero, an analogue of Prager style kinematic hardening is obtained. With Q=0, the flow rule provides a form of isotropic hardening in deformation space. More importantly, it is now possibly to "layer" the fracturing and plastic materials by choosing two separate loading functions F and f and writing

$$\dot{\sigma}_{ij} = \dot{\sigma}_{ij}^{E} + \dot{\sigma}_{ij}^{F} + \dot{\sigma}_{ij}^{P} \qquad (14)$$

with the various stress rates and change of stiffness following from equations 3, 10 and 13. In this way, then, a combined elastic-plastic-fracturing theory is obtained.

42. Somewhat similar approaches to incorporating the progressively fracturing solid into plasticity theory have been presented by Capurso (54) and Bazant (40) although without the convenience of working consistently in strain space. Bazant's paper is of particular interest as it shows that it is possible to obtain fair agreement with a wide range of test data on concrete using a kinematically hardening model. At present though, the formulation involves an inconveniently large number of experimentally determined coefficients and some simplification is required. In this, it could be helpful to include coupling between fracturing and plastic deformation beyond that implied by the parallel model and as suggested in para.17.

Time dependence and Endochronic theory

43. It has become conventional in dealing with creep of concrete to separate the time dependent creep and shrinkage, deformations from the other strains. Accordingly it would be in the spirit of the present work to replace the elastic stress component σ_{ij}^{E} with a component σ_{ij}^{VE} attributable to linear visco-elasticity. However it is doubtful whether this complication would be worthwhile in the analysis of behaviour under slow cyclic loading or earthquake conditions. Moreover, if some additional mode of energy dissipation is required, in addition to degradations of stiffness and plastic deformation, it would seem most convenient to include this in the analysis in the form of nodal damping in the equilibrium equations.

44. Although it is suggested that formal use of linear-viscoelasticity might well be avoided, it has been recognised that the structure of the visco-elastic stress strain relations allows an alternative approach to the treatment of load history and energy dissipation. In this, termed endochronic theory by Valanis (41), the stress strain relations have the same form as in linear visco-elasticity but with the exception that time is replaced by a monotonically increasing function which may involve strain, time, temperature etc. and so may be different at each point in a body. In these terms,

$$\sigma_{ij} = \int_0^z L_{ijkm}(z-\bar{z}) \frac{\partial \varepsilon_{km}}{\partial \bar{z}} d\bar{z} \qquad (15)$$

where the L_{ijkm} take the place of relaxation moduli and z is intrinsic time.

45. In applying his theory to metal plasticity Valanis (55) takes

$$z = \frac{1}{\beta}\log(1+\beta\zeta) \text{ with } \zeta = \int_0^t (\dot{\varepsilon}_{ij}\dot{\varepsilon}_{ij})^{\frac{1}{2}} dt \text{ and } \beta > 1$$

to define the intrinsic time scale. Following this, the assumption of isotropy and constant Poisson's ratio allows the L_{ijkm} to be written in terms of a single relaxation function G(z) and the initial values of Lame's constants λ and μ. Ie.

$$L_{ijkm}(z) = \left[\lambda\, \delta_{ij}\, \delta_{km} + 2\mu\, \delta_{ij}\, \delta_{jm}\right] G(z). \qquad (16)$$

By adopting a very simple form for $G(z) = e^{-az}$, Valanis obtained remarkably good agreement with experimental results from tests on copper and aluminium under complex strain histories including cyclic loading.

46. In the form described, the theory is attractively simple and has an advantage over plasticity in the elimination of the yield surface and the need to distinguish between loading and unloading. Bazant and his co-workers (42, 56) have extended and applied the endochronic concept to concrete with significant success in modelling behaviour. Unfortunately, much of the advantage and simplicity of Valanis' work has been lost in doing this as it has been found necessary to include specific criteria and corrective terms for loading, unloading and reloading. In this, as well as with the models involving plasticity, the question must arise as to how good a fit with experimental data is required for a model to be useful? The question is particularly relevant as it appears that if sufficient experimentally determined parameters are used, satisfactory agreement with experimental behaviour can be achieved with either plasticity or endochronically based theories.

Concluding Discussion

47. An attempt has been made to present a consistent viewpoint of material behaviour and the models used for its description. At present there are gaps in our knowledge and also difficulties in application. These suggest areas where further research is required.

48. First, it appears that there is a tendency for the model building process to become very detailed. It is doubtful whether this is justified in the context of analysis for earthquake conditions. The loading is somewhat arbitrary and the initial state of the material in the dam and foundation is also uncertain. Accordingly, it seems that there is a need to look for simpler descriptions than some that have been proposed. These should concentrate on broad phenomena and principally on energy dissipation. At present, it would appear that plasticity style models still have a lot to offer. However, whatever form of description is adopted, it is essential that further physical exploration of materials behaviour should be undertaken from a mechanics viewpoint. Tests must include unloading and reloading so that information on energy dissipation is obtained.

Also, the question of path dependence is crucial to the form of the model to be adopted and needs to be explored further.

49. Even at low stress levels, there is uncertainty on damping - what factors to include and how best to do this. Also, there is obvious convenience in using artificial values for damping in attempts to stretch linear analysis to situations where non-linear effects may not be altogether insignificant. When this is done, however, the influence of the material on the behaviour of the dam is not properly represented.

50. The role of damping in numerical analysis has also to be considered. With non-linear materials, the use of explicit time marching schemes becomes attractive. In these though, it is quite usual to include some damping even with time-independent behaviour in order to accelerate convergence. With time dependent loading, the choice of time step is particularly important as an unsuitable choice can introduce additional spurious damping into the system (57) Because of this, explicit time marching schemes seem to be best employed with materials descriptions involving time independent properties. If this view is adopted, there is an incentive to ignore linear damping and to include its effect with the other components of energy dissipation. The approximation involved is that of ignoring the effect of frequency on damping.

51. Nothing has been said about the problem of localisation of failure and the treatment of crack propagation. The approach to degradation using the progressively fracturing solid can be used equally well in tension and compression (47). However, the view provided is somewhat distorted. Failure of the model is progressive and so does not include the effects of the sudden release of strain energy that occurs with unstable crack propagation in a truly brittle material. Also the effects of degradation are spread throughout a region rather than localised at a crack. This leads to difficulties in assigning properties to the model material. If, for instance, values of energy dissipation per unit volume are used indiscriminately in specifying the model, the damping capacity of the structure could be over-estimated if failure is localised. Some practical guidance on these matters has already been provided by Clough and Zienkiewicz (58). However more research is required on techniques to model cracking and the inclusion of size effects. The use of formal fracture mechanics could be too complicated but an approach using a blunt crack band in a finite element analysis offers some prospect for further development (59).

52. Finally it is suggested that a consistent view of material behaviour is obtained in terms of energy. The total work done on a sample of material, less the kinetic energy, is considered as the sum of the stored strain energy U and components of dissipated energy. If the material is time independent, these may be considered to be due to micro-fracture D_F and friction or plastic deformation D_P. These quantities depend on the deformation path with different types of behaviour (from brittle to ductile) occuring according to the relative magnitudes of $\dot{U}$, $\dot{D}_F$ and $\dot{D}_P$. It is considered that the entire range of materials, rock, concrete and masonry is amenable to description in these terms.

REFERENCES

1. LANE R.G.T. General report on Question 51 seismicity and aseismic design of dams. Thirteenth Congress on Large Dams. New Delhi 1979, 669-725.
2. McHENRY D. and SCHIDELER J.J. Review of data on the effect of speed in mechanical testing of concrete. Symposium on speed of testing non-metallic materials. Philadelphia, 1956. ASTM Special Technical Publication STP185. 72-82.
3. HSU T.C., SLATE F.O., STURMAN G.M. and WINTER G. Microcracking of plain concrete and the shape of the stress strain curve. Proc. A.C.I. Vol.60, No.2, 1963, 209-224.
4. BRACE W.F. Microcracking in rock systems. Volume I, Paper 16. Structures, Solid Mechanics and Engineering Design. Ed. M. Teeni. John Wiley 1971, 187-204.
5. KOTSOVOS M.D. and NEWMAN J.B. Behaviour of concrete under multiaxial stress. Journal of the American Concrete Institute. Proceedings, Vol.74 No.9, Sept.1977, 443-446.
6. NEWMAN K. and NEWMAN J.B. Failure theories and design criteria for plain concrete. Volume 2 Paper 77, Structure, Solid Mechanics and Engineering Design. Ed. M. Teeni. John Wiley 1971, 963-995.
7. NEWMAN K. The structure and engineering properties of concrete in J.R. Rydzewski Ed. Theory of arch dams. Pergamon 1965. 683-712.
8. NEWMAN K. Criteria for the behaviour of plain concrete under complex states of stress. The structure of concrete. Ed. A.E. Brooks and K. Newman. Cement and Concrete Association. London, 1968, 255-274.
9. SINHA B.P., GERSTLE K.H. and TULIN L.G. Stress-strain relations for concrete under cyclic loading. Proceedings of the American Concrete Institute, Vol.61, No.2, 1964, 195-211.
10. BIENIAWSKI Z.T. Deformational behaviour of fractured rock under multi axial compression. Vol.1, Paper 50. Structure, Solid Mechanics and Engineering Design. Ed. M. Teeni. John Wiley 1971, 589-598.
11. SPOONER D.C. Progressive damage and energy dissipation in concrete in uniaxial compression. Ph.D. Thesis, University of London, 1974.
12. COOK D.J. and CHANDAPRASIRT P. Influence of loading history on the compressive properties of concrete. Magazine of Concrete Research, Vol.32, No.111, June, 1980, 89-100.
13. RUSCH H. Physikalische Fragen der Betonprufung (Physical problems in the testing of concrete). Zement-Kalk-Gips, Vol.12, No.1, Jan, 1959, 1-9. (Cement and Concrete Association Translation Cj 86).
14. SPOONER D.C. and DOUGILL J.W. A quantitative assessment of damage sustained in concrete during compressive loading. Magazine of Concrete Research, Vol.27, No.92, Sept, 1975, 151-160.
15. JAEGER J.C. and COOK N.G.W. Fundamentals of rock mechanics. Menthuen and Co, London, 1969.

16. KAPLAN S.A. Factors affecting the relationship between rate of loading and measured compressive strength of concrete. Magazine of Concrete Research, Vol.32, No.111, June 1980, 79-88.
17. ASHBEE R.A., HERITAGE C.A.R. and JORDAN R.W. The expanded hysteresis loop method for measuring the damping properties of concrete. Magazine of Concrete Research, Vol.28, No.96, Sept, 1976, 148-156.
18. JORDAN R.W. Tensile stress effects on damping. Concrete, Vol.11, No.3, March, 1977, 31-33.
19. JEARY A.P. Damping measurements from dynamic behaviour of several large multiflue chimneys. Proc.Instn.of Civil Engrs. Vol.57, 1974, 321-329.
20. JONES R. The effect of frequency on the dynamic modulus and damping coefficient of concrete. Magazine of Concrete Research. Vol.9, No.26, August 1957, 69-72.
21. COLE D.G. (1966). The damping capacity of hardened cement paste, mortar and concrete specimens. Proc. of a Symp. on "Vibration in Civil Engineering" Imperial College 1965. Ed. B.O. SKIPP, London, Butterworths, 1966, 235-247.
22. COLE D.G. and SPOONER D.C. The damping capacity of hardened cement paste and mortar in specimens vibrating at very low frequencies. Proc.Am.Soc. for Testing and Materials, Vol.65, 1965, 666-667.
23. SPARKS P.R. and MENZIES J.B. The effect of rate of loading on plain concrete. Magazine of Concrete Research. Vol.25, No.83, June, 1973, 73-80.
24. HUGHES B.P. and GREGORY R. Concrete subject to high rates of loading in compression. Magazine of Concrete Research. Vol.24, No.78, March, 1972, 25-36.
25. SAUCIER K.L. and CARPENTER L. Dynamic properties of mass concrete. Dynamic Geotechnical Testing, ASTM, Special Technical Publication STP.654, American Society for Testing and Mats. 1978, 163-178.
26. RICHART F.E., BRANDTZAEG A. and BROWN R.L. A study of the failure of concrete under combined compressive stress. Univ. of Illinois Engg. Experimental Station. Bulletin No.185, 1928, pp.102.
27. BRESLER B. and PISTER K.S. Failure of plain concrete under combined stresses. Transactions of the American Society of Civil Engineers. Vol.122, 1957, 1049-1059.
28. McHENRY D. and KARNI J. Strength of concrete under combined tensile and compressive stress. Proc. of the American Concrete Institute. Vol.54, April, 1958, 829-838.
29. GERSTLE K.H., LINSE D.H., BERTACCHI P., KOTSOVOS M.D., KO H.Y., NEWMAN J.B., ROSSI L.A. SCHICKERT G., TAYLOR M.A., TRAINA L.A. and ZIMMERMAN R.M. Strength of concrete under multiaxial stress states. Douglas McHenry Int.Symp. on Conc.& Conc.Structures. ACI publication SP-55 Oct, 1976, 103-131.
30. KUPFER H., HILSDORF, H.K. and RUSCH H. Behaviour of concrete under biaxial stresses. Journal of the American Concrete Institute. Proc. Vol.66, No.8, Aug, 1969, 656-666.
31. HUGHES B.P. and CHAPMAN G.P. The deformation of concrete and micro-concrete in compression and tension with particular reference to aggregate size. Magazine of Concrete Research. Vol.18, No.54, March, 1966, 19-24.
32. KUPFER H. The behaviour of concrete under multi-axial short term loading. Deutcher Ausschus fur Stahlbeton, Berlin, Heft 229, 1973.
33. KUPFER H.B. and GERSTLE K.H. Behaviour of concrete under biaxial stresses. Jour.Eng.Mech. Divn. Proc. of the American Society of Civil Engineers, Vol.99, No.EM4, Aug, 1973, 853-866.
34. SEED H.Bolton, DUNCAN J.M. and IDRISS I.M. Criteria and methods for static and dynamic analysis of earth dams. Numerical Analysis of Dams. Ed.Naylor, Stagg and Zienkiewicz. Proc.of a Conf, Swansea 1975, 564-588.
35. KOTSOVOS M.D. and NEWMAN J.B. Generalised stress-strain relations for concrete. Proc. of the American Society of Civil Engineers. Vol.104 No.EM4, Aug, 1978, 845-856. (Proc.Paper 13922).
36. KOTSOVOS M.D. and NEWMAN J.B. A mathematical description of the deformational behaviour of concrete under complex loading. Magazine of Concrete Research. Vol.31, No.107, June 1979, 77-90.
37. HUECKEL T. and NOVA R. On paraelastic hysteresis of soils and rocks. Bull de l'Acad. Polonaise des Sciences Serie.des sciences techniques. Vol.27, No.1, 1979, 49-55.
38. DOUGILL J.W. On stable progressively fracturing solids. Zeitschrift fur angewandte Mathematic und Physic (ZAMP) Vol.27, No.4, 1976, 423-437.
39. MAIER G and HUECKEL T. Non associated and coupled flow rules of elastoplasticity for rock-like materials. Int.J.Rock Mech.and Min.Sci. Vol.16, 1979, 77-92.
40. BAZANT Z.P. and KIM S.S. Plastic-fracturing theory for concrete. Jour.Eng.Mech.Div.ASCE, 1979. (In press).
41. VALANIS K.C. A theory of visco-plasticity without a yield surface - Archiwum Mechaniki Stosowanej. Vol.23, No.4, 1971, 517-551.
42. BAZANT Z.P. and BHAT P.D. Endochronic theory of inelasticity and failure of concrete. Jour. Eng.Mech.Divn. ASCE, Vol.102, No.EM4, 1976, 701-722.
43. BURT N.J. and DOUGILL J.W. Progressive failure in a model heterogeneous medium. Jour. E.M.Divn. ASCE, Vol.103, No.EM3, June, 1977, 365-376.
44. DOUGILL J.W., LAU J.C. and BURT N.J. Towards a theoretical model for progressive failure and softening in rock, concrete and similar materials. Mechanics in Engineering. Proc. ASCE-EMD. Conf.Univ.Waterloo Press, 1977, 335-355.
45. DOUGILL J.W. Some remarks on path independence in the small in plasticity. Quarterly of Applied Maths. Vol.33, No.3, 1975, 233-243.
46. Il'IUSHIN A.A. On the postulate of plasticity. Appl.Maths and Mechanics (PMiM), Vol.25, 1961, 746-752.
47. DOUGILL J.W. and RIDA M.A.M. Further consideration of the progressively fracturing solid. Jour.Eng.Mech.Divn. ASCE. (In press).
48. SANDERS J.L. Plastic stress-strain relations based on linear loading functions. Proc.2nd US Nat.Congress Appl.Mech. 1954, 455-460.
49. PRAGER W. A new method for analysing stresses and strains in work hardening plastic solids. J.Appl.Mech. Vol.23, 1956, 493-496.

50. ZEIGLER H. A modification of Prager's hardening hardening rule. Quart.App.Maths, Vol.17, No.1, 19 1959, 55-65.
51. MROZ Z. On the description of anisotropic work-hardening. J.Mech.Phy.Solids. Vol.15, 1967, 163-175.
52. HODGE P.G. Discussion of the paper by Prager. J.App.Mech. Vol.24, Sept, 1957, 482-483.
53. ARMEN J., PIFKO A.B., LEVINE H.S. and ISAKSON G. Plasticity. Ch.8. Finite Element Techniques. Ed.Tottenham H, and Brebbia C. Southampton Univ.Press, 1970, 209-257.
54. CAPURSO M. Extremum theorems for the solution of the rate-problem in elastic-plastic fracturing structures. Univ.of Bologna. Faculty of Engg, Technical Note No.33, ISCB, 1978, p.25.
55. VALANIS K.C. Observed plastic behaviour of metals vis-a-vis the endochronic theory of plasticity.
56. BAZANT Z.P. and SHIEH C-L. Constitutive behaviour of concrete: hysteretic fracturing endochronic theory. Jour.Eng.Mech.Div.ASCE. (In press).
57. CHRISTIAN J.T., ROESSET J.M. and DESAI C.S. Two and three-dimensional dynamic analyses. Numerical methods in Geotechnical Engg. Ed. Desai C.S. and Christian J.T., McGraw Hill, 1977
58. CLOUGH R.W. and ZIENKIEWICZ O.C. Finite element methods in analysis and design of dams. in Numerical Analysis of Dams, Ed.Naylor, Stagg and Zienkiewics. Univ.Swansea, 1975, 285-322.
59. BAZANT Z.P. and CEDOLIN L. Blunt crack band propagation in finite element analysis. Jour. Eng.Mech.Divn. ASCE. Vol.105, No.EM2, April, 1979, 297-315.

15 Failure and damage: earth and rock fill

D. A. HOWELLS, MA,FICE, Consulting Engineer, London

The stability of engineering structures may be studied either by means of stress analysis or by the limit equilibrium method. In the case of earthquake loading of earth and rock fill dams, the state of stress which might lead to failure depends on the response of the dam to the earthquake shaking. The failure is likely to be progressive and will take time. If the duration of the earthquake is less than the time needed for failure then damage to the dam, or perhaps even failure, after the earthquake is over may be expected.

INTRODUCTION

1. Engineers trying to analyse the behaviour of their constructions have long taken one or other of two approaches. The first has been to calculate the stresses produced within the structure by the external loads and to proportion the structure so that the stresses do not exceed permissible values derived from experience and the testing of materials. The second has been to consider the various forms of failure which have afflicted constructions of a particular class, to consider the equilibrium of the external loads and internal forces acting at the moment of failure and to proportion the structure so that these loads are greater by a factor than the loads which will act on the structure in specified circumstances.

2. The former approach was usually found more appropriate in traditional engineering practice but not in special cases such as the buckling of slender struts and even here it was common to disguise a limit equilibrium method to look as if it were one of stress analysis. On the other hand, the development of soil mechanics over the past half century has depended strongly on the limit equilibrium method. Over the years, both methods have been refined and associated with two theorems of plasticity theory relating to the lower and upper bound of the failure load.

3. In the case of earthquake loading of earth and rock fill dams, the state of stress which might lead to failure depends on the response of the dam to earthquake shaking at a stress level at which the behaviour may perhaps be treated as quasi-linear, but in the saturated parts of the dam the non-linearity may be enough to cause a build up of pore pressure. The failure is likely to be progressive and the duration of the earthquake may determine whether there is failure or damage short of the state of collapse.

BEHAVIOUR OF MATERIALS AND STRUCTURES

4. Stress analysis is based on the convention of continuity. The properties of the materials are supposed to be those of a point and the behaviour of a structure determined by integration over all the points in the structure. In fact, earth and rock fill dams, treated here as structures, are made of mineral particles of varying size pressing on one another and with voids between them which may be filled by either air or water. When properties of materials are deduced from the behaviour of test specimens the size and shape of these specimens are chosen with the intention that the hypothetical point properties of a continuum of material similar to the specimen may be fairly easily calculated. In fact, though, the behaviour of a test specimen is of comparable complication to that of a full scale engineering structure and calls for comparable detailed analysis. Tchalenko (ref. 1) has shown how similar shear zone structures are developed on the microscopic scale in the shear box test, on the intermediate scale in the Riedel experiment and on the regional scale in the earthquake fault.

5. Most of the materials in the shell of an earth or rock fill dam are compacted and may be expected to show shear brittle behaviour, but this may not be necessarily so for a clay core. These brittle materials show a markedly non-linear behaviour even at quite low stress levels before the peak strength is reached and are likely also to show an irreversible volume decrease depending in complicated manner on the three dimensional stress state. After the peak strength has been passed and at the large strains associated with the development of shear zones they are likely to be dilatant.

6. The behaviour described above is that of the mineral skeleton. In an earth or rock fill dam the shell materials may be dry or may be saturated. If a material is saturated, then the pre-peak compression will be associated with an increase in pore pressure and its irreversible component will cause, during earthquake shaking, an irreversible rise of pore pressure, thus decreasing the factor of

safety which depends on the effective stress state, while a fall in pore pressure will increase the factor of safety. These positive or negative pore pressures will be dissipated during a time which may be long or short relative to the duration of an earthquake depending on the permeability of the surrounding material: where negative pore pressures are developed in a shear zone the material immediately surrounding the shear zone is stressed to below peak strength and may itself be subject to positive excess pore pressures. The coupled behaviour of the mineral skeleton and the pore water is so complicated that they are normally thought of separately and uncoupled. In some cases this may be an oversimplification.

7. In estimating the hypothetical point properties of a continuous material subject to earthquake shaking allowance must be made for the fact that S waves are transmitted through the mineral skeleton of a saturated material while P waves are transmitted by the combined action of the skeleton and the pore water. Theoretical consideration of progressive or standing waves is very difficult unless the material properties are linearised in the form of stiffness and viscous damping.

8. The essence of the stress analysis approach is to calculate the load-deformation behaviour of complete assemblies by summation or integration of the stress-strain behaviour of the material. In the case of framed structures with a few major redundancies, it is relatively easy to visualise the way in which the force-deformation behaviour of the structure is derived from the stress-strain behaviour of the material, usually simple ductility, for example by means of the step by step process described by Massonnet and Save (ref. 2). As each plastic hinge forms the rest of the structure, now of lower stiffness, continues to behave elastically until the final hinge reduces the stiffness to zero. In the case of a plastic continuum, it is usual to replace the integration by summation over a number of finite elements.

9. Zienkiewicz et al. (ref. 3) treat among many other matters the case of simple slope of uniform material. In this case, the material is treated as elastic and visco-plastic, but may have different flow rules. The simplicity of the structure and the absence of constraint on volumetric deformation lead to strain rate contours which approximate closely to the failure mode which assumes a circular slip and may be investigated by limit equilibrium analysis. The factor of safety also approximates to that found from limit equilibrium.

10. For materials which show shear brittle behaviour, things are more complicated. They show quite markedly the tendency for shear deformation to become localised in shear zones. As the force acting on the structure increases, the shear strains increase and, because of the non-linear behaviour, these become significant as the peak strength is approached. For analytical exercises in which the material is assumed to be of constant strength throughout, this occurs at points of high stress concentration. In real materials, the strength varies from point to point and the initiation of large shear strains is likely to occur at a point where high imposed stress and low strength coincide.

11. Increasing shear deformation is then concentrated in a narrow shear zone which, in due course, extends to form a closed boundary on the surface of the structure and a slip surface within the structure. The shape of this slip surface when fully formed is the failure mode. Little is understood about the formation of such continuous shear zones. In some cases, it can be explained on a step by step basis comparable to that of the development of a mechanism in a framed structure described in paragraph 8. In others, it appears to take on a form leading towards a simple complete mechanism at an early stage. Rudnicki and Rice (ref. 4) suggest that localisation can be explained as an instability in the general constitutive description of inelastic deformation of the material. It is interesting to compare this approach with the approach of Bazant and Panula (ref. 5) whose model uses a set of concrete elements in parallel subject to compression and having random distribution of strength. They are concerned with the failure load of the system rather than the failure mode. This gap between the stress analysis approach to failure and the limit equilibrium method is the main cause of their continuing separate existence in relation to static loading. In the case of earthquake loading another complication arises. Up to the moment of failure the response of the dam to earthquake shaking is determined by stress analysis. At the moment of failure, the deformations in the shear zone become a dominant factor and the sliding mass has to be treated as a rigid body. So the parameters governing the shaking of the dam just before failure are frozen and taken into the limit equilibrium calculations as equivalent static loads.

FAILURE OF AND DAMAGE TO DAMS

12. The previous discussion deals with modern developments comparable to the "permissible stress" approach to engineering analysis. The other engineering approach, the study of failures to discover features they may have in common with a view to preventing them in future, is a much more rough and ready affair because it is covered by so many unquantifiable variables. Either by good luck or good judgment, the number of cases available for study is quite small. Dams are designed not to fail and the success of the designer hampers the empirical study of failure. Fairly large scale model tests on large shaking tables appear to offer the only way out of this dilemma. Statistics of the frequency of occurrence of a particular failure mode are likely to be strongly governed, inversely, by the attention paid to that failure mode by designers in the past. Design philosophy and construction practice change with time and we can be sure that those current

today will be superseded. The failure mode may be obscured by past-failure occurrences. There is a strong temptation to recognise in the failed mass failure patterns which have very little justification. Seed (ref. 6) in his 1979 Rankine lecture indicates the considerable effort that went into the investigation of the failure of the Lower San Fernando Dam and the identification of the form of failure.

13. From the practical point of view, failure may be defined as the loss of the ability of the dam to retain the reservoir at its design water level. In terms of theoretical soil mechanics it may be defined as the state at which the deformation ceases to be related to the load. Damage may be defined as any irreversible deformation other than that occurring before and during the first filling but also includes unacceptably large deformation during filling. Damage modes are not normally considered separately from failure modes but are rather less developed versions of the same deformation pattern.

14. On the other hand, one of the most commonly observed modes of earthquake damage to an earth fill dam is the appearance of cracks parallel to the crest of the dam and a short distance below it. They arise from the transverse extensional displacements of the dam caused by the amplified dam shaking. They may exist as damage features in their own right or they may act as initiators of other failure modes. It does not seem that they can be prevented, even for mild earthquakes, but they can be minimised by thorough compaction, setting up a high horizontal in-situ compressive stress not far below the surface.

15. Seed (ref. 6) lists nine possible ways in which an earthquake may cause failure of an earth dam: these are not repeated here. They all call for engineering decisions to minimise the failure risk but only those concerning the stability of slopes lend themselves to the application of engineering mechanics.

16. Potential liquefaction of the material of an earth fill dam is something to be avoided rather than analysed. In the case of older dams which may contain zones of potential liquefaction analysis may be helpful towards taking decisions about remedial work.

UPPER AND LOWER BOUNDS OF FAILURE LOADS AND MODES, INCREMENTAL COLLAPSE

17. The two approaches stress analysis and limit equilibrium analysis are parallelled in the study of theoretical plasticity of a continuum by the lower and upper bound theorems of plastic collapse. However, the uniqueness of plastic collapse loads has been proved only for ideal materials with somewhat unrealistic properties and has not been proved for shear brittle materials which usually form the shell and perhaps other parts of earth and rock fill dams. In particular relative displacements across shear zones may not be infinitesimal. Where uniqueness cannot be proved, the study of collapse loads must be closely related to detailed consideration of observed collapses while stress analysis must model step-by-step the actual loading pattern which may be expected to occur and against which the designer must make provision.

18. The concept of a collapse load at which the deformations are indeterminate has much in common with the question of the collapse load of a strut and it appears that the approximation principle used with Rayleigh's method may also apply to plastic collapse loads. This is that if the form assumed for the characteristic failure mechanism is a close approximation to the actual failure mechanism then the failure load calculated from it will be a very close approximation to the actual failure load. This is an assertion offered here without proof. But the limit equilibrium method should be used with an awareness of its limitations. Inherently, it gives higher estimates of the collapse load than the actual value and more approximations are introduced by the various detailed procedures for calculating the collapse load for any particular failure mode. The justification for limit equilibrium analysis must be empirical. The evidence is sparse and can easily be misinterpreted. The link with the upper bound theorem of plastic collapse is strong, but this should not lead to overconfidence in use.

19. During the development of the plastic theory of framed structures, the phenomenon of incremental collapse was studied. This dealt with the effects of cycles of loading, each of which stressed the structure into the plastic range. At high loads, less than the plastic collapse load, such a repetition of loading would lead to a repetition of plastic deformation in one direction culminating in collapse. At lower loads, but still in the plastic range, the development of plastic hinges did not go as far as collapse and a stable configuration was found. This is rather ambiguously referred to as "shake down" in the sense of "settle down". The main thrust of the studies was to find a method of applying the upper bound theorem to the calculation of the shake down load to avoid extremely long step-by-step computations. This work has not been applied in earthquake engineering largely because the coming of the computer has eased step-by-step calculations and the interest in assessing actual values of the deformations.

PROGRESSIVE FAILURE

20. A small step of increasing load may cause the stress at a point of a shear brittle material to become equal to the peak strength at that point. This may be because the point is a weak one or because of a concentration of stress. The decrease of stress resulting from the brittleness leads to an increase of stress at points nearby, together with an adjustment of deformations to give a new stable configuration. This takes a small finite time but no time dependent mechanism has yet been

formulated. A monotonic increasing load leading to the development of a shear zone requires a specific finite time to cause failure. If the time of application of the load is less than the time to failure this will not take place but the structure is weakened.

21. Related to this is the time required for the evolution of pore pressures in saturated materials during and after the earthquake loading. During the formation of a shear zone more and more of the material in the shear zone is dilating so that negative pore pressure is set up. The state of effective stress is thus safer than it would otherwise be. The negative pore pressure will suck in water from the surrounding material, which itself is in the pre-peak state and may well be subject to positive pore pressure, at a rate which depends on the permeability of this surrounding material.

22. It is not difficult to imagine circumstances in which a dam with a static factor of safety just above unity is weakened by the earthquake shaking and strengthened temporarily by negative pressures in the shear zone. After the earthquake is over the temporary strengthening dissipates and large irreversible deformations, even up to failure, take place.

DAMAGE

23. An earthquake shaking lasts for only a fairly short time and a structure which would fail if the same level of shaking were to continue may show considerable displacements, but not fail during a shorter earthquake. It is necessary to try to estimate what displacement to expect. Before this can be done it is necessary to choose realistic parameters to characterise potential future earthquakes for which the designer must provide. Peak acceleration alone is not enough. Takizawa and Jennings (ref. 7) have produced a two parameter formulation including duration defined from the energy flux-time graph which appears to be suitable for earth and rock fill dam applications. For analyses based on time histories of past earthquakes the problem does not arise.

24. Displacement analyses are in general based on representing the characteristic mode of failure as a block sliding upon an inclined plane. The model was first proposed by Newmark and the most recent refinements are presented by Sarma (ref. 8). The critical horizontal acceleration is determined for a critical slip circle for a particular structural section of an earth or rock fill dam, taking account of the variation of structural response with the height of the dam. For a particular earthquake accelerogram the excess above this critical acceleration will constitute a downslope driving force and a double integration will give the displacement. A number of calculations for combinations of different critical accelerations and earthquake traces indicates the trend of displacements.

25. The chosen mode of failure may be realistic for superficial slope failure. It may well represent a close approximation for deeper slips but calibration against observation of real earthquake behaviour augmented by large scale model experiments on a large shaking table would be needed before it can be accepted with confidence. In the case of superficial slope failure, relatively small scale model experiments by Goodman and Seed (ref. 9) tend to confirm the applicability of these calculations.

CONCLUDING REMARKS

26. Analytical techniques for identifying the important characteristics of material behaviour and of structures, a term which includes test specimens and earth and rock fill dams, have developed considerably over the past twenty years and have contributed to the understanding of the behaviour of dams subjected to earthquake. A particular need is for the validation of the approach to the calculation of displacements during an earthquake by large scale model tests and by using it as a framework for the observation of dams which are damaged but do not fail.

REFERENCES

1. TCHALENKO J.S. Similarities between shear zones of different magnitudes. Geological Society of America Bulletin, 1970, 81, June, 1625 - 1640.
2. MASSONNET C. and SAVE M. Calcul plastique des constructions. Centre Belgo - Luxembourgeois d'Information de l'Acier, Brussels, 1967.
3. ZIENKIEWICZ O.C., HUMPHESON C. and LEWIS R.W. Associated and non-associated visco-plasticity and plasticity in soil mechanics. Geotechnique, 1975, 25, 4, 671 - 689.
4. RUDNICKI J.W. and RICE J.R. Conditions for the localization of deformation in pressure-sensitive dilatant materials. Journal of the Mechanics and Physics of Solids, 1975, 23, 371 - 394.
5. BAZANT Z.P. and PANULA L. Statistical stability effects in concrete failure. Journal of the Engineering Mechanics Division, American Society of Civil Engineers, 1978, EM5, September, 1195 - 1212.
6. SEED H.B. Considerations in the earthquake-resistant design of earth and rock fill dams. Geotechnique, 1979, 29, 3, 215 - 263.
7. TAKIZAWA H. and JENNINGS P.C. Collapse of a model for ductile reinforced concrete frames under extreme earthquake motions. Earthquake Engineering and Structural Dynamics, 1980, 8, 2, 117 - 144.
8. SARMA S.K. Response and Stability of earth dams during strong earthquakes. U.S. Army Waterways Experiment Station, Vicksburg, 1979.
9. GOODMAN R.E. and SEED H.B. Earthquake-induced displacements in Sand embankment. Journal of the Soil Mechanics and Foundation Engineering Division, American Society of Civil Engineers, 1966, 92, SM2, 125 - 146.

16 Thirty years of research work on earthquake resistance of hydraulic structures in China

C. SHEN, and H. CHEN, Scientific Research Institute of Water Conservancy and Hydroelectric Power, China

This paper briefly discribes the main earthquake resistant research activities on hydraulic structures during the past thirty years since the founding of The People's Republic of China. It includes following topics such as: The engineering aspects of earthquakes in China; damages of hydraulic structures in severe earthquakes; main features in the first Earthquake Resistant design Code of China for Hydraulic Structures; up to date research works and future planning.

China is a country with high seismicity. As early as B.C. 1189 record of earthquake activity was found in history. According to imcomplete statistics up to date, 705 strong earthquakes with magnitudes greater than 6 took place from B.C. 780. The regions where earthquakes had occurred, almost covered the whole territory.

Subsequent to the founding of New China in 1949, recent years have witnessed more frequent earthquakes in China. There were 344 strong earthquakes with magnituses greater than 6, and 40 greater than 7.

Fig. 1 shows clearly the location of epicenters of strong earthquakes of magnitudes greater than 6. It reflects the peculiarities and areal distribution of seismicity in China.

All the earthquake's in China are tectonic earthquakes. The Chinese Bureau of Earthquakes organised some relevant Institutes to carry out extensive studies on historical earthquakes, geological conditions of seismicity, active faults, and recent seismic activities. They divided the whole country into 10 seismic zones and 30 seismic belts, and the seismic activities into 4 classes.

Class I: In this region, there were plentiful earthquakes with magnitudes greater than 7 and frequent earthquakes of magnitude greater than 6. Earthquakes of magnitude 8 were also found.

Class II: In this region, earthquakes of magnitude 6 were plendiful and frequent, some stronger than M=7.

Class III: In this region, earthquakes greater than M=6 were found with low frequency, and some were as strong as M=7.

Class IV: In this region, earthquakes of M=6 were found very few, and no earthquakes were found greater than M=7.

Most earthquakes in China are of shallow foci. The focal depth ranges from 10 to 45 Km. Some earthquakes of hypocenters deeper than 5oo km occurred in Prov. Heilongjiang and east Jiling.

The scale of intensity of earthquake used in China is divided into 12 degrees and has a value of relative acceleration for each degree similar to the Modified Mercalli scale widely used in many countries.

Fig. 2 shows the predicted intensity of earthquakes based on the above regionalization. In the design of dams and barrages one could find out from this diagram the earthquakes of where the structures are located.

When fundamental intensity is defined according to the engineering geology, hydrogeology, the importance of project as well as the property of rock foundation. In general, an increase or decrease by 1 degree may be effected as against the original intensity. For very large and important project concerned with the safty of lives and properties, 1 degree is added as a rule in the design for earthquake protection.

* * *

Before 1949, there were a few dams and barrages experienced strong earthquakes in China. After liberation, more than 84,000 reservoirs were built. They include 311 large reservoirs (storage capacity more than 100 million M^3) and 2200 medium reservoirs (storage capacity between 100 million M^3 and 10 million M^3). Large barrages are more than 200. There are scattered across the land, more chances to experience strong earthquake. However, no collapse of dam due to earthquake is found up to now.

Some typical examples of strong earthquakes greater than M=6 which caused damages to dams and barrages are tabulated below:

There are some examples of earthquakes induced by reservoirs impouding in China. The most famous and serious one is Hsinfengjiang Reservoir, which caused a strong earthquake of M=6.1, the details of which have been published elsewhere, no further account will be given here. Another 8 reservoirs had similar pheomna. Seven induced earthquakes occurred immediately after impounding. Only Danjiangkuo Reservoir(storage capacity 20.9 billion M^3, height of dam 97M) had a delay of 3 years. It impounded water in

1967 and some weak earthquakes were induced in 1970. It was only in Nov. 1973 that an of M=4.7 occurred. As the 8 examples are only of weak earthquakes, no enough attention was attracted, nor observation and analysis were made. However, there were some earthquakes which might not be induced by impounding of reservoirs.

These reservoirs are located on the seismically inactive regions, having tectonic faults. The earthquakes were not directly related to the capacities and depths of reservoirs. Although some reservoirs have large capacity, earthquakes occurred when the impounded water was not very deep. For instance, in the Hsinfengjiang Reservoir the earthquake began to be induced when the water depth was only 20 M. Some small reservoirs with only capacity of 15 to 20 million M^3 were also suffered from induced earthquakes.

The research on the mechanism of reservoir-induced earthquakes has made progress, but it still remains unclarified especialy the relationship between faults and water action are pending further investigations. The influence of rock type is quite ambigous. Seepage is serious in karst region, but direct influence cannot be confirmed.

Most dams in China are of earth dam. During strong earthquake rather serious damage might occur in this type of dam. According to the imcomplete statistics about 112 earth and rockfill dams had damages from 185. The highest earth dams high is Baisha at Miyuon County (H=66M). Most earth dams have height about 20 to 30 M. The maximum intensity is 10 degree.

The damages of earth and rockfill dams were next type in general:

1) Sliding: It is to be noted that the saturated sand and sandy gravel in shell of the dam happen to slide at low intensity. For example, during the earthquakes in Bohai, Haicheng & Tangshan. With the intensity of 6-7° there where 5 dams shell(saturated sand and sandy gravel)slided. The protective layer of sand and gravel of 2 dams also slided. The major cause, as we considered, was due to the liquefication of saturated sand and gravel. (Fig.3)

2) Cracks:. This is the most popular phenomenon among which longitudinal cracks might occur frequently sometimes they are also of transversal cracks at high seismic intensity. The widest crack amounts to 80 cm. at Douhe reservoir after Tangshan's earthquakes. If the unequal settement of foundation is quite remarkable transversal cracks often occured. Both the width and depth of cracks are closely related to thickness of alluvium, quality of construction, as will as the soil properties of earth dam.

3) Filtration: The discharge of filtration usually increased after earthquakes in a short duration. Sometimes muddy water might happen consequently. Afterward filtration could stop to continue automatically.

4) Settlement: Serious damage of earth dam always accociated with settlement. For instance, at Haicheng's earthquake the Sandaoling earth dam with central core (H=17M) with an intensity 8°. The settlement of the shoulder of dams was 20-22 cm. A longitudinal crack crossing the whole dam was found in the vicinity of upstream dam shoulder. (Fig.4)

5) Damage of Accessory Structure: Most of the parapet wall situated on the top of dam was made of masony work, which was a cantilever structure, being subject to larges acceleration during earthquake. It was found that many cases were under the threat of serious damage. For instance, almost half the length of parapet wall of Douhe Reservoir with a dam of 22 meters in hight was destroyed during the Tangshan's strong earthquakes having a intensity of 9°. (Fig.5)

General speaking, the concrete and rock masony dam, did not suffer from serious damage during these strong earthquakes. As we previously mentioned the buttress dam(H=102M) of Hsinfengjiang reservoir has a longitudinal crack with a length 82 M due to the concentration of dynamic stress under the intensity 8 degree. We considered that this crack hapened to be across all the section.

During Hsicheng's earthquake M=7.3 the

Location	Province	Date	Magnitude	Focal depth (Km)	Intensity at the epicenter (degree)
West part of Bachu County	Xinjiang	13 April 1961	6.8	20	9
East part of Longyao County	Hebei	8 March 1966	6.8	20	9
Between Ningjin and Dongwang Cunties	Hebei	22 March 1966	7.2	10	10
Bohai Gulf		18 July 1969	7.4	35	
Yangjiang County	Guangdong	26 July 1969	6.4	5	8
Tonghai County	Yunnan	5 Jan. 1970	7.7	5	10
Haicheng County	Liaoning	4 Feb. 1975	7.3	12	9+
Longling County	Yunnan	29 May 1976	7.3	20	9
Tangshan County	Hebei	28 July 1976	7.8	16	11
Songpan County	Sichuan	16 Aug. 1976	7.2	23	9
Liyang County	Jiangsu	9 July 1979	6.0	12	8

concrete gravity dam Shenwao (H=52.6 m) had a distance 75 Km from epicenter, the intensit was about 6 degree. During earthquake black water emerged from the drainage galleries at the bottom of dam. Six section crack opened 0.1-0.2 mm. and expansion joint opened 1-3 mm. 68 pieces of railings on the top of dam were brocken.

During Tonhai's earthquake the masony gravity dam of Redflag reservoir(H=35 M) had a intensity 7°. After earthquake 4 sections on the right bank were cracked, and water discharge was found to be as much as 20 l/s. Drainage gallery had some filtration.

During Tangshan's earthquake, there are 5 massony gravity dams caused damage. A dam 39.6 M in height and with an earthquake intensity 8° is a typical example. After earthquake, there were 4 transversal cracks on the bank of dam. Water filtrated. Another massony arch dam Siathinze(H=20.5 M) had an intensity 7° after earthquake on the bank apeared transversal crack. Between spillway and non-spill dam section transversal crack with a width of 2-3 mm also apeared and water filtrated.

Barrages had more damage during the strong earthquake. With referance to an incomplete statisties more than 100 small barrages were destroyed. Usually, less damage would be found on rock foundation, whereas more incidents would occur on soil foundation. Due to unequal settlement of foundation, the gate chamber of barrages was always suffered from earthquake seriously. Sometimes the load on the bridge acrossing the piers was very large and the sections of piers were comparatively small. During strong earthquake suport of piers would fail due to shearing , thus causing the collapse of bridge. For instance, during Tangshan's earthquake 20 small barrages out of 33 were seriously damaged,a barrage on the Ji Canal located in a region of 7° intensity, the piers were brocken off and the girders of the bridge were completely collapsed.

Beginning from the fifties the aseismic design of dams and barrages in China first followed the static theory. In the sixties the dynamic model tests and field vibration experiments in combination with preliminary dynamic analysis had been carried out to meet the need for the strengthning of Hsingfengjiang buttress dam after severe earthquake. Moreover the dynamic triaxial tests for the materials of earth dams have been undertaken at the same time. The first Earthquake Resistant Design Code for Hydraulic Structures in China was issued in 1978 for trial implementation. The main particularities of the code are as follows:

1. The analyses of dynamic behaviour and earthquake response of various types of hydraulic structures based on dynamic theory have been made, and the earthquake loads have been determined through summing up and simplefying the results of these analysis. The proposed distribution diagrams of earthquake loads over the height of various structures are simple for use and approximate to the actual response. According to the experiences drawn from earthquake damages and in the design of hydraulic structures the values less than the average of ground acceleration recorded during earthquakes have been adopted as a seismic factor in design. Special dynamic model test and analysis are required for very large and important water-retaining structures.

2. Some transient tension stresses in the concrete gravity and arch dam bodies are allowed when the earthquake resistance ability of these structures is checked, but the value of safety factor must be greater than 2.5. The allowable compression stresses in dynamic analysis may be 30% greater than that under the static loads.

3. The effect of elastic deformation of dam has been taken into account for determination of hydrodynamic pressure, but the compressibility of water is neglected.

4. The aseismic engineering measures is emphasized. On the basis of the experiences in engineering practice and in earthquake resistant design accumulated both in China and abroad some proposals were given.

The main organizations involved the research of earthquake resistance hydraulic structures in China are: Research Institute of Water Conservancy and Hydroelectric Power; Institute of Engineering mechanics, Academia Sinica; Research Institute of Development of Yangtze River; Nangjing Research Institute of Water Conservancy; Tsinghua University; Dalian Institute of Technology; East China Institute of Hydraulic Engineering etc.

The main research activities can be mentioned below:

1. Observation of strong motion. Since the sixties strong motion accelerogrophs have been installed in such large concrete dams as Hsingfengjiang, Fengman, Liujiaxia, Danjiangkou etc., as well as in such earth dams as Huangbizhang, Miyun, Guanting, Douhe etc.. In the great majority of cases the multi-channel optical recoding instruments made in China have been used. Some accelerograms of intensity IV-VI at dam sites and on the dam body proper were obtained in recent years.(Fig.6) On the basis of the recoded accelerograms, the ground motion at dam site was investigated and the dynamic method of analysis was checked as well.

With the accelerogram recorded at the dam site taken as ground motion imput, the acceleration response on the top of Hsingfengjiang buttress dam was calculated in order to check the dynamic analysis method. The result of comparision shows good agreement between the calculated response of dam and the actual recorded response.

2. Field vibration experiment. Field vibration experiements were carried out on the dams of various types. Usually the following types of exciting approaches are used, such as ambient tremble, explosion underground and under water, natural microearthquake, and mechanical exciters etc.. Now the exciters with frequency up to 20HZ have been produced. By utilizing the dynamic characteristics determined from the ambient tremble measurements and treating the recorded waves as the ergodic stationary random process, the

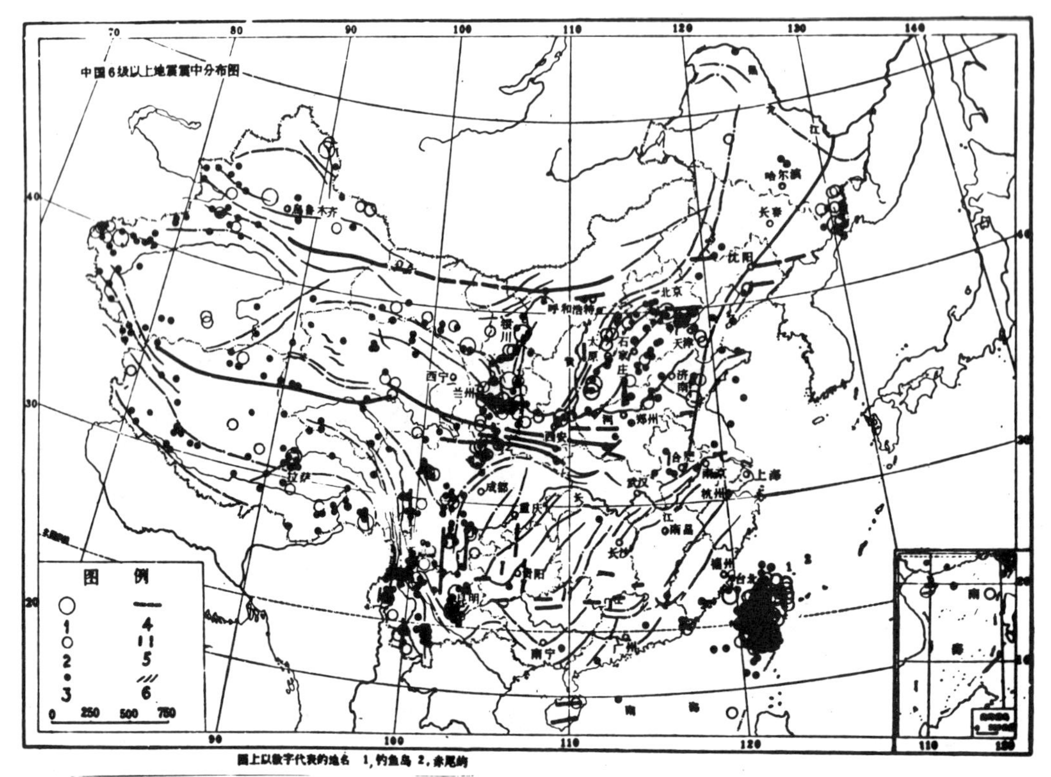

Fig. 1 Distribution of epicenters with M 6 in China
1. M=8-8.5 2. M=7-7.9 3. M=6-6.9 4. EW tectonic zone
5. NS tectonic zone 6. Other torsional structures

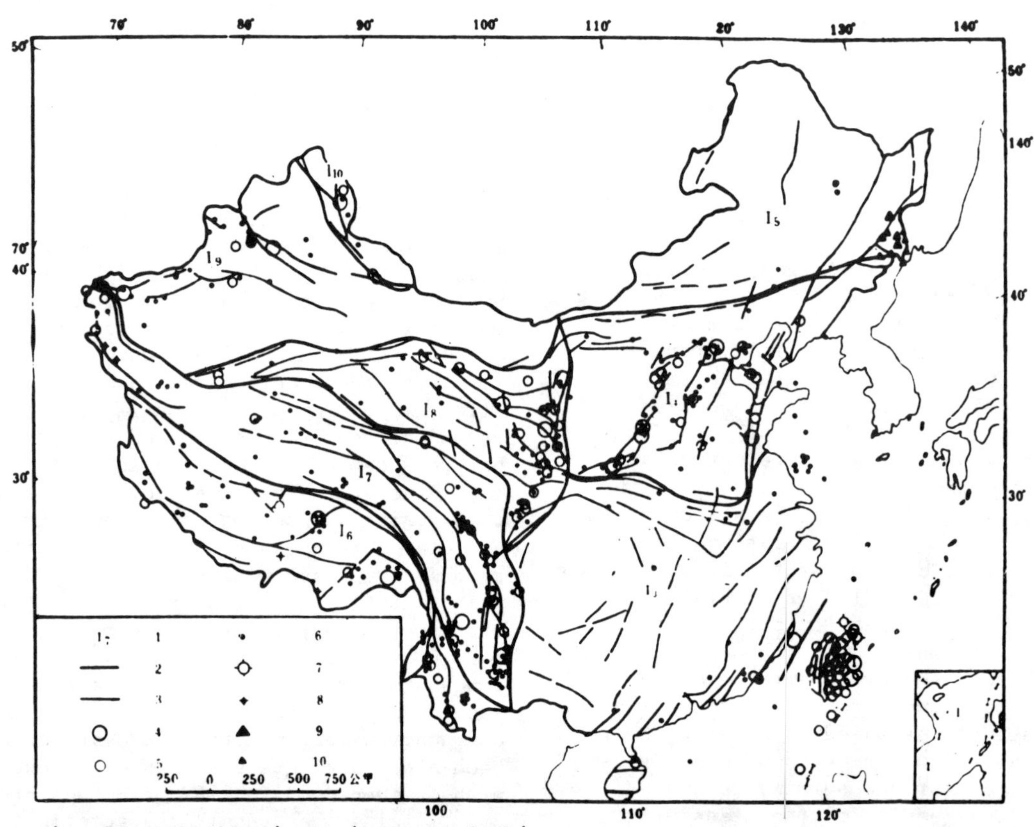

Fig. 2 Earthquake intensity map of China
1. No. Seismic zone 2. Boundary of seismic zone 3. Active faults (focal depth h 70 Km): 4. M 8 5. M=7-7.9 6. M=6-6.9(h=70-300 Km): 7. M=7-7.9 8. M=6-6.9(h=300 Km): 9. M=7-7.9 10. M=6-6.9

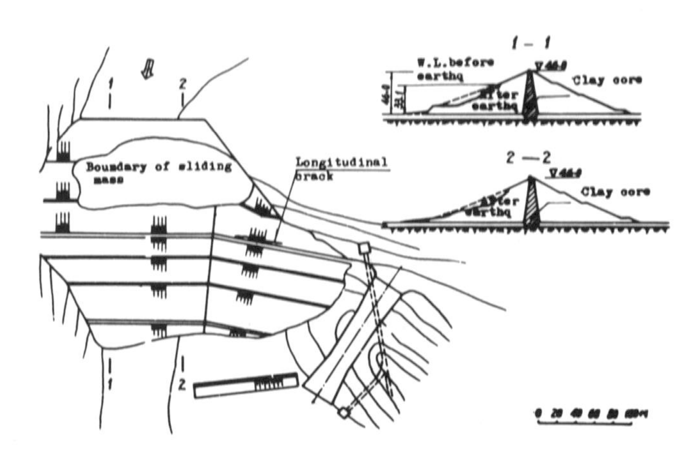

Fig.3 Sketh of earthquake damage of Shimen earth dam during Haicheng earthquake in 1975 (Earthquake intensity VII)

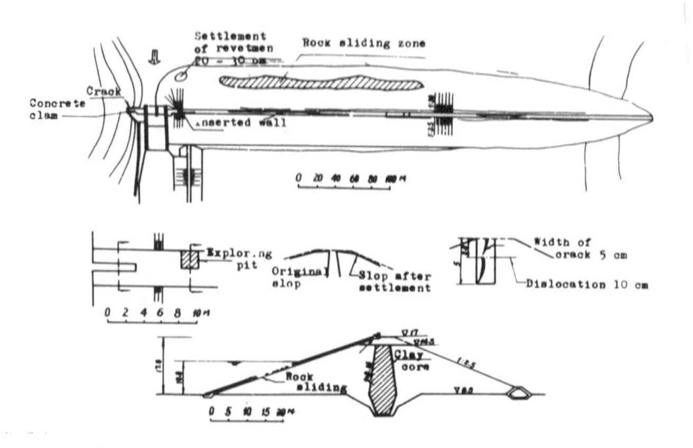

Fig.4 Sketch of earthquake damage of Sandaoling earth dam during Haicheng earthquake in 1975 (Earthquake intensity VIII)

Fig.5 Collapse of parapet on the top of Douhe dam during Tangshan earthquake in 1976 (Earthquake intensity IX)

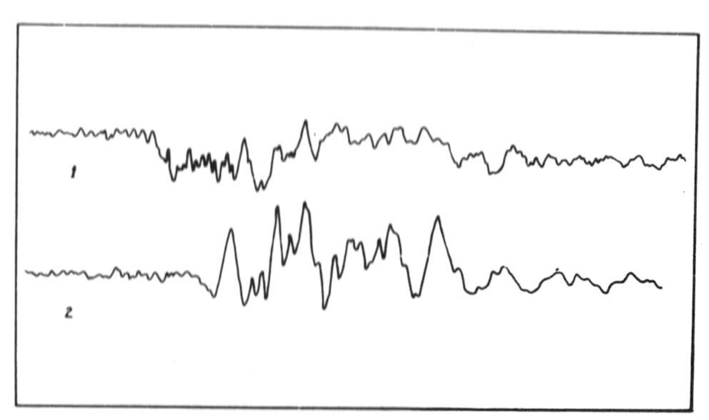

Fig.6 Accelerogram of Douhe Earth Dam in Tangshan

M=5.6 1979.9.2

1. Base of dam (0.1 mm/gal)
2. Top of dam (0.18 mm/gal)

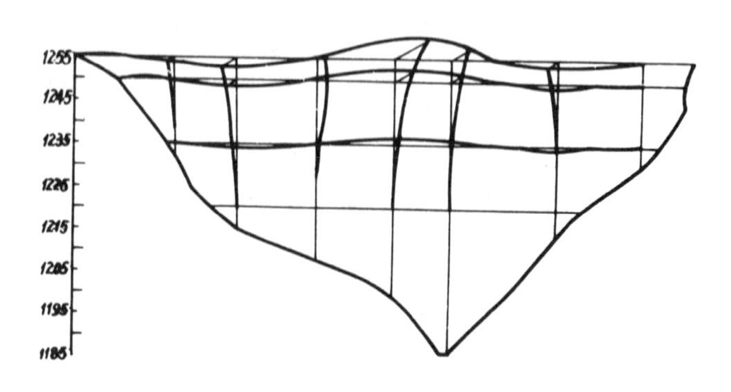

Fig.7 First Symmetrical mode shape of Hengshan arch dam (field vibration experiment) (Radial displacement, radial exciting)

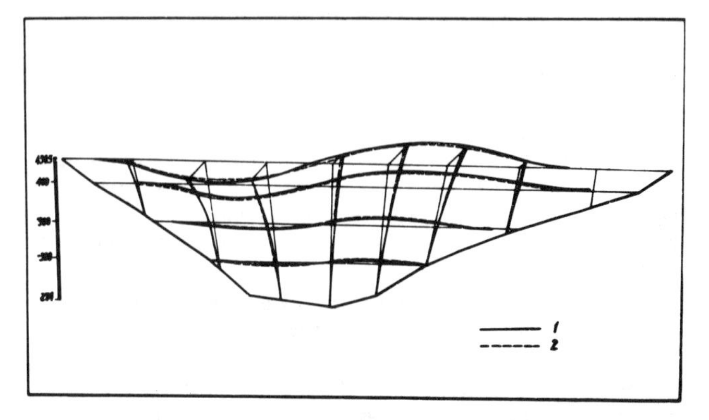

Fig. 8 First antisymmetrical mode shape of Baishan arch dam (3 dimentional dynamic calculation)

(Radial displacement)

1. With reservoir empty
2. With reservoir full

correlative function, power density spectrum and Fourier spectrum were calculated.

The mode shapes with reservoir empty of Hengshan double curvatural arch dam (H=69 m) are shown in Fig. 7. Results measured in field are in good agreement with those obtained from model tests in laboratory.

3. Model test in laboratory. A small-sized electromagntic shaking table simulating earthquake waves was instabled in 1966. Since the exciting force, load capacity and surface area of this shaking table are too small, to establish a medium-sized electro-hydraulic shaking table simulating earthquake waves for model tests of dams is now under consideration. Besides, the multiexciters technique is also applied to the fixed models of dams together with reservoirs.

The measurements made in tests involve frequencies, modes, dynamic stresses, hydrodynamic pressure etc.. The technique of three dimentional electric analogy is going to be used to determine the influence matrix of hydrodynamic pressure.

In order to meet the requirements of dynamic test with reservoir full, some special materials for model must be used. The brittle materials with low elesticity such as gypsum with iron filings, hard rubber containing lead filings, emulsion material with iron filings, epoxy resin with plasticine and iron filings etc. have been adopted in dynamic model tests. For all of these materials the dynamic simulality can be held and the accuracy in measurement can be satisfied.

In addition, the test with gelatin model using direct photographic technique has been performed to determine dynamic characteristics. The work on the leser hologram of vibration of dam model is being started, and an experiment of arch dam by this approach is also being carried out.

Recently a large scale model of Hsingfengjiang buttress dam of 4.5 m in height made of pumice concrete was built in the field. The earthquake response of the model dam was to be unveiled by the simulating earthquake explosion. A sluicing-siltation dam model of 14 m in height was also built in the field for the investigation of the liquefaction property of the dam by explosion.

Researches on the liquefaction property of the saturated sand and gravel have been conducted in connection with analyses of earthquake damage during Tangshan and others severe earthquakes. Apparators for two-direction cyclic triaxial test and simple shear test have been developed. The apparatus for cyclic torsional test and the large sized triaxial test mechine for testing gravelly materials are now under development.

4. Dynamic analysis. Dynamic analysis of dam using finite element method have been intensively conducted. More generally, the method of the mode superposition and step-by-step integration are adopted to solve the dynamic equation of dam. The general methods used to solve eigenvalue problems are the subspace iteralive procedure, orthogonal transform method and direct verse iterative procedure. The corresponding computer proframing has been made out. The studies of such problems as the dynamic analysis of arch dams and the effect of dam-reservoir interation are now in progressing. The researches on the effect of dam-foundation interaction and the nonlinear earthquake response of concrete dams have also been started. (see. Fig. 8)

As for the earthquake response analysis of earth dams, the limit equilibrium theory is used as a criterion to evaluate the incipient failure condition of the saturated sand under cyclic loading. Based on this consideration, the corresponding laboratory test and analytical methods are proposed. In the analysis of earthquake response of earth dam, the earth materials are assumed to be visco-elastic and the finite element method is employed. A preliminary computation scheme of dynamic analysis of the saturated sand mass by taking into account of the generation, dispersion and disipation of the pore pressure due to cyclic loading is outlined. The earthquake analysis of dams by using plasto-elastic constitution model is being studies.

In order to improve the earthquake resistant capability of the existing dams and barrages and to rehabilitate the earthquake damaged structures, aseismic engineering treatments and measures have also been studied. For example, the Hsingfengjiang buttress dams has been strengthened by backfilling concrete in the hollow space of the dam body; the gate operating frame on the top of Haihe barrage has been strengthened by oblique struts, the saturated sand lenses under the downstream foundation of Guanting earth dam has been compacted by vibroflotation.

However the following fact must be point out: Although a great deal of valuable data has been obtained from the field investigation and measurements on the damaged and undamaged dams and barrages which have been attacked by a series of severe earthquakes occurred in China in recent years, the scientific analysis and explanation of the various phenomena of earthquake damages are still inadequate. This is due to the insufficiency in research workers and necessary sceintific means. The firsthand materials are semetimes incomplete. There is still much work to be done.

Further research effort is to be stressed mainly in the following lines:

1. To intensity the work of observation on earthquake response of dams for collecting data to study the effect of the local condition of the dam site, non-uniform seismic wave imput within the foundation of dam, and dam-foundation interaction etc.. To make systematic field vibration tests on various types of large dams and barrages to accumulate data of vibration characteristics of these structures.

2. To explose the machanism of reservoir induced earthquake from the actual hydraulic projects and to find out the condition for its occurrance and the method of prediction.

3. To renew laboratory experiment equipments and work out the special computer programms of dynamic analysis of dams in order to carry out futher study of such problems as effect of complex foundation, seismic stability of the abutment of arch dams, liquefaction of earth dams, and nonlinear dynamic

analysis of various types of dams.

4. To continue researches on dynamic properties and stress-strain relationship of earth dam materials. To start the investigation of dynamic strength under complex stress condition, deformation properties and failure mechanism of hydraulic concrete.

5. To study the asismatic measures for rehabilitation of earthquake damage of dams and barrages along with actual projects.

References

1. Earthquake Intensity Maps of China. Beijing, 1979.

2. Deng Qi-dong et al: Principles and Methods of composing the Seismic Zoning Map of China " Acta Seismologica Sinica" Vol.2 No.1, 1980.

3. Hu Yuliang et al: Discussion on the Reservoir Induced Earthquakes in China and Some Problems Related to Their Origin "Seismoligy and Geology" Vol. 1. No. 4, 1979.

4. Songpan Earthquake in 1976, Beijing, 1979 .

5. Longling Earthquake in 1976, Beijing, 1979.

6. Shen Chung-kang, Chen Hou-chun et al., Earthquake Induced by Reservoir Impounding and Their Effect on Hsingfengjiang Dam, "Scientia Sinica, Vol. XVII No.2, 1974.

7. Earthquake Resistant Design Code for Hydraulic Structures (SDJ 10 - 78), 1978.

8. Chen Hou-chun et al., Dynamic Behavior and Earthquake Response of the Dam Section with Inserted Power Station of Fengshuba Dam, Proc. of 13th ICOLD, New Delhi, 1979.

9. Editorial Group of Earthquake Resistant Design Code for Hydraulic Structures, Earthquake Loads for Hydraulic Structures Proc. of 13th ICOLD, New Delhi, 1979.

10. Liu Lingyao Li Kueifen Bing Dongping Earthquake Damage of Baihe Earth Dam and Liquefaction Characteristics of Sand and Gravel Materials, Aug. 1979.

11. Wang Wen-shao, Strength, Liquefaction and Failure of Saturated Sands During Cyclic Loading, "Journal of Hydraulic Engineering" No.1 1980.

17 Soil amplified earthquake ground motions

J. N. PROTONOTARIOS, PhD, University of Patras and Greek Atomic Energy Commission

This paper investigates the effects of soil conditions on spectral characteristics of earthquake ground motions through: (1) Analytical procedures for a series of representative soil profiles, and (2) a study of site dependency on recent greek and romanian recorded motions. The results of the theoretical studies and the recorded data are compared with the prevailing views of the problems of soil amplification and site dependency of seismic motions and pertinent inferences are drawn.

INTRODUCTION

Site dependency of earthquake ground motions has been given an increasingly greater attention by seismologists and earthquake engineers during the last few years. The necessity for an orthological aseismic design of special structures like large dams or nuclear reactors has given further push to pertinent research. Most relative studies make extensive use of linear response spectra to characterize site dependent earthquake ground motions. Apart from the earthquake magnitude and the distance from the earthquake focus, local soil conditions have been reckognised as a most important factor influencing the spectral shape and the spectral values of an earthquake motion. Thus, it is already well established that large distant earthquakes result in motions with dampened high frequencies, whereas as close intermediate earthquakes produce motions near the epicenter with important high frequencies; pertiment attenuation relationships have been developed for ground motion characteristics like maximum ground acceleration, duration and predominant period (ref. 2,4). The effect of local soil conditions on intensity and frequency content of ground surface shaking has been also investigated the last few years. From the early research stages of this phainomenon it was realized that the main effect of soil layers is the filtering of the seismic motion, which propagates upwards from the rock base to the surface, through the soil medium, resulting in amplifying the frequencies which are close to the fundamental frequency of the soil profile. The soil effects upon earthquake ground motion characteristics have been investigated by means of purely analytical procedures utilizing methods of Soil Amplification Theory as well as by statistical processing of numerous actual strong motion records. Although the utilization of analytical methods has contributed greately to the problem, it should be stressed that their use may be followed by some important limitations: Site topography and its modelling, proper use of soil damping properties and most significantly uncertainties with regard to a suitable selection of a meaningful input base motion, might cause serious shortcomings for a reliable theoretical prediction. Thus drawing inferences from actual earthquake ground motions recorded at sites with diverse soil conditions is being considered at least equally important presently. The statistical processing of such data might lead to expected spectral shapes for different soil conditions, which could be a very useful guide for the prediction of specific site dependent spectra (ref. 5,10) and (ref.1,6,8).

This paper presents a number of response spectra, which were predicted theoretically for an equal number of cohesive and cohesionless soil profiles. It presents also a series of response spectra of strong ground motions recorded at sites of different soil conditions in Greece and Romania during strong earthquakes from 1972 until 1978. A correlation of spectral characteristics with soil conditions is attempted as well as a comparison of these spectra with mean site dependent spectra given by Seed and Hayashi et al (ref. 10,5). Finally some limitations of the procedure used to compute the theoretical spectra and their probable effects are discussed.

2. THEORETICALLY PREDICTED MOTIONS

2.1. Soil Profiles

Altogether, nine profiles were investigated to assess the influences of soil type and depth of profile upon the computed ground motions. For each soil type (Medium-soft clay, medium dense sand and very dense sand and gravel), three soil profiles of varying depths (50 ft., 100 ft., and 160 ft.) were used in the amplification studies. The three Clay profiles were designated as C-1, C-2 and C-3. The numbers 1,2 and 3 indicate the depth of the profiles as 50 ft., 100 ft. and 160 ft. respectively. Similarly, the medium dense sand and dense sand and gravel profiles were designated as MDS and DS respectively with the proper numbers following the abreviation profile names to indicate appropriate depth of the profiles.

The shear wave velocity of the Clay was taken 850 fps for low strains, corresponding to average in situ measurements of the MIT Campus Boston blue clay. The low strain shear modulii for the other two soil types, i.e. the medium dense sand and dense sand and gravel, were estimated from equation 1 (ref.9)

$$G = 1000\,K_2\sqrt{\bar{\sigma}_m} \qquad (1)$$

Following Seed's (ref.9) recommendations for a medium dense sand of relative density of about 75%, the value of K_2 was assumed to be equal to 61. Likewise, for the dense sand and gravel profiles of relative density of about 90% the value of K_2 was assigned to be equal to 120. In all the analyses, the shear velocity of rock for low strains was assumed to be equal to 6000 fps. Figures 1a, 2a and 3a present the clay, medium dense sand and dense sand and gravel profiles.

2.2. One-Dimensional Amplification Analysis

The computer program SHAKE-3 was used to conduct the amplification studies and compute ground response records for the nine profiles. The normalized shear modulus vs. strain and damping vs. stain curves suggested by Schnabel et al (ref.7) were employed to obtain strain compatible dynamic properties of the soil. An artificial earthquake record normalized to a peak accelaration of 0.053 g was used to study the nine previously described profiles. Figure 4 shows the acceleration response spectrum, for 5% damping of the artificial input record.

2.3. Results and Discussion

Figures 1.b, 2.b and 3.b present the acceleration response spectra, for 5% damping, for the three types of soils investigated: Boston blue clay, medium dense sand and dense sand and gravel, respectively. From these figures, the influences of profile depth and soil stiffness can readily be observed. For example, the greater the thickness of the soil deposit is, the larger is the fundamental period of the profile. The influence of profile depth is especially significant in the case of the clay profiles and less singificant in the case of dense sand profiles. In fact, increasing the depth of the Boston clay profile from 40 ft. to 160 ft. increases the fundamental period from about 0.4 to 1.6 seconds. Whereas, in the case of the dense sand profile, the fundamental period increases from 0.2 to about 0.7 seconds. Therefore, one can reasonably estimate the fundamental period of a "dense sand" deposit of up to 160 ft. thick to be in the range of 0.1 to 0.8 seconds. However, for soft deposits of considerable thickness, the estimation of the fundamental period may require more elaborate analysis. The fundamental period of rather shallow (less than 50 ft. thick) soft deposits may be approximately estimated from equation 2,

$$T = \frac{4H}{C_s} \qquad (2)$$

provided some "reasonable" allowance is made for reduction in the shear wave velocity depending upon the strength of the input record.
Another observation that can be made from these Figures is that the deeper the profile is, the smaller is the peak spectral acceleration. Here again this influence of the profile depth on the peak spectral accelerations is very predominant in the case of the soft deposits and less significant in the case of the stiffer deposit. This trend of reduction of the normalized peak spectral accelaration with increasing depth of prifile, is illustrated in Fig.5.
The response spectra of Figures 1, 2 and 3 show remarkably high spectral values in the range of the fundamental soil periods particularly for the shallow profiles. Such high values are met very seldom in actual recorded motions. The most reasonable explanations for this observation are: (1) The possibility of a higher damping capacity of the deposits than the used in the analysis, (2) The fact that the artificial used record contains relatively high spectral ordinates over a rather long range of periods (see Fig.4).
Table 1 summarizes the maximum ground accelerations and velocities for all the cases studied and presented in this chapter. The amplification ratios indicated in the table were obtained by dividing the maximum surface values by the maximum imput values.

Table 1. Summary of amplification Results

Soil Profil	Max. Ground Accel.		Max. Ground Vel.	
	Value	Ampl.	Value	Ampl.
C-1	0.105	2.0	0.261	1.6
C-2	0.066	1.2	0.290	1.8
C-3	0.052	1.0	0.352	2.2
DS-1	0.124	2.3	0.186	1.1
DS-2	0.140	2.6	0.218	1.3
DS-3	0.104	2.0	0.289	1.8
MDS-1	0.125	2.4	0.252	1.5
MDS-2	0.102	1.9	0.349	2.1
MDS-3	0.081	1.5	0.270	1.7

3. SITE DEPENDENT SPECTRA OF RECENT GREEK AND ROMANIAN EARTHQUAKES

3.1. Examined motions

Altogether, 16 records were investigated for this study. 14 motions, including the 1978 Thessaloniki records, were recorded at greek sites from 1972 to 1978. The Boucarest records of the 1977 great romanian earthquake were also used. Table 2 presents some basic parameters of the recorded motion, the sites and the corresponding earthquakes. The soil conditions of the greek sites, except Thessaloniki, vary from limestone (rock) to medium-stiff alluvium. Little information exists unfortunately for the soil profiles. The Thessaloniki record was recorded at the basement of the City Hotel, in the down town area of the lower city. The soil profile in this area of Thessaloniki consists of soft fill surface material up to 6 m thick overlying a well consolidated layer of deposits of unknown depth. The Bucharest record was obtained at the basement of the one-story frame building of the Building Research Institute, in the NE section of the city. The soil profile at this location consists of a surface layer of soft loess deposits up to 10 m thick, overlying 34 m of predominantly sand deposits of various densities, overlying in turn very deep deposits of clay and sand marl mixtures.
Figs. 6, 7 and 8 present normalized acceleration response spectra of the above motions, for 5% damping. All response values S_a in these spectra have been normalized by division over the

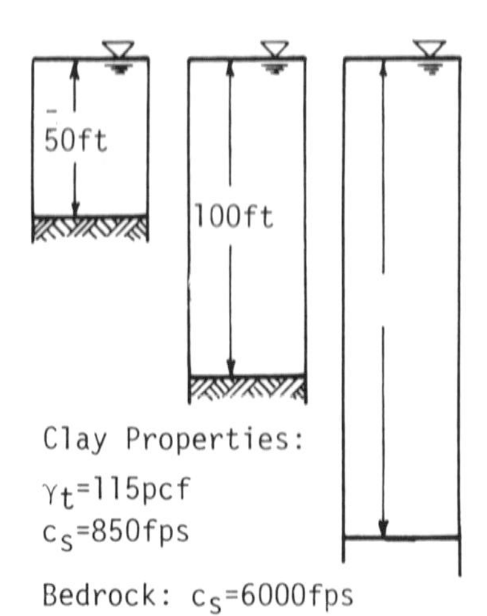

Fig. 1a. Clay Profiles

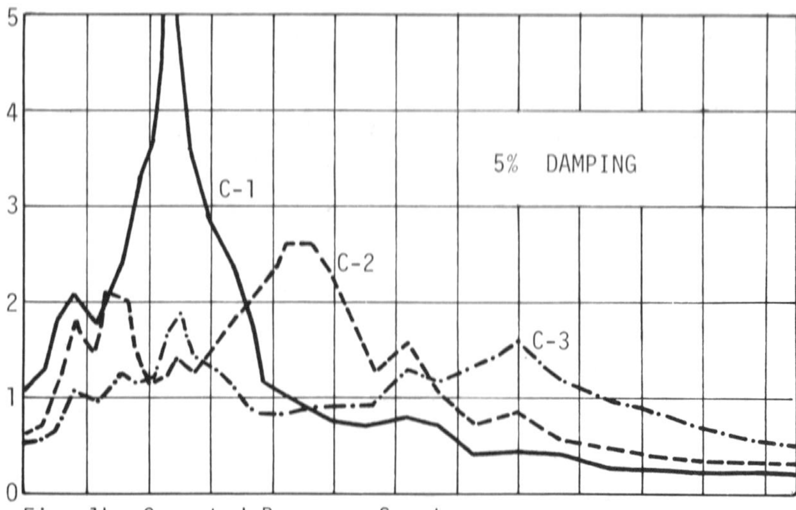

Fig. 1b. Computed Response Spectra

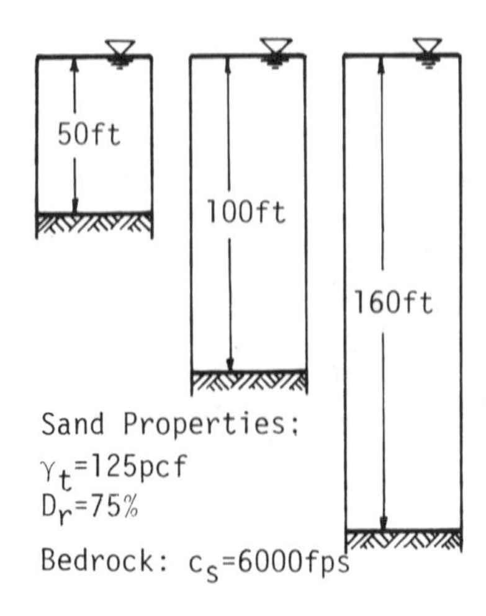

Fig. 2a. Medium Dense Sand Profs.

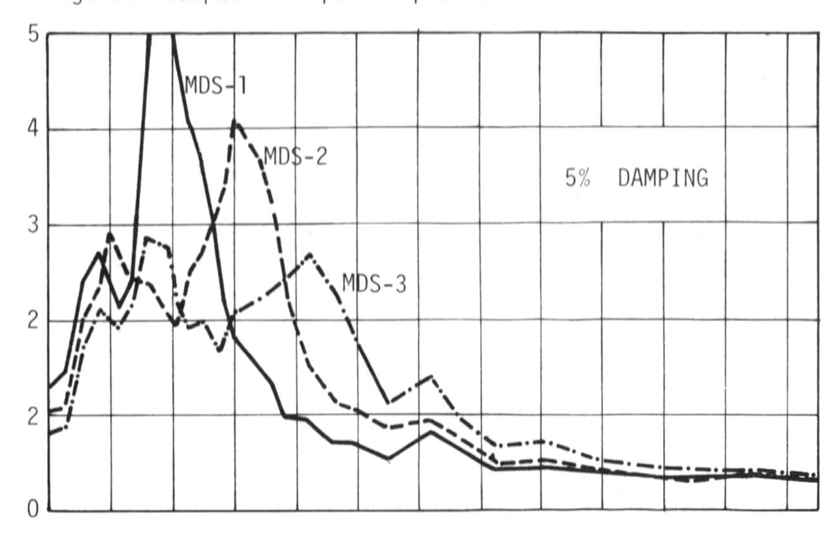

Fig. 2b. Computed Response Spectra

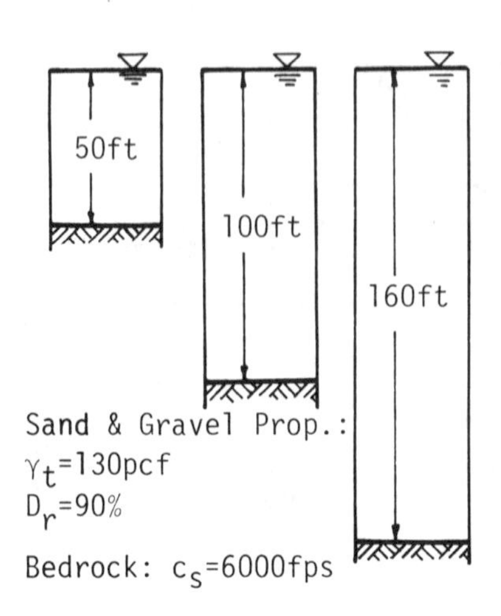

Fig. 3a. Dense Sand & Gravel Profiles

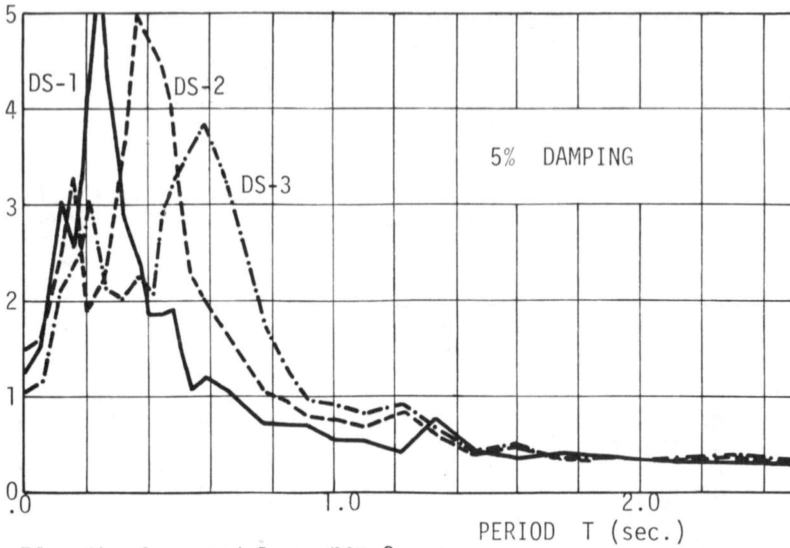

Fig. 3b. Computed Response Spectra

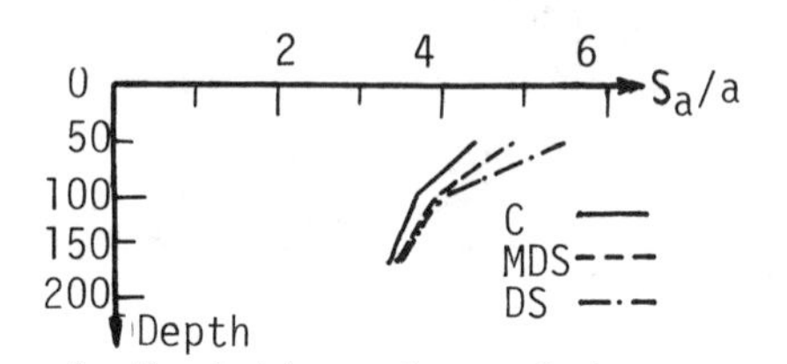

Fig. 4. Variation of max S_a/a with depth of profile

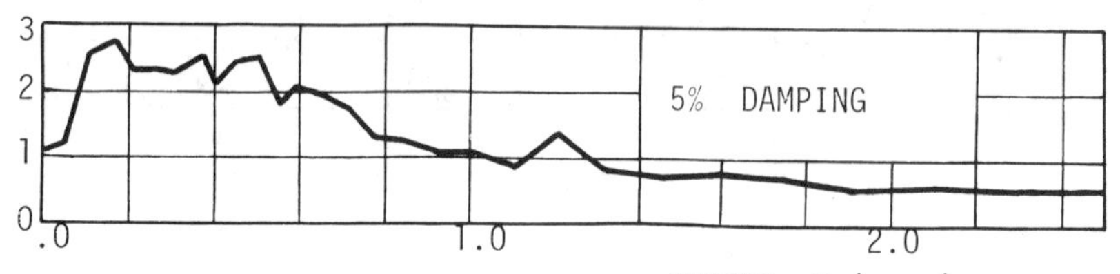

Fig. 5. Response Spectrum of Artificial Imput Motion

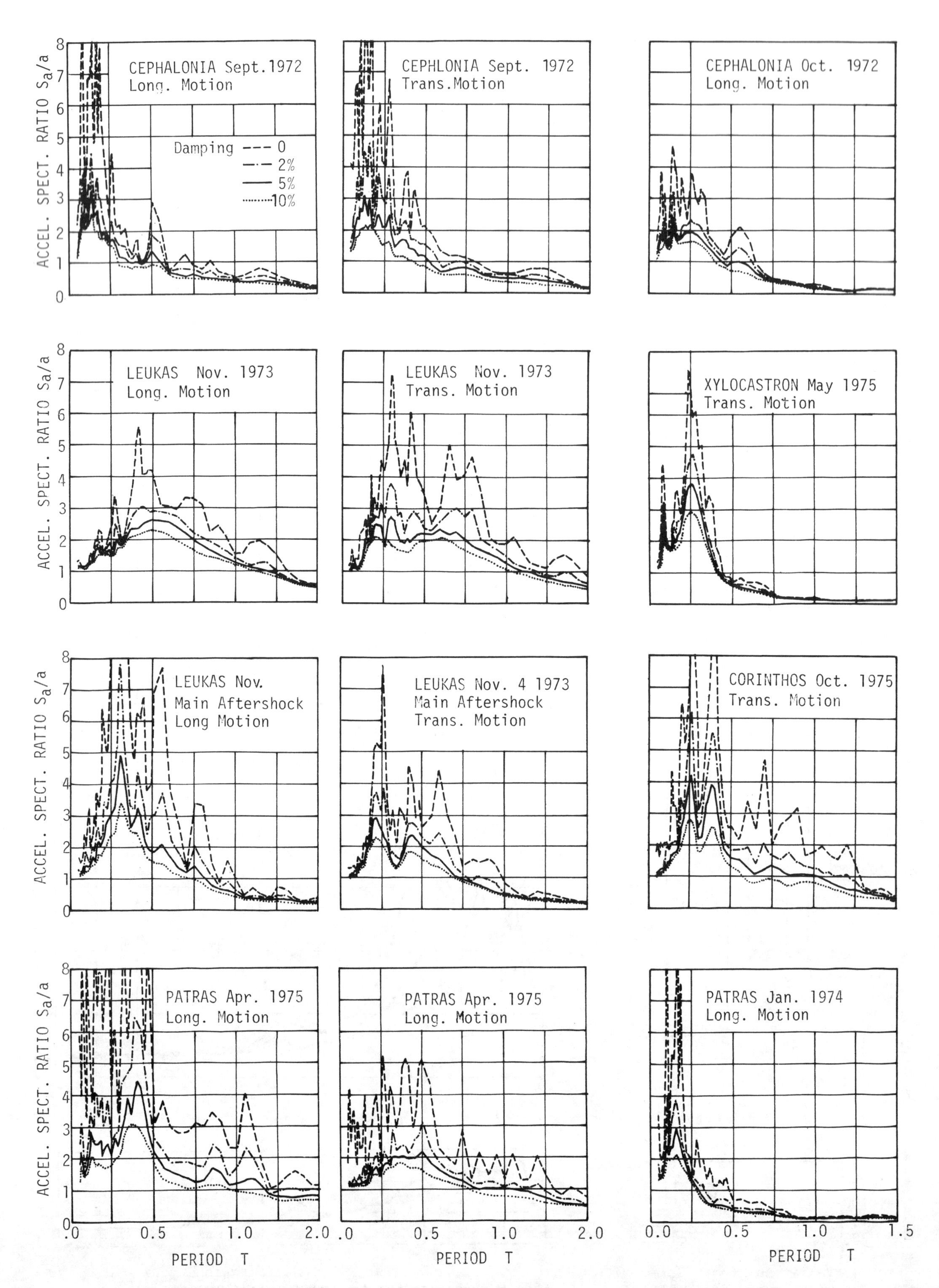

Fig. 6. Normalized Response Spectra of recent Greek Earthquake Ground Motions

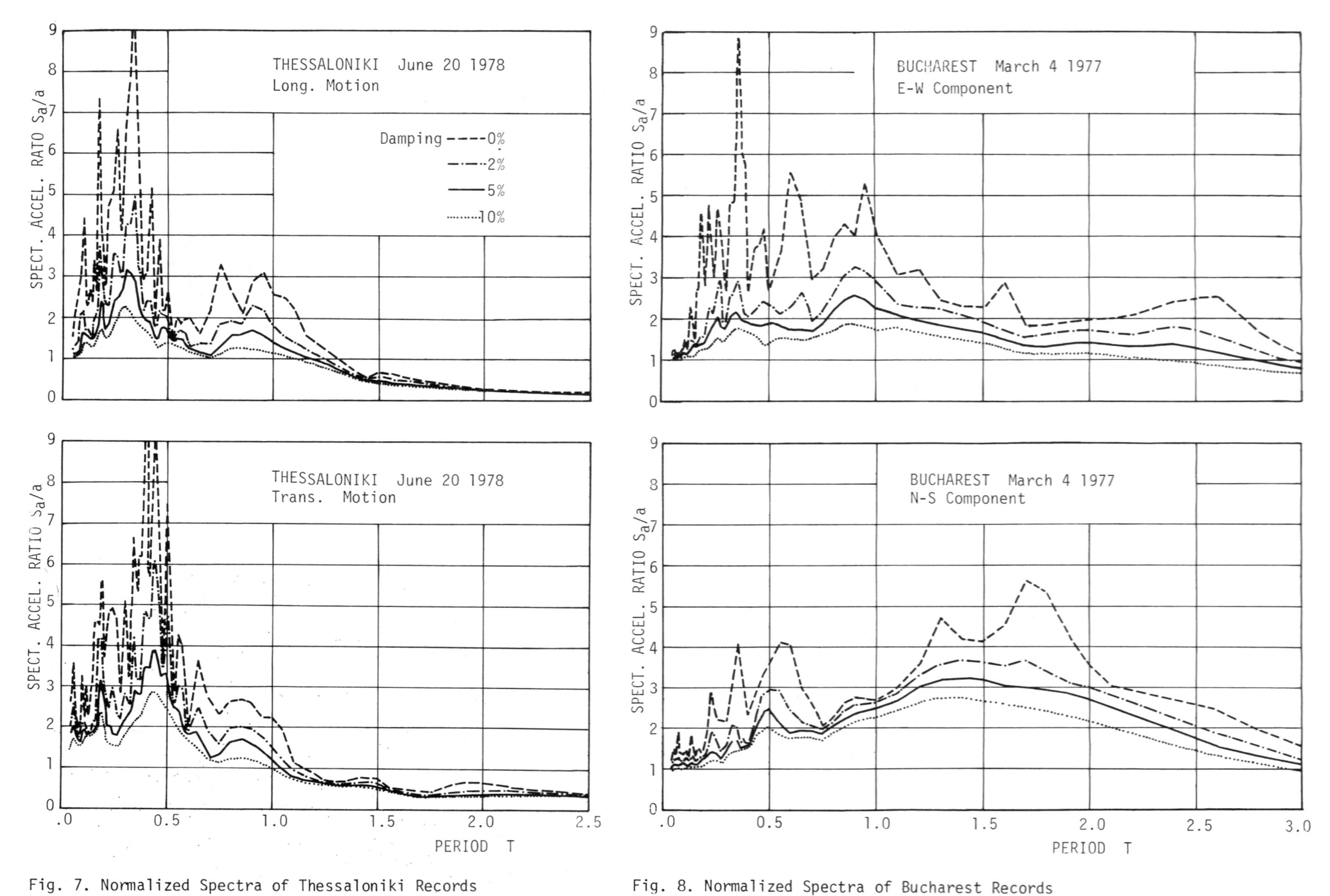

Fig. 7. Normalized Spectra of Thessaloniki Records

Fig. 8. Normalized Spectra of Bucharest Records

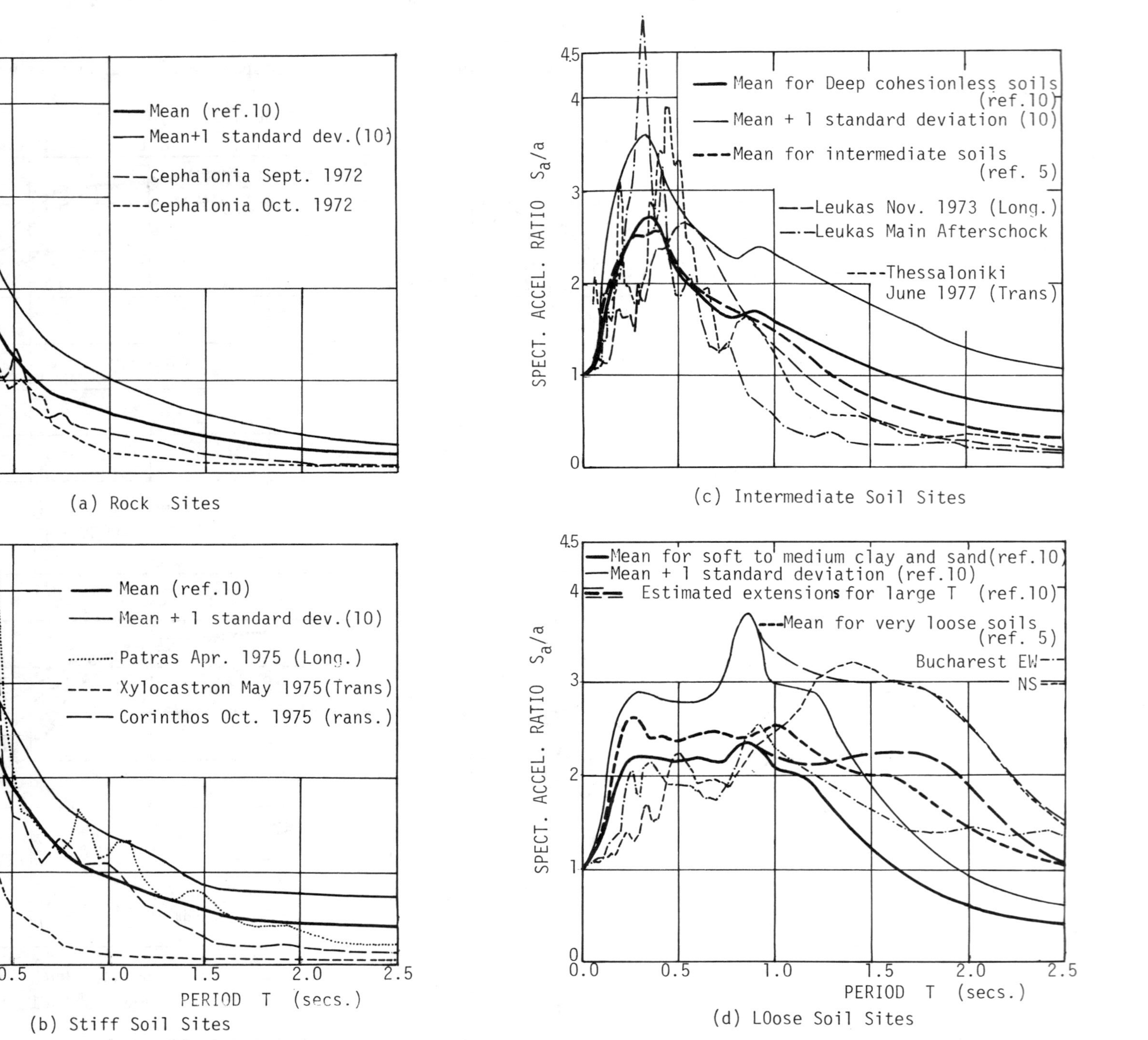

Fig. 9. Comparison of used spectra with mean site dependent spectra (ref. 5 and 10)

Table 2. Basic characteristics of used ground motions

EARTHQUAKE Date	Δ,h Km	M	Site Soil Cond.	Compon. Ident.	max a cm.sec^{-2}	max V cm.sec^{-1}	max d cm	Duration sec
CEPHALONIA	Δ=18	6.2	Limestone	Long.	170	8	1.8	6
Sept.17 1972	h=33			Trans.	120	5	1.7	8
CEPH. Oct.30 1970	Δ=22,h=13	5.5	Limestone	Long.	165	8.2	4.30	6
LEUKAS	Δ=20	6.0	Satur.	Long.	530	65	26.8	9
Nov.4 1973	h=13		Alluvium	Trans.	263	30	12	12.5
LEUKAS	Δ=12	5.0	Satur.	Long.	40	2.5	3.5	16
Nov.4 1973	h=35		Alluvium	Trans.	82	5.1	3.2	13
PATRAS, Jan.29 1974	Δ=17,h=31	4.4	Alluvium	Long.	41	1.5	0.6	7
PATRAS	Δ=32	5.5	Alluvium	Long.	19	2.0	1.55	14
Apr.4 1975	h=53			Trans.	50	3.37	1.83	14
XYLOC. May 75	Δ=15,h=36	4.3	Alluvium	Trans.	60	2.76	0.27	6.5
CORINTHOS Oct.75	Δ=16,h=24	5.0	Alluvium	Trans.	29	1.77	1.00	14
THESSALONIKI	Δ=28	6.5	Soft	Long.	151.4	18.2	6.4	15
	h=15		Alluvium	Trans.	141.9	14.3	5.7	11
BUCHAREST	Δ=166	7.2	Deep	E-W	174.5	75.1	10.6	14
March 4 1977	h=110		Alluvium	S-N	201.8	32.6	20.1	14

corresponding maximum ground acceleration a.

3.2. Discussion

Although it is rather difficult to draw definite inferences with regard to the effect of Soil conditions on the basic characteristics of the above greek earthquake ground motions, since little information about the soil profiles of the corresponding sites is known, some essential observations in relation with that matter can be pinpointed:

1. Expected influences of the soil conditions on the motions are apparent. These influences are more clear with regard to frequency content than spectral values. Thus motions recorded on rock sites (e.g. Cephalonia records, Fig.6a,b) or an stiff soil sites (e.g. Xylocastron record, Fig. 6d) have low predominent periods, whereas motions recorded on sites with less stiff and deep soil deposits (e.g. Leukas recond, Fig.6c, or Thessaloniki, Fig.7) present higher predominent periods.
2. Effects of magnitude and distance from the focus on the spectral characteristics may bring difficulties in assessing soil effects, which sometimes could be overriden by the former effects. This seems to be the case with the Patras records (Fig.6g,h), where it can be seen that spectral shapes are quite different for the same site due to different earthquakes. The motion from the more distant earthquake (Fig.6g) has clearly a higher predominent period than the motion from the closer one (Fig.6h).
3. Maximum spectral ratios S_{amax}/a vary considerably. Thus for the greek records they lie between approximately 2.7 and 4.8. They appear to have the highest values when intensity is relatively low and or when the predominent periods of the rock base motion coincide with the fundamental period of the soil profile (e.g. Leukas, Fig.6e or Patras, Fig.6g).
4. All greek records presented here present clearly very little energy in the low frequency (high period) range. The main reason for this must be the fact they are caused by rather close and medium-shallow earthquakes.
5. The combination of large distance and of soft and deep soil conditions cause the large predominent periods of high spectral values of the Bucharest motion.

Figure 9 presents a comparison of the examined spectra for 5% damping with the mean spectra derived by Seed (ref.10) and Hayashi (ref.5), for different categories of soil conditions. From these figures it can be seen that agreement is only relatively good and that it is better when compared with the japanese data than the american data. It should be stressed again that the main discrepancies appear in the high period range, where greek data fall considerably lower that the mean.

4. SUMMARY AND CONCLUSIONS

This paper has presented and commented (1) the results of amplification studies of a series of representative soil profiles, and (2) Response spectra of a series of recorded earthquake ground motions during recent greek and romanian earthquakes. The main conclusions of this study are:

1. One dimensional amplification analyses are very useful in predicting basic aspects of the influence of soil layers on the characteristics of earthquake ground motions. Factors, though, like soil damping parameters and input motion characteristics might bring some difficulties in the reliability of prediction specially with regard to the level of spectral values of the predicted motions.
2. Most characteristics of the presented greek and romanian records are consistent with established views of site dependency of earthquake ground motions including soil effects.
3. The available greek records show particularly low spectral values in the low frequqncy range, probably due to reasons of earthquake-

source mechanism and to the fact that they have been caused by relatively close intermediate magnitude earthquakes.

REFERENCES

1. AMBRASEYS N.M. Dynamics and Response of Foundation Materials in Epicentral Regions of Strong Earthquakes. Proc. 5th WCEE, Rome 1973.
2. BERNEUTER D.L. An overview of the Relations Earthquake-source Parameters and the Specification of Strong Ground Motion for Design Purposes. OECD Nuclear Energy Agency Report 49,521, Oct.1979.
3. CARYDIS P. Personal Communication. National Technical University, Athens, 1980.
4. DONOVAN H.C. Site Response Procedures for Major Buildings. EE 73-1, ASCE, Nation. Struct. Eng. Meeting, 1973.
5. HAYASHI S., TSUCHIDA H. and KURATA. Average Response Spectra for Various Subsoil Conditions. U.S.-Japan Panel on Wind and Seismic Effects UJNR, Tokyo, 1971.
6. PROTONOTARIOS J.N. Linear and Non-Linear Response to Site Affected Earthquake Ground Motios. MIT Ph.D. thesis, Cambridge 1974.
7. SCHNABEL P.B., LYSMER J. and SEED H.B. SHAKE-a Computer Program for Earthquake Response. University of California, Berkeley, EERC Report R72-12, 1972.
8. SEED H.B. "The Influence of Local Soil Conditions on Earthquake Damage. Soil Dynamics Proc. Spec.Session 2, 7th ICSMFE, 1969.
9. SEED H.B. and IDRISS I.M. "Soil Moduli and Damping Factors for Dynamic Response Analysis. University of California, Berkeley, EERC Report 70-10, 1970.
10. SEED H.B., UGAS C. and LYSMER J. Site Dependent Spectra for Earthquake Resistant Design. Bull. Seismological Soc. Am., Vol.66, No.1 p.p.221-243.

Session 2: Discussion

MATERIALS BEHAVIOUR WHEN SUBJECTED TO EARTHQUAKE

Mr D.A. Howells (introducing Paper 15). In paragraph 19 of my Paper I refer briefly to the 'shake down' of plastic structures. This derives from work such as that of Neal and Symonds, some twenty years ago on variable imposed loads but not with inertia loads. I find that this field is continuing a lively development, largely in Italy, and that formulations now take account of inertia loads. A review of the work of the past ten years on dynamic shake down is given by Cerradini in the June 1980 issue of the ASCE J. Appl. Mechanics, in a paper titled 'Dynamic shake down in elastic/plastic bodies.' This is a very theoretical and abstract statement of the position. An application to continuous beams and to frames by Corradi and Nova appears in Earthquake Enginrg and Struct. Dynam., 1974, No. 2, Oct-Dec, 139-155. This is obviously the same phenomenon as the very pragmatic approach to damage to earth and rockfill dams, initiated by Newmark and carried on by Seed and Sarma, which I refer to in paragraph 24. It would be an interesting exercise to try to present the latter in the highly abstract terms of the former.

My other comment is that, in paragraph 21, I refer to the contrast between the evolution of positive pore water pressure in the main body of the dam and the probable occurrence of negative pore pressure in the small shear zone over which failure is likely to take place. Empirical evidence for the former is given in Seed's account of the behaviour of Long Valley Dam. The analysis is probably helped by D.G. Gray (Finite element technique for two-dimensional consolidation. Proc. Instn Civ. Engrs, Part 2, 1980, Vol. 69, June, 535-542). This technical note includes a treatment of the Mandel-Cryer effect; with two such opposing tendencies at work simultaneously, the difficulties of interpreting observations are very great.

Professor H. Bolton Seed (introducing Paper 12). I would like to supplement my Paper with a few words about using 3-dimensional analysis on occasion to interpret the results of field observations. We can use such results to determine material properties.

In order to evaluate the seismic response of a dam it is necessary to know the shear modulus of the material comprising the dam, among other things. If we are dealing with a rockfill dam it is not an easy matter to determine the shear modulus of a rockfill material or material containing large cobbles in laboratory tests. The sample size required would be about four to five feet diameter and three times that in height. Nobody I know can perform tests on samples that large, especially under dynamic loading conditions. This means that, for many years, we have not known the true shear modulus of rockfill materials or cobbly materials, except by extrapolating data from smaller size particle tests conducted in the laboratory.

Under these circumstances it would seem that the best way to determine the true properties of rockfill or of large size particles in a dam is by observing the performance of a prototype structure under earthquake loading conditions. We had the opportunity to do this in the Oroville earthquake of 1965. As is often the case, Oroville Dam was built in what was thought to be a non-seismic region - one of the regions we talked about this morning - which suddenly became seismic, and had a profound effect on dam safety requirements throughout much of California. Furthermore, the records of Oroville Dam are themselves extremely interesting.

During the Oroville earthquake, the dam (750 ft high and having cobbly gravelly shells) was shaken by ground motions producing a maximum acceleration of about 0.1g at the base of the dam. The motions were amplified at the crest of the dam, producing accelerations of the order of 0.2 to 0.3g. We do not really know what the maximum crest motions were because the power supply failed at a critical stage of the shaking, so that two seconds of motion were obliterated from the records. We do know, however, as was mentioned by the Chairman, that the crest of the dam went on deforming long after the base motions had stopped. Thus in the latter part of the crest record the dam was essentially in free vibration and, therefore, one can determine from this part of the record the natural period of vibration of the structure. It is found to be about 0.8 seconds.

The shear modulus of cohesionless materials can be conveniently expressed by an equation of the form

$$G = 1000K_2\ (\sigma_m)^{\frac{1}{2}}$$

where K_2 is a soil property which is a function of the shear strain amplitude induced in the soil, the relative density of the soil and the grain size characteristics of the soil. At low strain levels K_2 has its maximum value, and this is referred to as K_2 maximum; it is a soil property characteristic of any cohesionless material, directly measurable as a property of that cohesionless material.

If the cross-section is now represented by a plain strain finite element model, knowing the natural period of the system is 0.8 seconds, it is possible to calculate the shear modulus of the material which comprises the system. In this way, one would arrive at the conclusion that K_2 maximum for this material is of the order of 320-350. Using such values for the K_2 max property in a plain strain finite element analysis calculation, we find that we can predict quite accurately the nature of the crest motions, given the base motions. However, this is not a true material property and hence it should be called a pseudo-K_2 max. What we are really doing in such a study is determining a material property which, when used in a plain strain analysis will predict the behaviour of a 3-dimensional system. Therefore, proceeding in this way, one can obtain a material property, but it is not a true property.

Many of the dams we deal with are 3-dimensional in nature. If we want to examine the response of these structures as they will behave in practice, we should use 3-dimensional analyses.

If one calculates the natural period of a dam of a given height (h) in a sloping valley, and compare it with the natural period of a dam of the same height in a plain strain condition, significantly different results may be obtained. One finds the natural periods are different and the plain strain system has a natural period which is longer than that of the 3-dimensional system. If we are to analyse results of this type, therefore, using plain strain systems, we would have to modify the calculation to take this fact into account.

Oroville Dam is a system of this type - the valley slopes are quite flat and the crest length is seven times the height. Nevertheless, even when the crest length is seven times the height, a difference of about 35% may still exist between the natural period of a plain strain system and the natural period of a dam in this kind of valley.

It follows that if Oroville Dam had a plain strain section, its natural period would not have been 0.8s but 1.35 times 0.8 which is about 1.08s. We would now have to determine the value of K_2max to produce this natural period and this would turn out to be 175. This is the real property of the soil. But if you use this value in a plain strain analysis you will, of course, predict an incorrect response. One would have to use this value in a 3-dimensional analysis to get the correct response.

I use this as an example to show the importance of the care required in interpreting field performance, to ensure that one arrives at the real properties of the soil when interpreting the results to determine material characteristics. It is quite easy, if one fails to make a 3-dimensional analysis of what is, in fact, a 3-dimensional system, to end up with an estimate of material properties which would be 100% in error from the calculations which have been made.

Professor A.J.L. Bolognesi (introducing Paper 11). My Paper refers to the types of soils and conditions which do not build up large pore pressures or cause significant strain loss due to earthquake shaking and associated effects. Up to three or four years ago there was unanimous consensus that a well-compacted gravelly shell did not develop pore pressures because, even if there were pore pressures in the laboratory tests due to the permeability of the shells, the pore pressures would dissipate as they built up. But in the last few years the computations and the theories developed by Professor Seed prove, in theory at least, that pore pressure should also be considered when we have these pervious shells. In order to stay within the limits of my Paper, if you have a gravelly shell you have three options: either you have a very pervious gravel or you make a consistent part of the upstream shell out of rockfill, which is possible under many construction conditions, or you have rock chimney drains to drain any pore pressures that might develop in the shells. I shall refer to this final case.

When the conditions referred to above apply and the foundations are satisfactory, then you may use the Newmark procedure analysis, with of course all the refinements which have recently been introduced, particularly the latest one by Makdisi and Seed (reference 6, Paper 11). Because of the title of the paper by Makdisi and Seed, when referring to this one would think of a simplified procedure, but the method can be as rigorous as anybody wants it to be.

If the conditions that I put forward apply, then all other things being equal, the critical value is the value that relates the yield acceleration (k_y) to the maximum average acceleration (k_{max}). It is easy to prove that with a solution on the safe side you can use, instead of the average acceleration, the accelerations along the slopes to make a balanced cross section.

What I have tried to do is to get the value k_y divided by the acceleration as constant as possible. Fig. 5 of my paper presents a conventional dam, with a 1:2 slope on the downstream side and a 1:2½ slope on the upstream side. The accelerations were computed by the QUAD-4 programme for the Taft earthquake scaled to maximum acceleration of 0.22g - you see how the acceleration multiplies near the crest. Because the critical value of the yield acceleration is along the slopes, as can be easily proved, the relationship k_y/u_{max} (z) is verified at points along the slopes. The dashed line S in Figs 5(b) and 5(c) of the Paper shows how this relationship varies when the slopes are constant. In both the upstream and downstream slope this relationship is lower on the upper part and higher on the lower part.

My proposal is, to get a balanced cross section for this case, to have this relationship as uniform as possible. For example, if the slopes in the upper part are flattened slightly then you can change the situation to the one shown by the solid lines. You improve the situation in the upper part and reduce the safety on the lower part, and you have a more balanced cross section. All things being equal, this relationship is the crucial factor in designing a balanced cross section. If the foundation is satisfactory it is unlikely that you will get a deep slide. Using this method, the first problem is to know how much the value $k_y/u_{max}(z)$ must be. Given value 1.0, it is almost certain that there will be insignificant displacements; it is also almost certain that one will over-design one's dam. You may do it for political reasons (as was mentioned this morning). If you have a dam near a city, where no one wants to see cracks, then you may go to a high number. But how much should the number be? If I want, for example, to make a balanced cross section, how much do I have to flatten the upper part of the slope? It all depends on choosing the appropriate value of the relationship $k_y/u_{max}(z)$. It looks as if the number 0.6 is a satisfactory value, if the design is based on accepting minor effects.

Having this relationship, one can compute the displacement. A great help can be obtained from studying all the cases that Makdisi and Seed have used in their simplified procedure. If anyone wants to, he can carry on the exact computation because they are limited to a small part of the upper part of the dam. For example, for the case presented in Fig. 5 of my Paper, when this relationship is of the order of 0.4, the displacement will be of the order of 0.25m.. If the number is increased to 0.6 the displacement will be of the order of 0.10m. This is computed by the Newmark method, as developed by Makdisi and Seed.

Paper 30 is a very interesting one, the results of which have also been commented on by Seed in Paper 12, who shed some light on this subject. Two dams are presented; they are symmetric - the upstream and the downstream slopes are exactly the same. With one of the dams, the Infiernillo Dam, there is no doubt that the pore pressure cannot build up. It is a rockfill, the properties of which have been known for many years. The second one is a gravel dam founded on gravel alluvium. Both are about the same distance from the epicentre. There is some difference but both are more or less about 100km from the epicentre. One important difference is the slopes. The upper part of the Infiernillo Dam has slopes of 1:1.75 and, according to the analysis that I presented (I used the analysis which has been published in Paper 11) the value of the relationship k_y/u_{max} is of the order of 0.5. One has to watch the period. I do not know the exact period of Infiernillo; I was informed that it must be about 1.5 seconds. The behaviour of the dam shows that if the relationship is of the order of 0.5 then you would expect minor problems. For example, with cracks alongside the crest: settlements of the order of up to 20cm, to which there is no objection but of course everybody concerned ought to know what is going to happen. In constrast, in La Villita Dam, the above relationship is practically 1.0, and the effect of the earthquake has been considerably less than at El Infernillo.

I think that this experimental information confirms the validity of the Newmark analysis procedure when applied to the conditions stated at the beginning of this discussion. When you have a problem, of multiplication of the acceleration, you must flatten the upper part of the slope; you can keep the same volume for the dam by steepening the lower part of the slope, as shown on Fig. 5 of my Paper.

Lastly, all the methods of analysis show that the critical slope is the upstream slope, the submerged slope, but in the particular case referred to in Paper 30 all the recorded damage is on the downstream side. Dr Resendiz said that this is because they did not record on the upstream side.

Professor R.T. Severn (Bristol University). Professor Seed commented on different occasions on the importance of the vertical component of acceleration, and also on the natural period of vibration, implying that there was only one. These two comments might be dangerous if taken together. The tallest rockfill dam in the UK is Llyn Brianne Dam - 95m high, and we have performed a series of prototype tests on this dam.

The difficulty of exciting a rockfill dam has been referred to. One has to ensure that the exciters themselves are securely fixed into the rockfill. That is quite a problem, requiring blcoks of concrete to be set into the rockfill, having volume in excess of $2m^3$. Our method of testing is somewhat different from others. There are two basic methods. One is to cover the structure with a large number of accelerometers, in which case the difficulty is one of calibration between the accelerometers, which is very important. If they are not calibrated properly the mode shapes measured are meaningless. Our approach takes more time. We have only two accelerometers which are moved around the dam. At the Llyn Brianne Dam there are berms on the downstream slope and we are able to make use of these with the travelling accelerometer. The force level is about 40kN total; the accelerometers are sensitive to $10^{-6}g$. We were not able to measure shear modulus from the real earthquakes, but we were able to measure shear modulus from surface measurements on the berms.

Professor Seed's comments on the correlation between such measurements at very low strains, and the higher strain values generated by an earthquake, are interesting.

There are few theoretical solutions on 3-dimensional rockfill dam work and the following results may therefore be of interest.
Fig. 1 shows that properties vary with depth, and that is very important. The Figure shows what the actual finite element mesh looks like. There are eight elements across the dam in this

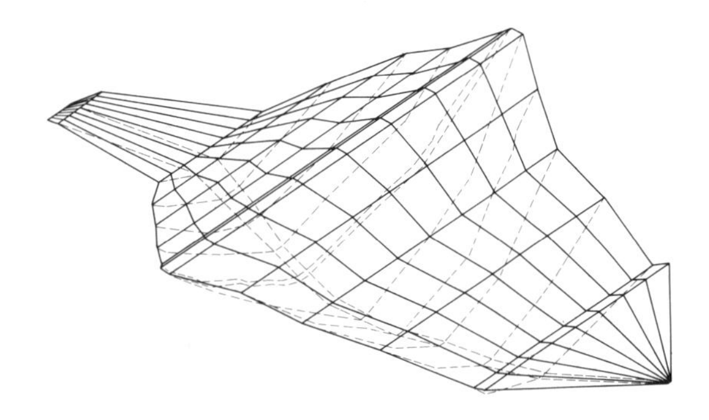

Fig.1. 3-dimensional mesh used for the dynamic analysis of Llyn Brianne Dam

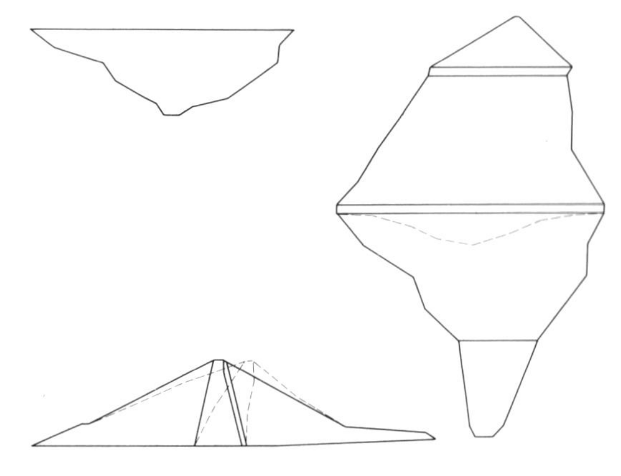

Fig.2. Mode shape 1; frequency = 3.45Hz

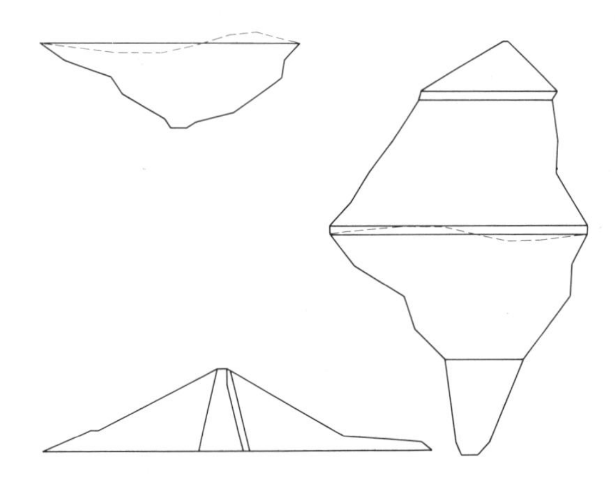

Fig.3. Mode shape 2; frequency = 3.99Hz

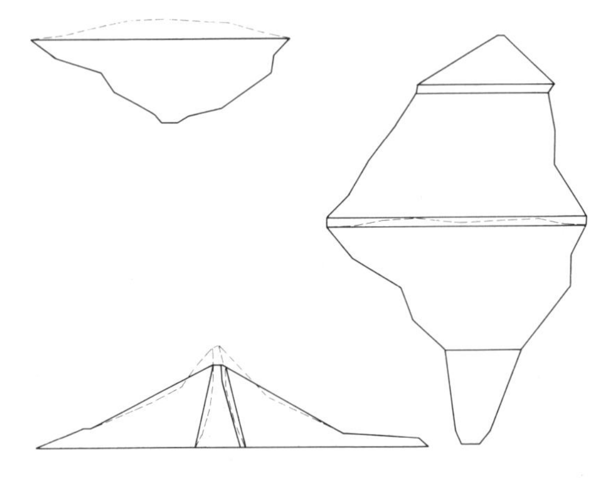

Fig.4. Mode shape 5; frequency = 5.08Hz

particular solution. An earlier solution had only four elements across the crest and, for those interested in finite elements, I think the results that we obtained from those two solutions are also important because it does matter, in the results one obtains, how sophisticated the element subdivision is. Some modes obtained with this finer mesh were not obtained with the coarser mesh, and could be missed entirely.

The first mode (Fig. 2) is satisfactory; it is a symmetrical upstream/downstream mode. The second mode (Fig. 3) is most interesting because this mode is not capable of being calculated by a 2-dimensional solution, and is what I refer to as a sloshing mode, in which the dam is moving from one abutment to the other, across the valley. There is hardly any upstream/downstream motion. The main motion is a depression on one side and an elevation on the other side. It is a sloshing mode which has a strong vertical component.

The mode is also interesting (Fig. 4). The frequencies are still within a sensible range - that is, within the significant part of the spectral curve. This is a predominantly vertical mode, as indicated on the elevation and on the cross section. It is a distinct vertical mode, at 5.08Hz; there is another vertical mode occurring at about 7.0Hz.

Of the measured modes the first frequency is 2.81Hz; with eight elements it is 3.45Hz across the valley; with four elements it is 3.66Hz. We have not measured the sloshing mode. The vertical component is important and it is there in the low frequencies. This frequency is 5.63Hz.

For the second point, I should like to comment on Professor Seed's work and Mr Baba's work - these two papers are interesting in that they give a somewhat different conclusion to the amplification of acceleration between the rock base and the crest. Professor Seed suggests that there is an amplification of between 1 and 2; Mr Baba suggested that it might be less than 1. I do hope that this discrepancy can be resolved.

Professor H. Bolton Seed. I do not think that we need to go to great lengths to reconcile the comments of Dr Baba and myself. In actual fact we are dealing with a non-linear problem. At low strains, such as those I talked about, with a very low base excitation, we find that soils tend to amplify the base motions while at higher strains you get into much softer materials characteristics and much more damping, so that the amplication decreases and may even be less than 1.0.

Regarding the natural period of a dam, I expect, because of the strain dependency of the shear modulus, that the natural period (or whatever we call the natural period) will be longer for stronger excitations. At very low amplitudes of motion it will be considerably lower than you would get for very strong earthquake motions. I do not think we have any disagreement at all.

Dr E.G. Prater (Swiss Federal Institute of Technology). As mentioned in Papers 7 and 9 the triaxial apparatus is frequently used, especially in commercial testing, to investigate cyclic stiffness characteristics at high shear strains. The results (secant modules and viscous damping) are then used in so-called 'equivalent-linear' dynamic response analyses. As is well known the shear modulus is not only strain dependent but varies also according to mean effective confining pressure. Most commonly a square root relationship is taken for the initial modulus at small strains ($\sim 10^{-4}$%). For accurate determination of the secant modulus and the area of the hysteresis loop it is necessary to maintain a constant mean effective stress, i.e.

$$\Delta\sigma_m' = (\Delta\sigma_1' + \Delta\sigma_2' + \Delta\sigma_3')/3 = 0$$

which requires that the cell pressure

$$\Delta\sigma_3 = -\Delta\sigma_d/3$$

since the change in $\Delta\sigma_1$ is $(\Delta\sigma_d + \Delta\sigma_3)$, $\Delta\sigma_d$ being the cyclic deviator stress. For saturated samples the changes in external volumetric stress are transferred directly to the pore water and the cyclic variation of cell pressure is not required. However, for moist or dry samples it makes an appreciable difference. In Fig. 5 the influence on the results for a gravel sample illustrates this point. It will be noticed that at relatively high axial strain the hysteresis loop has the form of a banana. Cycling the cell pressure improves matters but, unfortunately, does not make the loop symmetrical. This is probably due to the inherent difficulty of obtaining uniform strain conditions when going from compression to extension. For this reason in research programmes a more accurate test method may be required. Cycling the cell pressure remedies a common imperfection of technique, at the cost albeit of additional equipment.

Professor A.D.M. Penman (Building Research Station). The well known work of Professor Seed has had an effect which was reflected in the remarks of Professor Zienkiewicz; it can easily lead people to form the opinion that any soil when cyclically loaded for a sufficient length of time, will develop positive pore pressures and fail. I was interested to see that Professor Zienkiewicz accepted this in a straightforward way, and was prepared to generate programmes that would take account of this phenomenon.

I suggest that the phenomenon is not altogether a universal law, and there are many materials to which it does not apply, including some unsaturated fills. The proof we have is that many embankment dams stand up to earthquake forces very well, as has already been said. I suggest the reason is that we have rather differing conditions.

In the early days embankment dams were made in a very loose condition and we had hydraulic filled dams, such as Fort Peck, which were very susceptible to earthquake forces. Professor Zienkiewicz has illustrated to you the mechanism behind this behaviour. In order to get an increase in pore pressure the particles of fill material must rearrange themselves under the earthquake forces in such a way that the overall volume of material is reduced, the skeleton of the soil is partially collapsed, and the total stress is thrown on to the pore pressure, thereby reducing the effective stress and the stiffness or shearing strength of the soil.

If we are concerning ourselves with loose material, such as in hydraulic fill dams of the old days, or tailing dams of the present day, then there should be some apprehension. It is placed in such a loose condition that a small amount of extra force induced by earthquake can trigger off collapse, induce the high pore pressures we have talked about, and reduce the material to a liquid form - which can burst out, flow down the valley and so on.

In the majority of dams built today, so much energy is put into compacting the material in thin layers when it is being placed, there needs to be a certain threshold force coming from the earthquake before the structure breaks down and can possibly cause an increase in pore pressure. Therefore, for a given amount of fill that has been placed by modern machinery and is not saturated, there must be some threshold value or some magnitude of earthquake force which has to be exceeded before excess pore pressures are produced.

Another point I should like to make is that, the earthquake is usually of short duration, so that even if the threshold value is exceeded locally, and excess pore pressures develop, then there is an opportunity for them to dissipate, provided the earthquake shock does not go on

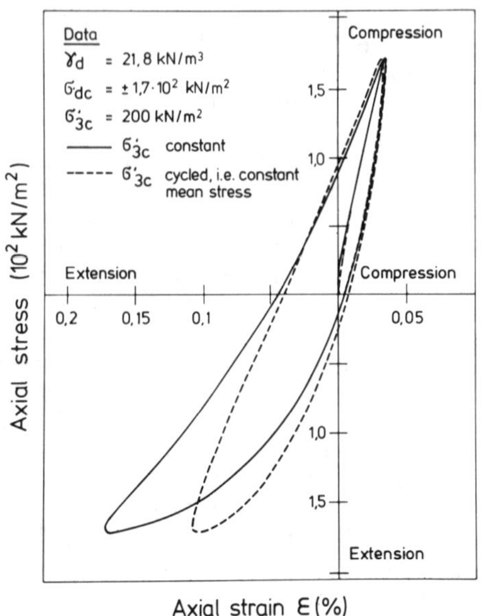

Fig.5

for too long. In the interval between successive earthquake shocks there may be, with certain fills, an opportunity for sufficient drainage to take place to bring the condition back to stability, so that it is not heading towards disaster but it is being checked in between the application of earthquake shocks.

My final point is that soil is a very good damper; that is why loose soil is used at the end of railway tracks so that if a train goes through the buffers it dives into the soil and the energy is absorbed by the soil. In the same sort of way, if we try to shake a mass of soil that forms an earth dam an enormous amount of energy is absorbed by the shear strains. If one applies strain to an element of soil one can deform it one way, but when the load is taken off it does not come back again. It stays there. So the built-in damping effect in the soil has a good stabilizing effect against the action of earthquake.

These factors would indicate that a modern earth dam, as the evidence appears to show, is perhaps better suited to resist earthquake forces than we might at first think.

Professor H. Bolton Seed. I am somewhat surprised by Professor Penman's suggestion that my work on cyclic loading of soils could 'lead people to form the opinion that any soil, when cyclically loaded for a sufficient length of time, will develop positive pore pressures and fail.' It is certainly not my personal belief that all soils are vulnerable to pore pressure generation and failure under the cyclic loading produced by earthquake or storms. In fact in the Rankine Lecture presented in London in 1979 and published in Geotechnique the same year I went to considerable lengths to differentiate between two types of soil from the point of view of their response to cyclic loading: (1) soils which do not build up pore pressures or lose significant strength as a result of cyclic loading; these included clays, dry or partially saturated soils, and some very dense saturated cohesionless soils; and (2) soils, such as medium dense saturated cohesionless sands or even relatively dense tailings materials, which can build up pore pressures and lose significant strength or resistance to deformations as a result of cyclic loading. The Lecture includes the statement: 'Since there is ample field evidence that any reasonably well-built dam constructed on a firm foundation can withstand earthquake shaking up to accelerations of 0.2g, from magnitude 6.5 to 7.0 earthquakes, with no detrimental effects, we should not waste our time analysing such problems....' The same paper points out: (1) that for dams constructed of clay soils experience has so clearly shown their ability to withstand the strongest earthquake shaking that only the simplest type of study is required to justify their seismic stability; and (2) in many rockfill dams the rate of pore pressure dissipation is so great that pore pressures could never build up to any significant extent during an earthquake.

These views are not dissimilar to those expressed by Professor Penman. My major concern with generalized statements of confidence in modern construction practices is that unexpected slope failures do occur from time to time as a result of earthquake shaking and it is well to avoid over-complacency about seismic stability problems. Nobody expected the slope failure in the Lower San Fernando Dam to threaten the lives of 80 000 people in the San Fernando earthquake of 1971, and the engineers involved certainly did not expect failure of the dams at the Mochi-Koshi tailings deposit during an earthquake in Japan in 1978. It would appear that most of the major failures of dams during earthquakes have resulted from failure by engineers to consider the possibility of pore pressure build-up in cases where such considerations are clearly warranted and necessary. I believe it is equally as important to be concerned about the possibility of such occurrences in cases where they may develop as to stress that such occurrences can not occur in all soils, as I am sure Professor Penman would agree.

Mr W.A. Wahler (W.A. Wahler Inc., U.S.A.). Professor Penman mentioned the problem of tailings dam stability. Tailings dams have been notoriously poorly designed in the past and many have failed in various ways, including liquefaction, induced by earthquake activity. However, these structures can be designed to be safe using principles that have become standard in the design of water dams such as flattening slopes, compaction, or zoning the structure. Saturated loose materials can be subject to liquefaction. Densification and/or saturation control can be used to preclude liquefaction failure. Current design practices control the saturation of the downstream slope area and/or densify this material to stabilize it against liquefaction. Traditionally stability has been treated as a slope problem but often it can be accomplished most economically by internal drainage or compacting a stabilizing buttress zone.

Mr B.L.Kutter (Cambridge University). A special package on the geotechnical centrifuge at Cambridge is being used to study the effects of earthquakes on model embankments. The centrifuge, described in detail in Professor A.N. Schofield's 1980 Rankine Lecture (Cambridge Geotechnical Centrifuge Operations. Geotechnique, 1980, Vol. 30, No. 3, 227-268), has a 4m working radius and is able to carry a 900kg payload up to 150g centrifuge acceleration. In order to impose lateral 'earthquake' accelerations on soil models, a special package was constructed that consists of a suspended container which is attached to a reaction mass via leaf springs. The springs are energized prior to flight by jacking them out relative to the container. The springs are held in place by a catch mechanism. When the soil specimen is spinning at the correct acceleration and after the soil water system has come into equilibrium, the catch mechanism is released and the container and reaction mass oscillate in a decaying sinusoidal fashion.

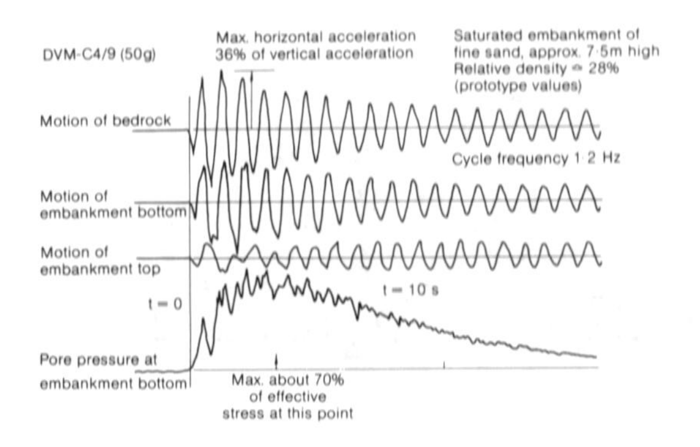

Fig.6. Model earthquake on centrifuged sand embankment (after Morris)

The earthquake package was designed and described by D.V. Morris. He also conducted a series of tests on sand embankments recording measurements of accelerations, pore pressures and deformation during model earthquake. A typical result is shown in Fig. 6. The bottom of the container is regarded as bedrock transmitting vibrations to the wet sand embankment. The maximum bedrock acceleration was 30% of the centrifuge acceleration. During the large earthquake pulses, the sand generated positive pore pressures - measured by one transducer to be 70% of the effective stress. The acceleration of the sand adjacent to the bedrock was measured and is shown to be very similar to the bedrock motion. However, the acceleration of the crest was heavily attenuated since the partially liquefied embankment was not strong enough to transmit the large accelerations to the crest. After the excess pore pressures had dissipated the crest motion became re-coupled with the base motion and the embankment vibrated almost as a rigid body.

Research is now being conducted to observe the behaviour of clay model slopes during simulated earthquake. Sliding block calculations, as proposed by Newmark, were used to predict dynamic displacements with varying degrees of success. One of the assumptions of sliding block calculations is that a distinct slip surface forms during yield. To investigate this assumption, lead powder threads were injected into the specimens before testing. Then post-test X-ray radiography was used to reveal the internal deformation patterns. Fig. 7 shows a sample that was subjected to one earthquake with a peak acceleration of 47% of the vertical acceleration. In prototype terms the crest moved 42cm - about 20% of the height of the slope. While a short rupture surface appeared at the toe, and tension cracks were visible at the surface, the central portion of the sample distorted over a wide band, not a distinct surface.

Figure 8 is a radiograph of a later sample. The sample was first subjected to earthquake and deformation occurred on the wide band that is seen to intersect the left side of the photograph. After the earthquake the centrifuge acceleration was increased until a 'static' slip occurred on the very distinct circular arc that intersects the top surface of the specimen. This Figure emphasizes a difference in mechanism between static deformations (which occur on a distinct surface) and dynamic deformations (which occur on a wide band).

There are many soil properties which influence their tendency to form rupture surfaces. But in this case the reason for the difference between dynamic and static mechanism is that static (constant) loading failure occurs when the factor of safety is one, so that soil on only the most critical slip surface is subjected to failure stresses and deformation is confined to this surface; in contrast, dynamic loading can produce factors of safety much less than one for a short time. During this time deformation occurs on a band of slip surfaces which all have factors of safety less than one.

In spite of the discrepancy between assumed and observed mechanisms, sliding block displacement predictions were reasonable. However, the fact that predictions are highly sensitive to the assumed value of soil strength had to be taken into account. To demonstrate this point, consider an example: a sliding block with a yield acceleration ($k_y g$) of 0.1g rests on an inclined 2:1 cohesive slope. A single, horizontal, sinusoidal acceleration pulse of amplitude $k_m g$ at a frequency of 1.0Hz is applied to the base of the model. Now consider the error in displacement predictions during the single acceleration pulse if the predictions are based on a 5% overestimate of shear strength (to obtain soil parameters of this accuracy would be very lucky). For a large sine pulse with k_y/k_m = 0.2 one would predict a displacement of 400mm and would observe 460mm, a 15% unconservative error. For one small sine pulse with k_y/k_m = 0.9 a displacement of 1.0mm would be predicted and 17mm would be observed, a 1600% unconservative error. Relative errors tend to be larger for small pulses that barely exceed the yield acceleration. This fact makes it difficult to test and compare analytical techniques based on existing field evidence of measured displacements during relatively small earthquake. The sensitivity to strength of dynamic displacements is the nature of the problem, not of the analytical technique, and therefore should not be considered a drawback of sliding block type calculations.

In conclusion, the relative importance of all assumptions of an analytical technique should be kept in mind. The assumption that a distinct failure plane forms, albeit inaccurate, may be less significant than inaccurate characterization of soil strength. Centrifuge modelling allows one to conduct a large number of tests helping us to keep these considerations in the proper perspective.

Professor J.W. Dougill (introducing Paper 14). My main interest in materials is in the descriptions of behaviour that provide the link between experimental results and structural performance. In particular, my work is concerned with theoretical models which give the connection between stress and strain for various materials.

In the analysis of dams, the situation with concrete is very similar to that for soil.

Fig.7. Earthquake deformation of a clay model

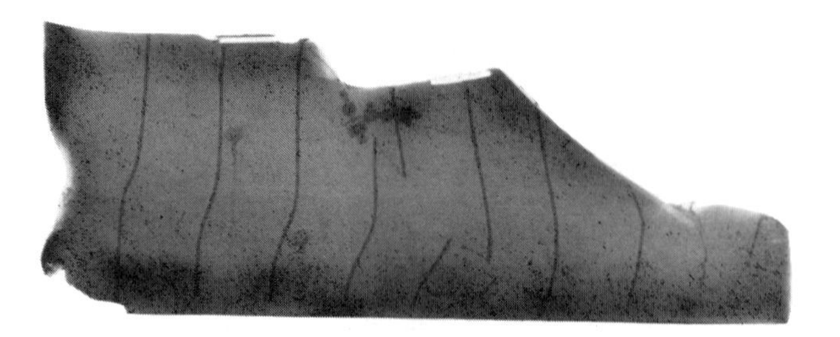

Fig.8. Static slip surface superimposed on earthquake deformations

We have well developed methods of analysis together with a range of materials descriptions ready to be used. I have commented on these in my Paper: they include basic linear models, plasticity style models, including those giving some account of degradation and softening, and the more recent endocronic descriptions of material behaviour.

The main factor that holds back further development and application is a lack of data on materials performance. In particular we are short of results relevant to loading conditions in earthquakes and for dams which can be put in a form suitable for use with theoretical descriptions.

Where should people put the emphasis in further research on materials behaviour? Consider first the use of linear models for situations in which the strain amplitudes can be taken to be small. Here the analysis is based on linear elasticity with, in some cases, the addition of damping. At present with concrete it seems that damping is included in an almost arbitrary fashion and certainly without consideration of real material behaviour. Of course, there are gaps and uncertainties in the available data on damping. These include the effect of normal stress and frequency, the role of free water in the concrete and (if one attempts to stretch the range of application of the linear models) the effect of small scale cracking. Most results are for unidirectional loading when, in general, results are needed for combined stress states. Again, in using linear models there is a fundamental difficulty in relating results to failure conditions and assessing the strength of a structure; unless behaviour is ideally brittle, it is difficult to associate ideas of strength with the use of a linear model. Essentially any load factor used is intimately connected with the method of linear analysis. In what way can rate effects, for instance, and a possible mechanism of failure be brought into this procedure?

In considering non-linear behaviour, it is of interest that there has been more useful feedback from earth and rockfill dams under earthquake conditions than from mass concrete structures. This is because not many concrete dams have suffered significant damage under earthquake conditions. This is very satisfactory but does mean that there have been few opportunities of investigating failures and comparing the results of different analyses with real behaviour observed in a major structure. Because of this, there are uncertainties to do with the effectiveness of our modelling procedures. Some of these arise because our information on the behaviour of non-linear dissipative materials is effectively testing machine dominated. The load paths used in testing are those built into the testing machine and we have not yet been able to explore materials under fully general stress combinations. Also, most models of material behaviour leave out degradation - the change in stiffness that occurs with increasing strain amplitude. This could be important in affecting response under dynamic loading particularly when there are a few cycles of loading which lead to fairly large strains in parts of the structure.

These problems seem to be common to soil, rock and concrete together with the question of softening and localization of failure and the possibility of breakdown of the analysis when yield or degradation occurs in one place accompanied by large deformations.

Lastly, size may have an effect of its own that needs to be taken into account in analysis and design. Here the effect of size occurs in the phenomena being examined and also in describing the material properties that are to be put into the analytical model. This last is a significant and relevant problem for both the analyst and the materials engineer.

Dr P. Bertacchi (introducing Paper 7). Some further information is required in connection with part B of Paper 7 regarding the behaviour of concrete under dynamic loading. An important document should have been included. This is a review prepared on behalf of the Materials Committee on Large Dams, by the national USA Committee, by Mr Graham (Water and Power Resources Service). The report concerns the effect of the rate of loading on strength of concrete in different conditions of ageing and moisture. Other more recent research, quoted in the report, deals with the critical cyclic testing and stress/strain relationships under seismic conditions. The mechanics of failure under slow/applied cycles and sustained compression loading is being presented in the study.

The behaviour of concrete when subjected to slow cycles and very sustained loads has a particular effect because the failure might occur at lower levels than with the static load.

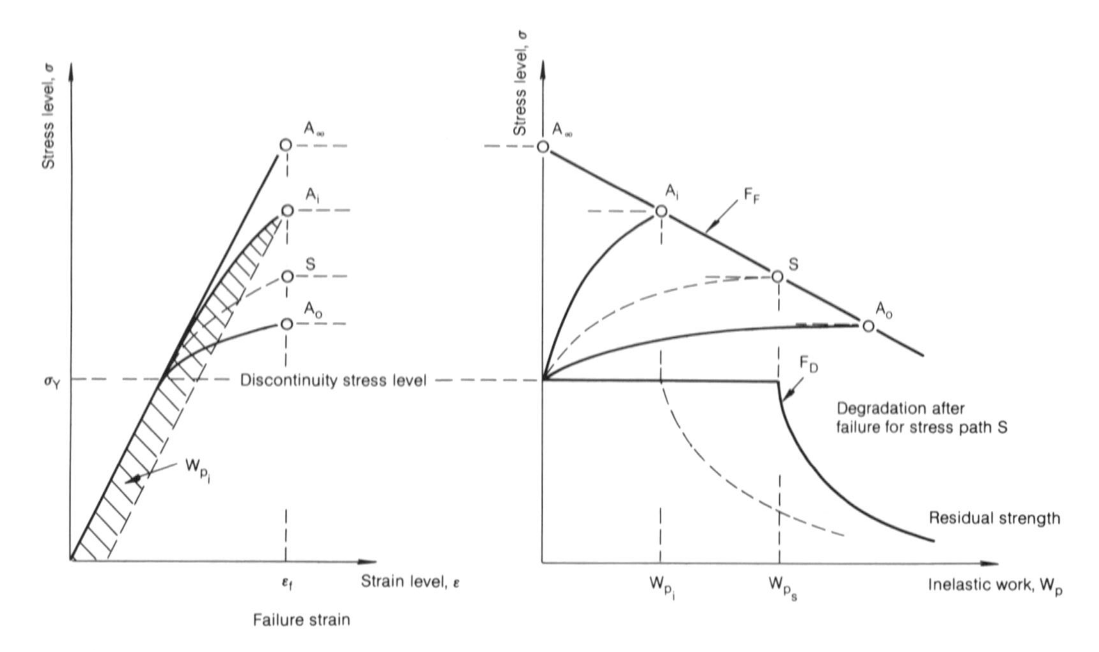

Fig.9

As the percentage of the final maximum load is increased, one sees that the lower is the number of cycles supported by the material. It depends on when the load is over the limit of the formation of micro-cracks or over the limit of self-propagation of the same micro-cracks. Another observation regarding the possibility of determining the damping of concrete through laboratory tests, already commented on by Mr Dougill, is the very good work of Ashbee and Jordan of the Central Electricity Laboratory in Leatherhead. The examination of the form and surface of the hysteresis loop is considered. A very recent work, not yet published by Mr Jordan, examines the effect of the stress, frequency, curing, mix and age on the damping of the concrete.

Dr N. Bicanic (Faculty of Civil Engineering, Zagreb). If Professor Dougill did not use enough words to suggest that you read his Paper, I would urge that you do it. It is an excellent review of the material characteristics which occur in concrete under high rates of loading, and which should be included in numerical models. I would like to quote two statements in the Paper. 'It is becoming more generally recognized that our capacity to perform extensive high speed computation with model materials has outstripped our knowledge of material behaviour itself', and 'It seems there is a need to look for simpler descriptions'. The numerical model we are suggesting follows a similar philosophy i.e. it attempts to represent basic phenomena with simple descriptions. The basic phenomena in direct and cyclic compression and tension tests of concrete under fast loading that are readily recognized by many investigators (Figs. 2. and 3, Paper 20) are: significant stress rate dependence of failure strength for both tensile and compression tests; failure strain remains almost constant, irrespective of the rate of loading; in the low cycle fatigue test, there is a pronounced hysteresis effect with an envelope curve in the stress/strain space, which can be taken as almost equivalent to the stress/strain curve for the low stress rate in the monotonic case.

The suggested model is an elasto-viscoplastic of concrete behaviour (Fig. 4, Paper 20) where the fluidity parameter, allowing for inelastic straining, is stress rate dependent. For the description of material behaviour two surfaces are used The discontinuity stress level, denoted in Fig. 9 as σ_Y is equivalent to the discontinuity surface in the multiaxial stress space. For a very slow stress rate the failure is reached at the point A_o, and if one had infinitely fast load rate, the failure is reached at the point A_∞. Here we take into account the fact that observed failure strain is more or less constant irrespective of the stress rate. If one plots the same information in a different co-ordinate space, relating inelastic work (dissipated energy) to the failure stress level, points A_o, A_i and A_∞ come into different positions. With infinite load rate the failure is reached with no inelastic work done, whereas for a very slow rate point A_o is reached. If, for simplicity the discontinuity stress level is kept constant in the prefailure regime and the linear fit is used to connect failure points, the surface F_D is constant and the failure surface F_F is shrinking, depending on the amount of inelastic work done. The discontinuity surface may harden with inelastic work, but the failure surface in this context starts from its maximum position and can only shrink due to the damage done to that particular stress point. The failure can be reached in any position on the co-ordinate W_p, depending on the stress level reached. After failure the discontinuity surface F_D is degraded to some residual strength.

The generalization of the model is described in Paper 20, using simple surface descriptions based on relatively reliable uniaxial test data. The prediction of failure in cyclic test was satisfactory, and the model does produce the main effects in the prefailure regime.

18

A simplified method for the earthquake resistant design of earth dams

S. K. SARMA, BTech, PhD, Imperial College, London

There are four stages in the design of earth dams for earthquakes. (a) Determination of earthquake input data. (b) Response analysis of the dam to the input data and determination of Seismic Coefficients. (c) Determination of the critical seismic acceleration. (d) Determination of the consequences when the response seismic coefficients show larger values than the critical acceleration. A simple reasonable way of determining these quantities is provided.

ABSTRACT
There is a great imbalance of knowledge in engineering seismology and earthquake engineering. Methods of analysis are highly developed and amenable to rational analysis. However, the rest of the problem, namely the input loading of the earthquake and the behaviour of foundation and fill materials, is at a primitive stage and little understood at this time.

It is the uncertainty and lack of reliable data for the selection of the appropriate design parameters at a particular site and for a particular earth fill dam, that dictate the use of a design method which will be of an accuracy comparable to that of the input data.

The purpose of this note is to present a simple engineering method for the design of earth fill dams to resist earthquakes with acceptable damage levels, using a fail-safe technique.

The basic principle in this type of analysis is that under strong earthquake forces, soil masses will deform and settle, and the proposed method aims at determining how much they will distort and establishing what deformations are acceptable in the design. The first stage is the selection of earthquake input data.

The steps to be followed are:

Seismicity study of the region in terms of active tectonics and unified magnitudes and reliable hypocentral location. Uncertainties in the determination of these parameters must be known, (8), (9).

Seismic zoning of the region and thorough geological study of the immediate vicinity of the site for Quaternary tectonic or secondary ground deformations.

Cumulative frequency distribution of earthquakes of different magnitudes determined for the different zones identified. This distribution should not necessarily be linear, and an upper bound magnitude should be assessed to truncate it. (5), (10).

Adoption of attenuation laws for near- and distant-fields. These laws must be up-to-date and if possible, reflect the type of faulting or source mechanism predominant in the region. (2), (3), (7).

Seismic hazard analysis, in its simplest form, may be carried out, following a model similar to that derived by Cornell (1968). Design ground accelerations, velocities and durations must be expressed in terms of probabilities of exceedance for the life-time of the structure. The design values to be chosen must reflect the relative importance of components of the structure and its appertainances.

Bedrock motions should be expressed in terms of ground velocity rather than ground acceleration. This is because peak velocity is related to the energy flux of the earthquake (14) whereas peak acceleration does not appear to have any such relationship. Therefore, peak ground velocity is a more reasonable parameter for designers.

Once the peak ground velocity has been selected, at least 3 or 4 existing ground motion records of different frequency characteristics may be used after they are normalised to the same design ground velocity. For example, for a design peak velocity of 45cm/sec we may select 4 records such as

	Peak Acceleration	Peak Velocity	Predominant Period
Koyna earthquake (1967) Long Component.	.63g	32cm/sec	.15 sec
El Centro (1940) N-S Component	.31g	35cm/sec	.5 sec
Parkfield (1966) ST.2.	.52g	67cm/sec	.6 sec
San Fernando (1971) S-16E Pacoima Dam	1.03g	110cm/sec	.4 sec

These records may be normalised to peak ground velocities of 45cm/sec. i.e.

Scale Factor	Peak Velocity	Peak Acceleration	Predominant Period
45/32	45cm/sec	.89g	.15sec
45/35	45cm/sec	.40g	.5 sec
45/67	45cm/sec	.35g	.6 sec
45/110	45cm/sec	.42g	.4 sec

The above four normalised record may then form the basis of the analysis.

The next stage is the analysis of the response and stability of the dam.

In order to analyse the stability of an earth dam during an earthquake, the following information is needed:

(a) The inertia forces that will be generated in the dam during an earthquake..

(b) The resistance of the dam against these forces along with the pre-existing static forces.

(c) The possible consequences, when the resistance of the structure is not sufficient to withstand these forces temporarily, allowing the development of deformations.

(a) Inertia Forces

The inertia forces to be generated during an earthquake will depend on i) the geometry of the dam and its foundations, ii) the material properties and iii) the earthquake time history.

A simplified model for the assessment of these forces is to construct a simple mathematical model, assuming elastic properties for the material. There are solutions already available (1), (4), (16), (18), which should be adequate for the simplified design. These are based on the following assumptions.

i) Geometry: (a) An untruncated wedge resting on bed rock or an untruncated wedge resting on a layer, overlying bed rock (figure 1).
(b) The wedge is symmetrical about the central axis (Y axis).
(c) Very long compared to height ($L > 4h$).
(d) Very flat slopes ($< 1:1\frac{1}{2}$) so that the vibration takes place in shear only.

For earth dams, these assumptions are not too restrictive. Consequently, the height of the dam (h) and the depth of the foundation layer (d) below are the only geometrical parameters involved in the analysis.

ii) Material Properties: It is assumed to be elastic, homogeneous with energy dissipating capacity. We know that the soil is inelastic and also non-homogeneous. For small vibrations, when the strain induced is small, it behaves almost elastically. For large vibrations, when the material strains into the inelastic zone, an estimate of the energy loss due to non-linearity can be taken care of by the proper use of damping. Damping is assumed to be of viscous type mainly due to analytical convenience and it represents an equivalent damping lumping the energy loss in radiation through the bed rock and in internal friction. A 20% viscous damping appears to be a reasonable value in strong earthquake vibrations. The necessary elastic properties of the dam and the layer are their shear wave velocities and the densities. Since in most cases the densities are more or less equal shear wave velocities (S) become the operative parameters.

The analysis shows that the most convenient parameter for the vibrational analysis of the dam are its fundamental period and the damping coefficient.

The fundamental period of the dam layer system is given by (16):

$$T_1 = \frac{2\pi}{\bar{a}_1} \cdot h/S_1$$

Where h is the height and S_1 is the shear wave velocity in the dam; $\bar{a}_1$ is a parameter which depends on the properties of the foundation with respect to the dam. Two parameters m and q can be defined which are

$$m = \frac{S_1\rho_1}{S_2\rho_2} \qquad q = m(H - h)/H$$

Where ρ is the density of the material. Figure 2 gives $\bar{a}_1$ as a function of the two parameters m and q. The higher mode periods are given by $T_n = \frac{2\pi h}{\bar{a}_n S_1}$ and the numbers $\bar{a}_n$ are tabulated in Sarma (1979).

Given the earthquake input data and the fundamental period of the dam with m and q, the response of the dam can be assessed from which the accelerations in the dam may be obtained as a function of the depth from the crest and time. The response point acceleration is then scanned to find the absolute maximum value at a given height. However, these maximum accelerations which vary with depth are not of much use since they do not occur at the same time or even in the same direction (4). It is an average seismic coefficient, therefore, that should be used which gives the net maximum earthquake load within a potential failure mass. The average seismic coefficient (K_m) when multiplied by the weight of the mass with a failure surface gives the maximum total earthquake load on the mass. This is obtained by assuming a failure surface and by computing the total load on this mass at different times and then by finding the absolute maximum value of the total load.

Simplified potential slip surfaces can be defined in advance (4), (18) as shown in Figure 3 and spectral curves for a given earthquake record can be obtained as a function of the fundamental period of the dam and the depth of the slip surface from the crest. Figure 4 shows such spectral curves, which are normalised to the maximum ground acceleration level (16). Figure 4a gives the maximum point acceleration whereas 4b and 4c give average seismic coefficients corresponding to slip surfaces shown in figures 3a and 3b. Similar spectral curves for other earthquake records and m and q values are available in Sarma (1979)

As such, for a given dam, we may prepare curves as shown in figure 5, which give the design average seismic coefficients at different levels of the dam.

The resistance of the dam is best defined by the critical acceleration which is needed to bring the factor of safety or the possible failure surface to one.

It is obvious that the critical acceleration depends on dynamic strength parameters and pore pressures. In the case of loose sand this cyclic loading may even bring about liquefaction of the sand.

If the cyclic pore pressure can be predicted in advance, these can be taken into account in any of the existing stability analysis methods to determine either the factor of safety or the critical acceleration.

Sarma (1975) defines a dynamic pore pressure parameter A_n which is analogous to Skempton's A parameter but dependent on the number of cycles. These parameters can be obtained from laboratory test results, as shown in figure 6 (17), and a limit equilibrium method of analysis of slip surfaces may be used wherein the A_n parameters are utilised to determine the critical acceleration as a function of the number of cycles (16). In fact it is not necessary to do the analysis for many cycles. A 20 cycle A_n parameter analysis should be sufficient to check the safety of the structure. Results of such an analysis are shown in Figure 7, wherein the A_n parameters are arbitrarily chosen. In this analysis, the assumption is made that the total earthquake load acts on the centre of gravity of the constituent mass segments. The increasing acceleration towards the top of the dam suggests that the resultant of the earthquake load acts at a higher point than the mass centre of gravity. However, the application of the load on the mass centre of gravity is on the safer side. We should determine the critical acceleration for surfaces at different heights of the dam and not find the one most critical, figure 8.

Consequences of failure

Since the instantaneous acceleration during the earthquake may be large enough to reduce the factor of safety below one, surfaces of discontinuity may be produced and displacements may develop along such slip surfaces.

The displacement of the sliding mass can be estimated roughly by using sliding block technique, (12), (15),(21). Sarma (1975) has shown that the displacement due to an earthquake record depends on the maximum value of the acceleration ($K_m g$) in the record, the predominant period of this record (T) and the ratio of the critical and the maximum acceleration (K_c/K_m), figure 9.

Since in the case of the dam, we are looking at the time history of the average seismic coefficient, it is difficult to know the predominant period. In general, the predominant period will be that of the earthquake record itself or it will be one of the mode periods of the dam. It should be remembered that when a surface of discontinuity is produced, the period of vibration of the structure will be reduced. But by how much, will only be a guess. We can, therefore, safely assume that the predominant period will be the largest of the values mentioned above. Using the graph (figure 9), the displacement can then be estimated. Table 1 gives an estimate of such displacement computed for slip surfaces shown in figure 8. It is obvious that the computed displacements for large A_n values are unacceptable. However, it should be noted that the displacements in this case are computed on the basis of the residual strength as well as on arbitrarily chosen A_n values. If it is assumed that the residual strengths are applicable at the end of the earthquake, then the factor of safety, at the post seismic stage, is greater than one.

TABLE 1

DISPLACEMENTS

STEADY STATE SEEPAGE

y/H		K_m	K_c	K_c/K_m	$4x_m/K_m gT^2$	x_m (cm)
		iii)	T = 0.5 sec			
0.2		1.0	.28	.28	1.2	75
0.4		0.98	.28	.29	1.1	67.4
0.6		0.88	.30	.34	.7	38.5
0.8		0.74	.33	.45	.18	8.3
1.0(a)		0.62	.45	.73	.03	1.2
1.0(b)	A_n	0.62	.19	.31	1.0	38.8
$1.0(b_1)$	0.5	0.62	.14	.23	1.5	58
$1.0(b_2)$	1.0	0.62	.09	.15	3.0	116
$1.0(b_3)$	1.5	0.62	.06	.10	4.5	175

CONCLUSIONS

In conclusion, it appears that the simplified method does give a reasonable approach to the problem of earthquake resistant design of earth dams. What we need now is field experiences of measured or computed displacements with rigorous analysis using non-linear properties including change in pore pressures in time, so that simplified computed displacements may be correlated to actual displacements.

ACKNOWLEDGEMENTS

The author gratefully acknowledges Professor Ambraseys for reading through the manuscript.

REFERENCES

1. AMBRASEYS, N. (1960) The seismic stability of earth dams. Proceedings, 2nd World Conference Earthquake Engineering, Vol. 2, pp 1345-1363, Tokyo.
2. AMBRASEYS, N. (1975) Trends in engineering seismology in Europe. Proceedings 5th European Conference Earthquake Engineering, Vol. 3, pp 39-52, Istanbul.
3. AMBRASEYS, N. (1978) Preliminary analysis of European strong motion data, 1965-1978, part II. Bulletin of European Association Earthquake Engineering. Vol. 4, 1, pp17-37, Skopje, Jugoslavia
4. AMBRASEYS, N. and SARMA, S.K. (1967) The response of earth dams to strong earthquakes. Geotechnique, 17, pp 181-213.
5. BATH, M. (1979) Some aspects of global seismology. Tectonophysics, 54, T1-T8.
6. CORNELL, A.C. (1968) Engineering seismic risk analysis. Bulletin of the Seismological Society, America, 58, 5, pp 1583-1606.
7. ESTEVA, L. and VILLEVERDE, R. (1974) Seismic risk, design spectra and structural reliability. Proceedings 5th World Conference Earthquake Engineering, 2, pp 2586.
8. KARNIK, V. (1969) The seismicity of European area, Part 1, D. Reidel, Dordrecht, Holland.
9. KARNIK, V. (1971) The seismicity of European area, Part 2, D. Reidel, Dordrecht, Holland.
10. LOMNITZ, C. (1974) Global tectonics and seismic risk. Elsevier, Holland.
11. MARTIN, G., FINN, W.D.L., SEED, H.B. (1975) Fundamentals of liquefaction under cyclic loading Journal Geotechnical Engineering Division, ASCE, 101, GT5, pp 423-438.
12. NEWMARK, N. (1965) Effects of earthquakes on dams and embankments. Geotechnique, 15, 2, pp 139-160.
13. SARMA, S.K. (1968) Response characteristics and stability of earth dams during strong earthquakes. Ph.D. Thesis, London University.
14. SARMA, S.K. (1971) Energy flux of strong earthquakes. Tectonophysics, 11, pp159-173.
15. SARMA, S.K. (1975) Seismic stability of earth dams and embankments. Geotechnique, 25, 4, pp 743-761.
16. SARMA, S.K. (1979) Response and stability of earth dams during strong earthquakes. Misc. paper GL-79-13. Geotechnical Laboratory, U.S. Army Engineer Waterways Experiment Station, P.O. Box 631, Vicksburg, Miss. 39180.
17. SARMA, S.K. and JENNINGS, D. (1980) A dynamic pore pressure parameter A_n. Proceedings of the International Symposium on soil under cyclic and transient loading. Swansea, January, pp 295-298.
18. SEED, H.B. and MARTIN, G. (1966) The seismic coefficients in earth dam design. Journal Soil Mechanics and Foundations Division ASCE, 92, SM3.
19. SEED, H.B. (1967) Earthquake resistant design of earth dams. Canadian Geotechnical Journal, IV 1, pp 1-27.
20 SEED, H.B. (1979) Considerations in the earthquake resistant design of earth and rockfill dams. Geotechnique, 29, 3, pp 215-263.
21. AMBRASEYS, N. (1974) Dynamics and response of foundation materials in epicentral regions of strong earthquakes. 5th World Conference on Earthquake Engineering, Vol. 1, pp CXXVI-CXLVII, Rome.

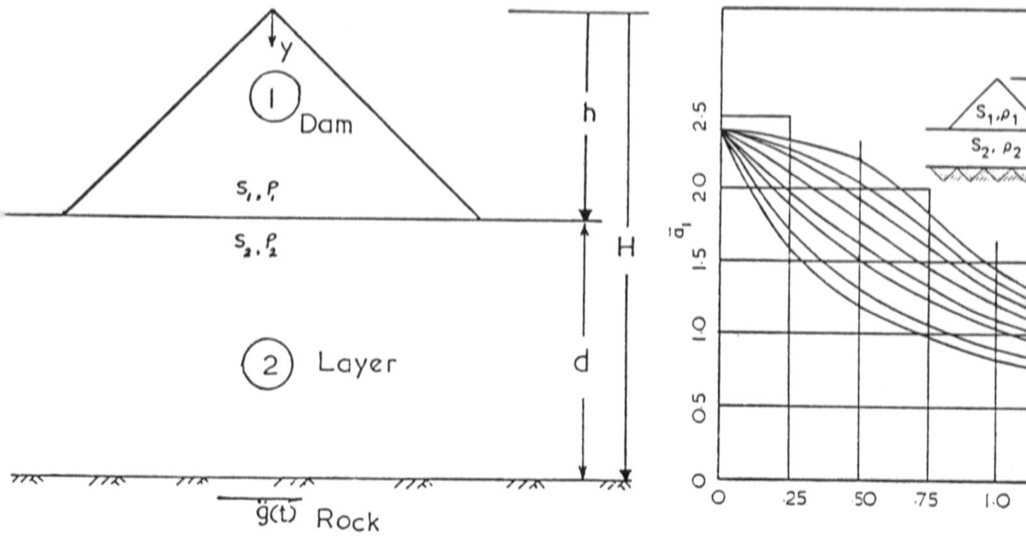

Figure 1. Geometry of dam-foundation system

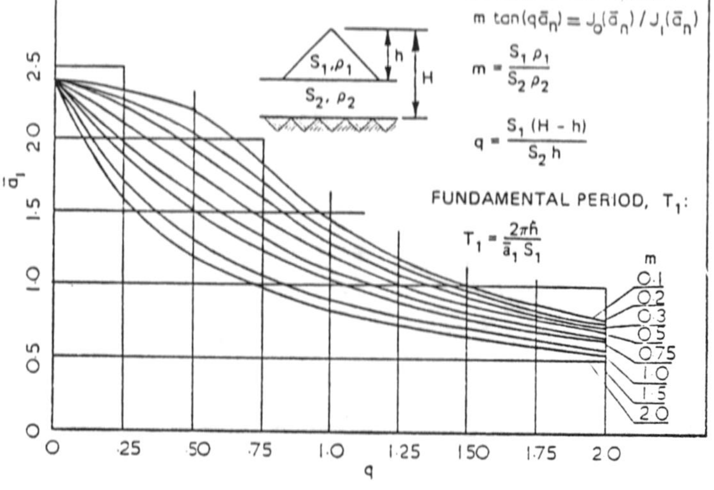

Figure 2.. Fundamental period of dam-foundation system

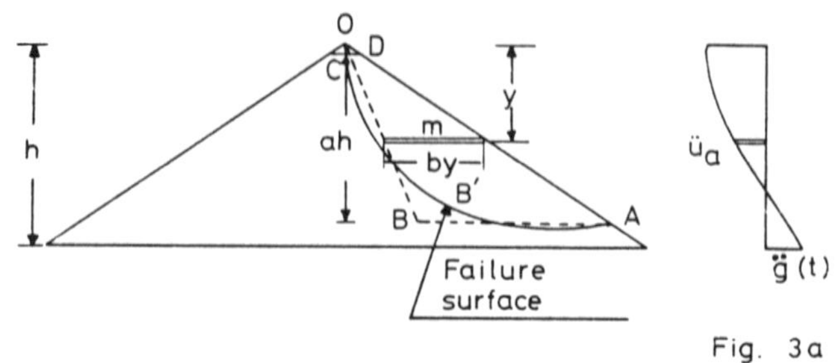

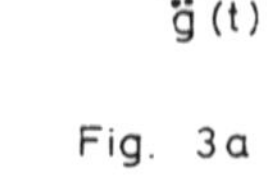

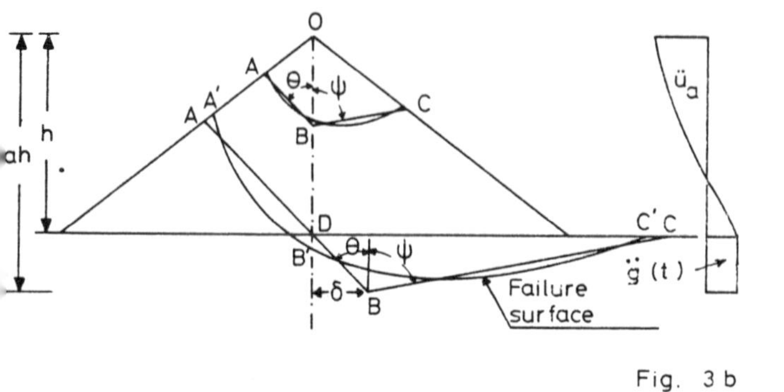

FIGURE 3: SIMPLIFIED POTENTIAL SLIDING SURFACES

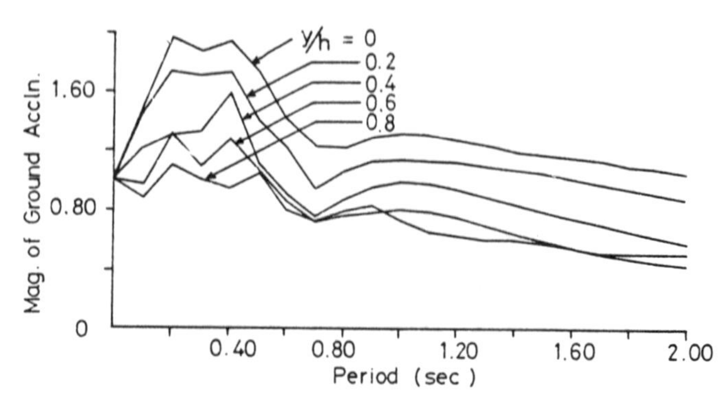

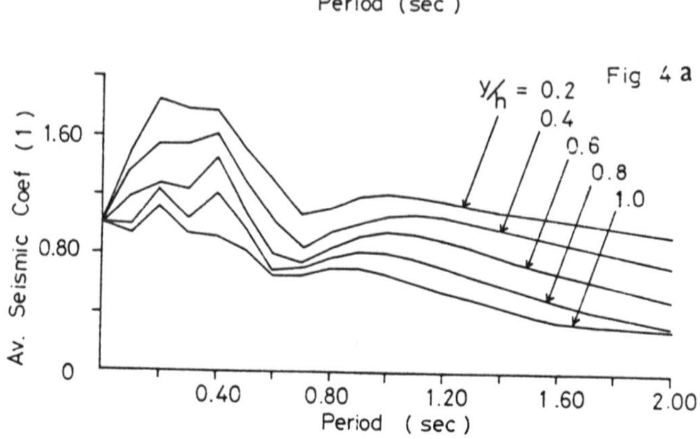

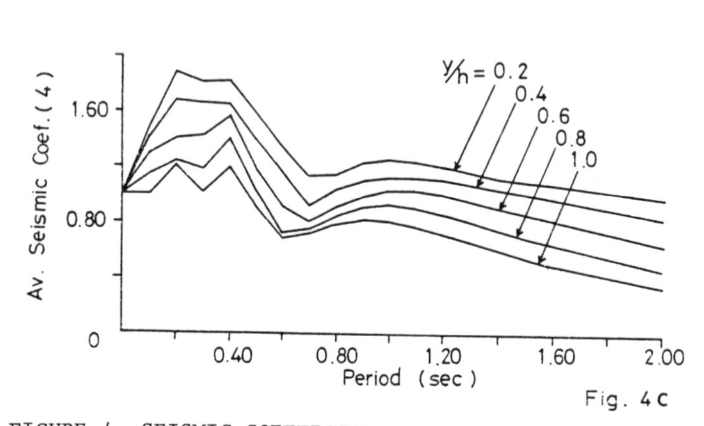

FIGURE 4. SEISMIC COEFFICIENTS SPECTRA

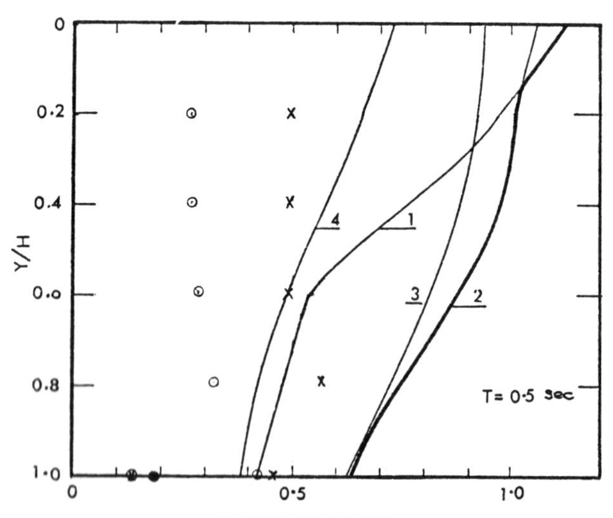

(Numbers on the curves give type of earthquake record)

Critical accn.

- × End of Construction Stage
- ⊗ " " " " (Residual Strength of Foundation)
- O Steady State seepage Stage
- ● " " " " (Residual Strength of Foundation)

Figure 5: Design seismic coefficients and critical accelerations.

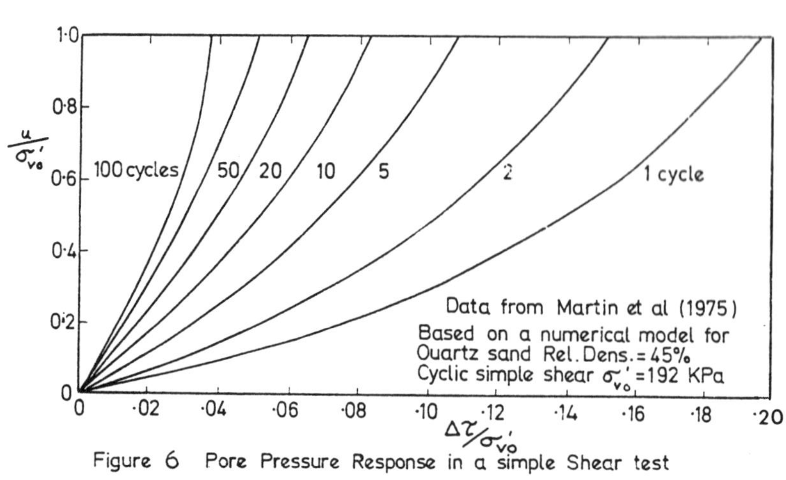

Figure 6 Pore Pressure Response in a simple Shear test

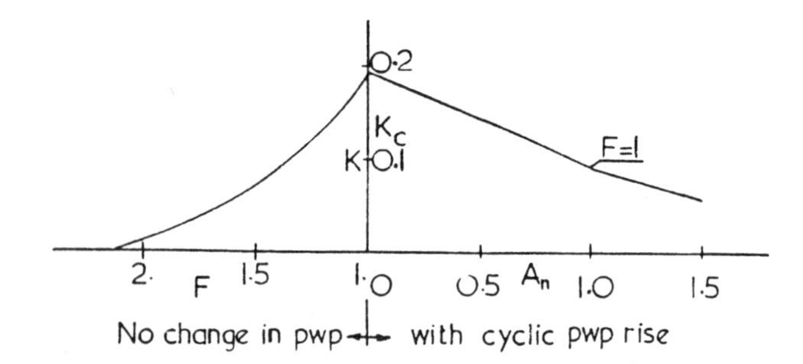

Figure 7: Stability analysis of slip surface 5

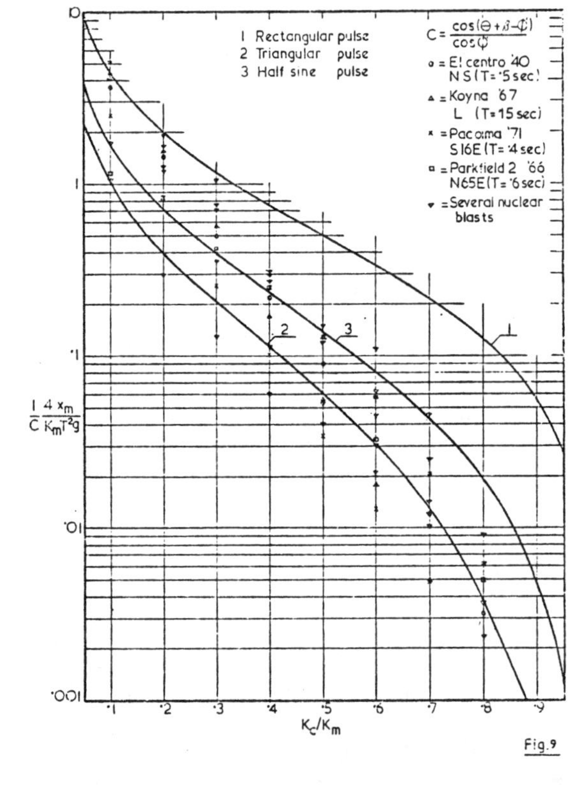

Fig.9

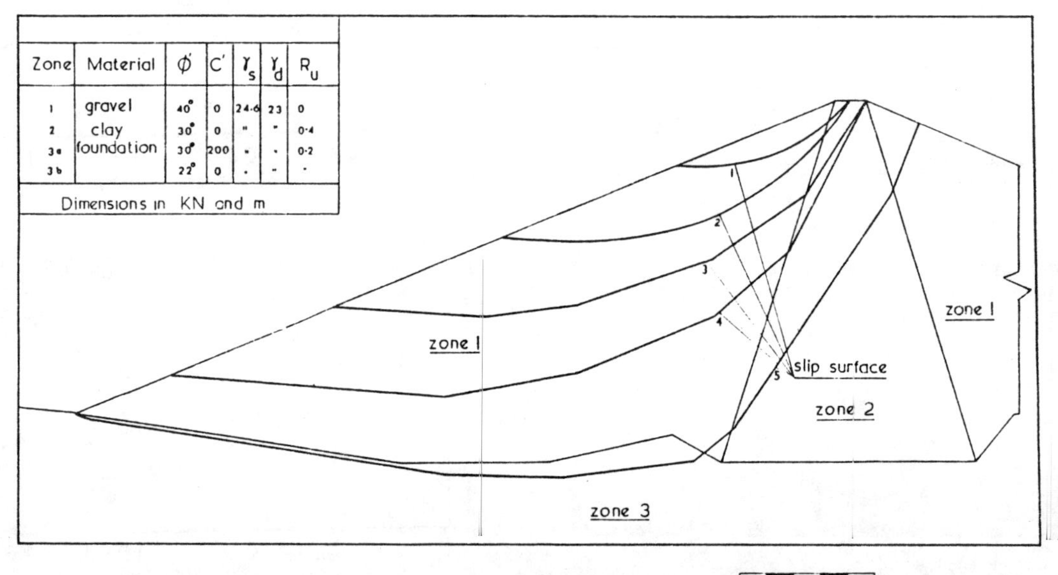

Zone	Material	φ'	C'	γ_s	γ_d	R_u
1	gravel	40°	0	24.6	23	0
2	clay	30°	0	"	"	0.4
3a	foundation	30°	200	"	"	0.2
3b		22°	0	"	"	"

Dimensions in KN and m

Figure 8: Typical dam section

19 Seismic analysis and design considerations for concrete dams

K. J. DREHER, BSc MSc, Water and Power Resources Service (formerly, US Bureau of Reclamation) Denver

Current techniques for analyzing and evaluating the response of concrete dams to earthquakes are summarized. After discussing appropriate descriptions for fault displacement and vibratory ground motion, certain factors essential to the analysis of concrete dams are considered. These include hydrodynamic interaction, foundation interaction, and damping. Selected results from the analysis of a gravity dam are then presented. The differences between two different formulations for hydrodynamic interaction and the effects of idealized reverse fault displacement are shown. Finally, criteria for evaluating the seismic performance of concrete dams are briefly considered.

INTRODUCTION

1. Although there has not been a catastrophic failure of a concrete dam to date due to the occurrence of an earthquake, the safety of concrete dams located in seismic areas continues to receive and deserves a substantial amount of study. The reasons for this are the lack of recorded ground motion from rock sites in the near field of large magnitude earthquakes, the lack of recorded structural performance data from concrete dams subjected to severe earthquakes, and the inability to conclusively determine the earthquake-producing potential of many faults. However, continued study of these factors has brought about an increase in the level of earthquake ground motions considered appropriate to ensure adequate safety at many damsites. As a result, the performance evaluation of concrete dams in regions where large earthquakes might occur has become increasingly difficult to assess.

2. Fortunately, most earthquakes do not bring about ground motions large enough to cause damage to concrete dams. Fig. 1 shows a suggested limit of earthquakes which a well-designed, well-constructed concrete dam in the western United States may be assumed to resist without significant damage when fault displacement is not a factor. To evaluate the safety of a concrete dam subjected to earthquakes above this limit, dynamic response analyses must usually be performed. The purpose of this paper is to summarize current techniques for accomplishing such analyses for both earthquake ground acceleration and fault displacement. Selected results from two-dimensional analyses of a gravity dam for two formulations of hydrodynamic interaction and idealized fault displacement are included, and the evaluation of results obtained from numerical analyses are briefly discussed.

APPROPRIATE PARAMETERS FOR ANALYSIS

3. Before the seismic safety of a major dam can be assessed, comprehensive seismological studies must be performed to establish the magnitude and location of all earthquakes which could significantly affect the structure. Geological features which could have a pronounced affect on potential ground motions should also be identified and the potential for surface faulting assessed. The outcome of these studies is not always definitive and conclusions may be largely based on the judgments of seismologists and geologists. While such judgments are necessary and usually appropriate, specified parameters may be based on insufficient or excess conservatism.

Fault displacement

4. Although no tectonic fault displacement is known to have occurred in the foundation of a concrete dam, a well-designed, well-constructed concrete dam can be expected to safely withstand limited fault displacement. This is not meant to imply that it is generally acceptable to site concrete dams in locations where surface faulting is likely. Rather, there are individual cases where the probability of faulting is extremely remote and the foundation conditions are such that allowing exposure to potential faulting can be considered. In fact during the 1930's, Morris Dam, which is a 330-ft (100-m) high gravity structure near Los Angeles, California, was constructed with an "open joint" above the trace of a fault to allow for movement in case displacement should occur (ref. 1).

5. To analyze a concrete dam for fault displacement, the geometry of the fault along with the amplitude and sense of potential movement must be defined. Data describing joints in the foundation rock including orientation, spacing, continuity, and shear strength, are also required such that the effective strengths of the jointed rock mass can be estimated. Although the processes of faulting and seismic wave propagation can not physically be separated, the assumption that fault displacement occurs prior to the generation of vibratory

ground motion simplifies the analysis and is conservative.

Vibratory ground motion

6. The easiest means of specifying the time history of ground acceleration for a dynamic response analysis is to obtain accelerograms recorded on rock at similar causative fault distances during earthquakes of similar magnitude to those established for the site under consideration. However, there are only a relatively small number of accelerograms recorded on rock for moderate to large earthquakes, and records for a prescribed magnitude and fault distance having appropriate intensity, frequency content, and duration, do not usually exist. One way of overcoming this difficulty is to bracket the appropriate ground motion characteristics and perform analyses using sets of scaled or unscaled accelerograms from several different earthquakes. This can significantly increase the cost of accomplishing a seismic response study and may create difficulties when evaluating the results. Another technique is to "piece together" accelerograms using various parts of scaled or unscaled historical recordings. While only one set of accelerograms results for each earthquake, the frequency content of such motions may be inappropriate.

7. Response spectra. Currently, the best method for determining appropriate ground acceleration input is to develop response spectra representing earthquakes established for the site. Rather than defining the time history of ground acceleration directly, a response spectrum shows the extent any single-degree-of-freedom structure, with some assumed level of damping, would respond to the ground motion. Thus, a displacement, velocity, or acceleration response spectrum indicates both the intensity and frequency content of the ground motion and, with duration, serves as a basis for defining that ground motion.

8. There are several methods which can be used to develop a response spectrum for a particular earthquake based on response spectra computed from historical accelerograms. The primary differences between these methods are the parameters used to quantify the intensity of the ground motion being specified and whether site-independent or site-dependent spectra are developed. The most commonly used intensity parameter is peak ground acceleration. However, peak ground acceleration is not by itself a meaningful indicator of ground motion producing significant structural response, particularly near the causative fault. Since the peak ground acceleration is a high frequency pulse which is normally beyond frequencies of structural response, peak acceleration is only indicative of structural response when it is in proportion to that part of the ground motion which does cause significant response. A better indicator of ground motion severity for most structures is peak ground velocity. An even better measure is the area under the velocity response spectrum curve between the periods of vibration most important to the response of the structure being analyzed. This parameter is termed spectrum intensity (ref. 2) and can be used in developing response spectra as described in ref. 3.

9. The first methods proposed for developing response spectra were based on standard spectrum shapes determined from historical accelerograms recorded over a wide range of geological and seismological conditions and were independent of the characteristics at any particular site. With additional data from historical earthquakes, it is now apparent that spectral shapes are largely influenced by site conditions such as whether the site is composed of rock or soil. Since concrete dams are founded on rock, only site-dependent spectra developed for rock sites should be used to formulate ground motions for seismic design and analysis. If site-independent spectra are used, the level of computed structural response may be substantially greater than the level that should be predicted, especially when the site is close to the causative fault.

10. In addition to conditions of soil or rock, there may be other site-dependent factors such as local geological features, topography, type of faulting, etc., which could significantly influence ground motion at a particular site. Thus, these factors should also be considered during the development of site-dependent response spectra. Although the effects of these characteristics can be theoretically determined to some extent, there is presently insufficient data to verify required assumptions in all but a few cases. Consequently, these effects are usually not included and when they are, their definition is based largely on judgment.

11. Time histories. With an appropriate response spectrum defined, a historical accelerogram can be selected according to how well its computed response spectrum matches the specified response spectrum with or without scaling. If a suitable match cannot be obtained, an artificial accelerogram having a particular duration can be generated on a computer such that its response spectrum is nearly identical to the prescribed spectrum (ref. 3). For a two-dimensional analysis of a gravity dam, an additional accelerogram for vertical excitation is usually required. A three-dimensional analysis of an arch or gravity-arch dam normally requires an additional horizontal accelerogram as well. If suitable historical accelerograms are not available, artificial time histories can again be generated and are usually scaled to have some fraction of the intensity assigned the primary horizontal component.

12. The accelerograms selected or developed for a seismic response analysis are usually assumed to act uniformly across the site. Furthermore, the maximum horizontal component is usually assumed to occur in a direction normal to the axis of the dam at the maximum

section. Although these assumptions generally introduce additional conservatism, there is rarely enough data at a particular site about the propagation of seismic waves to allow more precise specification.

Deterministic and probabilistic analyses

13. The approach to formulating input for seismic design and analysis described above is deterministic. That is, specific input is developed for a particular earthquake occurring at a particular location. At the present time, however, it is not possible to determine the exact magnitude and location of future earthquakes at many sites. Hence, the usual approach is to determine near upper-bound magnitude, location, fault displacement, and ground motion characteristics. Although each parameter may individually be credible, their simultaneous occurrence may not be credible. The probabilistic approach is an attempt to quantify the likelihood of occurrence for these various parameters separately and in combination. Unfortunately, the probability distributions of most earthquake characteristics are either unknown or poorly known because of the lack of adequate data. Also, the interdependence of various earthquake characteristics is not precisely known at present, and accurately determining probabilities of occurrence for various combinations of characteristics is difficult. Even if completely reliable probabilistic determinations of fault displacements and accelerations could be made, criteria setting forth acceptable levels of occurrence probability or risk do not generally exist. Also lacking is a unified probability-based approach whereby the earthquake specification, ground motion definition, and structural response analysis are compatibly evaluated. Consequently, the bulk of analyses for critical structures, including dams, continues to be deterministic in whole or in part.

METHODS OF ANALYSIS

14. Descriptions of the theory for dynamic structural response computations and the various methods used to analyze concrete dams are presented in many publications. Rather than reiterate this information, the present discussions consider selected factors essential to the analysis of concrete dams and sample results from two-dimensional analyses of a gravity dam. These discussions are directed towards the finite element method of analysis because it is currently the most widely used technique for predicting the seismic response of concrete dams. Simpler techniques are usually limited to psuedo-static approximations of inertia loading and thus are inappropriate in terms of providing realistic estimates of structural response. Other techniques, such as trial load and related methods, have capabilities for performing psuedo-dynamic response spectrum or simplified time history analyses. Although there are many instances when such analyses are appropriate, these methods usually do not have suitable representations of hydrodynamic or foundation interaction for large seismic excitations, flexible provisions for varying material properties, and capabilities for modeling fault displacements.

Hydrodynamic interaction

15. Various investigators have shown that the effects of hydrodynamic interaction are significant. The most widely used technique to account for reservoir effects is the Westergaard "added mass" approach (ref. 4). Assuming a horizontal harmonic ground motion acting on a rigid dam of infinite extent, Westergaard showed that hydrodynamic pressures are opposite in phase to the ground motion and therefore can be quantified as equivalent inertia forces. Although this concept has been utilized in the analysis of many concrete dams throughout the world, it neglects coupling between the response of a flexible dam and its reservoir. This is not significant for excitations at frequencies below the natural frequency of the reservoir. In fact, hydrodynamic pressures measured during model studies of an arch dam subjected to low-level harmonic excitations have shown reasonable agreement with the pressure distribution calculated using Westergaard's technique (ref. 5). However, when the frequency of the excitations is such that significant response of the reservoir occurs, neglecting coupling with the dam can introduce considerable error.

16. Over a period of several years, Chopra and Chakrabarti (ref. 6) developed a technique whereby a two-dimensional reservoir-dam system is divided into two substructures. The flexible dam substructure is represented as an assemblage of finite elements and the reservoir substructure as a semi-infinite continuum governed by the wave equation. The response of the total system is computed by combining the complex frequency response function of the hydrodynamic forces with modal frequency response functions of the dam and calculating the response to arbitrary excitation through Fourier integration. While this approach is much more realistic than Westergaard's technique in most cases, the approach is currently limited to gravity dams having vertical upstream faces and does not account for actual reservoir geometries. A comparison of the hydrodynamic pressures and resulting structural response computed using both the Westergaard and substructuring techniques is presented in a later section.

17. The development of complex frequency response functions for arch dam-reservoir systems with general boundary conditions has not been completely successful to date. One approach which is being tested in the United States (ref. 7) involves dividing the submerged face of an arch dam into a number of subsurfaces as defined in a finite element model. For each subsurface, the face is assumed to be vertical and the curvature, water depth, and angle to the abutment, are assumed to be constant. The complex frequency response functions of the hydrodynamic forces acting on each subsurface are summed and

combined with the modal frequency response functions of the dam. This procedure is similar to that described above for gravity dams and also uses Fourier integration to compute response to arbitrary excitation. While satisfactory results have been obtained for simple cases, further studies including comparisons with results from laboratory model studies are required to confirm the validity of this technique for the solution of general arch dam-reservoir systems.

18. Other methods for realistically accounting for hydrodynamic interaction include finite difference and finite element representations of the reservoir. Although these techniques can be used to idealize reservoirs having arbitrary geometry and various boundary conditions, additional computer resources are required which may exceed those remaining after formulating a finite element model of a dam and its foundation. This is particularly likely for arch dam systems where these methods would otherwise be especially useful. Consequently, the application of these techniques has thus far been relatively limited.

Foundation interaction

19. The interaction between a dam, reservoir, and foundation are intertwined and should not be analyzed separately. Conceptually, the foundation can be treated as a substructure and complex frequency response functions determined. These expressions could then be combined with those describing hydrodynamic forces and modal responses of the dam. While this seems feasible for two-dimensional gravity dam systems, extending this concept to three-dimensional arch dam systems currently appears intractable.

20. The current practice is to treat foundation-dam interaction separately from reservoir-dam interaction. Although the foundation of a gravity dam may usually be considered as a semi-infinite elastic continuum, the foundations of both gravity and arch dams are usually modeled as part of finite element idealizations. The principal error caused by this approach is the introduction of fictitious boundaries which are treated as rigid supports (ref. 8). Thus, vibration waves in the modeled foundation are reflected back towards the dam rather than propagating continuously away from the dam. There are two ways to reduce the effect of these boundary reflections. First, "quiet boundary" formulations may be used to "absorb" vibration waves at the boundary. Unfortunately, such formulations are only partially successful in eliminating boundary reflections in either two- or three-dimensional problems (ref. 9). The second and more satisfactory approach is to model enough foundation such that spurious vibrations are absorbed by internal damping. Extending the modeled foundation the maximum height of the dam upstream and downstream in radial directions, one-half the maximum height tangentially from the abutments, and the height of the dam as it varies from left to right in depth, has been found to produce satisfactory results.

21. When the foundation is formulated as part of a finite element model, the mass of the foundation must be specified to be consistent with the definition of input ground acceleration. Normally, the input motion is defined to occur at the base of the structure along the top surface of the foundation, and structural response is computed relative to this surface. However, if the mass of the foundation is realistically specified, the structural response will be calculated relative to the base of the modeled foundation. This can cause significant errors since the displacements along the base of the dam will not be zero. To prevent these errors and ensure that calculated accelerations along the base of the dam are equal to and in phase with the ground acceleration used as input, the mass of the modeled foundation should be specified to be zero. Unfortunately, an idealized foundation model without mass does not totally model foundation-dam interaction. An alternate approach would be to develop base rock motions for use as input rather than surface motions. But few direct observations of such motions have been made and little analytical effort has been expended to compute base rock motions from given surface motions.

Damping

22. The primary energy loss mechanism currently assumed in the analysis of concrete dams is viscous damping. The term viscous damping implies that damping forces are proportional to the velocity response of the structure and thus are frequency dependent. Although damping forces may actually be frequency independent, approximating damping forces as viscous usually produces reasonable results.

23. Equivalent viscous damping constants have been determined experimentally. Shaking tests using low-level excitations have been performed on concrete dams throughout the world, and damping ratios of 2 to 5 percent of critical are frequently reported. However, damping ratios as high as 10 percent of critical have been measured during higher levels of excitation (ref. 9). These tests indicate that energy losses increase as displacements become large which is a reasonable result since large displacements are accompanied by small openings of vertical contraction joints. A significant dependence on other factors such as reservoir elevation or frequency of vibration has not been reported. Therefore, damping ratios of 2 to 10 percent presently appear reasonable for most concrete dams. The choice for a particular seismic analysis should be based primarily on the expected level of displacement and may require adjustment after an analysis has been performed.

Linear and nonlinear analyses

24. Evaluations of the seismic response of concrete dams are largely based on linear elastic analyses. However, when computed stresses or strains exceed elastic limits, evaluations must be based on nonlinear analyses, qualitative data, and judgment.

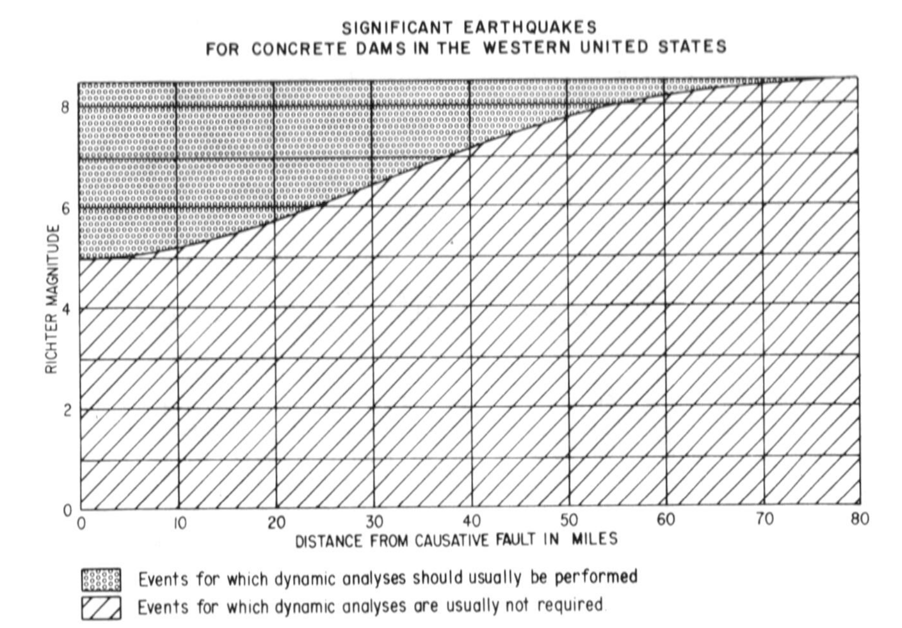

Fig. 1. Significant earthquakes for concrete dams

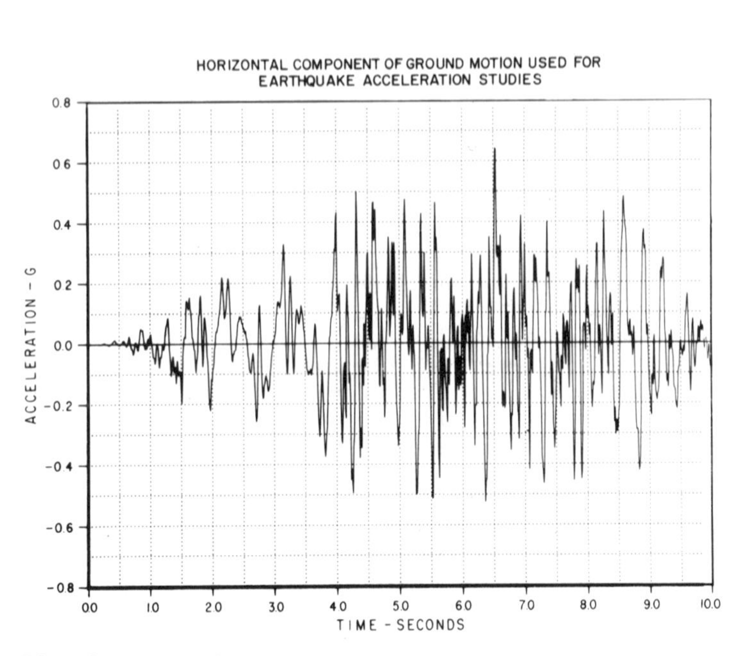

Fig. 2. Upper bound M_L 6.5 accelerogram

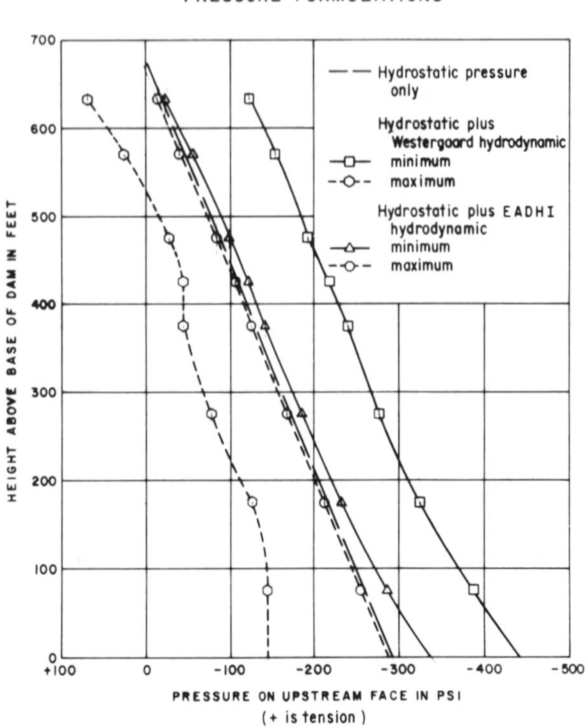

Fig. 3. Comparison of hydrodynamic pressures

Although further verification and refinement is needed, nonlinear material models developed specifically for concrete are available for use in finite element analyses. One such model (ref. 10) allows for material weakening under increasing compressive loads. Triaxial failure envelopes are used to account for multiaxial stress conditions and to define failure in tension and crushing in compression. There are also provisions to model post-cracking and crushing behavior including unloading, reloading, and deactivation of tensile failure planes. Using this model, analyses for fault displacement effects can be performed for both gravity and arch dams provided that the fault displacement is treated as a static loading condition. Selected results from such an analysis for a gravity dam are shown and discussed in the following section. Nonlinear dynamic response analyses are theoretically possible and will become practical when the extensive computer resources required and computational expense become reasonable.

Sample analyses of a gravity dam

25. Sample analytical results are presented for a 680-ft (210-m) high gravity dam. The earthquake considered is a local magnitude 6.5 occurring at a shallow focal depth immediately adjacent to the dam producing severe ground accelerations and reverse fault displacement of approximately 1 ft (0.3 m) underneath the dam. An upper-bound estimate of the maximum ground acceleration component is shown in Fig. 2. This accelerogram was artifically generated such that its response spectrum matches a prescribed site-dependent spectrum. A vertical accelerogram was also generated for the analyses using the same site-dependent spectrum scaled down by a factor of 0.6.

26. Inertia loading and hydrodynamic effects. Two different techniques were used to account for hydrodynamic interaction in conjunction with five-mode linear time history analyses of a plane strain finite element model of the maximum section of the dam and its foundation. Westergaard's added mass technique was used along with the general purpose computer program SAPIV (ref. 11) in one case, and in the other, the program EADHI (ref. 12) was used which incorporates the frequency domain substructuring technique developed by Chopra and Chakrabarti. Fig. 3 shows a comparison of the maximum and minimum total water pressures on the upstream face with the hydrodynamic portions computed using these two formulations. The reservoir water surface elevation was assumed near the top of the dam, and both horizontal and vertical accelerations were considered as described above. Although the general shapes of the bounded pressure distributions are similar, significantly larger variations from hydrostatic pressure are predicted with Westergaard's added mass approach. Also, negative pressures are computed to occur on the upper portion of the upstream face with Westergaard's method. This indicates that in the analytical model, the reservoir at times "pulls" the dam upstream which physically cannot happen. These differences have pronounced effects on the stresses calculated in the dam as shown in Fig. 4. The stresses shown are the maximum principal tensions computed at any instant of time during the analyses on both the upstream and downstream faces. The stresses include the effects of gravity, hydrostatic pressure, hydrodynamic pressure, and inertia loading for damping equal to 10 percent of critical. A comparison of maximum principal compressive stresses shows differences similar to those in Fig. 4. Although maximum instantaneous stresses provide only a partial indication of the structural response, this comparison shows that Westergaard's technique is generally excessively conservative for the case considered. For lower height dams, or for ground accelerations which are less severe or having a different frequency content, the differences between Westergaard's technique and more precise formulations may not be as large. Similar comparative studies are currently being performed for arch dams of various geometrical shapes. Although conclusive results have not been obtained, it is expected that differences attributable to hydrodynamic formulations will be at least as large as for gravity dams.

27. Reverse fault displacement. The finite element model used to assess the effects of 1-ft (0.3-m) reverse fault displacement together with gravity and water loads is shown in Fig. 5. The dam, jointed foundation rock, and fault were represented with nonlinear plane strain elements and the initial analysis performed with the computer program ADINA (ref. 10). For the case studied, the fault was assumed to have a strike perpendicular to the plane of the maximum section extending from the base of the modeled foundation to an area near the midpoint of the base of the dam. The modeled fault corresponds to a 5-ft (1.5-m) wide zone and was treated as a nonlinear elastic orthotropic material. Both the foundation and the dam were modeled with the nonlinear concrete model described in paragraph 24. The concrete was assumed to have a compressive strength of about 8,000 lb/in^2 (55.1 MPa) and a tensile strength of 500 lb/in^2 (3.45 MPa). The jointed foundation rock was assumed to have a tensile strength of less than 100 lb/in^2 (0.69 MPa) based on fracture mechanics considerations. The fault displacement was accomplished with concentrated forces applied parallel to the fault by means of one-dimensional stiffness elements along the right edge and base of the modeled foundation downstream of the fault.

28. In the analysis, the gravity load was applied first, followed by the water load, and then the fault displacement. Each of these loadings was applied incrementally, and resulting tensile cracks at gauss points are portrayed in fig. 5 with small dashes. The application of gravity and water loads results in small areas of anomalous cracking in the foundation caused by boundary effects, lumped gravity loading, and the behavior of the anisotropic fault. This cracking is shown in

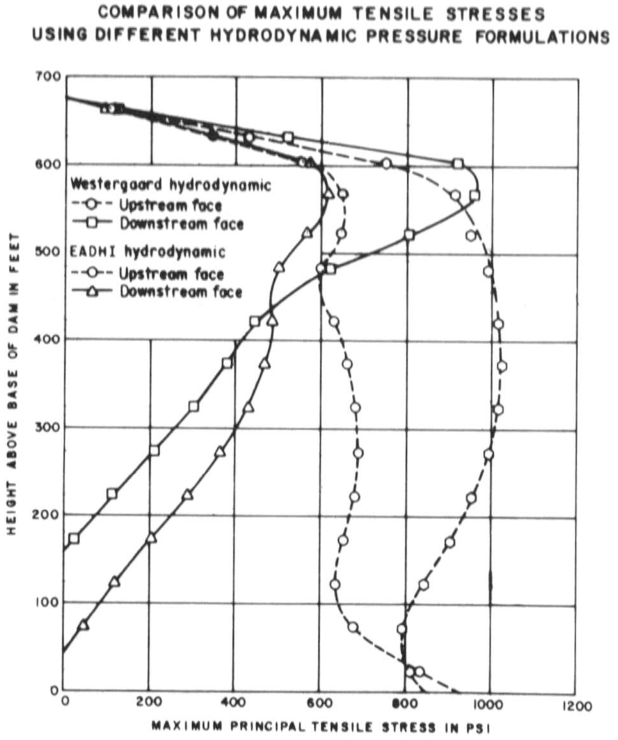

COMPARISON OF MAXIMUM TENSILE STRESSES
USING DIFFERENT HYDRODYNAMIC PRESSURE FORMULATIONS
Westergaard hydrodynamic
Upstream face
Downstream face
EADHI hydrodynamic
Upstream face
Downstream face
HEIGHT ABOVE BASE OF DAM IN FEET
MAXIMUM PRINCIPAL TENSILE STRESS IN PSI
0
100
200
300
400
500
600
700
0
200
400
600
800
1000
1200

Fig. 4. Comparison of maximum tensile stresses

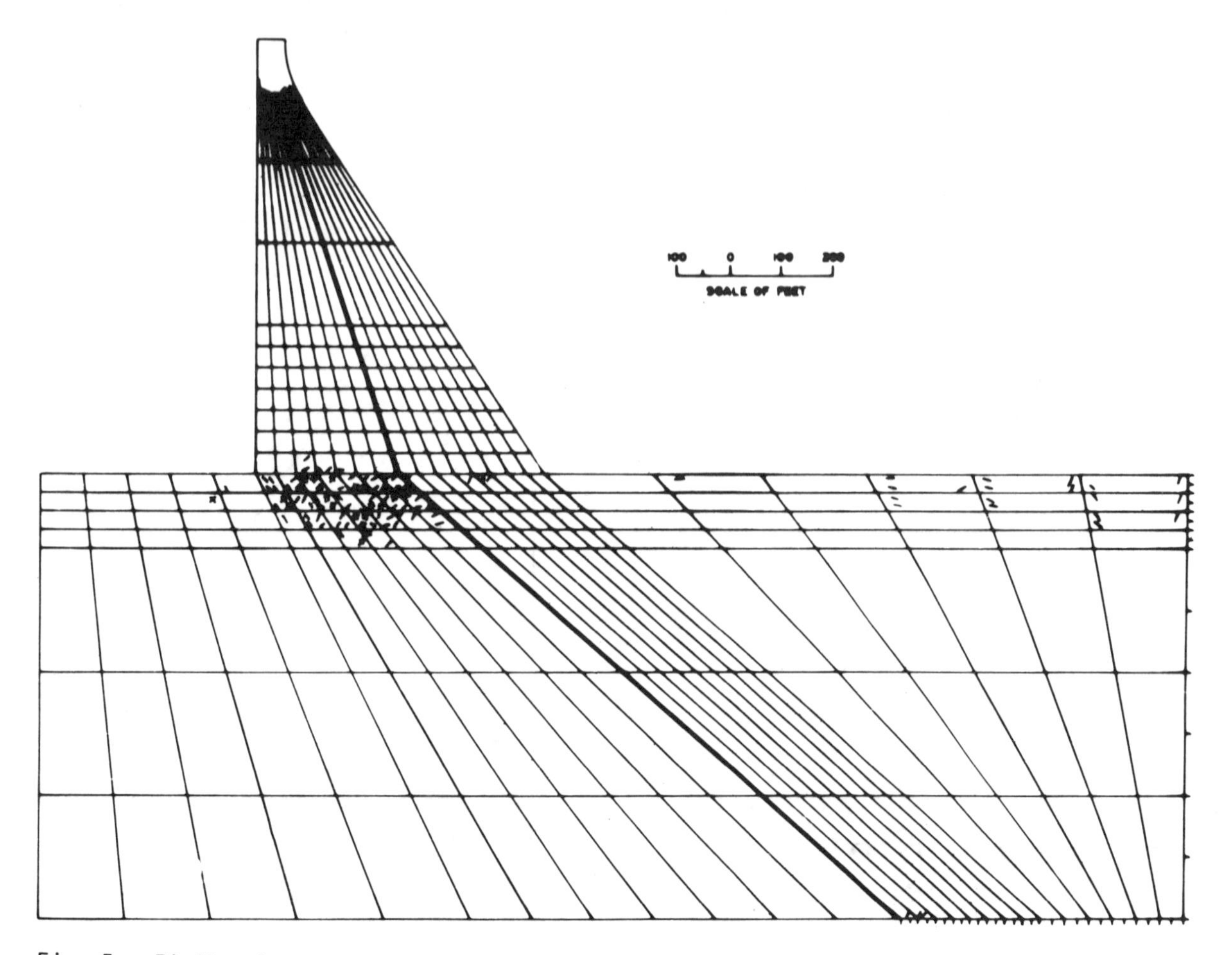

Fig. 5. Finite element mesh for reverse fault displacement studies

Fig. 5 on the left side of the model near the surface, deep in the foundation near the fault, and at isolated locations under the downstream half of the dam. This spurious behavior of the numerical model is overshadowed by the tensile cracking due to fault displacement which is largely restricted to the foundation under the upstream half of the dam. Since the foundation rock is already jointed, the results indicate that the upstream part of the dam tends to lift off the foundation as it is thrust upward, and joints in the foundation rock open upstream of the fault. Fig. 5 also shows minor cracking within the base of the dam. However, this cracking cannot physically occur because the necessary tension cannot be transmitted by the jointed foundation rock.

29. These results show that for reverse displacement on the idealized fault, the dam lifts without cracking. Consequently, stresses can be determined by means of simplified linear elastic studies in which the foundation elements underneath the dam where tension is indicated are iteratively softened until foundation support with fault displacement is provided by zones of compression only. The results of these studies are shown in Fig. 6 and Fig. 7 and indicate that the primary effects of the fault displacement considered are stress concentrations in the vicinity of foundation support.

30. For cases where the fault contacts the base of the dam near the downstream toe, the dam remains in contact with the foundation downstream of the fault, and joints in the foundation rock again open upstream of the fault. The weight of the dam allows it to tilt upstream slightly maintaining contact with the foundation under the upstream face. The effects of this condition are also stress concentrations near the areas of foundation support. Similarly, if the fault displacement was normal rather than reverse, two modes of two-dimensional behavior could occur depending on the orientation of the fault. Although stress distributions would be somewhat different following normal fault displacement, the magnitude of the stress concentrations would not be significantly greater than those caused by reverse fault displacement.

31. Studies are currently underway in which reverse faulting in a direction perpendicular to that considered above and various orientations of strike-slip faulting are modeled. Three-dimensional studies of fault displacements underneath arch dams are also planned.

32. In order to adequately assess the effects of fault displacement, it is imperative that the stability of the foundation also be evaluated along with the structural stability of the dam. As a result of fault displacement, loading on the foundation will be redistributed, and seepage conditions together with pore pressures could be significantly altered.

33. Combined effects of fault displacement and inertia loading. If the fault displacement is assumed to occur prior to the generation of

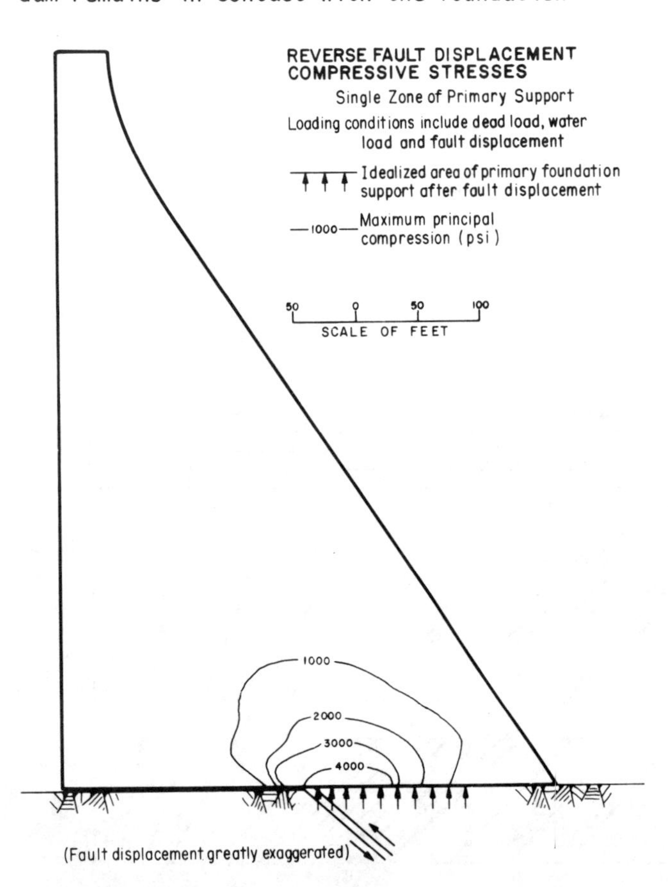

Fig. 6. Compressive stress concentration

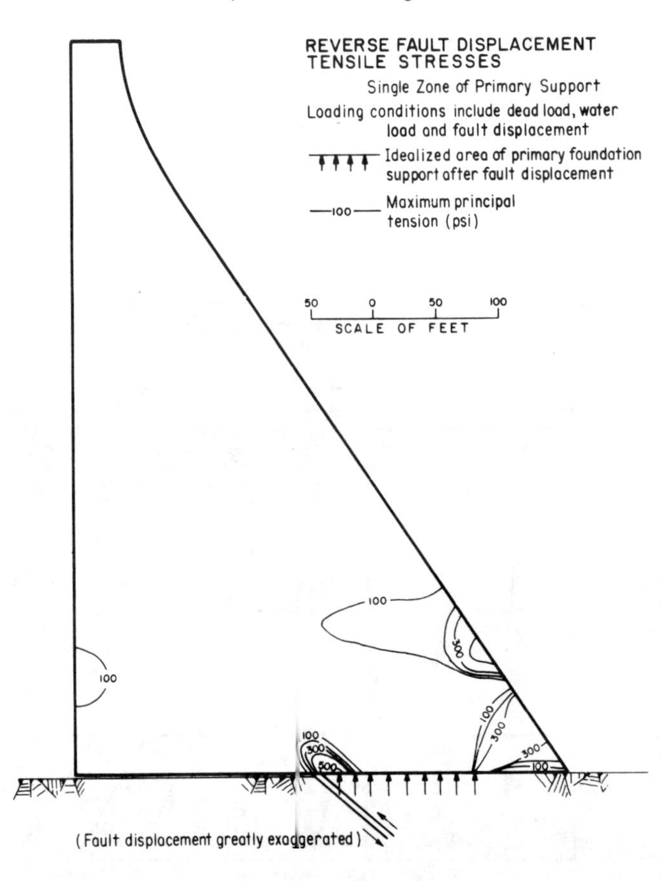

Fig. 7. Tensile stress concentrations

vibratory ground motion, and if the dam remains linear elastic as in the above example, then the combined effects of fault displacement and inertia loading can be accounted for by performing dynamic response analyses using a linear finite element model with softened foundation elements as described in paragraph 29. The stresses computed in this manner for the combined condition may be predominantly less than those determined for vibratory ground motion alone because of the reduced stiffness of the dam-foundation system after fault displacement. The reduced stiffness is accompanied by a reduction in the natural frequencies of the system. Since the intensity of ground shaking close to causative faults generally increases with frequency in the frequency range of most concrete dams, a reduction of the natural frequencies generally indicates that the level of dynamic response will be reduced as well.

RESPONSE EVALUATION AND OTHER CONSIDERATIONS

34. Currently, there are no strict criteria which can be used to fully evaluate the performance of concrete dams subjected to severe earthquakes. Although traditional criteria setting forth factors of safety based on concrete strengths are frequently used, they do not provide a basis for assessing overall safety of a concrete dam. For example, computed tensile stresses from linear elastic analyses may exceed the tensile strength of the concrete. However, even if cracking should occur, structural failure or instability will not necessarily develop either during or after the seismic event. Unfortunately, nonlinear dynamic response analyses of cracked dams are currently not practical because of limited data concerning triaxial stress-strain relationships up to failure and the inordinate computer resources needed. Consequently, it is common practice to require that the prevalent level of computed tensile stress be no greater than the tensile strength of the concrete. However, it is usually acceptable to allow tensile stresses exceeding the tensile strength of the concrete at a limited number of locations, for brief instants of time, for earthquakes having remote probabilities of occurrence. Computed compressive stresses, on the other hand, should not be allowed to exceed the compressive strength. Significant damage to concrete is more likely to result from compressive failure than from tensile failure, and concrete dams, especially arch dams, rely on the capacity to withstand compression more than tension to safely resist static loads. For gravity dams, stability against sliding and overturning may also need to be considered. Although, instantaneous inertia forces may be large enough to cause instantaneous instability, these forces are so transitory that actual sliding or overturning failure of a well-designed dam is unlikely. However, if the potential exists for tensile cracking through the entire thickness of a gravity or arch dam, the orientation of such cracking may promote sliding instability and must be considered.

35. If the seismic performance of a proposed concrete dam is judged to be unsatisfactory, there are several design modifications, in addition to simply increasing the strength of the concrete, which may be beneficial. The shape of the dam should be reviewed and revised as necessary to minimize any abrupt changes in stiffness or mass. Sensitivity studies might be considered to estimate the optimum distributions of these parameters. If the extent of cracking in an arch dam is unacceptable, reshaping may also be used to increase the compressive stresses caused by static loads and reduce the cracking in the areas of concern.

REFERENCES

1. LOUDERBACK G.D. Faults and engineering geology. Symposium on application of geology to engineering practice, Geological Society of America, 1950, November, 125-150.
2. HOUSNER G.W. Intensity of earthquake ground shaking near the causative fault. Proceedings, third world conference on earthquake engineering, New Zealand, 1965, 3, 94-115.
3. TARBOX G.S., DREHER K.J., and CARPENTER L.R. Seismic analysis of concrete dams. Transactions of the thirteenth international congress on large dams, New Delhi, 1979, 2, 963-994.
4. WESTERGAARD H.M. Water pressures on dams during earthquakes. Proceedings, American Society of Civil Engineers, 1931, November, 1303-1318.
5. NORMAN C.D., CROWSON R.D., and BALSARA J.P. Dynamic response characteristics of a model arch dam. Report N-76-3, U.S. Army Engineers Waterways Experiment Station, Vicksburg, Mississippi, 1976, March.
6. CHAKRABARTI P. and CHOPRA A.K. Earthquake analysis of gravity dams including hydrodynamic interaction. International journal for earthquake engineering and structural dynamics, 1973, 2, October-December, 143-160.
7. ADAP2. PMB Systems Engineering, Inc., San Francisco, California, 1977.
8. CLOUGH R.W. and ZIENKIEWICZ O.C. Finite element methods in analysis and design of dams, part c. Proceedings of an international symposium on criteria and assumptions for numerical analysis of dams, Swansea, 1975, September.
9. HATANO T. Aseismic design criteria for arch dams in Japan. Proceedings, ninth international congress on large dams, Istanbul, 1967, 4, 1-17.
10. BATHE K.J. ADINA, a finite element program for automatic dynamic incremental nonlinear analysis. Report 82448-1, Acoustics and vibration laboratory, Massachusetts Institute of Technology, Cambridge, 1977, May.
11. BATHE K.J., WILSON E.L., and PETERSON F.E. SAPIV, a structural analysis program for static and dynamic response of linear systems. Report EERC 73-11, Earthquake engineering research center, University of California, Berkeley, 1974, April.
12. CHAKRABARTI P. and CHOPRA A.K. A computer program for earthquake analysis of gravity dams including reservoir interaction. Report EERC 73-7, Earthquake engineering research center, University of California, Berkeley, 1973, June.

20 Computational models for the transient dynamic analysis of concrete dams

O. C. ZIENKIEWICZ, FRS, and E. HINTON, PhD, University College, Swansea; and N. BIĆANIĆ, PhD, and P. FEJZO, Dipl Ing, Gradevinski Institut, Zagreb

With the progress in the development of the finite element method and the simultaneous increase of computer speed and capacity, full non-linear, step by step analysis of structures subject to earthquakes can be carried out today at reasonable cost. It is thus feasible to check the performance of dam designs subject to extreme conditions of earthquake and to ascertain the degree of damage incurred providing the behaviour of the materials can be adequately modelled. In this paper we present briefly some essential features of a model used in the analysis of concrete dams, together with numerical studies of a typical case. In particular attention is focussed on a new rate dependent viscoplastic concrete model.

INTRODUCTION

1. Nonlinear, finite element analysis of structures subject to earthquake shock is today possible due to advances made in computer technology and numerical methods. With suitable modelling of the material behaviour, reasonably accurate predictions of permanent deformation and other damage can be made and the engineer can assess the performance of his structures subject to extreme earthquakes. Clearly in the design of dams such considerations are of extreme importance in view of the damage potential. The object of this paper is thus to

(a) present some general techniques currently available, and
(b) to discuss typical models of material behaviour adopted for concrete dams in recent studies.

ANALYSIS TECHNIQUE

2. A typical problem of dam earthquake analysis is illustrated in Fig. 1. Here, as shown, the dam, fluid and foundation regions are "discretised" using a similar type of element throughout. We shall not discuss the details of the procedures and formulation here as they can be found elsewhere, (refs. 1-3). Instead we shall simply state the basic outline of the equations necessary. Thus, if the primary unknowns are the displacements

$$\underline{u}^T = [u_x, u_y, u_z] \qquad (1)$$

and these are "discretised" by the shape functions N_i and nodal displacement parameters $\bar{u}_i$ as

$$\underline{u} = \sum_{i=1}^{n} N_i \bar{\underline{u}}_i = \underline{N}\,\bar{\underline{u}} \qquad (2)$$

we can write the full equilibrium statement in an approximate form (which represents a set of n-ordinary differential equations suitable for the study of dynamic problems) as

$$\int_\Omega \bar{\underline{B}}^T \underline{\sigma}\, d\Omega + \underline{M}\,\ddot{\bar{\underline{u}}} + \underline{C}\,\dot{\bar{\underline{u}}} = \underline{f} \qquad (3)$$

In the above equation $\bar{\underline{B}}$ is a matrix defining strain increments as

$$d\underline{\varepsilon} = \underline{B}\, d\bar{\underline{u}} \qquad (4)$$

and this can be related to large displacement (Lagrangian) forms. The matrices $\underline{M}$ and $\underline{C}$ are the mass and damping matrices respectively and these can be easily computed (although insertion of a damping matrix represents, as is well known, some arbitrary features).

3. The earthquake input is contained in the specified force vector $\underline{f}$ and if there is no other external loading, we may write

$$\underline{f} = -\underline{M}\,\ddot{\bar{\underline{u}}}_g \qquad (5)$$

where $\ddot{\underline{u}}_g$ is the vector of earthquake accelerations. Furthermore, if damping is neglected, the acceleration at time "t" is approximated by a central difference approximation

$$\ddot{\underline{u}}_t = \frac{1}{(\Delta t)^2}\left(\bar{\underline{u}}_{t+\Delta t} - 2\,\bar{\underline{u}}_t + \bar{\underline{u}}_{t-\Delta t}\right), \qquad (6)$$

and the mass matrix $\underline{M}$ is lumped i.e. diagonalised (a successfully used procedure is proposed in ref. 4), system (3) uncouples to a system of independent equations (7) which for each degree of freedom "i" has the form

$$\bar{u}^i_{t+\Delta t} = -\frac{(\Delta t)^2}{M^i}\left[\int_V \underline{B}^T \underline{\sigma}(t)\, dV\right]^i + \\ + \left(2\bar{u}^i_t - \bar{u}^i_{t-\Delta t} - (\Delta t)^2\, \ddot{u}_g(t)\right) \qquad (7)$$

Here $\left[\int_V \underline{B}^T \underline{\sigma}(t)\, dV\right]^i$ indicates that integration is only to be performed over the domain which contributes to the internal force for a particular degree of freedom "i".

4. To complete the formulation appropriate material behaviour laws must be specified and

then the solution of the equation system (7) coupled with the system of constitutive equations (see below) can be accomplished in a step by step manner using explicit processes which are generally very economical and relatively easy to implement even on small sized computers.

5. Some general problems of analysis are mentioned briefly below.

- The water retained by the dam can be discretised in the same manner as the elastic solid by assigning to it a very low shear modulus (ref. 2).
- It is quite important that proper radiation boundary conditions be imposed at "infinite" boundaries. A new and powerful technique for radiation isolation was recently developed and is presented in ref. 5.
- To ensure stability in explicit computations it is necessary to restrict the time step Δt to about one-half of the time taken for a compression wave to travel the distance between any two nodes of the mesh.

MODELLING OF CONCRETE BEHAVIOUR

6. Most of the concrete models developed so far are for use in static analysis under monotonic loading conditions and are based on data collected during static experiments on concrete specimens (refs.6-8). However, experimental data for plain concrete subjected to dynamic loading (refs. 9-12), although scarce and obtained from uniaxial tests only, clearly shows that

- the strength of plain concrete and its initial elasticity modulus depend on the straining (loading) rate, (Fig. 2)
- the failure strain remains almost constant for any rate of loading for a particular concrete (Fig. 2)
- the cyclic compressive loading produces a pronounced hysteresis effect in the stress-strain curve (Fig. 3)
- the stress-strain curves under compressive load histories possess an envelope curve which may be considered unique and identical to the stress-strain curve under the constant strain rate test (Fig. 3).

7. Obviously for dynamic loading conditions strain rate sensitive models are needed and it is only recently that some have been reported (refs. 13,14). A new strain rate and stress history dependent numerical model for plain concrete is now proposed. The model is essentially a modification of Perzyna's elasto-viscoplastic model and the identification of the model parameters is based on uniaxial experiments by Hatano and co-workers (refs. 11,12,15,16).

8. Firstly some basic expressions from the viscoplastic theory are presented. Fig. 4 illustrates the viscoplastic rheological model. At any instant the total strain can be written as

$$\underline{\varepsilon} = \underline{\varepsilon}^{e} + \underline{\varepsilon}^{vp} \tag{8}$$

with $$\underline{\varepsilon}^{vp} = \int_t \dot{\underline{\varepsilon}}^{vp}\, dt \tag{9}$$

Rates of stress and elastic strain are related through the expression

$$\dot{\underline{\sigma}} = \underline{D}\, \dot{\underline{\varepsilon}}^{e} \tag{10}$$

Viscoplastic behaviour is initiated when the yield surface defined as

$$F = F(\sigma_1, \sigma_2, \sigma_3) = 0 \tag{11}$$

is reached.

9. The rate and direction of viscoplastic strain are defined by a flow rule usually of the form proposed by Perzyna (ref. 17)

$$\dot{\underline{\varepsilon}}^{vp} = \gamma < \Phi\ (F) > \frac{\partial Q}{\partial \underline{\sigma}} \tag{12}$$

where the plastic potential Q defines the direction of viscoplastic strain and the fluidity parameter γ and an appropriately chosen function $<\Phi(F)>$ controls the amount of viscoplastic straining. In a case when the plastic potential Q is made equal to the yield surface F, expression (12) can be written as

$$\dot{\underline{\varepsilon}}^{vp} = \gamma < \Phi\ (F) > \frac{\partial F}{\partial \underline{\sigma}} \tag{13}$$

which is an associative viscoplastic flow rule. By combining (13) and (10) the constitutive relationship can be written as

$$\dot{\underline{\varepsilon}} = \dot{\underline{\varepsilon}}^{e} + \dot{\underline{\varepsilon}}^{vp} = \underline{D}^{-1}\, \dot{\underline{\sigma}} + \gamma < \Phi\ (F) > \frac{\partial F}{\partial \underline{\sigma}} \tag{14}$$

Discretized in time (9) can be approximated by

$$\underline{\varepsilon}^{vp}_{t+\Delta t} = \underline{\varepsilon}^{vp}_{t} + [(1-\beta)\dot{\underline{\varepsilon}}^{vp}_{t} + \beta\dot{\underline{\varepsilon}}^{vp}_{t+\Delta t}]\Delta t \tag{15}$$

with β ranging from zero (Euler explicit forward integration) to one (implicit backward integration). In the former case viscoplastic strain at the end of the time increment is

$$\underline{\varepsilon}^{vp}_{t+\Delta t} = \underline{\varepsilon}^{vp}_{t} + \Delta\underline{\varepsilon}^{vp}_{t} = \underline{\varepsilon}^{vp}_{t} + \dot{\underline{\varepsilon}}^{vp}_{t}\ \Delta t \tag{16}$$

10. In the present model (ref. 18) material behaviour is described using two surfaces in stress space - the discontinuity surface F_D and the bounding failure surface F_F (Fig. 5). During inelastic straining, when a stress point S is outside the discontinuity surface F_D, both surfaces F_D and F_F change in a manner which depends on the amount of accumulated damage expressed as inelastic work W^p. Surface F_D also depends on the memory parameter κ which is related to the post-failure regime. Hence

$$F_D(\underline{\sigma}, W^p, \kappa) = 0 \tag{17}$$

$$F_F(\underline{\sigma}, W^p) = 0 \tag{18}$$

Perzyna's viscoplastic associative flow rule (13) is modified by making the fluidity parameter γ dependent on the rate of the elastic strain, i.e.

$$\dot{\underline{\varepsilon}}^{vp} = \gamma\ (\dot{\underline{\varepsilon}}^{e}) < \Phi\ (F) > \frac{\partial F}{\partial \underline{\sigma}} \tag{19}$$

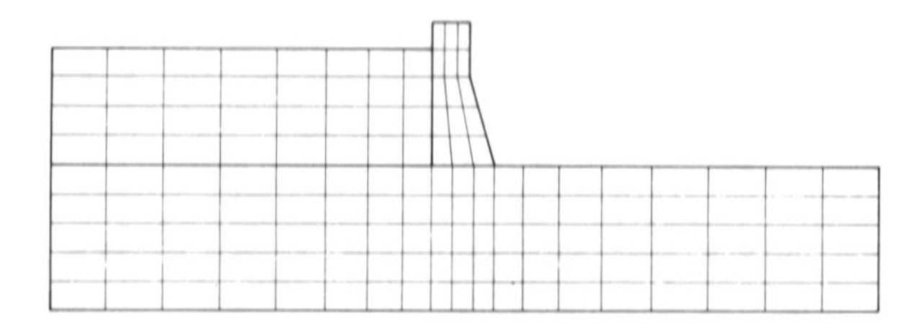

Fig. 1 Finite element mesh of a dam-fluid-foundation system

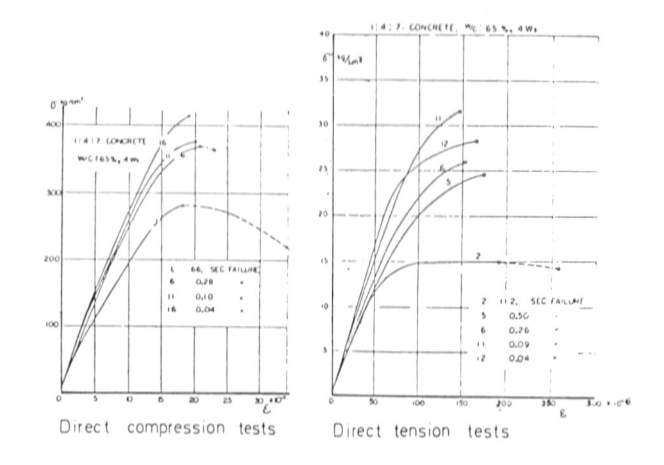

Fig. 2 Uniaxial test results (Hatano)

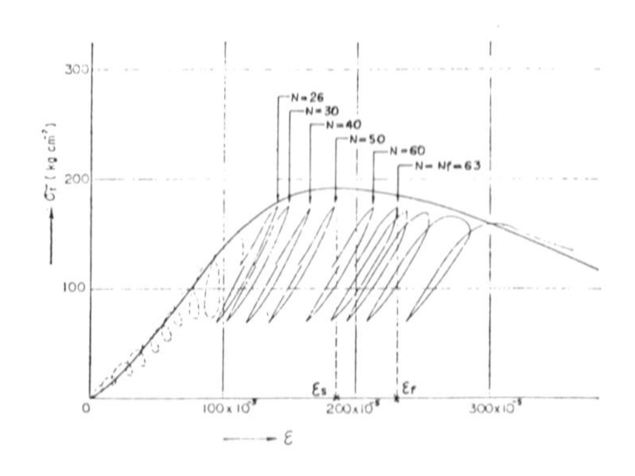

N : Number of loading cycles
N_f: Value of N at failure
ε_s: Strain at failure in static compression test
ε_f: Total strain at failure in fatigue test

Fig. 3 Low cycle fatigue test results (Hatano)

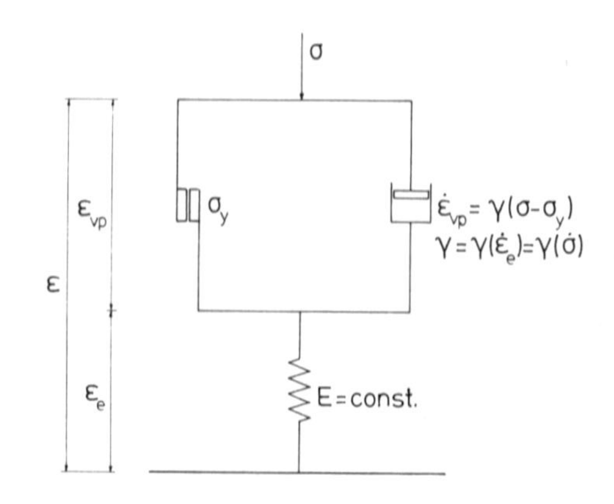

Fig. 4 Viscoplastic rheological model

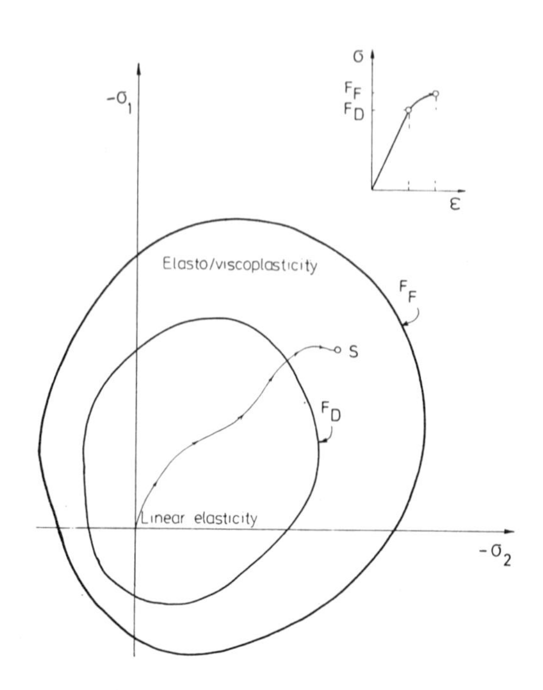

Fig. 5 Surfaces for the description of material behaviour

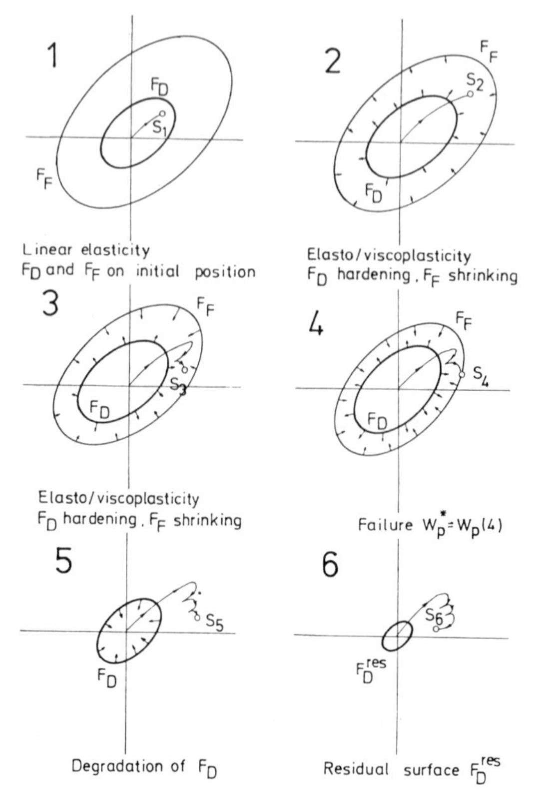

Fig. 6 Changing pattern of the discontinuity and failure surfaces

Since only uniaxial properties (f_{cu} and f_t) are available, expressions (17) and (18) can be rewritten as

$$F_D(\underline{\sigma}, f^D_{cu}(W^p,\kappa),\ f^D_t(W^p,\kappa)) = 0 \qquad (20)$$

$$F_F(\underline{\sigma}, f^F_{cu}(W^p),\ f^F_t(W^p)) = 0 \qquad (21)$$

where

$f^D_{cu}(W^p,\kappa)$, $f^D_t(W^p,\kappa)$ define the change of the discontinuity stress level in compression (f^D_{cu}) and in tension (f^D_t) as a function of W^p and κ

$f^F_{cu}(W^p)$, $f^F_t(W^p)$ define the change of the failure stress level in compression (f^F_{cu}) and in tension (f^F_t) as a function of W^p

W^p is the inelastic work (dissipated energy) density defined as

$W^p = \int_t \underline{\sigma}^T\ \underline{\dot{\varepsilon}}^{vp}\ dt$

and κ is a memory parameter defined below.

11. The basic concept of the model is that both surfaces F_D and F_F change with the accumulated inelastic work. The bounding failure surface F_F is initially defined by the uniaxial stress levels which correspond to brittle fracture $\bar{f}^F_{cu}$ and $\bar{f}^F_t$ i.e. to failure stresses which would be obtained with infinite load rates and no inelastic strains. The surface F_F changes (shrinks) in a manner which depends on the amount of dissipated energy - the change is controlled through the changes of $f^F_{cu}(W^p)$ and $f^F_t(W^p)$. The discontinuity surface F_D is initially defined through the uniaxial stress levels $\bar{f}^D_{cu}$ and $\bar{f}^D_t$ which correspond to the first departure from linear elasticity. The surface F_D is changing (hardening or softening) in a manner which depends on the amount of dissipated energy - the change is controlled through changes in $f^D_{cu}(W^p)$ and $f^D_t(W^p)$. Up to the moment when the stress point reaches the failure surface F_F the rate of viscoplastic straining is governed by the distance of the stress point from the discontinuity surface F_D. After failure takes place the surface F_F loses its significance and the post-failure regime controlling parameter κ is 'switched-on' to control the degradation of the discontinuity surface. Parameter κ is defined as the post-failure dissipated energy density i.e.

$$\kappa = W^p - W^{p*} \qquad (22)$$

in which W^{p*} is the dissipated energy density at a moment of failure. From the above considerations it is clear that the bounding failure surface F_F serves as a monitoring device which enables modelling of the fact that the failure stress depends on the accumulated damage. Fig 6 summarizes the previous discussions.

12. It remains to define the changes of surfaces F_D and F_F in terms of the dissipated energy i.e. to determine the changes of uniaxial parameters upon which these surfaces depend. The change of failure surface parameters is defined as

$$f^F_{cu}(W^p) = \bar{f}^F_{cu}\ c_F(W^p)$$
$$f^F_t(W^p) = \bar{f}^F_t\ t_F(W^p) \qquad (23)$$

The change of discontinuity surface parameters is defined as

$$f^D_{cu}(W^p,\kappa) = \bar{f}^D_{cu}\ c_D(W^p)\,d_c(\kappa)$$
$$f^D_t(W^p,\kappa) = \bar{f}^D_t\ t_D(W^p)\,d_t(\kappa) \qquad (24)$$

Here

$\bar{f}^F_{cu}, \bar{f}^F_t, \bar{f}^D_{cu}, \bar{f}^D_t$ are initial values of uniaxial parameters

$c_F(W^p), t_F(W^p)$ are compressive and tensile failure functions

$c_D(W^p), t_D(W^p)$ are compressive and tensile discontinuity functions

$d_c(\kappa), d_t(\kappa)$ are compressive and tensile degradation functions (d_c=1, d_t=1 in the pre-failure regime)

The change of uniaxial parameters with dissipated energy is illustrated in Fig. 7.

13. A Mohr-Coulomb surface (the only one which has the correct principal characteristics for concrete behaviour and which can be defined using only two uniaxial test parameters f_{cu} and f_t) has been chosen for the description of both the discontinuity and the failure surface. Graphical representation of both surfaces in the prefailure and post-failure regimes is given in Fig. 8.

Identification of material model parameters

14. The identification of material model parameters is based on experiments by Hatano and coworkers (refs. 11,12,15,16) who tested three different concrete mixes by applying monotonic uniaxial loading to failure in compression and tension and compressive periodic loading with different amplitudes and frequencies. Results obtained for a concrete mix 1-4-7 are shown in Figs. 2 and 3.

15. Before the evaluation of the material model parameters a choice of surface parameter functions has to be made. A simple set of such functions is shown on Fig. 9, i.e. the discontinuity surface is assumed to remain constant up to failure after which it degrades to the residual collapse surface and the failure surface parameters depend linearly on the dissipated energy.

16. With surface parameter functions chosen, material parameters are evaluated using monotonic test results and are then used to predict behaviour in low cycle fatigue tests. Results are then compared and parameters adjusted in order to fit both monotonic and low cycle fatigue test results. The procedure for the evaluation of the model parameters from monotonic test results is illustrated on Fig. 10 and briefly outlined below.

1. Establish load histories assuming constant

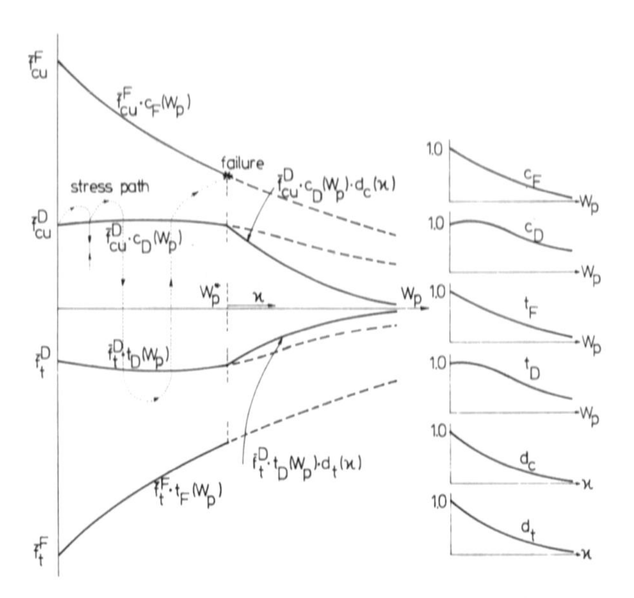

Fig. 7 Change of surface parameters with dissipated energy - general form

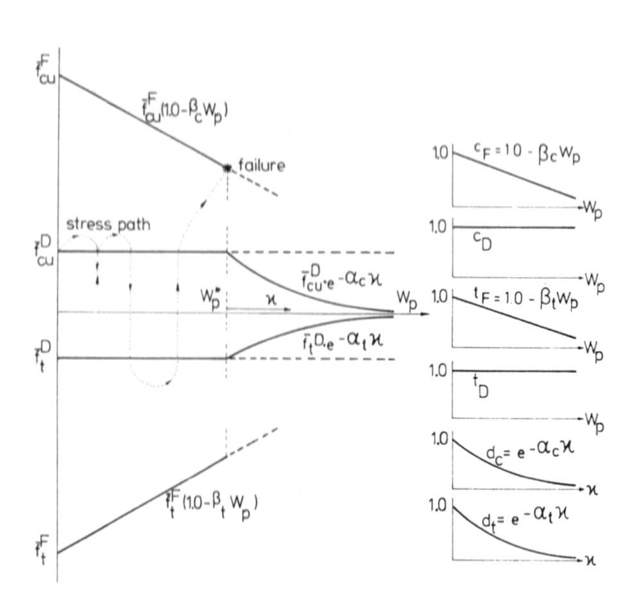

Fig. 9 Change of surface parameters using a set of simple functions

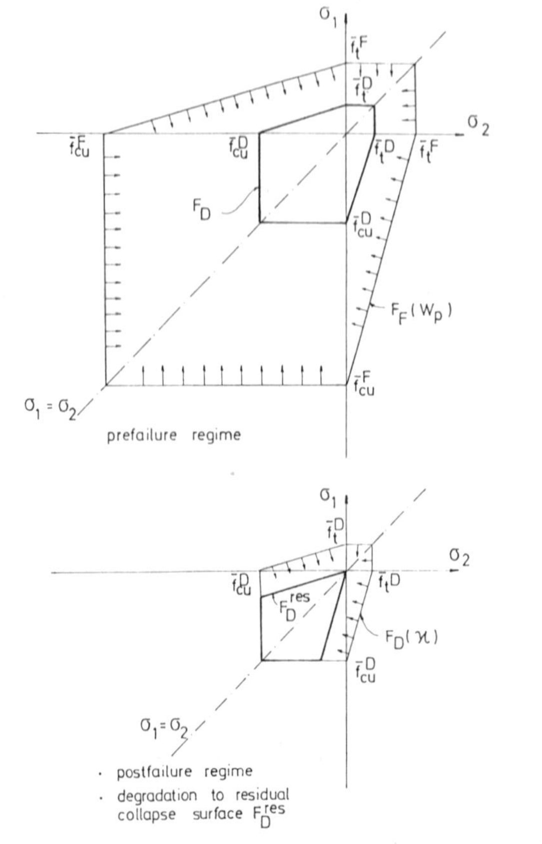

Fig. 8 Mohr-Coulomb surface before and after failure (collapse)

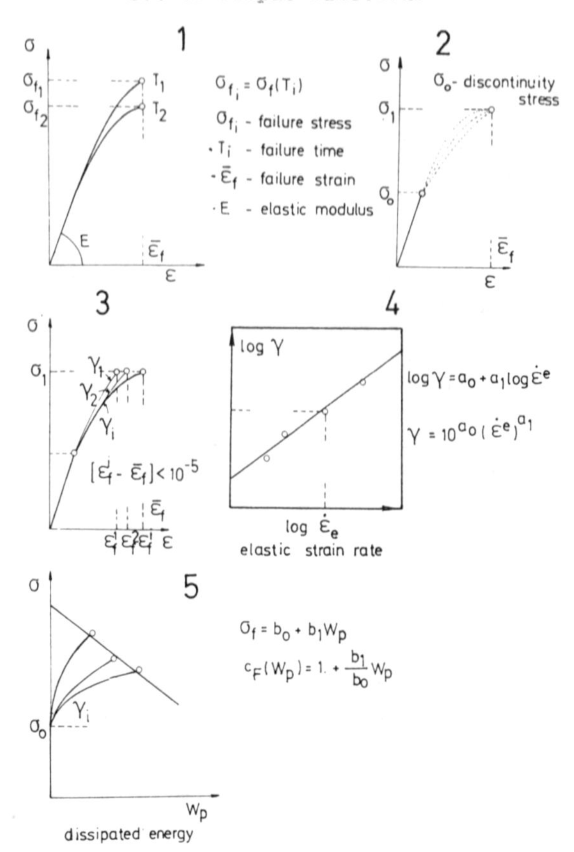

Fig. 10 Evaluation of model parameters

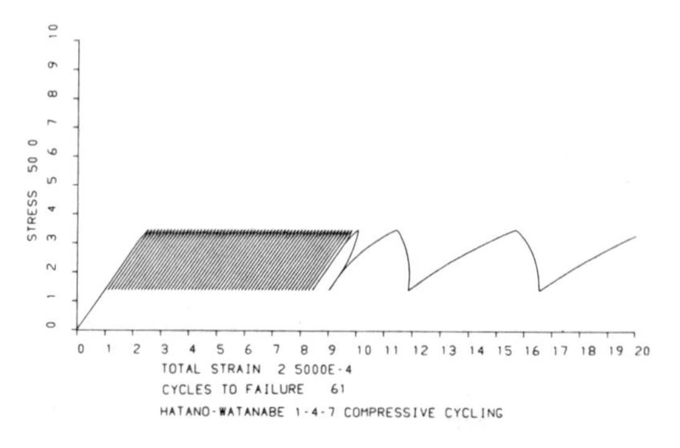

Fig. 11 Computed low cycle fatigue test results

load rate and using Hatano's relations between failure times T_is and failure stresses σ_{fi}s

2. Choose discontinuity stress level σ_o (at approximately 30% of the static strength)
3. Determine the dependence of the fluidity parameter γ on the elastic strain rate $\dot{\varepsilon}^e$
4. Determine the relationship between the fluidity parameter γ and the elastic strain rate $\dot{\varepsilon}^e$. This is reasonably well defined by

$$\log_{10}\gamma = a_o + a_1 \log_{10}(\dot{\varepsilon}^e)$$

or

$$\gamma = 10^{a_o} (\dot{\varepsilon}^e)^{a_1} \tag{25}$$

5. Use expression (25) to repeat the monotonic tests and plot the accumulated dissipated energy for every stress history against the failure stress. The linear fit obtained is given as

$$c_F(W^p) = 1 + \left(\frac{b_o}{b_1}\right) W^p \tag{26}$$

and is compatible with the failure surface parameter function described above. (Admittably a linear fit is more appropriate for moderate and fast strain rates than for all rates including slow ones). Results of a low cycle fatigue test predicted by using the evaluated material parameters for a concrete mix 1-4-7 are presented in Fig. 11.

17. To implement this material model in a standard elasto-viscoplastic finite element computer program and use it in a 2-D or 3-D analysis, it only remains to define a measure of the elastic strain rate upon which a fluidity parameter γ is dependent. Deviatoric strains cause most damage to concrete, so the rate of the second elastic deviatoric strain invariant has been chosen to be that measure. Thus

$$\dot{J}_2(\underline{\varepsilon}_t^e) = \frac{1}{\Delta t}\left[J_2(\underline{\varepsilon}_t^e) - J_2(\underline{\varepsilon}_{t-\Delta t}^e)\right] \tag{27}$$

and the fluidity parameter γ is then defined as

$$\gamma_t = 10^{a_o} (\dot{J}_2(\underline{\varepsilon}_t^e))^{a_1} \tag{28}$$

18. The algorithm for transition from a state at time "t" with known displacements ($\underline{u}_t$), viscoplastic strains ($\underline{\varepsilon}_t^{vp}$) and dissipated energy (inelastic work) density (W_t^p) to a state at time "t+Δt" is shown on a flow diagram in Fig. 12.

SUBSTITUTE SPECTRA

19. In any step by step analysis the computation cost depends heavily on the duration of the dynamic excitation and several attempts (refs. 19,20) have been made to produce short duration substitute accelerograms which would in some sense model the true longer duration phenomena. For linear analysis such short duration accelerograms are satisfactory providing they exhibit a similar frequency response spectrum to the original parent accelerogram.

20. Although this condition is not sufficient in nonlinear approaches, satisfactory results have been obtained using a suitably designed "combisweep" accelerogram (ref. 18). "Suitably designed" means that the substitute accelerogram is not only site and design-parent accelerogram dependent, but that it also takes the structural characteristics into account. In other words, there is a different combisweep accelerorgram for a different structure built on the same site and having the same design response spectrum.

21. From the results obtained it appears that in practice use of such substitute accelerograms is strongly justified at least in the preliminary design stages.

NUMERICAL EXAMPLE

22. With the proposed material model incorporated in a standard elasto-viscoplastic finite element computer program, a nonlinear seismic analysis of a Koyna dam section is now performed. The section is modelled using a very coarse mesh of quadratic eight-noded isoparametric elements and the mass is lumped as described in ref. 4. A real 10 sec-accelerogram (Koyna transversal component) and its short duration substitute 2.23 sec-accelerogram (Koyna combisweep) are used (Fig. 13). The results of these analyses (Fig. 14) compare well, especially in predicting the permanent deformation.

CRACKING OF CONCRETE

23. So far there has been little mention of the tensile behaviour of the concrete. Cracks, a dangerous phenomenon in any structure, pose a special and as yet undetermined risk for dams because of the destabilisation influence of the water entering the cracks under high hydrostatic pressure.

24. The concrete model described above provides a rather crude representation of concrete behaviour in tension. A model involving a cohesionless material with no tensile strength and only a limited compressive residual strength after failure, while being realistic for the post-failure state after failure in compression, clearly underestimates the residual strength after failure in tension. In reality, when failure in tension occurs, in only one direction at a time, the tensile strength in that direction is thereafter reduced to zero, but the compressive strength in that direction and both the tensile and compressive strengths in orthogonal direction(s) are hardly affected at all. Moreover, due to aggregate interlock, some residual shear strength (dependent on the crack opening) is also retained.

25. Modelling of post-cracking behaviour is by now almost a standard procedure (refs. 21,22,23) with two distinctly different approaches adopted. The more convenient crack "smearing" (refs. 24,25) approach is at present much more frequently used than the more complicated and time consuming approach in which cracks are introduced through mesh redefinition (refs. 26, 27). In the context of the present concrete material model research is currently in progress using the "smearing" technique coupled with a strain state based criteria for failure in tension while retaining the existing

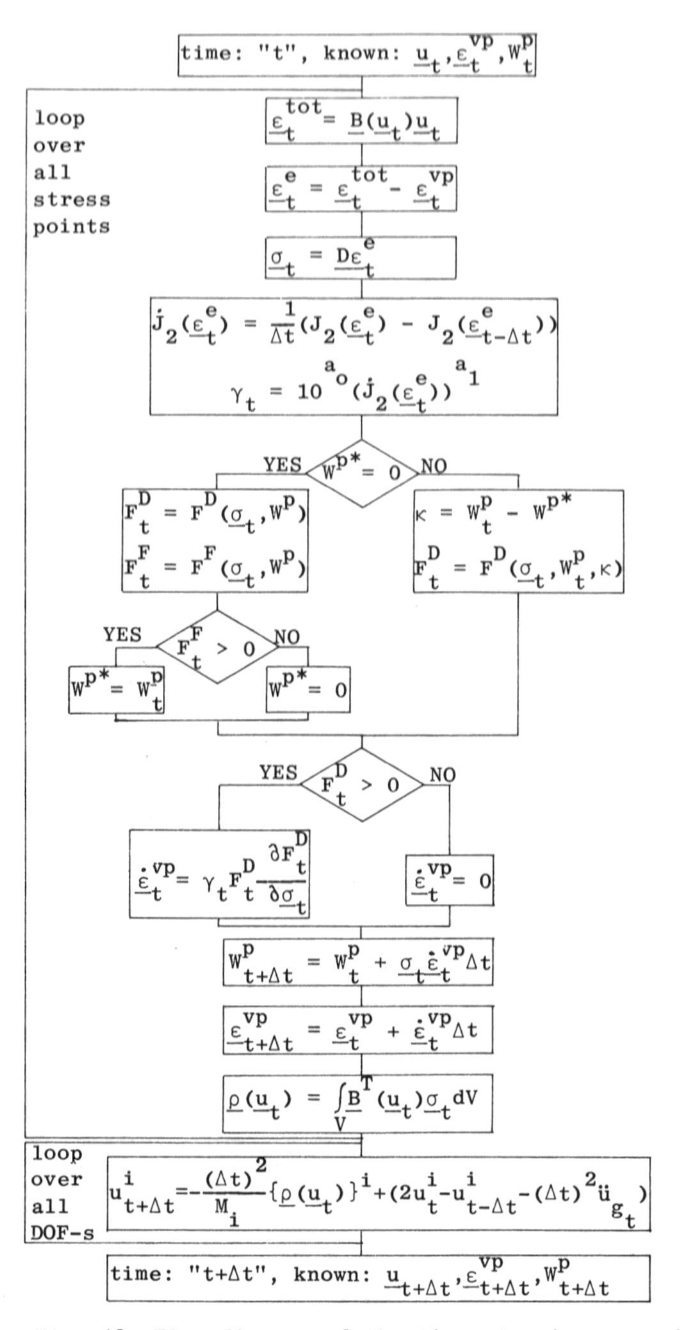

Evaluate total strain

Evaluate elastic strain by subtracting viscoplastic strain from total strain

Evaluate stresses corresponding to el. strain

Evaluate fluidity parameter as a function of the second elastic deviatoric strain invariant (a_o & a_1 are material constants)

If at a stress point the failure surface has not been reached previously ($W^{p*}=0$), define current discontinuity and failure surfaces. If the failure surface has been reached ($W^{p*} \neq 0$), define the current discontinuity surface and evaluate postfailure dissipated energy density κ

Check whether the failure surface has been reached in this step and if so ($F_t^F>0$), memorize dissipated energy density at the moment of failure W^{p*}

Check whether the discontinuity surface has been reached in this step and if so ($F_t^D>0$), evaluate viscoplastic strain rate

Evaluate dissipated energy density at t+Δt

Evaluate viscoplastic strain at t+Δt

Evaluate elements of the global elastic restoring force vector

Evaluate displacements at t+Δt

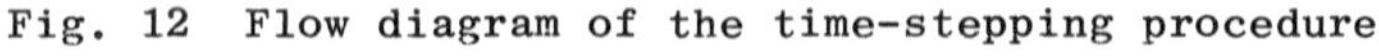

Fig. 12 Flow diagram of the time-stepping procedure

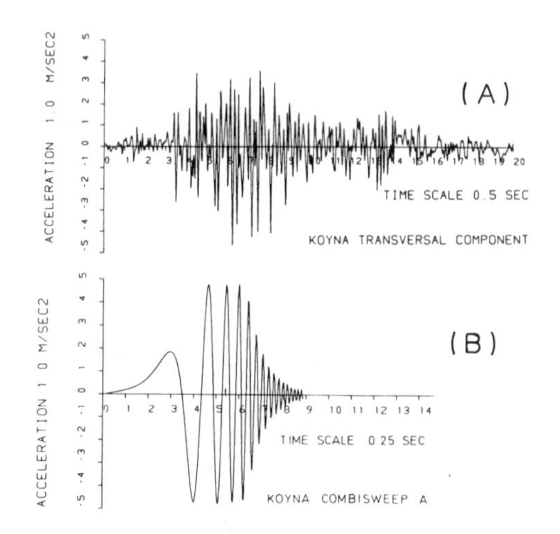

Fig. 13 A real accelerogram (A) and its substitute (B)

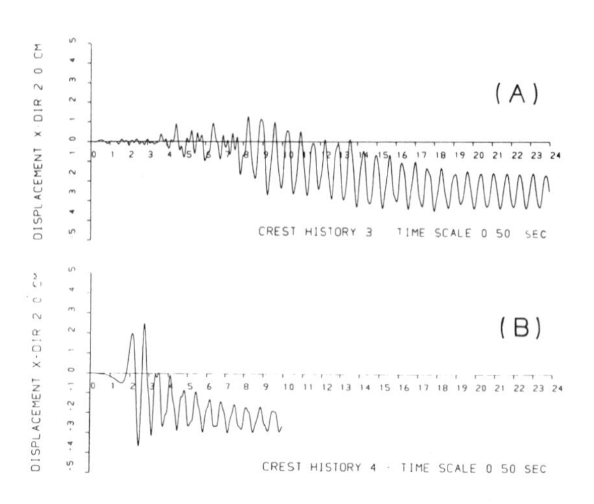

Fig. 14 Koyna dam analysis results
(A) using a real accelerogram
(B) using a substitute combisweep accelerogram

criteria for failure in compression.

CONCLUDING REMARKS

26. A procedure for the nonlinear, finite element analysis of concrete structures subjected to earthquakes has been described. A new elasto/viscoplastic strain rate and stress history dependent numerical model for plain concrete behaviour was presented in detail. Using this model and real and substitute earthquake accelerograms, an analysis of the Koyna dam section was performed.

27. Although results obtained are satisfactory, there clearly exists a need for further research and experimental studies, notably in

(i) The modelling of concrete behaviour and particularly of its behaviour in tension. In this context a study of possible pore water pressure development in cracks is needed to clarify the question of additional risks.

(ii) The reduction of the computational costs by improving numerical techniques. Here coupling of explicit and implicit integration methods promises much more economy (refs. 28,29).

ACKNOWLEDGEMENT

The financial support granted by the Građevinski Institut (Civil Engineering Institute) and the Samoupravna interesna zajednica za naučni rad (Selfmanaged interest group for scientific work) both from Zagreb, Yugoslavia, to N. Bićanić and R. Fejzo is hereby gratefully acknowledged.

REFERENCES

1. ZIENKIEWICZ O.C. The finite element method. McGraw-Hill, London, 1977.
2. SHANTARAM D., OWEN D.R.J. and ZIENKIEWICZ O.C. Dynamic transient behaviour of two- and three-dimensional structures including plasticity, large deformation effects and fluid interaction. Earthquake Engineering and Structural Dymamics, 1976, 4, Oct. 561-578.
3. CLOUGH R.W. and PENZIEN J. Dynamics of structures, McGraw-Hill, New York, 1975.
4. HINTON E., ROCK T. and ZIENKIEWICZ O.C. A note on mass lumping and related processes in the finite element method. Earthquake Engineering and Structural Dynamics, 1976, 4, Jan. 246-249.
5. ZIENKIEWICZ O.C., BETTESS P. and KELLY D.W. The Sommerfeld (radiation) condition on infinite domains and its modelling in numerical procedures. Proceedings, 3rd Int. Symp. on Computing Methods in Applied Science and Engineering Pt.1, IRIA, Paris, 169-203.
6. OTTOSEN N.S. Constitutive model for short-time loading of concrete. Proceedings A.S.C.E., 1979, 105, EM1, Feb. 127-137.
7. DARWIN D. and PECKNOLD D.A. Nonlinear biaxial stress-strain law for concrete. Proceedings A.S.C.E., 1977, 103, EM2, Apr. 229-241.
8. KUPFER H.B. and GERSTLE K. H. Behaviour of concrete under biaxial stresses. Proceedings A.S.C.E., 1973, 99, EM4, Aug. 853-866.
9. SINHA B.P., GERSTLE K.H. and TULIN L.G. Stress-strain relations for concrete under cyclic loading. Journal ACI, 1964, 61(2), Feb. 195-211.
10. KARSAN I.D. and JIRSA J.O. Behaviour of concrete under compressive loadings. Proceedings A.S.C.E., 1969, 95, ST12, Dec. 2543-2563.
11. HATANO T. Dynamical behaviours of concrete under periodical compressive load. Central Research Institute of Electric Power Industry, C-6104, Tokyo, 1962.
12. HATANO T. and WATANABE H. Fatigue failure of concrete under periodic compressive load. Transactions JSCE, 1971, 3(1), 106-107.
13. NILSSON L. Impact loading on concrete structures. Chalmers University of Technology, Göteborg, Publication 79:1, 1979.
14. PAL N. Seismic cracking of concrete gravity dams. Proceedings A.S.C.E., 1976, 102, ST9, Sept. 1827-1844.
15. HATANO T. and TSUTSUMI H. Dynamical compressive deformation and failure of concrete under earthquake load. Proceedings 2nd WCEE, 3, Tokyo, 1960, 1963-1978.
16. HATANO T. Relations between strength of failure, strain ability, elastic modulus and failure time of concrete. Central Research Inst. of Electric Power Industry, C-6001, Tokyo, 1960.
17. PERZYNA P. Fundamental problems in viscoplasticity. Advances in Applied Mechanics, 1966, 9, 243-377.
18. BIĆANIĆ N. Nonlinear finite element transient response of concrete structures. Ph.D. Thesis, C/Ph/50/78, University of Wales, Swansea 1978.
19. JOHNSON G.R. and EPSTEIN H.I. Short duration analytic earthquake. Proceedings A.S.C.E. 1976, 102, ST5, May, 993-1000.
20. WANG W.Y.L. and GOEL S.C. Prediction of maximum structural response by using simplified accelerograms. Proceedings 6th WCEE, 3, New Delhi, 1977.
21. DODGE W.G., BAZANT Z.P. and GALLAGHER R.H. A review of analysis methods for prestressed concrete reactor vessels. ORNL-5173, Oak Ridge National Laboratory, 1977.
22. WEGNER R. Finite element models for reinforced concrete. Formulations and Computational Algorithms in Finite Element Analysis (U.S. - Germany Symposium). M.I.T. Press, Cambridge, Mass., 1977, 393-439.
23. BERGAN P.G. and HOLAND I. Nonlinear finite element analysis of concrete structures. Computer Methods in Applied Mechanics and Engineering, 1979, 17/18, 443-467.
24. PHILLIPS D.V. and ZIENKIEWICZ O.C. Finite element non-linear analysis of concrete structures. Proceedings of the Institution of Civil Engineers, 1976, 61(2), Mar., 59-88.
25. BATHE K.J. and RAMASWAMY S. On three-dimensional nonlinear analysis of concrete structures. Nuclear Engineering and Design, 1979, 52, 385-409.
26. NGO D. and SCORDELIS A.C. Finite element analyses of reinforced concrete beams. Journal ACI, 1967, 64, Mar., 152-163.
27. NILSON A.H. Nonlinear analysis of reinforced concrete by the finite element method. Journal ACI, 1968, 65, Sept., 757-766.
28. HUGHES T.J.R. and LIU W.K. Implicit-explicit finite elements in transient analysis. J. of Applied Mechanics, 1978, 45, June, 371-378.
29. BELYTCHKO T. and MULLEN, R. Mesh partition of explicit-implicit time integration.Ref.as (22)

21 Earth dam analysis for earthquakes: numerical solution and constitutive relations for non-linear (damage) analysis

O. C. ZIENKIEWICZ, FRS, K. H. LEUNG, MSc, E. HINTON, PhD, and
C. T. CHANG, University College, Swansea

This paper discusses (a) The essential nature of soil behaviour under repeated loading and discusses here briefly appropriate constitutive relations. (b) Numerical formulations and solution techniques for dynamic soil problems coupled to pore water flow. (c) Presents some results of transient non-linear analysis for typical problems of earth dams and foundation layers in which permanent deformation and liquefaction occur under earthquake shocks.

INTRODUCTION

1. The ever present need for designing safe structures and the prevention of possible catastrophies has focussed much attention on the performance of earth dams in earthquake motion. Such "near disasters" as have occurred recently at the San Fernando dams and elsewhere justify our concern. It is the major object of this paper to indicate to the engineering practitioner an analysis procedure capable of predicting the extent of permanent damage which can occur when the foundation of the dam is subject to a given prescribed earthquake shock.

2. Clearly, we must be able to ensure that no major failure will happen under all probable combinations of earthquake motion but under certain extreme conditions a limited damage may be permissible. In a report now under preparation the ICOLD committee on Seismicity will be concerned with the matter of predicting earthquake motions and their probabilistic distribution at a given site. In another report of the Analysis and Design committee the subject of the earthquake magnitudes which should be realistically considered is discussed. These matters although of extreme importance are omitted from the present paper and we assume at the outset that the maximum motions to be resisted are known (at least in character at the bedrock level).

3. In static analysis of dams two possible approaches exist to predict unacceptable deformation or failure. In the first of these, "limit equilibrium concepts" are used ignoring the deformability characteristics of the material and concentrating on the collapse mechanism. Here only its strength properties need to be considered, and with a good estimate of these, reasonably accurate predictions of complete collapse can be found.

4. In the second approach appropriate "constitutive" relations for the material are first established and a complete solution for the deformations is obtained under all loads by an approximate solution of the "field" equations. Such solutions are only feasible by an application of a numerical discretization process (such as that provided by the finite element method) and the use of powerful computers.

5. Non-linear analysis approaches provide a more realistic and accurate estimate of static behaviour right up to the point of collapse than are possible with the first, "limit equilibrium" approach (albeit at higher costs generally).

6. In the dynamic problem the two alternative approaches do not exist. The reasons for this are that:

(1) The loading is itself dependent on the deformations and cannot be, a priori, predicted.

(2) The material strength is often dependent on the rate of straining and on its deformation history, and finally;

(3) The duration of loading is often so short that even if "failure" occurs the total deformation suffered by the dam may be insignificant.

7. In earthquake analysis, therefore, we cannot avail ourselves of 'limit equilibrium' methods to predict the extent of damage. In the same vein, linear analysis using idealized material properties is not in general applicable.

8. For important structures and for extreme intensity earthquakes we recommend that full non-linear dynamic analysis be carried out.

9. Before any analysis can be attempted it is necessary to:
(1) Model suitably the material behaviour.
(2) Establish the general behaviour pattern and governing differential equations of the problems.
(3) Derive the appropriate numerical discretization processes and their possible computer solution.

10. All these problems will be discussed in some detail in the present paper but particular attention will be focussed on the soil behaviour.

11. Even under static loads the description of soil behaviour is not easy and much research which has gone into this subject has not yet arrived at a universal, quantitative, model which is generally acceptable. Nevertheless reasonably accurate models are today available which can predict, with sufficient accuracy, collapse conditions and non-linear, permanent, deformations occurring under load (ref.1).

12. The situation becomes more complex when fluctuating loads occur, such as may be expected in the earthquake response of earth dams. Here two new phenomena are encountered. Referring to drained behaviour of the soil we note that

(1) Permanent shear strains occur after each cycle of stress application and that

(2) Permanent volume contraction (densification) occurs after each cycle of loading (except for extremely dense materials).

Of the two effects, the second is the one of greatest importance in practice where saturated, undrained, or partially drained behaviour predominates.

13. It is this densification phenomenon that is responsible for the increases of pore pressure when cyclic load is applied to a sample in an undrained state and which accounts for the large displacements occurring when the pore pressures build up to the value of the mean effective, compressive, stress. When this happens incipient liquefaction (ref.1) is reached and the material is at a point of yielding. Some dilatancy which occurs in sands during plastic flow counterbalances at this stage the pore pressure increase phenomena and complete liquefaction follows only when such dilatancy is exhausted after a considerable shearing strain. Figure 1 shows such behaviour in a typical cyclic 'shear box' test carried out on a sand recently (ref.2) and shows the relative insignificance of cumulative shear strains before the onset of liquefaction and the following dramatic increase of strains.

14. Clearly it will not be an easy matter to derive a model capable of reproducing fully all such phenomena. In the first part of this paper we shall therefore review some existing models and introduce a simple one which is capable of quantitative predictions concerning the pore pressure increase evaluation.

15. The basic generally applicable differential equations, which describe the behaviour of of porous media under draining conditions have been described elsewhere (ref.3) and it was shown under what circumstances undrained behaviour can be assumed. For a typical structure in which sands or silts comprise an important part such assumptions are not generally adequate and we shall show how drainage can be readily included in analysis.

16. The numerical discretization details necessary for a non-linear solution of the coupled fluid/soil system will be introduced and the paper will conclude with some examples of computation and a discussion of the failure of the San Fernando dams (refs.4,5) where liquefaction of fill and subsequent large deformation was nearly responsible for a major catastrophe.

CONSTITUTIVE LAWS AND LIQUEFACTION

17. Some constitutive laws for static behaviour. All constitutive relations for soil in which pore pressure changes play an important role are most conveniently defined in terms of effective stress variables. Undrained properties follow uniquely from such descriptions.

18. In rate independent soils, i.e. when creep strains can be considered as negligible, the constitutive law for a non-linear material must be specified as an incremental relation between the changes of effective stress $d\underset{\sim}{\sigma}'$ and the changes of that part of strain which we consider to be directly related to stress i.e. $d\underset{\sim}{\varepsilon}^{\sigma}$. *

19. Thus we can write generally

$$d\underset{\sim}{\sigma}' = \underset{\sim}{D}\, d\underset{\sim}{\varepsilon}^{\sigma} \qquad (1)$$

where D is the tangent modulus matrix. Such a tangent modulus matrix will in general be a function of effective stress level, straining history etc.

20. Many alternative models defining the constitutive law of the type given in equation (1) are available. Amongst these we have the classes of

(1) Non linear elasticity (hyper-elasticity)

(2) Plasticity or viscoplasticity

(3) Hypo-elasticity

(4) Endochronic theory

21. All have their proponents and antagonists - and each shows some philosophical or computational merits. If the application of loads on a structure is non-monotonic, non-linear elasticity can lead to entirely erroneous results and therefore such models are discarded a priori. As hypo-elasticity and endochronic theory are basically alternative descriptions of the plasticity (or visco-plasticity) phenomena,

---

* In the present paper we shall use a vector/matrix notation for stresses, strains and related quantities. Thus

$\underset{\sim}{\sigma}^T \equiv [\sigma_x, \sigma_y, \sigma_z, \sigma_{xy}, \sigma_{yz}, \sigma_{xz}]$, $\underset{\sim}{u}^T = [u_x, u_y, u_z]$ etc.

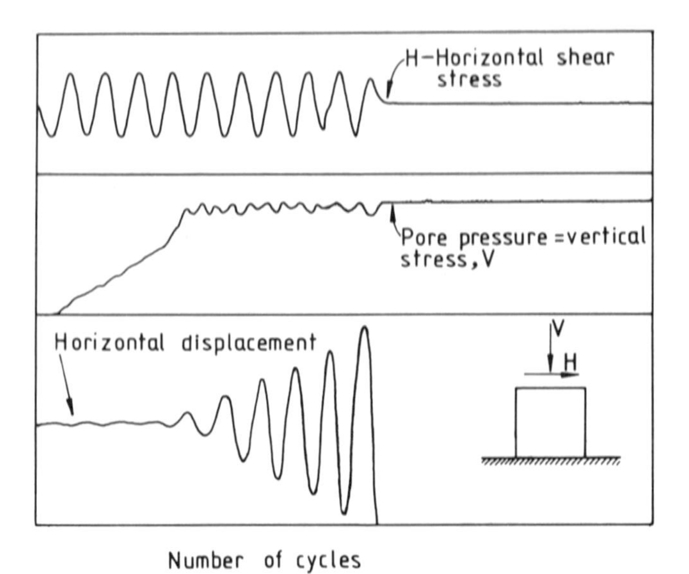

FIGURE 1 Typical behaviour of a sand under cyclic shear stress (undrained)

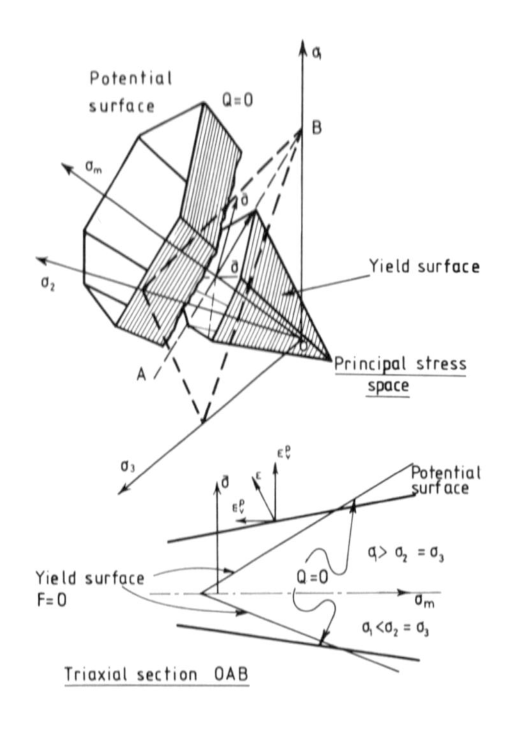

FIGURE 3 Model B - Ideal non-associated plasticity with Mohr-Coulomb yield surface

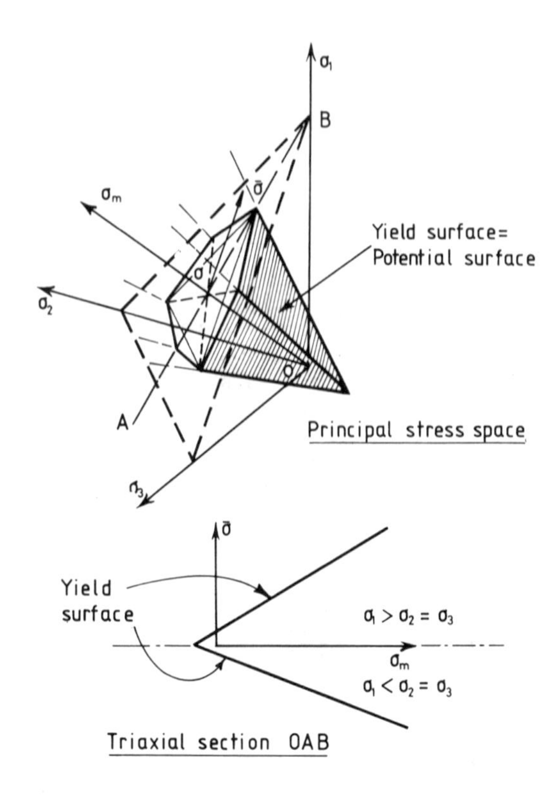

FIGURE 2 Model A - Ideal associated plasticity with Mohr-Coulomb yield surface

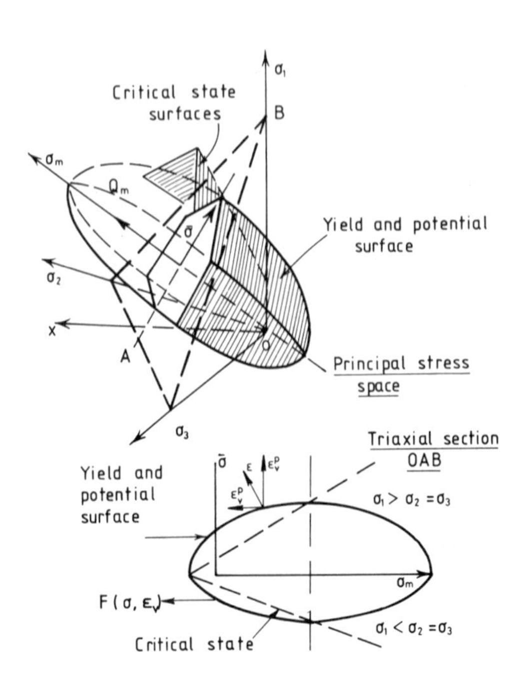

FIGURE 4 Model C - Strain hardening critical state associated plasticity with Mohr-Coulomb critical state surface

we shall entirely concentrate on the plasticity model.

22. This is defined by

(a) a yield surface

$$F(\underset{\sim}{\sigma}', \underset{\sim}{\varepsilon}_p) = 0 \qquad (2)$$

which limits the field of elastic deformation. Thus elastic straining only can occur if

$$F < 0 \qquad (3a)$$

and when

$$F = 0 \qquad (3b)$$

both plastic and elastic straining can be present. Note that $\underset{\sim}{\sigma}'$ represents the effective stresses and $\underset{\sim}{\varepsilon}_p$ the plastic strains.

(b) To complete the plasticity definitions the directions of plastic straining have to be defined by a plastic potential

$$Q\,(\underset{\sim}{\sigma}', \underset{\sim}{\varepsilon}_p) = 0 \qquad (4)$$

such that we can write

$$d\underset{\sim}{\varepsilon}_p = \lambda \frac{\partial Q}{\partial \underset{\sim}{\sigma}'} \qquad (5)$$

where λ is a proportionality constant.

(c) Finally it is assumed that the total, stress dependent, strain increment $d\underset{\sim}{\varepsilon}^{\sigma}$ can be divided into elastic and plastic parts, i.e. that

$$d\underset{\sim}{\varepsilon}^{\sigma} = d\underset{\sim}{\varepsilon}_e + d\underset{\sim}{\varepsilon}_p$$
$$d\underset{\sim}{\varepsilon}_e = \underset{\sim}{D}_e^{-1}\, d\underset{\sim}{\sigma} \qquad (6)$$

$\underset{\sim}{D}_e$ represents the matrix of elastic constants.

23. Plasticity rules are associative if

$$F = Q \qquad (7a)$$

and non-associative if

$$F \neq Q \qquad (7b)$$

The definition of yield, potential and hardening rules suffice to determine the incremental $\underset{\sim}{D}$ matrix of equation (1). Details of the necessary algebraic operations are presented elsewhere (ref.29).

24. In attempting to model a soil by plasticity theory we note at the outset that the best known characteristic of soils is that of their critical behaviour characterised by an envelope of Mohr's circles usually approximated by a drained cohesion c' and friction angle ϕ' values. At such stress states, shown in the space of principal stress of Figure 2 as a pyramid (ref.1), deformation can continue without appreciable stress changes. Clearly it is therefore natural to associate this critical surface with the form of plasticity model used.

25. Three models of plasticity of a relatively simple kind have been used with success in static problems:

(A) In this ideal, classic, plasticity is assumed with the critical Mohr-Coulomb surface playing the role of the yield surface $F(\underset{\sim}{\sigma}') = 0$. Further, fully associative behaviour is assumed with Q=F. This model is shown in Figure 2.

(B) This model is similar to (A) but with a non-associative flow rule assumed in which Q has a similar form to that of F but with ψ replacing the friction angle ϕ'. This rule for flow allows a smaller dilatancy to be imposed on the material during flow (with the most frequent assumption being $\psi=0$ at which no dilatancy occurs). This model is shown in Figure 3.

(C) The last model is the modified critical state model based on original work of Roscoe and his followers (refs.6-8) and amended by Zienkiewicz et al. (ref.9) to include a Mohr-Coulomb type critical surface in three dimensional stress space as shown in Figure 4. With this model strain hardening depending on the plastic volumetric strain

$$-(\varepsilon_{ii})_p = -\underset{\sim}{M}^T \underset{\sim}{\varepsilon}_p \qquad (8)$$

is assumed but a fully associative flow rule is adopted. All loading paths starting in the elastic domain for such a material will show either 'hardening' or 'softening' behaviour but will, at large strains, converge to the critical surface.

26. Numerical experiments reported elsewhere (refs.1,9,10) lead one to the following conclusions applicable to static foundation or embankment analysis:

Static drained behaviour - normally consolidated materials. Models (A), (B) and (C) give almost identical collapse loads and, with some adjustment of elastic constants, very similar displacement behaviour.

Static undrained behaviour of normally consolidated materials. Model (B) with ψ=0 and model (C) give a very similar collapse and displacement performance, but model (A) shows no collapse due to negative pressures developed on continuing dilatancy during yield, and finally

Static undrained behaviour - over-consolidated Here only model (C) is capable of dealing with the over-consolidation phenomenon - and results recently obtained indicate excellent agreement with experiment (ref.1)*.

*Although the overconsolidative, drained behaviour has not been explicitly investigated we note that on failure, strain-softening will occur with the residual strength values being the same for models (C) and (B). We would thus expect no advantage to arise for model (C) in estimating final strengths.

27. We see that this is not the case for cyclic (or generally variable) load histories as within an ideal (or isotropically hardening) yield surface no amount of stress cycling can produce any permanent strains.

28. Modifications of static plasticity for cyclic (or variable) loads. To reproduce the essential features of cyclic (or variable) load response the constitutive models discussed in previous sections need to be modified or replaced. Two main lines of attack present themselves.

29. In the first the basic formulation used for static problems is replaced by a new one capable of accommodating the additional effects. Here for instance the critical state model discussed in the previous section could be augmented by additional kinematically hardening yield surfaces as suggested by Mroz et al. (refs. 11-13) or alternative plasticity models can be introduced. The work of Ghaboussi (refs.14,15), Pender (ref.16), Nova (ref.17) and others (refs. 18,19) follows in this category. The addition of cumulative effects is natural in the context of endochronic models and here some success has been reported (refs. 20,21) necessitating however the use of numerous measured parameters.

30. In the second approach a simpler philosophy is adopted. The static model with its well-tested structure is retained and the new effects are made by addition. Such an approach has the advantage of simplicity and the most direct use of experimental evidence and for this reason we have singled out its use in the present paper.

31. In the introduction we have already stated that the most important feature of cyclic strain response is that of the cumulative densification which is responsible for such phenomena as liquefaction (refs.22-24) and loss of strength. We shall therefore concentrate our efforts on including this aspect in our model.

32. In the previous section we have concluded that for static behaviour studies the non-associative plasticity (Model B) or the 'critical' state models (Model C) were applicable. Now we need to include an additional accumulation of strain $\underset{\sim}{\varepsilon}^{o}$ which is volumetric nature, i.e.

$$\underset{\sim}{\varepsilon}^{o} = \underset{\sim}{m}\, \varepsilon_{v}^{o} \tag{9}$$

and which is caused by the history of elastic and plastic straining.

33. Before considering detailed formulas for this it is essential to investigate the effect of such a volumetric strain on the changes of pore pressure under undrained conditions. Such conditions are characteristic of standard tests conducted under cyclic loads and allow the simplest correlation between ε_{v}^{o} and the measured pore pressure changes. In the next section we shall discuss this in detail, indicating the basic mechanism of liquefaction.

34. The effects of densification strain and some empirical models. Retaining the elastic plastic constitutive law of equation (1) and noting that the total strain $\underset{\sim}{\varepsilon}$ is given on the sum of $\underset{\sim}{\varepsilon}^{\sigma}$ and $\underset{\sim}{\varepsilon}^{o}$ we can write

$$d\underset{\sim}{\sigma}' = \underset{\sim}{D}\,(d\underset{\sim}{\varepsilon} - d\underset{\sim}{\varepsilon}^{o}) \tag{10}$$

in which we assume that $d\underset{\sim}{\varepsilon}^{o}$ has been independently determined.

35. The definition of effective and total stresses and their link with pore pressure gives

$$d\underset{\sim}{\sigma} = d\underset{\sim}{\sigma}' - \underset{\sim}{m}\, dp \tag{11}$$

and if undrained behaviour of a sample is considered

$$dp = d\hat{p} = -\frac{K_f}{n}\, \underset{\sim}{m}^{T}\, d\underset{\sim}{\varepsilon} \tag{12}$$

36. Let us now investigate the behaviour of a sample of soil illustrated in Figure 5 on which the total stress does not change ($d\underset{\sim}{\sigma} = 0$) and where $d\underset{\sim}{\varepsilon}^{o} = \underset{\sim}{m}\, d\varepsilon_{v}^{o}$ occurs due to some extraeous causes. From equations (10) and (11) we can write

$$\underset{\sim}{m}\, d\hat{p} = D(d\underset{\sim}{\varepsilon} - \underset{\sim}{m}\, d\varepsilon_{v}^{o}) \tag{13}$$

37. From equations (12) and (13) after eliminating $d\underset{\sim}{\varepsilon}$ we can obtain a simple relation

$$d\hat{p} = -\beta\, d\varepsilon_{v}^{o} \tag{14}$$

where

$$\beta = 1/(\frac{n}{K_f} + \frac{1}{K_T}) \simeq K_T \text{ if } K_f >> K_T \tag{15}$$

and the tangent bulk modulus of skeleton is given as

$$K_T = 1/\underset{\sim}{m}^{T}\, \underset{\sim}{D}^{-1}\, \underset{\sim}{m} \tag{16}$$

38. This mechanism of densification causing an increase of pore pressure is fundamental to our understanding of the phenomena and we note that by virtue of equation (14) it is immaterial whether we specify with primary variable or simply the undrained pressure increase $\hat{p}$.

39. Typical tests on sand under cyclic loading are reproduced in Figures 6a and b (ref.25) in which

$$\theta = |\bar{\sigma}|/\sigma'_{mo} \tag{17}$$

with $|\bar{\sigma}|$ being the fluctuating cyclic shear (deviatoric invariant) and σ'_{mo} the mean effective stress at the start of the loading.

40. As we have already mentioned, the development of the densification strain $\underset{\sim}{\varepsilon}^{o}$ must be related to the total strain history (and the level of the stresses). With this in mind a parameter ξ is defined such that

$$d\xi = \sqrt{de_{ij}\, de_{ij}} \qquad (18)$$

with e_{ij} representing deviatoric strain components. In some earlier work (ref.25) we show that a good correlation with experiment can be defined by taking an expression

$$d\varepsilon_{v}^{o} = \frac{-A}{1 + B}\, d\kappa \qquad (19)$$

where

$$d\kappa = e^{\gamma\theta}\, d\xi \qquad (20)$$

Here the constants A, B and γ define the characteristics of cyclic response.

41. In Figure 7 a correlation obtained by the above expressions is shown for the sand whose behaviour was given in Figure 6.

42. Other formulations have been proposed (ref.26) and we must observe that development of ε_{v}^{o} is very much dependent on the initial density of soil. It is clear that if the soil is packed to its maximum density no further densification is possible and thus $A \to 0$.

43. While the search for an optimal, all embracing, expression continues we note that by equation (14) we can interchangeably use ε_{v}^{o} or $\hat{p}$ in all computations. The use of the latter is convenient as it represents a most direct connection between experiment and subsequent calculations. Seed et al. (refs.27,28) show that for a wide variety of sands the relative pore pressure rise $\hat{p}/\sigma'_{mo}$ can be related to θ and to the number of stress cycles in a manner shown in Figure 8.

44. If we note that during a single cycle we can compute $\Delta\xi$ as twice the total absolute range of shear strain, it is easy to show that for uniform cycles

$$\xi = N|\Delta\bar{\sigma}|2/G = N\theta\, \sigma'_{mo}\, 2/G \qquad (21)$$

where N is the number of cycles and G the average shear modulus. The curves of Figure 7 can then be amended to compute $d\hat{p}$ directly for non-cyclic strains.

45. Thus, we now have a reasonably good representation of the cyclic pore pressure (or ε_{v}^{o}) increase rate which can be used in non-linear analysis.

NUMERICAL FORMULATION

Basic relationships: Discretized forms and time stepping

46. The differential equation governing dynamic phenomena in soils in which drainage occurs is now considered. The constitutive relations and definitions are

$$d\underset{\sim}{\sigma} = d\underset{\sim}{\sigma}' - \underset{\sim}{m}p \qquad (22a)$$

$$d\underset{\sim}{\sigma}' = \underset{\sim}{D}\,(d\underset{\sim}{\varepsilon} - d\underset{\sim}{\varepsilon}^{o}) \qquad (22b)$$

$$d\underset{\sim}{\varepsilon} = L\, d\underset{\sim}{u} \qquad (22c)$$

where in general $\underset{\sim}{D}$ and $\underset{\sim}{\varepsilon}^{o}$ depend on stresses, strains and their history in the manner already discussed, $\underset{\sim}{\varepsilon}$ represents total strains, $\underset{\sim}{u}$ displacements and $\underset{\sim}{L}$ an appropriate differential operator. Overall equilibrium can be expressed by the equation

$$\underset{\sim}{L}^{T}\underset{\sim}{\sigma} + \rho\,\underset{\sim}{g} = \rho\ddot{\underset{\sim}{u}} \qquad (23)$$

and finally the porous fluid flow equations (defining the mass balance) are

$$-\underset{\sim}{\nabla}^{T}\underset{\sim}{k}\,\underset{\sim}{\nabla}p + \underset{\sim}{m}^{T}\dot{\underset{\sim}{\varepsilon}} + \underset{\sim}{\nabla}^{T}(\underset{\sim}{k}\,\rho_f\,\underset{\sim}{g}) = \underset{\sim}{\nabla}^{T}\underset{\sim}{k}\,\rho_f\,\ddot{\underset{\sim}{u}} - \frac{n\,\dot{\underset{\sim}{p}}}{K_f} \qquad (24)$$

with $\underset{\sim}{k}$ being the permeability, ρ_f and K_f respectively the fluid density and bulk modulus and η the porosity. The above system with appropriate boundary conditions on displacements (or on total tractions) and pore pressures (or their normal gradients) can represent closely most dynamic phenomena encountered - and must be solved with the non-linear characteristics already discussed (and possibly a permeability which depends on strains).

47. To achieve a solution, the initial step usually involves a finite element discretization. In this the displacements $\underset{\sim}{u}$ are described in terms of the nodal values $\bar{\underset{\sim}{u}}$ as

$$\underset{\sim}{u} = \underset{\sim}{N}\,\bar{\underset{\sim}{u}} \qquad (25)$$

with a similar discretization for pressures

$$\underset{\sim}{p} = \bar{\underset{\sim}{N}}\,\bar{\underset{\sim}{p}} \qquad (26)$$

where $\bar{\underset{\sim}{p}}$ is the vector of pressure nodal values. In equations (25) and (26), $\underset{\sim}{N}$ and $\bar{\underset{\sim}{N}}$ are appropriate shape functions.

48. Following standard discretization procedures which the reader can find in appropriate texts, e.g. ref. (29), we arrive simply at a semi-discrete system

$$\int_{\Omega} \underset{\sim}{B}^{T}\underset{\sim}{\sigma}'\, d\Omega + \underset{\sim}{M}\,\ddot{\bar{\underset{\sim}{u}}} - \underset{\sim}{Q}\,\bar{\underset{\sim}{p}} = \underset{\sim}{f} \qquad (27)$$

with

$$d\underset{\sim}{\sigma}' = \underset{\sim}{D}\,(\underset{\sim}{B}\, d\bar{\underset{\sim}{u}} - d\underset{\sim}{\varepsilon}^{o}) \qquad (28)$$

representing equations (22) or (23) and

$$\underset{\sim}{H}\,\bar{\underset{\sim}{p}} + \underset{\sim}{S}\,\dot{\bar{\underset{\sim}{p}}} + \underset{\sim}{Q}^{T}\dot{\bar{\underset{\sim}{u}}} - \hat{\underset{\sim}{M}}\,\ddot{\bar{\underset{\sim}{u}}} = \hat{\underset{\sim}{f}} \qquad (29)$$

representing equation (24).

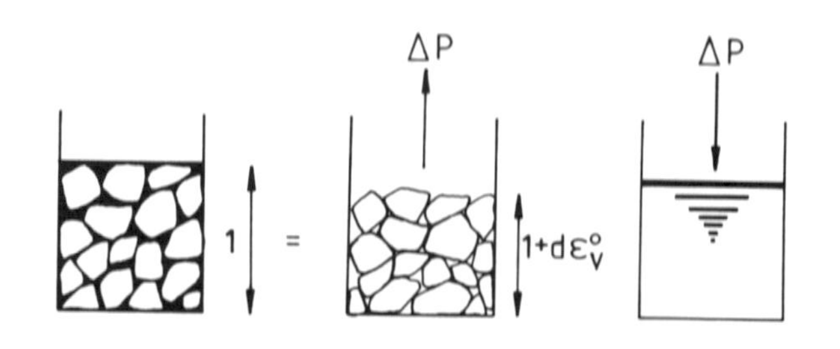

FIGURE 5 The mechanism of the development of pore pressure due to skeleton contraction - $d\varepsilon_v^0$

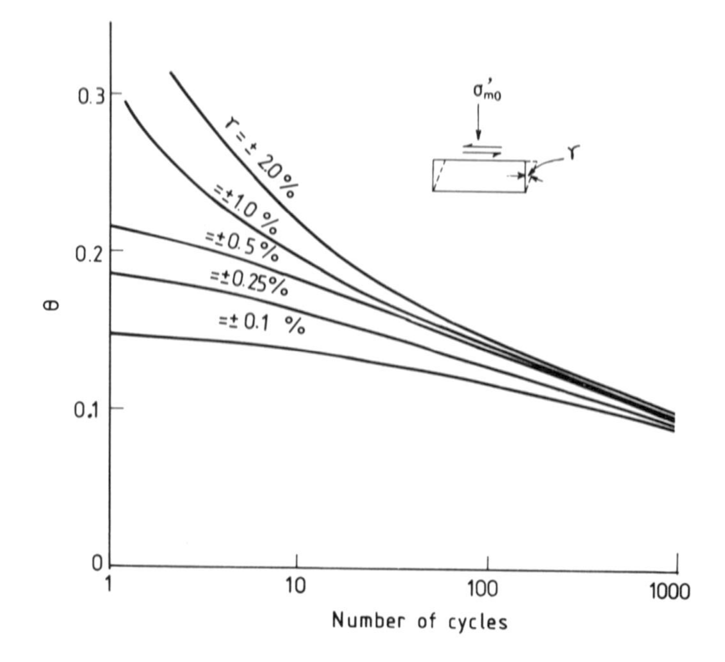

FIGURE 6a Development of cyclic shear strains during undrained cyclic simple shear loading (N.G.I. sand)

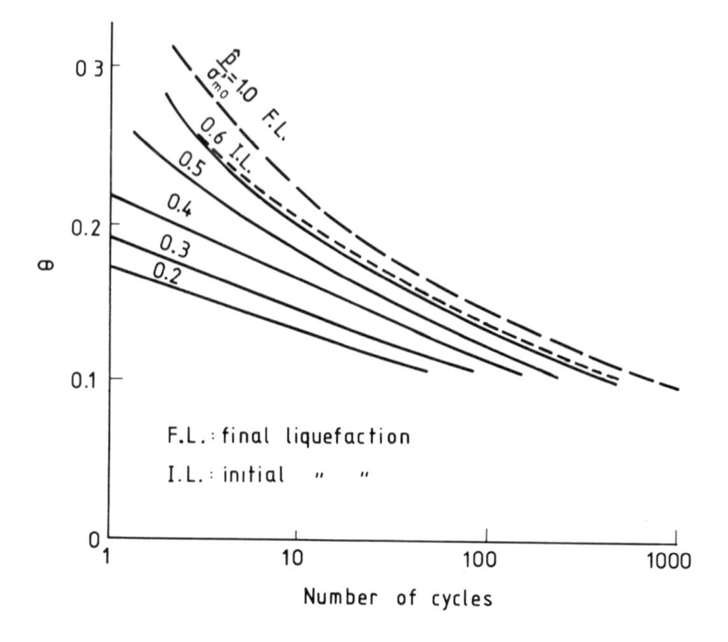

FIGURE 6b Pore pressure build-up during undrained cyclic simple shear loading (N.G.I. sand)

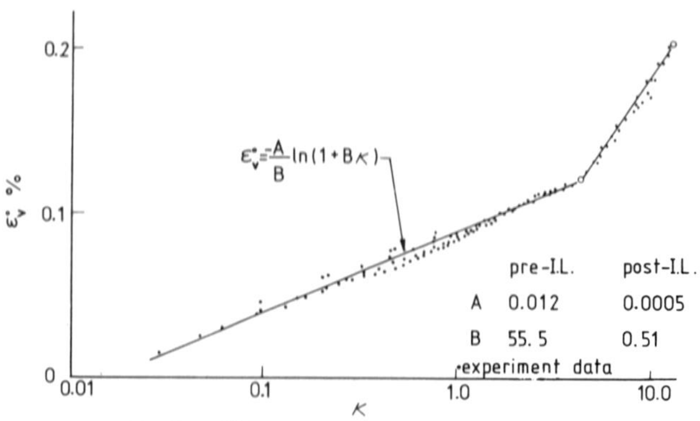

FIGURE 7 Densification strains verus κ (N.G.I. sand)

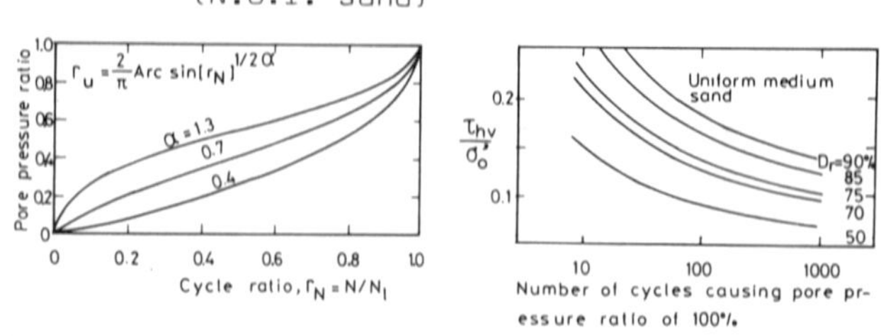

FIGURE 8a Theoretical expression for undrained rate of pore water pressure build-up

8b Typical undrained cyclic loading test data (after Seed, ref.28)

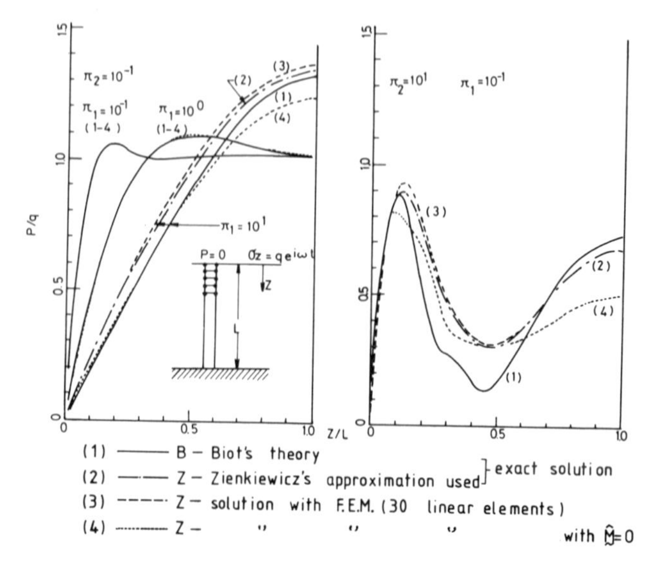

FIGURE 9a,b A comparison of the exact and the finite element solution

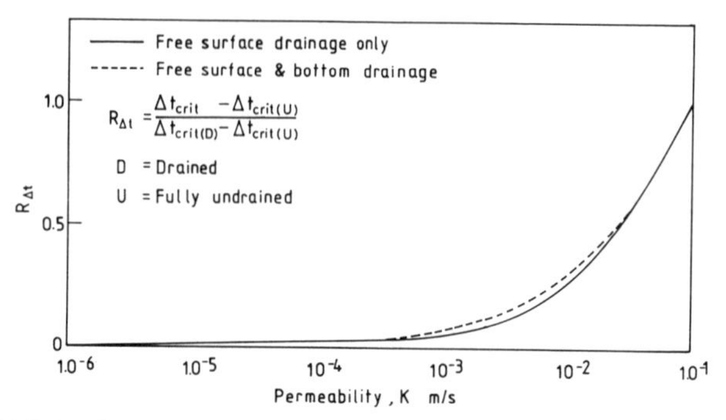

FIGURE 9c Critical time step versus permeability of the layer problem

49. Note that $\underset{\sim}{B}$ is the standard strain matrix and $\underset{\sim}{M}$ the mass matrix defined by components as

$$\underset{\sim}{B}_i = \underset{\sim}{L}\,\underset{\sim}{N}_i \qquad (30a)$$

$$\underset{\sim}{M}_{ij} = \int_\Omega \underset{\sim}{N}_i^T \rho \,\underset{\sim}{N}_j \, d\Omega \qquad (30b)$$

and the other matrices are

$$\underset{\sim}{Q}_{ij} = \int_\Omega \underset{\sim}{B}_i^T \underset{\sim}{m} \,\bar{\underset{\sim}{N}}_j \, d\Omega \qquad (31a)$$

$$\underset{\sim}{H}_{ij} = \int_\Omega (\underset{\sim}{\nabla}\, \bar{\underset{\sim}{N}}_i)^T \underset{\sim}{k}\, \underset{\sim}{\nabla}\, \bar{\underset{\sim}{N}}_j \, d\Omega \,. \qquad (31b)$$

$$\hat{\underset{\sim}{M}}_{ij} = \int_\Omega \bar{\underset{\sim}{N}}_i^T \underset{\sim}{\nabla}^T \underset{\sim}{k}\, \rho_f \,\underset{\sim}{N}_j \, d\Omega \qquad (31c)$$

$$\underset{\sim}{S}_{ij} = \int_\Omega \bar{\underset{\sim}{N}}_i^T \, n/K_f \,\bar{\underset{\sim}{N}}_j \, d\Omega \qquad (31d)$$

50. The system naturally has to be supplemented by appropriate plasticity relations for computing the $\underset{\sim}{D}$ matrix and expressions such as those given by equations (4), (18-20) for computing the densification strains $\underset{\sim}{\varepsilon}^o$.

51. Although certain 'symmetries' exist in the equation system, the primary variables $\bar{\underset{\sim}{u}}$ and $\bar{\underset{\sim}{p}}$ have a different structure (and indeed different physical units) and therefore standard time stepping algorithms are not easy to apply. It is therefore best to adopt a 'staggered' solution process in which

(a) Equation (27) is solved for the displacement changes $\Delta\bar{\underset{\sim}{u}}$ using some "extrapolated" value of $\bar{\underset{\sim}{p}}$

and

(b) Equation (29) is solved for $\bar{\underset{\sim}{p}}$ using the now available values of $\Delta\bar{\underset{\sim}{u}}$.

52. Such staggered processes have been used successfully for other "coupled" physical problems and a very full analysis of their characteristics with regard to stability and accuracy has recently been provided by Park and Felippa (refs. 30,31).

53. Many alternative time stepping procedures are obviously available and a number of alternative schemes for second order equations (such as equation (27)), are presented in Chapter 21 of reference 29 (viz. also reference 32).

54. For our purposes we find it convenient to use the explicit, central difference scheme for equation (27) which we shall write

$$\underset{\sim}{M}(\bar{\underset{\sim}{u}}_{n+1} - 2\bar{\underset{\sim}{u}}_n + \bar{\underset{\sim}{u}}_{n-1})/\Delta t^2 + \left(\int_\Omega \underset{\sim}{B}^T \underset{\sim}{\sigma}\, d\Omega\right)_n - (\underset{\sim}{Q}\,\bar{\underset{\sim}{p}})_n = \underset{\sim}{f}_n \qquad (32)$$

where n+1, n, or n-1 denote a set of consecutive time points separated by increments Δt of time.

55. We note that in equation (32) we have simply extrapolated $\bar{\underset{\sim}{p}}$ as the value of this vector at time n.

56. As initial values of $\bar{\underset{\sim}{u}}$ and $\dot{\bar{\underset{\sim}{u}}}$ as well as $\bar{\underset{\sim}{p}}$ are given, i.e. values of $\bar{\underset{\sim}{u}}_n$, $\bar{\underset{\sim}{u}}_{n-1}$ and $\bar{\underset{\sim}{p}}_n$ are available at the start of computation (n=0), equation (32) can be solved for $\bar{\underset{\sim}{u}}_{n+1}$.

57. This solution is almost trivial if $\underset{\sim}{M}$ is a lumped, diagonal matrix. Various schemes for such diagonalisation are discussed in reference 29 and throughout the numerical computation used here such schemes were employed allowing the programs to be implemented on small computers with minimal storage requirements.

58. It is well known that the central difference scheme is conditionally stable and requires

$$\Delta t \leq \Delta t_{crit} \qquad (33)$$

where the value of Δt_{crit} is governed by the largest eigenvalue of the system in the linear cases. Thus if equation (27) is linearized (with the elasticity matrix considered constant) and the term $\underset{\sim}{Q}\,\underset{\sim}{p}$ is neglected, we find that

$$\Delta t_{crit} \simeq h/c \qquad (34)$$

$$c = \sqrt{\frac{K_T}{\rho}}$$

in which h is the size of the smallest element and c is the compression wave velocity determined by the tangent modulus K_T of the material. As in plastic processes a 'softening' of such a modulus tends to occur and it therefore suffices to use

$$K_T = K_e \qquad (35)$$

where K_e is the elastic bulk modulus of the (drained) material. As equation (27) is coupled with equation (29) through the pore pressure $\bar{p}$ where the value of $\bar{\underset{\sim}{p}}$ depends on the permeability of the material and the free drainage boundary condition of the system respectively, it is apparent that the Δt_{crit} is also subjected to the influence of these factors. Therefore, the critical time step Δt_{crit} is governed by the stability of the system of coupled equations and has a value ranging from Δt_{crit} (undrained) to Δt_{crit} (drained).

59. Once $\bar{\underset{\sim}{u}}_{n+1}$ is determined, equation (29) can be used for determination of $\bar{\underset{\sim}{p}}_{n+1}$. Now we can compute, approximately

$$\ddot{\bar{\underset{\sim}{u}}}_n = (\bar{\underset{\sim}{u}}_{n+1} - 2\bar{\underset{\sim}{u}}_n + \bar{\underset{\sim}{u}}_{n-1})/\Delta t^2 \qquad (36)$$

$$\dot{\bar{\underset{\sim}{u}}}_n = (\bar{\underset{\sim}{u}}_{n+1} - \bar{\underset{\sim}{u}}_n)/\Delta t$$

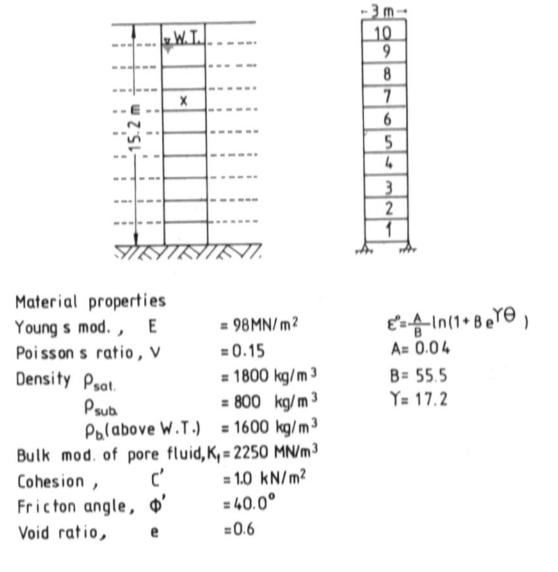

FIGURE 10 Saturated soil stratum

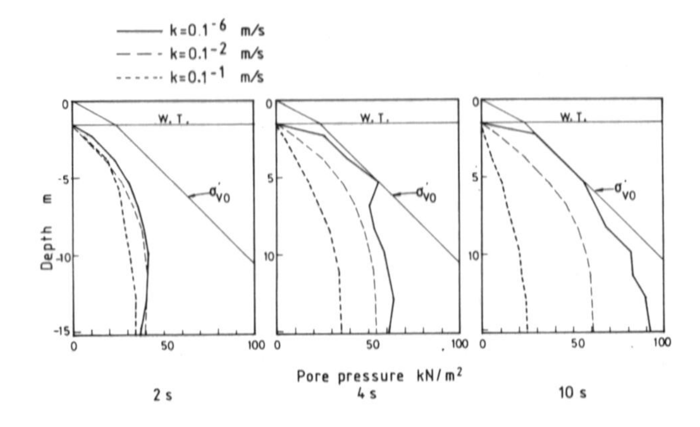

FIGURE 11a Build-up of excess pore water pressure at 2,4,10 sec. from the start of motion

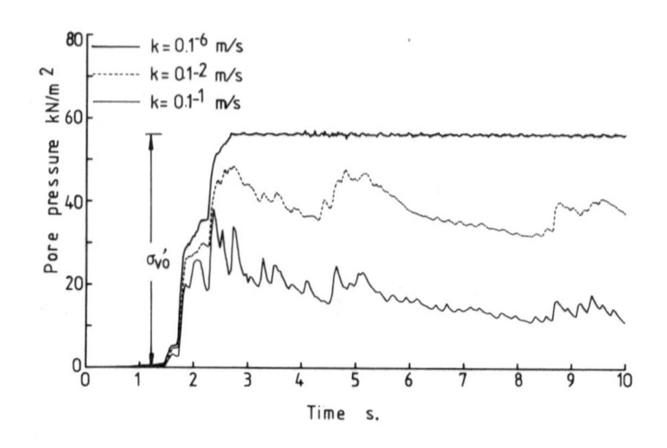

FIGURE 11b Build-up of excess pore water pressure with respect to time in layer 7 due to El Centro earthquake of May, 1940 scaled to 0.1g at base

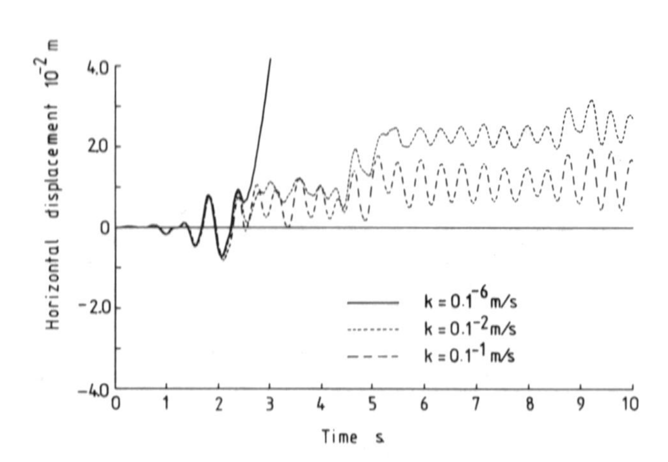

FIGURE 12 Surface horizontal displacement time history of horizontally layered saturated soil subjected to El Centro earthquake of May, 1940 scaled to 0.1g at the base

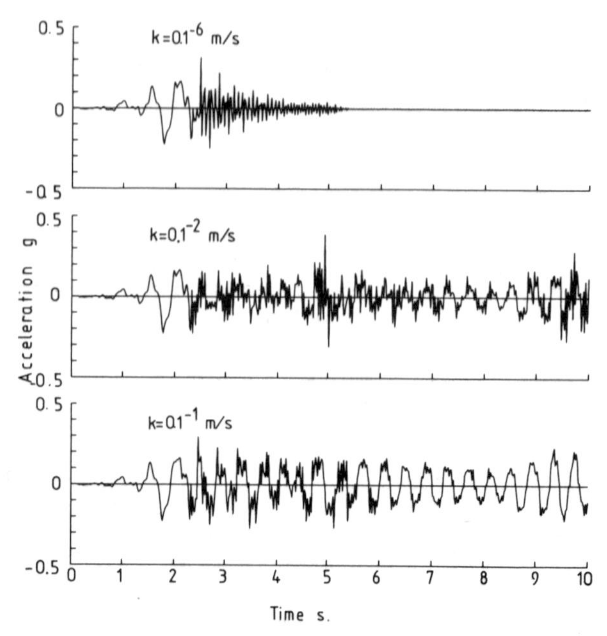

FIGURE 13 Surface horizontal acceleration time history of a horizontally layered soil subjected to El Centro earthquake of May, 1940 scaled to 0.1g at base.

and introduce an appropriate time marching scheme for $\bar{p}$. The whole family of such schemes can be written as

$$H(\theta \bar{p}_{n+1} + (1-\theta)\bar{p}_n) + S(\bar{p}_{n+1} - \bar{p}_n)/\Delta t =$$

$$\hat{f}_n - Q^T \dot{\bar{u}}_n + \hat{M} \ddot{\bar{u}}_n \qquad (37)$$

with schemes for $\theta \geq 1/2$ being unconditionally stable, and $\theta=0$ leading (on suitable lumping of the matrix S) to an explicit computation mode.

60. As the compressibility matrix S can, on occasion, become very small in value, the critical time steps for the explicit scheme can become very low so it is essential to adopt unconditionally stable schemes for this part of the operation, if the same time step as that used for the solution of equation (32) is to be retained. We have thus adopted a simple backward difference scheme with $\theta=1$ which has the advantage of avoiding all oscillation (other schemes with $\theta = 2/3$ for instance may prove to be even better and numerical experiments are proceeding here).

Simplification of undrained behaviour

61. If undrained behaviour is assumed - and this is a tenable assumption when the permeability k is low or if the frequencies to be considered are high - then a simplification occurs. We note from equation (24) that (with k=0) we have simply

$$-\eta \dot{p}/K_f = m^T \dot{\varepsilon} \qquad (38)$$

or $$-\eta \, dp/K_f = m^T d\varepsilon$$

Now, the constitutive relation can be written in total stress terms (using equations (22a-22c) as

$$d\sigma = \bar{D} d\varepsilon - D d\varepsilon^o \qquad (39)$$

with

$$\bar{D} = D + m \frac{K_f}{\eta} m^T \qquad (40)$$

The discretized equations of equilibrium now involve only $\bar{u}$ and can be written as

$$\int_\Omega B^T \sigma \, d\Omega + M \ddot{\bar{u}} = f \qquad (41)$$

with

$$d\sigma = \bar{D} B \, du - D \, d\varepsilon^o \qquad (42)$$

62. This, once again, can be solved by time stepping processes and an apparent simplification occurs as $\bar{p}$ solution is no longer needed. An explicit scheme for such a solution has been considered by Zienkiewicz et al. (ref.25) and used with considerable success since. However, we must note that now the critical time step is governed by the undrained compression modulus $\bar{K}_T$ (and the corresponding compression wave velocity $\bar{c}$), i.e.

$$\Delta t \leq \Delta t_{crit} \simeq h/\bar{c}; \quad \bar{c} = \sqrt{\frac{\bar{K}_T}{\rho}} \qquad (43)$$

63. As the bulk modulus of the pore fluid, e.g. water, is of the order of 2500 MN/m^2, an undrained analysis requires a very small value of Δt_{crit} and is therefore expensive. The Δt_{crit} for a solution based on the full $\bar{u} - \bar{p}$ form of the previous section with a permeability which tends to zero has a lower bound value equal to Δt_{crit} (undrained) and is found to vary depending on the boundary conditions. We have found in some of our studies that the maximum Δt_{crit} of a partially drained analysis with a permeability which tends to zero is equal to 3.2 times of the Δt_{crit} of an undrained analysis and hence under such conditions there is no economic advantage in adopting an undrained rather than a partially drained analysis procedure.

Test example

64. As several exact solutions for full Biot's equations as well as for the approximate form given by equations (22-24) have been obtained analytically (ref.33), it is convenient to test transient analysis computations on such examples.

65. The problem used concerns a layer of material of depth L placed on an impermeable base with free surface drainage. On the surface of this layer a vertical, periodic total stress q is applied at a period T. Taking a porosity $\eta = 1/3$ and $\rho_f/\rho = 1/3$, the solution is found to depend on two nondimensional parameters

$$\pi_1 = \frac{2}{\pi} k\rho \frac{T}{(\hat{T})^2}$$

$$\pi_2 = \pi^2 \left(\frac{\hat{T}}{T}\right)^2 \qquad \hat{T} = \frac{2L}{\bar{c}} \qquad (44)$$

66. In Figures 9a and b we show a comparison between exact periodic solutions for pressure amplitudes and results of numerical computation which was carried out using 30 bi-linear isoparametric quadrilateral elements in the layer and taking the time stepping solution to the point where almost steady state response was achieved. The excellent agreement can be noted for the type of approximation which we have here.

67. In the same figure we also show results of a numerical computation carried out omitting the $\hat{M}$ term in equation (29). We note that, at higher values of π_1, the error introduced is quite substantial and we do not recommend that this term should be generally suppressed, although at low permeabilities its importance is small.

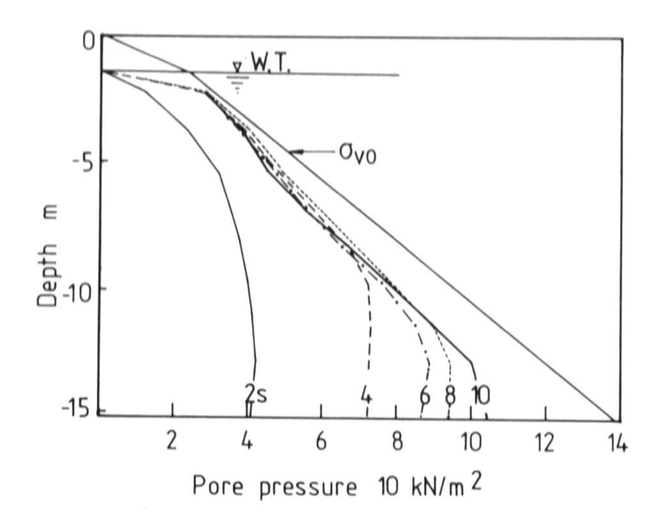

FIGURE 14a Build-up of excess pore water pressure at 2 sec. intervals from the start of motion (Model C - critical state model with x=100.0)

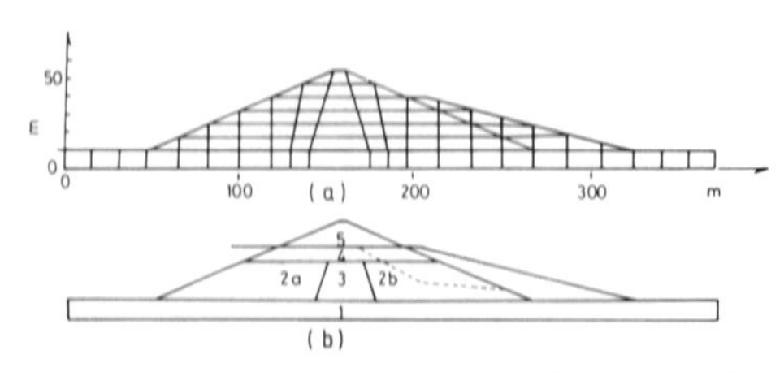

	Material zone	Elastic modulus, E MN/M²	V	Φ' Degree	C' KN/M²	Unit weight T/M³	β = A/B (1·Be^{γθ}) A	B	γ
1	Alluvium	200	0.4	38	10	2.09			
2a	Hydraulic fill - sand	90	0.41	37	10	2.02	2914	1904	4.6
2b	" " "	110	0.41	37	10	2.02-sat 1.71-dry	2914	1904	4.6
3	Clay core	90	0.41	37	10	2.02			
4	Ground shale-hydraulic fill	90	0.41	37	10	2.02sat 1.71dry			
5	Rolled fill	60	0.3	25	126	2.0			

FIGURE 16 The simulation of the Lower San Fernando Dam. (a) finite element mesh, (b) material zones and material properties.

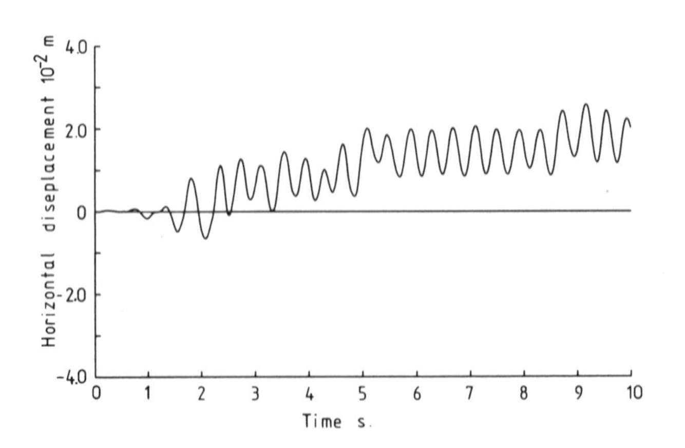

FIGURE 14b Surface horizontal displacement time history of a horizontally layered soil subjected to El Centro earthquake of May, 1940 scaled to 0.1g at base (Model C - critical state model with x=100.0)

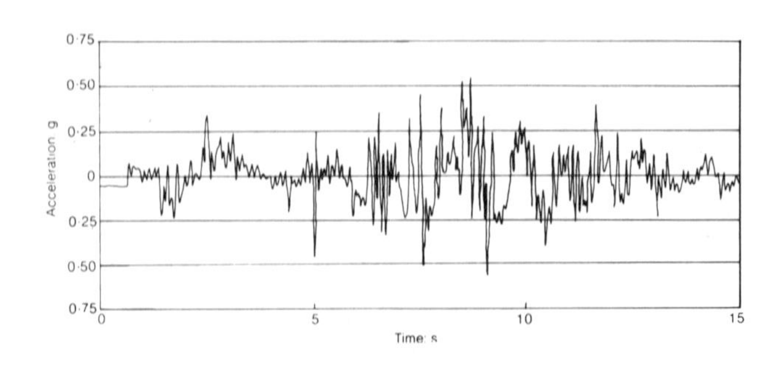

FIGURE 17 Scott's acceleration record for San Fernando Dam (after Seed, ref.5)

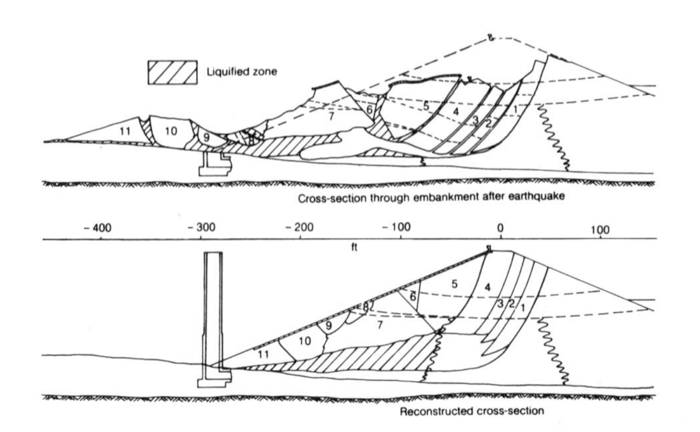

FIGURE 15 Cross-section through slide area and reconstructed cross-section of the Lower San Fernando Dam (after Seed, ref.36).

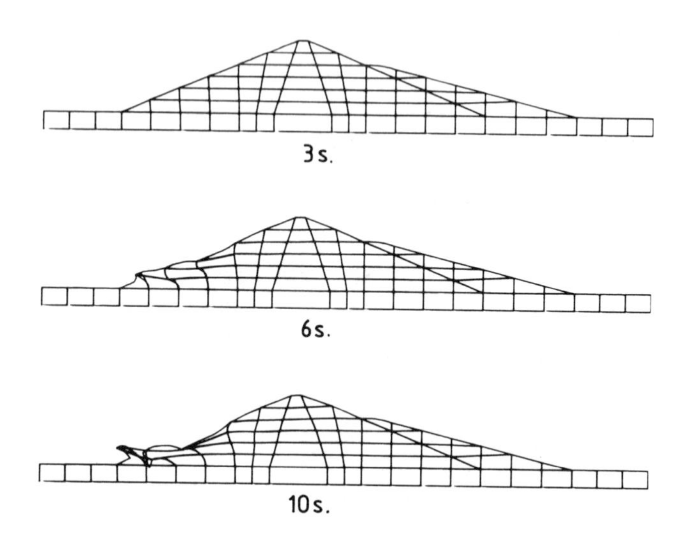

FIGURE 18 Deformed mesh of the dam at 3,6,10 sec. from the start of motion (displacements x 2.0)

68. In Figure 9c we show the relationship between the value of Δt_{crit} and the permeability. The increase of Δt_{crit} corresponding to the change of permeability is noticeable. In order to illustrate the effect of free drainage boundary conditions, a pervious boundary is imposed at the bottom of the layer and the result is shown in the same figure.

69. The computer program in which the various features of analysis are incorporated is named LIQ2 and includes simple facilities for amendment of constitutive laws etc.. The basic elements used are

(a) the four node, isoparametric, quadrilateral element.

(b) the eight node, serendipity, isoparametric quadrilateral element.

(c) the nine node, Lagrangian isoparametric quadrilateral element.

Mass lumping procedures described elsewhere (ref.29) are used, and the performance of elements (b) or (c) using 'reduced' integration is optimal.

APPLICATION

General Remarks

70. The formulation presented, together with the constitutive laws introduced, allows any problem of earthquake response to be studied quantitively. Naturally, the results will only be as good as the physical data supplied and the idealization used for the constitutive model - irrespective of the degree of accuracy of the numerical representation.

71. In the examples that follow we deal with sands or silts subjected to various earthquake shocks. In all of the examples the simple densification model introduced in the previous section (equations (18-20)) is used in conjunction with ideal, non-associative plasticity (Model B) or the critical state model (Model C). In real sands more complex models are generally needed as a certain degree of dilation will occur during plastic deformation (near the critical state). In a recent study, we find that by introducing a small degree of dilation in a non-associative plasticity model, a higher resistance against liquefaction is predicted. However, this will be affected by the degree of confinement of the system. Generally, the effect of dilatancy will be more pronounced in a two-dimensional problem than in a one dimensional problem. If the effect of dilatancy is ignored the main effects will be more clearly illustrated and the solution will overestimate the danger of liquefaction and thus be on the safe side.

72. Again we should remark that the strain hardening critical state model will be less prone to liquefaction (some of the plastic dilatancy counteracting the densification strains) and therefore in the study of the Lower San Fernando dam such a model is not used.

A layer of pervious sand subjected to a horizontal earthquake shock

73. A horizontal soil layer of depth 15.2m is subjected to a base motion of the N-S component of the El Centro Earthquake, May, 1940 with the maximum acceleration scaled to 0.1g. The soil is modelled as a simple elastic-plastic material with zero dilatation (Model B) and the values of the densification strain parameters A, B and γ based on ref.25 are adopted in this study. The relevant results of the solution are shown in Figures 10-13.

74. The same problem was first studied by Finn et al. (refs. 24,34) using a one-dimensional program called DESRA and more recently another approximate study was carried out by Seed and Martin (ref.28) using programs MASH (ref.35) and APOLLO (ref.38). Unfortunately, there is no field pore pressure data available at present to provide a check on the numerical solution of any analysis of this type. However, our approach did manage to reproduce the basic features that one can expect during the generation and dissipation of pore pressure caused by an earthquake. For example, there is a 'cut off' of surface motion after the liquefaction of any layer occurs. It is of interest to note that similar behaviour can be observed in the results of the analyses by Finn et al.

75. In Figure 14 we show similar results for the undrained case but here the critical state model (Model C) is used. The model employed has a full elliptical yield surface and a normal consolidation condition is assumed. As anticipated earlier, complete liquefaction does not occur in this case.

A simulation of the Lower San Fernando Dam behaviour

76. In this fully two-dimensional example all features of the analysis capability are demonstrated and we show that an approximate simulation of the behaviour of this dam together with an estimate of the permanent deformation occurring during the earthquake can be achieved.

77. The details of the dam and its failure in the San Fernando earthquake are described and analysed in reports prepared by Seed et al. (refs.4,5) as well as his Rankine lecture of 1979 (ref.36). As a result of the earthquake, the dam slid in the upstream shell by good fortune retaining sufficient freeboard with a low reservoir level to prevent a disastrous spillage. A figure of a cross section through the slide area and its reconstruction reproduced from reference 36 is shown in Figure 15.

78. The data for the present analysis were taken from reference 5 and reinterpreted in terms of the densification strain. Appendix A lists some of this information which is also summarised in Figure 16. This figure also shows the mesh of isoparametric 8-node elements used in the analysis. It should be noted that in the analysis, the zones above the original

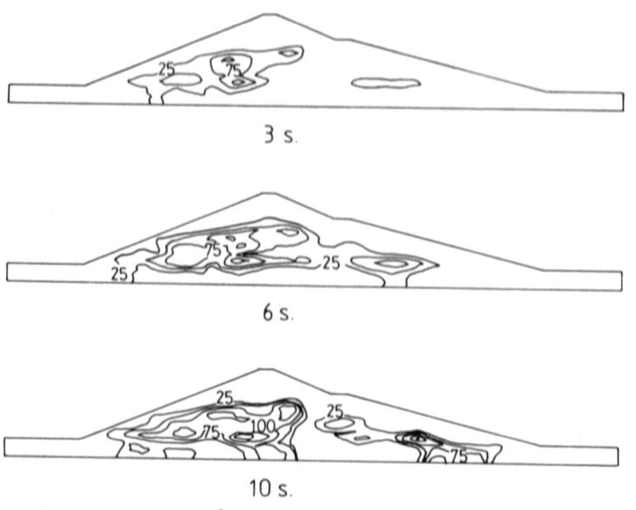

Contour unit = kN/m²
Contour spacing = 25 kN/m²

FIGURE 19 Contours of build-up of excess pore water pressure at 3,6,10 sec. from the start of motion

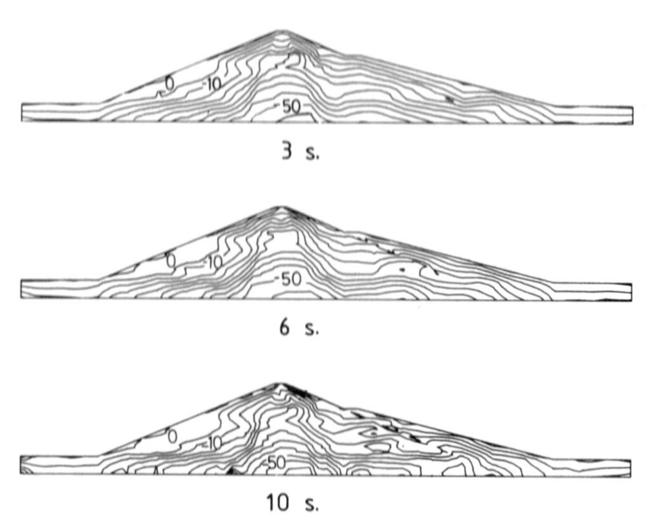

Contour unit = 10 kN/m²
Contour spacing = 50 kN/m²

FIGURE 20 Vertical effective stress contour at 3,6,10 sec. from the start of motion

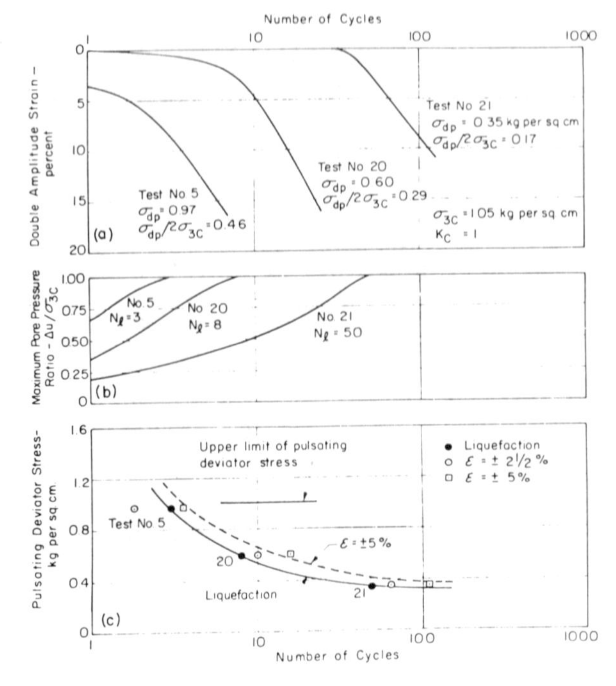

FIGURE A1 Results of cyclic load tests on isotropically consolidated samples of hydraulic sand fill - Upper San Fernando Dam (after Seed et al, ref.5)

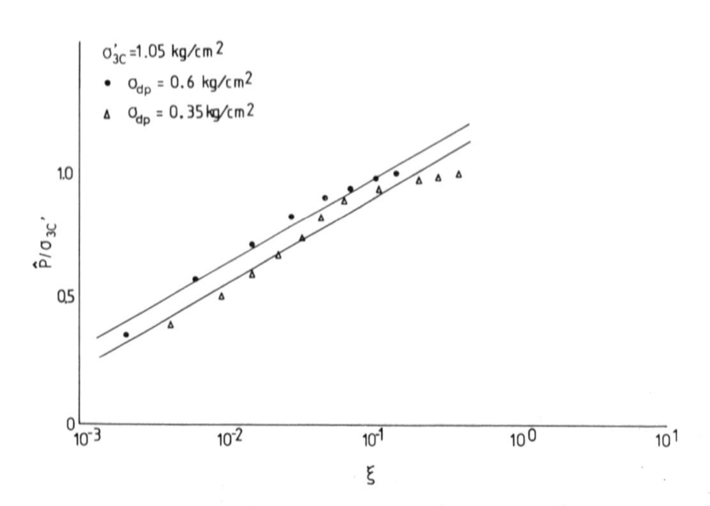

FIGURE A2 Excess pore water pressure $\hat{P}$ versus the total strain path ξ

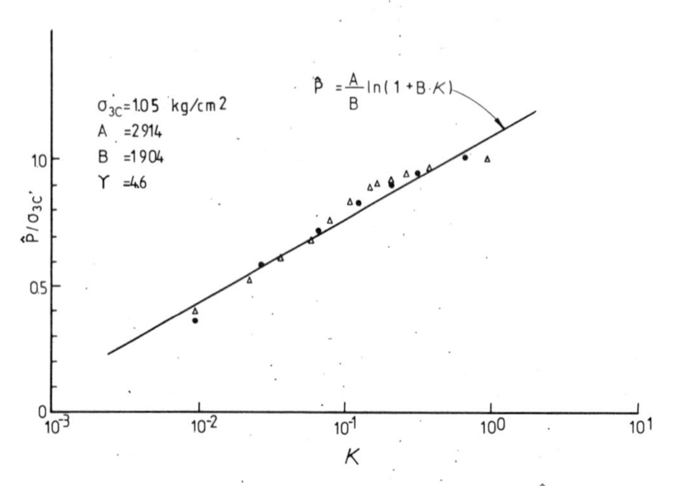

FIGURE A3 Excess pore water pressure $\hat{P}$ versus κ

phreatic surface are assumed 'dry' and the pore pressure there is simply set to zero.

79. Initial stresses and pore pressures are computed using an elasto-viscoplastic soil mechanics program described in reference 39. The analysis is carried out by a 'one step gravity turn on' approach and the stages of construction of the dam are not taken into account. After determining the initial stresses, the dam is subjected to a base motion which is developed from the abutment seismoscope record of the Lower San Fernando Dam by R.F.Scott.The same base motion was used by Seed et al. (refs.4,5) and is shown in Figure 17. The analysis is carried out for the first 10 sec. and the deformed mesh, contours of the build up of excess pore pressure and contours of the vertical effective stress at various times during the earthquake passage are shown in Figures 18, 19 and 20 respectively. The analysis shows that;

(1) Noticable permanent deformation occurs at the upstream slope after 10 sec. from the start of motion.

(2) Less excess pore pressure is predicted in the inner region of the upstream shell than in Seed's analysis.

There are two possible explanations for this behaviour. Firstly, permanent deformations occur at the upstream toe of the dam and as a result decrease the magnitude of horizontal vibration of the inner part of the upstream shell, and therefore produce less densification strain and hence less pore pressure. Secondly, small deformation theory is employed in the present analysis and the volumetric strain due to the nonlinear contribution is not taken into account.

80. Seed (ref.36) concluded that most of the motion of the slide occurred some time after the passage of the main shock due to a redistribution of pore pressure. Our computation stops before such a redistribution can take place. Indeed this poses certain problems as the formulation in principle includes all non-linear consolidation effects. To include this stage of computation, two alternative approaches are being studied;

(a) A reduction of the time scale for post earthquake phenomena and a continuation of the explicit computation on a different time scale resulting from adjusted permeability values

and

(b) the use of an implicit, stable process with a subsequent increase of the time step length.

The reader will however observe that all the essential forms of behaviour have been modelled and the computation has permitted permanent and significant movement due to liquefaction to be observed.

Concluding remarks

81. The paper shows that with the present stage of development of numerical methods and with the current understanding of liquefaction phenomena it is possible to carry out a full non-linear analysis to determine whether any permanent displacement will occur in a dam or its foundation after the passage of an extreme earthquake motion.

82. Clearly when improved models of the material become available they can be readily inserted into the analysis - but at the present stage of development it is found that relatively simple elastic-plastic models including densification are adequate.

83. Research however continues on development of improved numerical algorithms for combining post-earthquake behaviour studies as well as on the improvement of basic models to take account of effects of multi directional shaking such as that studied recently by Seed et al. (ref.37).

REFERENCES

1. ZIENKIEWICZ,O.C. Constitutive laws and numerical analysis for soil foundation under static, transient or cyclic loading. Proceedings of Second International Conference on Behaviour of Offshore Structures (BOSS), London, 1979.

2. NASSER,S. NEMAT Private Communication, 1980.

3. ZIENKIEWICZ,O.C. and BETTESS,P. Soil and other saturated porous material under transient, dynamic condition, general formulation and validity of various simplified assumptions. Soil Under Cyclic and Transient Loading, Editors O.C.Zienkiewicz and G.N. Pande, John Wiley (to be published).

4. SEED,H.B., IDRISS,I.M., LEE,K.L. and MAKDISI,F.I. Dynamic analysis of the slide in the Lower San Fernando Dam during the earthquake of Feb.9, 1971. J. Geotech. Engng. Div., Proc. Am. Soc. Civ. Eng., 1975 101, GT9, 889-911.

5. SEED,H.B., LEE,K.L., IDRISS,I.M. and MAKDISI,F.I. Analysis of the slides in the San Fernando Dam during the earthquake of Feb. 9, 1971. EERC Report No. EERC73-2, University of California, Berkeley, Calif. 1973.

6. ROSCOE,K.H., SCHOFIELD,A.N. and WORTH,C.P. On the yielding of soils. Geotechnique 1958, 8, 22-53.

7. ROSCOE,K.H. and BURLAND,J.B. On the generalised stress/strain behaviour of 'wet' clay. in Engineering Plasticity, Editors J. Hayman and F.A. Lockead, Cambridge University Press, 1968, 535-609.

8. SCHOFIELD,A.N. and WORTH,C.P. Critical State Soil Mechanics. McGraw Hill, 1968.

9. ZIENKIEWICZ,O.C., HUMPHESON,C. and LEWIS, R.W. Associated and non-associated visco-plasticity and plasticity in soil mechanics. Geotechnique, 1975, 25, 671-689.

10. ZIENKIEWICZ,O.C., HUMPHESON,C. and LEWIS, R.W. A unified approach to soil mechanics including plasticity and viscoplasticity. in Finite Elements in Geomechanics, Editors G. Gudehus, John Wiley,1977

11. MROZ,Z., NORRIS,V.A. and ZIENKIEWICZ,O.C. An anisotropic hardening model for soils and its application to cyclic loading. Int. J. Num. and Analytical Meth. Geomech. 1978, 2, 203-221.

12. MROZ,Z., NORRIS,V.A. and ZIENKIEWICZ,O.C. Application of an anisotropic hardening model in the analysis of elastic-plastic deformation of soils. Geotechnique, 1979, 29, 1-34.

13. MROZ,Z., NORRIS,V.A. and ZIENKIEWICZ,O.C. Elastic plastic and viscoplastic constitutive models for soils with application to cyclic loading. - to be published.

14. GHABOUSSI,J. and DIKMEN,S.U. Liquefaction analysis of horizontally layered sands. J. Geotech. Engng. Div., Proc. Amer. Soc. Civ. Engng, 1978, 104, GT3, 341-356.

15. GHABOUSSI,J. and MOMEN,H. Modelling and analysis of cyclic behaviour of sands. in Soils Under Cyclic and Transient Loading, Editors O.C. Zienkiewicz and G.N. Pande, John Wiley - to be published.

16. PENDER,M.J. Cyclic mobility - a critical state model. Proceedings of the Int. Symposium on Soils Under Cyclic and Transient Loading, Swansea 7-11, Jan.1980 325-335.

17. NOVA,R. A constitutive model for soil under monotonic and cyclic loading. in Soils Under Cyclic and Transient Loading. Editors O.C. Zienkiewicz and G.N. Pande, John Wiley - to be published.

18. CARTER,J.P., BOOKER,J.R. and WORTH,C.P. A critical state soil for cyclic loading. in Soils Under Cyclic and Transient Loading, Editors O.C. Zienkiewicz and G.N. Pande, John Wiley - to be published.

19. DAFALIAS YANNIS,F. Bounding surface formulation of soil plasticity. in Soils Under Cyclic and Transient Loading, Editors O.C. Zienkiewicz and G.N. Pande, John Wiley - to be published.

20. BAŽANT ZDENĚK,P., ANSAL,A.M. and KRIZEK,J. Endochronic models for soils. in Soils Under Cyclic and Transient Loading, Editors O.C. Zienkiewicz and G.N. Pande, John Wiley - to be published.

21. VALANIS,K.C. and READ,H.E. Recent development and application of the endochronic theory to the behaviour of soils. Presented in the Int. Symposium on Soils Under Cyclic and Transient Loading, Swansea, January 1980, 7-11.

22. SILVER,M.L. and SEED,H.B. Volume changes in sands during cyclic loading. J. Soil Mech. Found. Div., Proc. Am. Soc. Civ. Eng., 1971, 97, 1171-1182.

23. MARTIN,G.R., FINN,W.D.L. and SEED,H.B. Fundamentals of liquefaction under cyclic loading. J. Geotech. Eng. Div., Proc. Am. Soc. Civ. Eng., 1975, 101, 423-438.

24. FINN,W.D.L., LEE,K.W. and MARTIN,G.R. An effective stress model for liquefaction. J. Geotech. Engng. Div., Proc. Am. Soc. Civ. Eng. 1977, 103, 517-533.

25. ZIENKIEWICZ,O.C., CHANG,C.T. and HINTON,E. Non linear seismic response and liquefaction. Int. J. Num. and Analytical Meth. in Geomechanics, 1978, 2, Issue No.4, 381-404.

26. NASSER NEMAT,S. and SHOKOOH,A. Densification and liquefaction of sand in cyclic shearing. Central American Conference on Earthquake Engineering, El-Salvador, January 1978.

27. SEED,H.B., MARTIN,P.P. and LYSMER,J. Pore-water pressure changes during soil liquefaction. J. Geotech. Engng. Div., Proc. Am. Soc.Civ. Eng. 1976, 102, GT4, 323-346.

28. MARTIN,P.P. and SEED,H.B. Simplified procedure for effective stress analysis of ground response. J. Geotech. Engng. Div., Proc. Am. Soc. Civ. Eng. 1979, 105, GT6, 734-758.

29. ZIENKIEWICZ, O.C. The Finite Element Method, McGraw-Hill, London, 1977.

30. PARK,K.C. Partitioned transient analysis procedures for coupled field problems - stability analysis - to be published.

31. PARK,K.C. and FELIPPA,C.A. Partitioned transient analysis procedures for coupled field problems - accuracy analysis - to be published.

32. ZIENKIEWICZ,O.C. Finite elements in the time domain. State of art survey of finite element methods - to be published, Proc. Am. Soc. Mech. Eng. 1980.

33. ZIENKIEWICZ,O.C., CHANG,C.T. and BETTESS,P. Drained, undrained, consolidating and dynamic behaviour assumptions in soils. Limits of validity - to be published.

34. FINN,W.D.L., MARTIN,G.R. and LEE,M.R.W. Comparison of dyanmic analyses for saturated sands. presented at the 1978 Conference on Earthquake Engineering and Soil Dynamics, Pasadena, Calif.

35. MARTIN,P.P. and SEED,H.B. MASH - A computer program for the non-linear analysis of vertically propagating shear waves in horizontally layered soil deposits. EERC Report No. UCB/EERC-78/23, Univ. of Calif. Berkeley, Calif., Oct. 1978

36. SEED,H.B. Considerations in the earthquake-resistant design of earth and rockfill dams. 19th Rankine Lecture of the British Geotechnical Society, Geotechnique, 1979, 29 215-263

37. SEED,H.B., PYKE.R.A. and MARTIN,G.R. Effect of multi directional shaking on pore pressure development in sands. J. Geotech. Engng. Div., Proc. Am. Soc. Civ. Eng.,1978, 104, 27-44

38. SEED,H.B., MARTIN,P.P. APOLLO - A computer program for the analysis of pressure generation and dissipation in horizontal sand layers during cyclic or earthquake loading. EERC Report No. UCB/EERC-78/21, Univ. of California, Berkeley, Calif., Oct. 1978

39. ZIENKIEWICZ,O.C. and HUMPHESON,C. Visco-plasticity- A general model for description of soil behaviour. in Finite Elements in Geomechanics ,Editors, C.S. Desai and C.Christian, McGraw Hill, 1977

ACKNOWLEDGEMENT

The financial support granted by the Science Research Council of U.K. under Research Grant GR/2602.8 and GR/A/H17/3 to C.T. Chang and K.H. Leung is hereby gratefully acknowledged.

APPENDIX A

The evaluation of the densification parameters

The densification strain parameters (equation (19)) used in the simulation of the Lower San Fernando Dam behaviour are evaluated from the results of the cyclic triaxial tests on isotropically-consolidated samples of the hydraulic fill of the Upper San Fernando Dam (ref.5). The reason for adopting the Upper Dam data in this study is that the required information for the Lower Dam is not available from reference 5. Tests show that the resistance to liquefaction which is defined as the maximum cyclic deviator stress required to cause a certain amount of strain in a number of cycles , for the Lower Dam is found to be slightly higher than the value of the Upper Dam. Therefore the use of the Upper Dam data should provide conservative results.

Some results from reference 5 are shown in Figure A1. The strain amplitude and maximum deviator stress of each load cycle can be used to evaluate the value of θ, $d\xi$ and ξ by the use of equations (17) and (18). The relationship between the excess pore pressure $\hat{p}$ and ξ is shown in Figure A2. Finally, the values in the

$\hat{p}$ - ξ plane can be transfromed into the $\hat{p}$ -κ plane by employing equation (20) and the result is shown in Figure A3. The parameters A, B and γ in Figure A3 are used in the present study.

22

Seismic studies in relation to the design of Alicura dam and appurtenant works

B. GILG, Dr sc techn, Department for Hydropower, Irrigation and Water Treatment of Electrowatt Engineering Services Limited, Switzerland

For the Alicura dam site and storage lake area extensive seismic investigations have been performed. Since the location of this future power plant is relatively close to important epicenter-concentrations in Chile, detailed studies were needed in connection with the various types of earthquakes occuring in the southern part of the south american continent. Besides the historical events, which allow a certain extrapolation with respect to the determination of a maximum possible earthquake, the problem of induced seismic shocks is also raised and the corresponding investigations are described. Finally the importance of a well functioning bottom outlet is pointed out.

INTRODUCTION

One of the important rivers flowing on the Argentinean side of the Andes from SW towards NE is the Rio Limay. Its origin is the lake of Nahuel Huapi (San Carlos de Bariloche) and its main tributary is the Collon Cura rising at different springs around the volcano Lanin (3740 m.asl).

At a distance of 110 km downstream from San Carlos the river has created a 250 m wide gorge of a depth of 150 m at which place a large power plant will be implemented. The corresponding characteristics are the following : (Fig.1)

Hydrology

Catchment area	6'980 km^2
Average annual inflow	8'324 Mio m^3
Average discharge	265 m^3/sec
Design flood inflow	3'026 m^3/sec

Reservoir

Total storage	3'500 Mio m^3
Surface area	60 km^2

Dam

Type	Rock/Earthfill
Hight above foundation	130 m
Crest length	880 m
Crest level	710 m.a.s.l.
Volume	13 Mio m^3

Spillway

3 radial gates	10 x 14 m^2
Capacity	3'000 m^3/sec

Bottom outlet

2 slide gates	3.5 x 2.2 m^2
Capacity	580 m^3/sec

Waterways

Headrace canal	length	450 m
4 penstocks	length	240 m
	diameter	6.8 m

Power house

Average rated gross head	121 m
Turbine (Francis) discharge	4 x 262 m^3/sec
Installed capacity	4 x 250 MW

GENERAL EARTHQUAKE SITUATION

The gorge of Alicura is situated at 40.8° S and 70.7° W, i.e. at a distance of about 100 km from the chilean border. Thus the location is at the same latitude as the southern part of one of the most active seismic zones of the American continent, namely the Peru-Chile trench, along which the Nazca plate, a plate of the oceanic lithosphere is being thrust downward beneath the South American continental plate. The convergence of these two plates is a result of sea floor spreading along the East Pacific rise and along the Mid-Atlantic ridge.

Seismological data compiled by Gutenberg and Richter (1954) for the period of 1906 to 1944 show that numerous large earthquakes of shallow (0-70 km) and intermediate (71-300 km) focal depth occurred along the western margin of South America as far south as latitude 37° S. Earthquakes with focal depths exceeding 300 km (deep-focus earthquakes) are found east of the Andes from the equator to latitude 37° S. South of latitude 37° S, earthquakes are mostly shallow with some sporadic intermediate focal depth earthquakes. (Fig. 2)

Several studies based on recent earthquake data have helped to define the nature of the deep crustal structure and the seismic zones beneath the western continental margin of South America. A detailed examination of available hypocenter data indicates that the spatial distribution of earthquakes is quite different at various latitudes and that the nature of the Benioff (subduction) zone varies along the continental margin. A focal mechanism study of earthquakes in Chile during the period 1962-1970 indicates that "subduction of the oceanic plate at any given time is taking place in discrete and localized episodes and that the lithospheric slab itself is broken into a series of tongues that are absorbed independently and quite differently from one latitude to the next or even at one depth as opposed to the other". (Stauder 1973)

Of the large and destructive historical earthquakes that have occurred along the western margin of South America, the most significant for the project site is the May 22, 1960 earthquake of southern Chile. This seismic event consisted of a sequence of major shallow focus earthquakes that included a main shock of magnitude 8.5.

In the vicinity of the project site, preliminary regional geologic and tectonic studies have recently been carried out or are presently being undertaken. Among the available literature is the preliminary report (1978) on the geology and natural resources of Neuquen Province presented at the 7th Argentinian Geologic Congress of April 1978. The structural study was based in part on interpretation of Landsat imagery. This analysis indicates a complex structural setting and the existence of important structural lineaments which in most cases divide the province into different structural units or elements. These lineaments are interpreted as old lines of weaknesses in the Preliasic (Pre-Jurassic) basement which may or may not have been reactivated to affect the younger rock units. The project site is located within a structural block referred to as the Collon Cura Basin. This old tectonic depression is bounded by the basement rocks of the Alumine Batholitic Belt to the west and north and the Sanico Sill to the east and is separated from them by major fractures trending N 15° W.

In the immediate project site area, the geologic studies have been conducted in relatively fine detail to develop an understanding of the geologic and tectonic history of the project area, particularly with regard to fault activity. The oldest rock unit at the project site is the Paso Flores Formation of Upper Triassic to Lower Jurassic age. This formation consists of interbedded sandstone and conglomerate laid on the basement igneous rocks. At the damsite, the Paso Flores Formation comprises the entire bedrock foundation of the dam and consists of in interbedded sandstone, claystone and some limestone. The Paso Flores Formation was subjected to deformation including shearing, folding, and faulting during Upper Cretaceous Cordilleran orogeny and a subsequent Early Tertiary Cordilleran orogeny.

EARTHQUAKE - CLASSIFICATION

Based on the findings described in the previous paragraph, an earthquake classificiation was established. 4 families of seismic events were distinguished :

- Earthquakes of various magnitudes (up to 8.5) with epicenters along the Peru-Chile trench and with shallow focal depths (0-70 km). The width of this epicenter zone is about 300 km, extending from the Andes to the trench, which is situated 100 - 150 km offshore in the Pacific.
- Earthquakes of various magnitudes with epicenters in the Andes and eastward of them on the Argentinean territory having intermediate focal depths (70-300 km).
- Small to medium magnitude earthquakes with epicenters in the nearer and farther vicinity of Alicura dam site and relatively shallow focal depths (0-90 km).
- Possible induced earthquakes with epicenters in the area of the dam site and the future storage lake.

The first two types of earthquakes have the same origin, i.e. movements along the separation plane between the South American plate and the subjacent Nazca plate. Since this plane is inclined from W to E, the focal depth is contineously increasing from the Peru-Chile trench towards the Argentinean territory. At the Alicura site the depth might be at 100 km or more. Therefore even a very strong earthquake of this type can by no means provoke at the surface intensities higher than 8 - 8.5. The corresponding ground acceleration would then amount to about 10 - 20 % g.

The third type of seismic shocks is related to shallow earthquakes within the reach of the South American plate, which may be the consequence of deeper movements. Therefore these events are developing a smaller amount of energy and are showing magnitudes 6.5. The focal depths are relatively small so that the intensities might come up to similar values as in the case of deep earthquakes.

The forth type of earthquakes is caused by a possible reactivation of faults generated in former geological times. Their origin might also be an increase of movements along faults, which are presently only provoking microseismic events. The prediction of such impacts is rather difficult and requires two kinds of special studies :

- Elaboration of a fault register near the dam site and in the area of the storage lake
- Recordings of microseismic events in the Alicura zone during at least one year

On the other hand it is very useful to compare the situation of Alicura with other damsites,

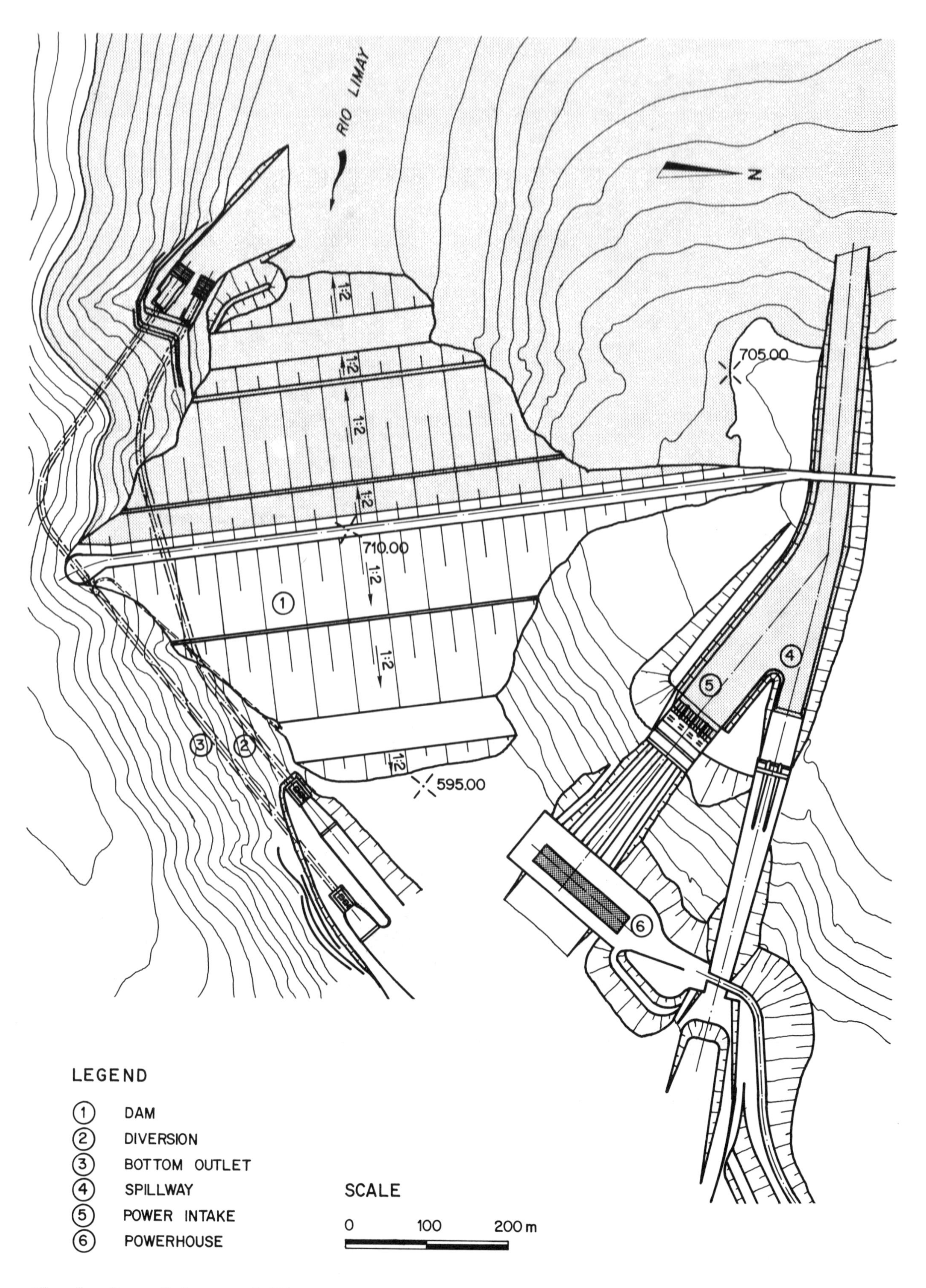

Fig. 1. General layout of Alicura power plant

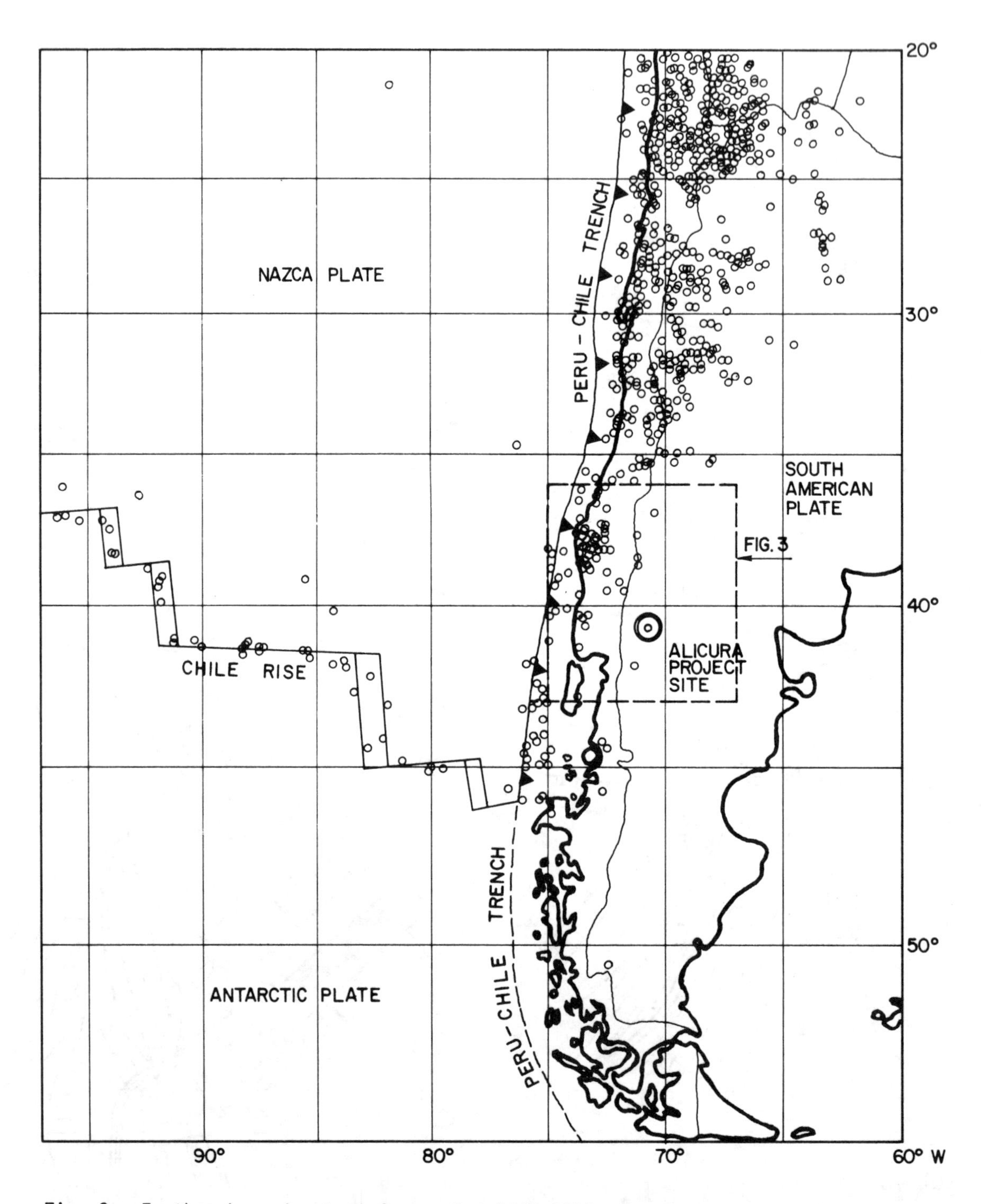

Fig. 2. Earthquake epicenters for period 1962-1969

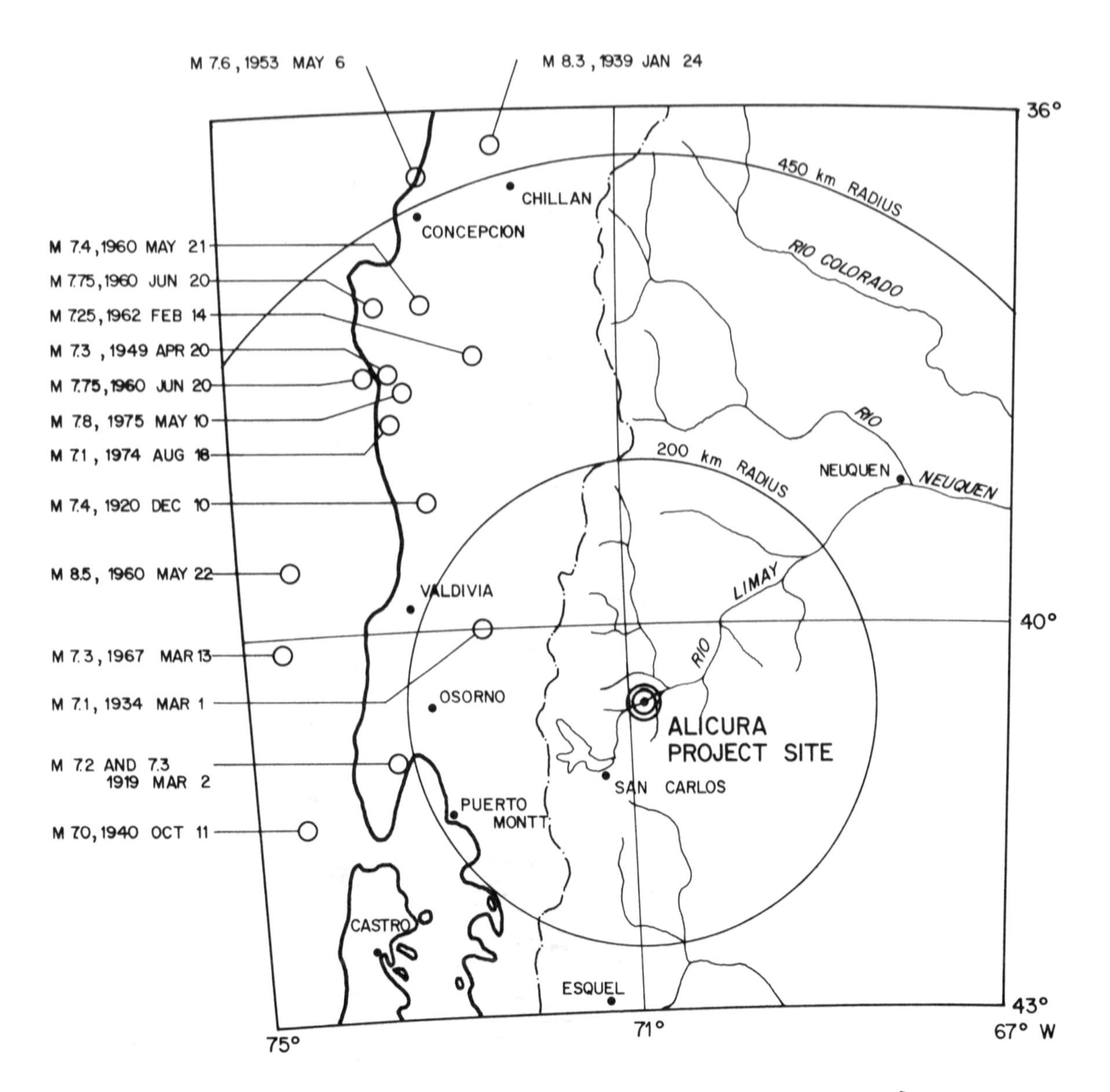

Fig. 3. Regional seismicity map with epicenters of major earthquakes M $\geq$7.0

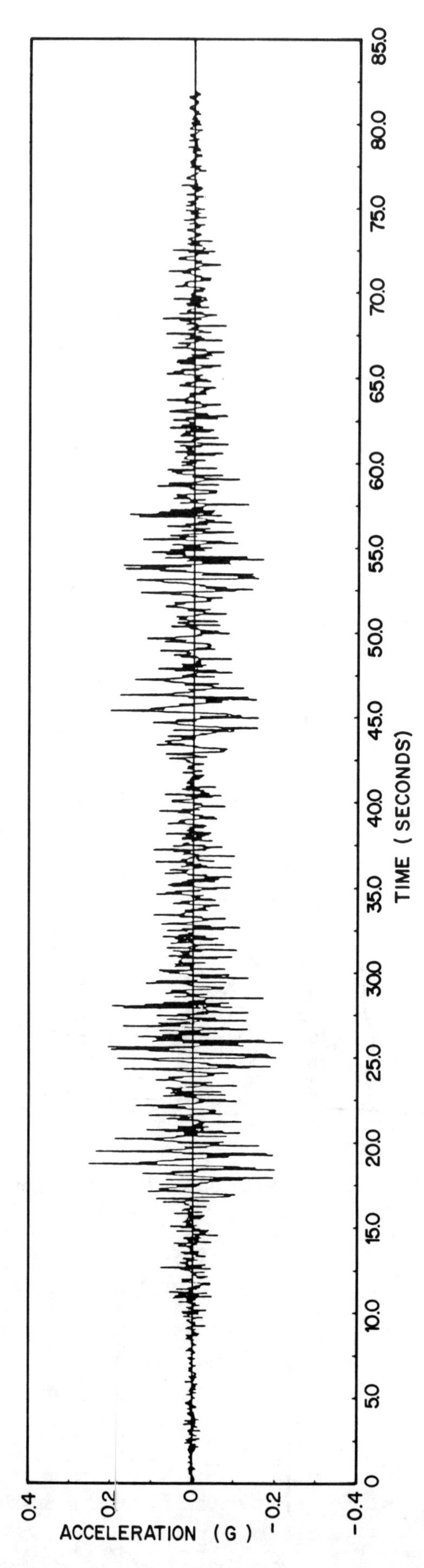

Fig. 4. Accelerogram of a M=8.5 distant earthquake (140 kms)

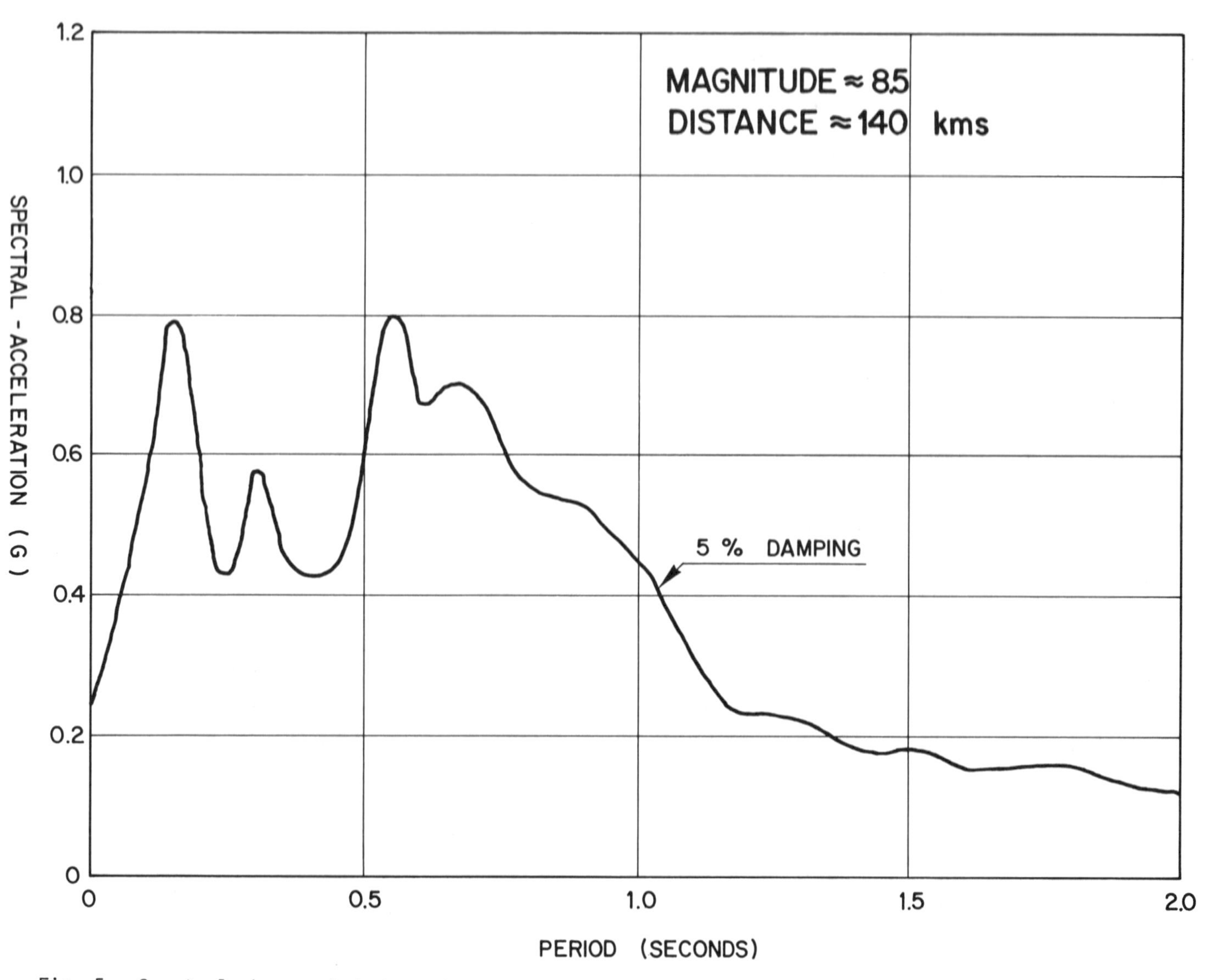

Fig. 5. Spectral shape related to the earthquake of Fig. 4

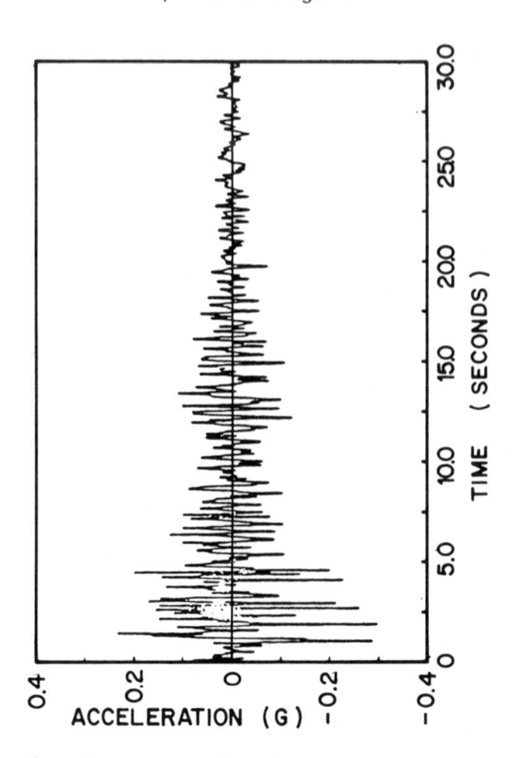

Fig. 6. Accelerogram of a deep-seated local earthquake

where induced earthquakes really took place in the past, in order to find out possible similarities.

EVALUATION OF THE EARTHQUAKES ALONG THE OVERTHRUST PLANE BETWEEN THE SOUTH AMERICAN AND THE NAZCA PLATES

It is well known, that some of the most catastrophic earthquakes which have destroyed human settlements, took place in the zone of the Peru-Chile trench. Nevertheless, when analysing the special situation of Alicura damsite, the biggest concentration of epicenters is located north of the 36° S latitude and therefore in a distance of more than 500 km from the future power plant. These events can therefore be neglected, since even a 8.5 magnitude earthquake is to be considered as practically inefficacious, once the distance of its epicenter exceeds 450 km.

A second and smaller group of epicenters is located between the cities of Conception and Puerto Mont, and this zone belongs to the area laying inside of a circle around Alicura, the radius of which is 400 km. (Fig. 3)

The haviest earthquakes in this area were :

Time	Magnitude	Depth	Epicenter	Dist.of Alicura	Int. at A.
16.12.1575	8.5	>10 km	73.6°/39.8°	250 km	7
24.12.1737	7.8	>10 km	73.6°/39.8°	250 km	6
07.11.1837	8.0	>10 km	73.6°/39.8°	250 km	6-7
02.03.1919	7.3	40 km	73.5°/41.0°	220 km	5
10.12.1920	7.4	>10 km	73.1°/39.0°	255 km	5-6
01.03.1934	7.1	120 km	72.5°/40.0°	155 km	4
11.10.1940	7.0	>10 km	74.5°/41.5°	315 km	4-5
20.04.1949	7.3	70 km	73.4°/38.0°	355 km	4
22.05.1960	8.5	60 km	74.5°/39.5°	330 km	6
20.06.1960	7.75	60 km	73.6°/38.0°	370 km	4
14.02.1962	7.25	45 km	72.5°/37.8°	330 km	4-5
13.03.1967	7.3	36 km	74.6°/40.1°	320 km	5
18.08.1974	7.1	36 km	73.4°/38.4°	320 km	5
10.05.1975	7.8	6 km	73.2°/38.1°	335 km	5-6

The preceeding table is demonstrating that in the last 500 years the corresponding family of earthquakes has never provoked a higher intensity than 7 at Alicura damsite. The intensity with a return period of 50 years is about 6 and the intensity 4 is reached practically every 2 years.

By extrapolation we obtain the intensity of 9 as an absolute maximum, which is corresponding to a ground acceleration of the rock mass in Alicura of about 20 % of the gravity.

EVALUATION OF THE CATEGORY OF EARTHQUAKES WITH EPICENTERS WITHIN 200 KM DISTANCE OF ALICURA DAM SITE AND SHALLOW TO INTERMEDIATE FOCAL DEPTHS

This family of seismic events has magnitudes 6 and the recordings are not yet very ancient. Nevertheless a summarizing table gives some indications on the activity during the last 25 years.

The main data are the following :

Time	Magnitude	Depth	Distance fr.Alicura	Intensity at Alicura
10.02.54	4-5	100 km	165 km	2 - 3
28.07.55	4-5	10 km	75 km	4
05.01.56	4-5	10 km	80 km	4
22.05.60	4-5	10 km	165 km	3
07.06.60	4-5	10 km	110 km	4
24.07.61	4-5	10 km	60 km	4
15.08.63	4.1	58 km	105 km	2
19.12.63	4.3	33 km	180 km	2
24.05.64	4.0	33 km	135 km	2
16.07.65	4.1	54 km	200 km	2
29.08.65	3.9	98 km	95 km	1.5
07.05.67	4.3	98 km	165 km	1.5
26.11.67	4.2	39 km	150 km	2
19.08.69	5.0	14 km	145 km	3.5
25.08.69	4.7	13 km	185 km	3
30.11.70	4.7	161 km	120 km	2
01.02.72	4.8	138 km	125 km	2
01.08.72	4.4	121 km	85 km	2
02.09.75	5.1	33 km	145 km	3.5

Although the time of recordings is very limited, we can estimate the 20 years return period intensity at 4 and the 2 years intensity at 2, which allows an extrapolation for the corresponding maximum possible intensity of this type of earthquake at 9 once more. This value is also obtained, when we assume a seismic event with the magnitude M = 6.5, in a depth of 10 km and a distance of 8 km from the dam site. Such a seismic event can reasonably be considered as a possible maximum.

INDUCED EARTHQUAKES

It is a well known fact, that the filling of a storage lake may produce earthquakes even at a dam site, which has never been considered before as being located in a seismic area.

However such events must be related to some geological features which are favoring the dynamic behaviour of the rock.

Earthquakes induced by reservoir loading are caused by changes in effective stresses produced in rocks below the reservoir. These are related to percolation of water from the reservoir into fissures and pores in the rock, causing increases in pore or fissure water pressure. For reservoirs where there are pre-existing faults, two possible triggering mechanisms have been proposed by Beck and Housner :

1) an increase in shear stress across a fault due to the water load, and
2) a decrease in shear strength of a fault, either due to the water pressure decreasing the effective normal stress, or due to the physical and chemical alteration by the water

of the properties of the material in the fault zone.

Before entering into the details of the fault system study performed at the Alicura reservoir site, a short review is given of the different cases, where induced earthquakes have taken place or where earthquakes most probably were caused by induced seismicity : Table Nr 1.

Today the situation at Alicura can be characterized as follows :

- A very extensive investigation of the various fault systems over an area of 2000 km^2 did not show the presence of any fault line where seismic activities have taken place in the historical past.
- The prehistoric activity goes at least back to the later Pliocene (2 million years ago) and can even not be proved with certainty.
- The microseismic recordings which are being undertaken for the last 4 months with the help of 4 stations have given some indications on seismic events of low magnitudes varying between M = 1.5 and 2.1. However, it was not possible to detect the accurate location of the epicenter.

All these results let assume that the Alicura reservoir site does not belong to the very sensitive places with respect to induced earthquakes.

Thus the magnitude of a possible seismic event would by no means exceed 5.7, as it was for instance recorded at the Oroville dam site. Probably it would even be smaller, and it can not be excluded that finally no induced earthquake will occur at all.

A pessimistic estimate can therefore be obtained by assuming a seismic event of M = 5.7 having a focal depth of 5 km. The corresponding epicenter intensity would then amount to I_o = 9, thus being practically the same value as for the other types of earthquakes investigated here before.

TYPICAL ACCELEROGRAM AND ITS DURATION

Besides the estimation of the possible maximum seismic intensity at a given site, the determination of a typical earthquake accelerogram and the corresponding duration of the shaking is a very difficult undertaking. Therefore until today very often the NRC-accelerogram is used not only for the dynamic analysis of nuclear power plants but also for dams and appurtenant works.

However such a proceeding is not very satisfactory since the local characters of the event are neglected.

For Alicura dam site it was finally decided, to distinguish between 2 kinds of seismic shocks, corresponding :

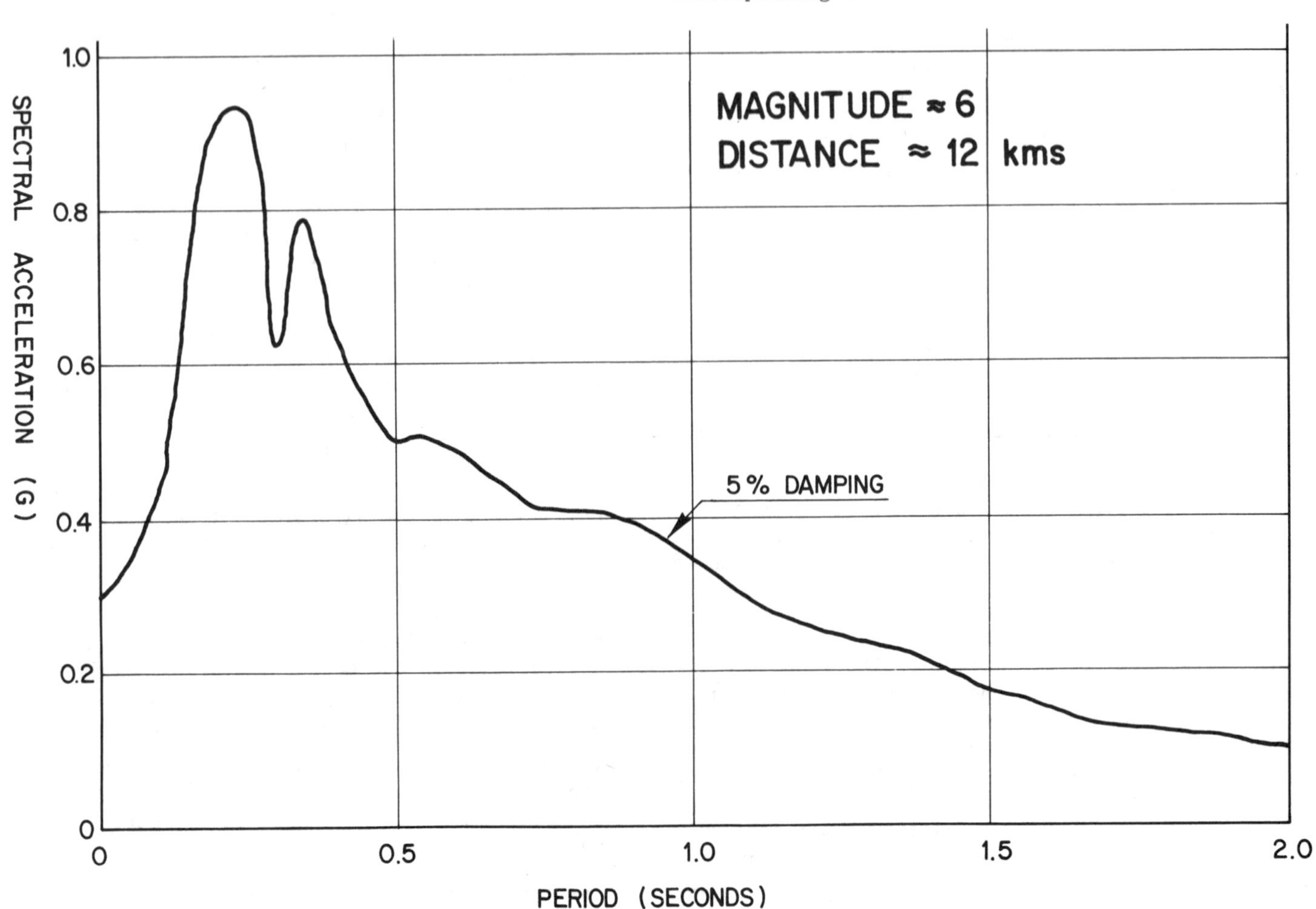

Fig. 7. Spectral shape of a local earthquake

1) to a relatively deep-seated local earthquake producing groundmotions at the site of $I_0 \simeq 8$, with maximum acceleration of 0.25 g and a duration of 20-30 seconds

2) to a distant earthquake (D = 150 km) with magnitude 8.5, a maximum acceleration of 0.20 g and a duration of 60 seconds (> 0.05 g)

Fig. 6 and 4 show one possible accelerogram for each of these cases as an example. It is likely that other kinds of similar time histories will be used in the analysis of the Alicura dam, in order to find out the one with the strongest response.

The analysis of the stability and the deformation of the dam body and the rock-slope as well as the determination of a possible deformation of the alluvial foundation will be based on both types of the seismic events. The same procedure will be used for the foundation stability of the appurtenant works on the rock of the valley flanks.

On the other hand for the design of concrete and steel structures, the natural period of which is less than 0.5 seconds, it was decided to perform the dynamic analysis by use of a NRC spectrum slope scaled to a zero period ordinate of 0.25 g.

SOME DETAILS OF THE DYNAMIC ANALYSIS

Complementary to the methods which are usually applied for dynamic computations, some special conditions will be considered :

- On the downstream side of the left abutment, where the inclined rock layers could move on pelite seams, an assessment of such movements is made with the aid of a Newmark-type displacement analysis.
- The loss of strength and the resulting instability in the alluvial foundation materials underlaying the up- and downstream shells of the Alicura dam, is investigated by a dynamic analysis, for which the results of cyclic loading tests on representative soil samples are taken into account.
- For the dynamic response analysis of concrete structures as spillway, powerhouse, penstocks and others, the use of standard procedures is adequate.
- High concrete structures (as e.g. intake towers) are not recommended. Therefore the intake structure of the diversion tunnel, in which the bottom outlet is placed, shows only a reduced hight of about 25 m, as it is necessary for constructional and hydraulic reasons. The fact that the upstream part of the tunnel between intake and gates can no more be controlled, is of course a severe disadvantage and therefore a strong concrete lining in combination with radial grouting from the tunnel is foreseen on the entire length.

FINAL REMARKS

As pointed out in the previous paragraphs the elaboration of the basic criteria for the dynamic analysis and design of Alicura dam and power plant was a very complex work. It has to be emphasized that the one or other speculative assumption had to be made because of the well know fact that in the field of seismology normally the quantity of meaningful measuring data is relatively scarce.

Therefore it is unavoidable that there remains some uncertainty especially in the domain of induced earthquakes. With all the security measures taken a rare unforeseen seismic event might occur during the lifetime of the scheme. Therefore a relatively rapid drawdown of the reservoir should always be possible and this is one of the most significant reasons for the installation of a bottom-outlet in every dam, which creates a major lake in a populated area.

REFERENCES

1. BECK J.L. and HOUSNER G.W (1977), Oroville Reservoir California and Earthquakes of August 1, 1975; in Proceedings, 6th World Conference on Earthquake Engineering, India.

2. CASTANO J.C., MEDONE C.A. and CARMONA J.S. (1970), Superficie de Focos de Sismos, su Distribution su Liberacion de Energia y un Analisis Estadistico del Proceso al Sur de los 12o de Latitud Sur, in Proceedings, Conference on Solid Earth Problems, Buenos Aires, October 26-31, 1970

3. GUTENBERG B. and RICHTER C.F. (1954), Seismicity of the Earth and Associated Phenomena (2nd ed.), Princeton University Press, Princeton, New Jersey, 322 p.

4. GUTENBERG B. and RICHTER C.F. (1956), Earthquake Magnitude, Intensity, Energy, and Acceleration, Seismological Society of America Bulletin, Vol. 46, No 2.

5. PLAFKER G. (1972), Alaskan Earthquake of 1964 and Chilean Earthquake of 1960: Implications for Arc Tectonics, Jour. Geophys. Res., V. 77, p.901.

6. QUARTINO B. Dr. (1979), Personal communication May 24, 1979, Buenos Aires, Argentina.

7. RAMOS V.A. (1978), "Estructura", in Relatorio Geologia y Recursos Naturales del Neuquen, VII Congreso Geologico Argentina, 9 al 15 de Abril de 1978.

8. RODRIQUEZ R., CABRE R., and MERCADO A. (1976), Geometry of the Nazca Plate and Its Geodynamic Implications, in the Geophysics of the Pacific Ocean Basin and Its Margin, Geophysical Monograph 19, Amer. Geophys. Union, Washington, D.C.

9. STAUDER W. (1973), Mechanism and Spatial Distribution of Chilean Earthquakes with Relation to Subduction of the Oceanic Plate, Jour. Geophys. Res., V.78, p. 5033.

10. SYKES L.R. and HAYES D. (1971), Seismicity and tectonics of South America and adjacent oceanic areas (abstr); Geol.Soc.America, V.3, p. 206.

23 Natural frequencies and response characteristics of gravity dams

D. ALTINISIK, MSc, and R. T. SEVERN, DSc FICE, University of Bristol

Many natural frequency values of gravity dams were presented at XIII ICOLD, and others have been obtained by the present authors. Here they are compared with results from a formula proposed by Chopra and Corns (ref 2). General indications are that the formula is not sufficiently accurate for satisfactory prediction of stresses, whereas 3D calculations, which include the foundation elasticity, agree well with such prototype measurements as exist. The paper also presents results for asynchronous excitation of gravity and rockfill dams, indicating that the effect is to reduce dynamic magnification by a large factor compared with the normal synchronous values.

INTRODUCTION

1. The XIII ICOLD Congress in New Delhi discussed (Question 51) 36 Reports on aseismic design, many concerned with gravity dams. The sophistication of analysis varied from 2D with rigid base, to 3D with foundation included, and some reports supplied valuable results of natural frequency measurements on prototypes. Chopra and Corns (ref 2) provided an approach for preliminary design, making use of simple formulae and design curves, but pointing out that for final design a full dynamic analysis is required. It is with the Simplified Approach that the first part of this paper is concerned, comparing the results which it produces with those presented in the ICOLD Reports, and also with the results obtained by the present authors, particularly for Baitings dam where prototype measurements have been made.

2. The second part of the paper studies the problem of multiple support excitation of dams, especially gravity and rockfill. Here recognition is made of the finite time which an earthquake takes to travel across a dam, and the consequent need to input different accelerations to different parts of the base.

PART 1. ASSESSMENT OF THE SIMPLIFIED APPROACH

Principles of the Simplified Approach (ref 2)

3. It is assumed that all significant response of a gravity dam to an earthquake occurs in the first mode, the natural frequency of which can be calculated from the simple formula $(E)^{\frac{1}{2}}/12H$, where E is the dynamic Young's moduls in kpa units, and H is height in metres. To account for the effect of the reservoir, a modifying factor is taken from prepared curves. This frequency is then used to determine an appropriate spectral acceleration (S_a) from published data, thereby allowing the distribution of horizontal dynamic force to be determined. From beam theory, with the dam behaving as a cantilever, stresses are then determined. Neglect of the second (and higher) modes should be noted. As will be shown, this Simplified Approach gives a fundamental frequency appreciably too large in all cases where it has been properly assessed, leading to the erroneous conclusion that the second mode is of such high frequency as to be outside the significant range of the spectral curves.

4. Spectral curves, even when averaged over a number of earthquakes, have pronounced peaks, and this makes it important for frequency values to be assessed accurately. In the Simplified Approach the frequency formula derives from 2D calculations, which not only assume a rigid base, but also that the dam is built in a rectangular valley of infinite extent. In practice valley shapes vary considerably, being generally described as triangular, trapezoidal or irregular. It is certainly difficult to characterise a dam by a single height value.

Analysis of frequency values presented at the XIII ICOLD Congress

5. It will be shown that stresses in the vulnerable upper part of gravity dams are assessed with sensible accuracy by the Simplified Approach if the fundamental frequency itself is accurate. Careful scrutiny of such values, presented at XIII ICOLD for dams throughout the World, is therefore of value. These results are collected in Table 1, where values in parentheses are given by the Simplified Approach; f_1 and f_2 are the first and second natural frequencies.

6. Starting with Crystal Springs dam (ref 7) the small ambiguity between the stated height and that appearing on the drawing makes no difference to the conclusion which will be reached, although it does indicate the difficulty of precise definition of height. The valley is trapezoidal, with little opportunity for plane-strain conditions to develop. The 3D calculations are very valuable because they include foundation elements having a dynamic modular ratio (E_r/E_c) of 0.2 in one set, and

0.05 in the other. The former agree more closely with measured values, and it is these that are given in Table 1. The 3D calculations were also made for both reservoir empty and full conditions; the prototype measurements were, however, made for a reservoir level at about mid-height, which is closer to empty than to full in considering hydrodynamic effects. Considering Table 1 for the empty reservoir, the Simplified Approach gives 12.50 Hz for f_1, whereas 3D calculations with modular ratio 0.2 gives 6.62 Hz. For full reservoir, corresponding values are 8.85 and 5.97 Hz. The measured frequency values, obtained by spectral analysis of ambient vibrations, gave 6.10 Hz for the f_1, which should be compared with a calculated value in the range 6.62-5.97 Hz, though nearer to the former. However, from results given in ref. 7 only a small change is required in the modular ratio to bring the values in congruence. The second mode is also of interest. 3D calculations give 10.12 and 8.94 Hz for the empty and full reservoir, whereas measurement gave 8.40 Hz, for a modular ratio of 0.2, for the near-empty condition. Two noteworthy features of this second mode are that it is an upstream-downstream antisymmetrical mode which would not be produced by 2D calculations. Second, its frequency is lower than f_1 from the Simplified Approach, and would correspond to a greater value of S_a. A final point arising from this valuable work at Crystal Springs, relates to the validity of using'fixed' rather than'radiating' boundaries in the foundation. The former appears adequate if a suitable modular ratio is used, and so long as the boundaries are sufficiently distant from the dam. This has been borne out by other work (ref 10).

7. The Izvorul Montelui (ref 1) and Fenshuba (ref 6) dams can be considered together. The very reasonable agreement in f_1 between 2D calculations and the Simplified Approach, merely indicates the truth of the fact that the formula for f_1 used in the latter is a best-fit formula for many such 2D calculations. A rubber model was used for the measured values, and that too was fixed at the base.

8. Luijiaxia dam (ref 8) is particularly interesting. Its valley profile has a distinct step, so that one part has approximately one-half the height of the other part. It was also constructed in 10 monoliths. grouted after application of hydrostatic load. Again, the f_1 value for 2D calculations and the Simplified Approach are in good agreement; but for this dam, 3D calculations have also been made (regrettably without foundation), and can be seen to differ from the Simplified Approach values by a large amount for both empty and full reservoir conditions. When full, the measured value of 5.56 Hz is lower than the calculated value 6.17 Hz, but this difference would certainly be reduced if foundation elements were introduced into the calculations.

9. For 3 Italian dams (ref 3) at the bottom of Table 1, it is tantalising that 3D computations were only made for Alpe Gera, to compare with the prototype measurements, and even they were only performed for a rigid-base. Noting that the valley profile is triangular, it is even more difficult to select a characteristic height for the Simplified Approach formula. If maximum height is chosen, f_1 = 2.72 Hz, which is much lower than f_1 = 4.22 from the 3D calculation with rigid base. The corresponding meas-

Table 1: Natural frequencies of gravity dams presented at XIII ICOLD 1979

Dam $E_c:E_r$ (GPA)	Height (m) Valley-Shape		Natural Frequencies (Hz) Calculated Empty		Calculated Full	Measured Empty	Measured Full
Crystal Spring	38.7	f_1	6.62 (12.50)		5.97 (8.85)	6.10	
33.1 : 6.2	Trapezoidal	f_2	10.12	3D	8.94	8.40	
Izvorul-Munteli	127	f_1	3.75 (3.60)		(2.60)		
30 : ∞	Trapezoidal	f_2	5.74	2D			
Fenshuba	93	f_1	5.20 (4.85)		4.33 (3.52)	4.35	
29.4 : ∞		f_2	10.00	2D	9.17	8.33	
Luijiaxia	147	f_1	3.28 (3.11)		2.82 (2.25)		
30 :	Irregular	f_2	7.81	2D	7.04		
		f_1	6.71		6.17		5.56
		f_2	8.13	3D	7.52		
Alpe Gera	174	f_1	4.22 (2.72)		(1.97)	3.47	
32.37 : ∞	Triangular	f_2	5.59	3D		4.72	
Campo Moro	97	f_1	(4.88)		(3.53)	6.24	
32.37 : ∞						8.73	
Morasco	59	f_1	(8.02)		(5.80)		6.49
32.37 : ∞	Irregular						

ured value of f_1 is 3.47, a value towards which the 3D value would reduce if foundation elements were included. Evidently, within the varying height of the dam some value could be chosen which would enable the Simplified Approach to give the correct value, but this does not present a satisfactory general approach.

Baitings gravity dam. Frequency analysis

10. A comprehensive series of prototype tests and calculations have been made for the Baitings gravity dam (UK). The dam is 53.8 m high, slightly curved in plan, and has a shallow trapezoidal profile. Prototype tests were made using an eccentric-mass exciter system (ref 10), and during testing reservoir level was 3m below spillway. For convenience, this condition is described as'full'. From Table 2 it is seen that 3 natural frequencies were measured below the 10 Hz upper limit of the exciters. Making the assumption that the frequency of a gravity dam in a trapezoidal valley is inversely proportional to its height, it is seen that these measurements are sensibly consistent with the Crystal Springs results in Table 1. From refs 7 and 10 it can also be seen that similar mode-shapes were obtained. Referring to the 2D calculations for Baitings, using maximum height and dynamic E = 26 GPa, f_1 agrees closely with the Simplified Approach for both empty (8.57 : 8.50 Hz) and full (7.56 : 7.27 Hz) conditions, but the measured value for the 'full' condition was 5.0 Hz. Continuing with the 2D results, introduction of foundation elements brings the fundamental frequency towards 5.0 Hz, but it is not until 3D elements are introduced that good agreement is obtained when the ratio of rock/ concrete dynamic moduli is 0.2. This gives 5.08 Hz compared with the measured 5.0 Hz. The two measured higher modes, at 6.1 and 8.0 Hz, are also in good agreement with the 3D calculations if the modular ratio is taken at 0.2. . Because of their mode-shapes (ref 10), these higher modes are not produced by 2D calculations. They are important, however, as may be seen by reference to spectral curves. For example, for S16E San Fernando 1971, the S_a at 6.1 is considerably larger than that at 7.56 Hz, and in the Simplified Approach only the latter is considered.

Baitings gravity dam. Stress Analysis

11. Following the determination of f_1 the second step in the Simplified Approach obtains dynamic forces with the help of a S_a value taken from published curves; the third step applies these horizontal forces to the dam behaving as a cantilever. A comparison is made in Fig. 1 between vertical direct stresses(σ_y) obtained in this way (A), and those derived from 2D finite element calculations (B and C) with fixed base, and using S16E San Fernando 1971. Stresses on upstream and downstream faces are equal and opposite in the Simplified Approach for the same height because they are derived from beam theory, so only one curve (A) is required. Both with and without water the agreement may be described as adequate for the upper part of the dam, particularly on the downsteam (B) face, but not so elsewhere. The stress values given are for f_1 only. Contributions from higher modes were calculated, but, as previously noted, the frequency of these higher modes, calculated using 2D elements and with fixed base, is such as to generate very small S_a values, and small stresses therefore. To check possible contributions from a vertical acceleration, S16E San Fernando 1971, factored by one-half, was applied vertically at the base. The resulting σ_y (D) was almost equal on the two faces, and relatively small. As expected, stresses due to the same earthquake are altered when foundation elements are introduced into 2D calculations, and again when 3D elements are used. These various results for σ_y on the upstream face are collected in Fig. 2. It should be noted that there were only 3 elements in the height for the 3D solution, which is sufficient to give accurate frequencies, but somewhat inaccurate stresses.

Analysis of the Owen Falls gravity dam

12. Although prototype measurements have not been made at Owen Falls, 2D calculations are available (ref 11), which support the conclusions derived from Baitings. Table 3 is a summary of values of f_1, with values in parentheses from the Simplified Approach. Clearly the latter agree with finite element values for rigid base, but differ for an elastic foundation, even with Er/Ec = 1. No 3D calculations have been made for Owen Falls. Fig. 3 gives σ_y stresses on the upstream face: curves A are for the Simplified Approach; B from 2D finite elements with rigid base; C for 2D finite elements with Er/Ec = 1. With reservoir empty, an elastic foundation lowers the frequency to a value corresponding to a higher S_a, and therefore increases the stress. A similar effect is produced by the reservoir. It is not suggested that any of these frequencies, produced by 2D calculations, are correct for Owen Falls, simply that dynamic stress is dependent on the values used.

Table 2: Baitings gravity dam - natural frequencies E_c = 26 GPa H = 53.8m

Hz	Er/Ec = ∞ Empty	Er/Ec = ∞ Full	Er/Ec = 1.0 Empty	Er/Ec = 1.0 Full	Er/Ec = 0.2 Empty	Er/Ec = 0.2 Full	Measured
Mode 2D 1	8.57(8.50)	7.56(7.27)	6.48	5.56	3.98	3.39	
2D 2	19.74	17.02	11.88	11.37	5.79	5.74	
Mode 3D 1 Sym			7.85		5.08		5.0
2 A/Sym			10.00		6.86		6.1
3D 3 Sym			11.77		7.79		8.0

Conclusions

13. For gravity dams constructed on rigid foundations, in wide valleys of constant height, the Simplified Approach is adequate for preliminary aseismic design. But for the majority of gravity dams, available evidence indicates a true fundamental frequency different from that given by the Approach. Varying height, three dimensional behaviour, and interaction with foundation, all contribute to this difference, which is important in view of the large change in S_a caused by these frequency differences. Analysis presented here also recognises that natural frequencies greater than the first will exist for gravity dams higher than about 50 m, which lie within the significant regions of the S_a curves, therefore giving appreciable response. These frequencies are, of course, ignored by the Simplified Approach.

Part 2. MULTIPLE SUPPORT EXCITATION

General Considerations

Calculation of response of a dam to an earthquake usually assumes that every point of the base experiences the same acceleration at any instant. Since the speed with which the pulse travels is finite, perhaps as low as 1000 m/s, this assumption is clearly incorrect, and it will be useful to know how different accelerations at different parts of the base change the response from that produced by uniform acceleration. The outline of the theory has been presented in ref. 9, but since it is not well-known, an explanation is given here in an Appendix.

Multiple support excitation of Baitings dam

15. The results presented here are for the 2D plane-strain, rigid-base case, but the conclusions are valid generally. As stated in the Appendix, when the whole base is given the same acceleration, the r-vector expresses a unit rigid-body movement of the whole dam. When a single point on the base is accelerated, with all other points stationary, the required r-vector expresses the deformed shape of the dam consequent on this single degree of freedom having unit displacement, with all other base points fixed. To illustrate the effect of multiple support excitation, the base of Baitings dam is divided into two halves, as in Fig. 4, and it is intended that S16E San Fernando be applied first to half A, with B fixed, and then to B, with A fixed. For the first of these excitations the calculated r-vector shape is given in Fig. 4(a) by the dashed line, and for the second excitation in Fig. 4(b). The two shapes are plotted in opposite senses from the original position only for clarity. As seen from Fig. 4, vector (b) is more than twice vector (a) at the crest.

vector (a) at the crest.

16. Fig. 5 gives the σ_y stress on the upstream face of Baitings with reservoir empty. Curve A is for the A-half only of the base (Fig 4) subjected to S16E San Fernando, and curve B when B-half only is subjected to the same input. The curve A + B must clearly be the same as that corresponding in Fig. 1. The important conclusion from this simple illustration may be stated as follows. The A + B curve assumes the same S_a to be active over the whole base; in fact, because the input is travelling across the base, some parts will experience a lower acceleration value than S_a, by virtue of the definition of S_a. Thus, taking only the two halves of Fig. 5, the real σ_y is less than A + B. A more detailed analysis of this kind is given below.

Multiple support excitation of a rockfill dam

17. Fig. 6 is a typical rockfill dam cross-section, 100 m high, 410 m at the base and 10 m wide at the crest. For this analysis realistic E-values have been taken for core and rockfill, varying with depth as shown. For the multiple support analysis, the base was sub-divided first into four equal regions, and then into seven. The speed of the S16E San Fernando and Port Hueneme inputs (v) was taken first as 1000 m/s and then as 1500 m/s. The analysis process consisted of the evaluation of S_a and the time (t_a) at which it occurred in the earthquake record. This S_a is to be applied to part 1 of the base. If the centre of part 2 is distance d from the centre of part 1, then part 2 is deemed to experience that part of the input occurring a time d/v before t_a, and the corresponding value of the Duhamel integral is obtained, which is then applied to part 2. In a similar way, ground inputs for all the sub-divided regions are obtained. The entries in Table 4 are those of dynamic magnification factor at the crest, with multiple support excitation values compared with corresponding values produced by uniform ground acceleration. In all cases the former is appreciably smaller than the latter, and this could be a significant factor in the ability of rockfill dams to withstand major shocks.

APPENDIX. THEORY OF MULTIPLE-SUPPORT EXCITATION. (see also ref. 9)

Introduction

18. Such difficulties as arise in this problem do so because conventional analysis of structures subjected to ground vibration, in which all ground points experience the <u>same</u> acceleration, assume that in the dynamic equations of equilibrium,

Hz	Er/Ec = ∞		Er/Ec = 1		Er/Ec = 0.5	Er/Ec = 0.2
	Empty	Full	Empty	Full	Empty	Empty
Mode 1 2D	14.49(14.90)	11.21(12.00)	11.36	8.43	9.68	7.07

Table 3 : Owen Falls Dam - Natural frequencies E_c = 35 GPa H = 27 m

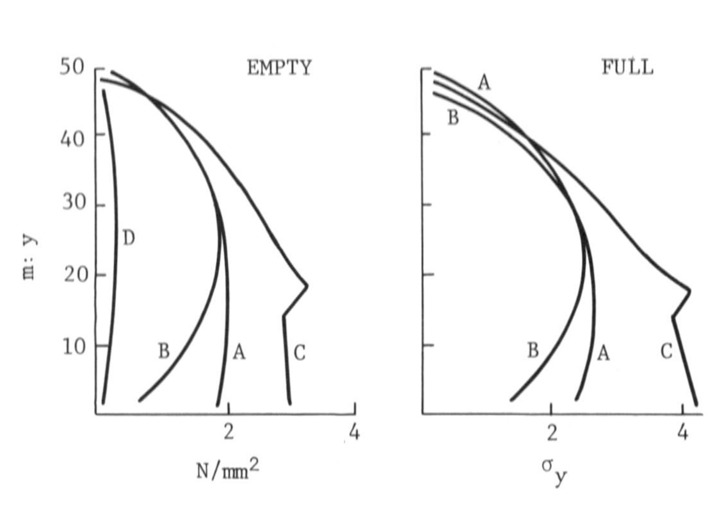

Fig. 1. Baitings dam σ_y stresses. 2D rigid base

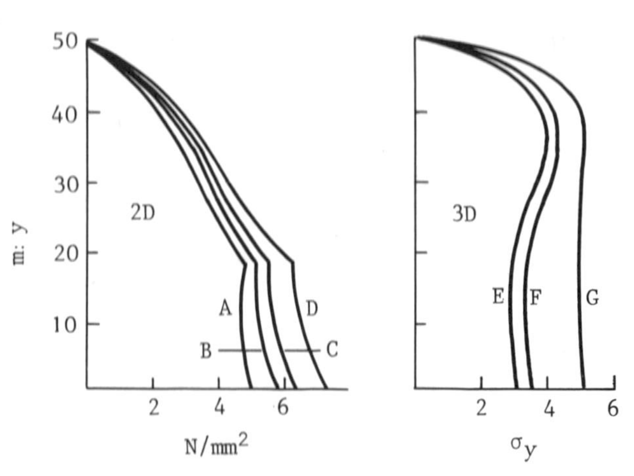

Fig. 2. Baitings dam σ_y on upstream face. For A, B, C no water and E_r/E_c = 1, 0.5, 0.2
For D, with water and E_r/E_c = 1. For E, F, G, no water and E_r/E_c = 1, 0.5, 0.2

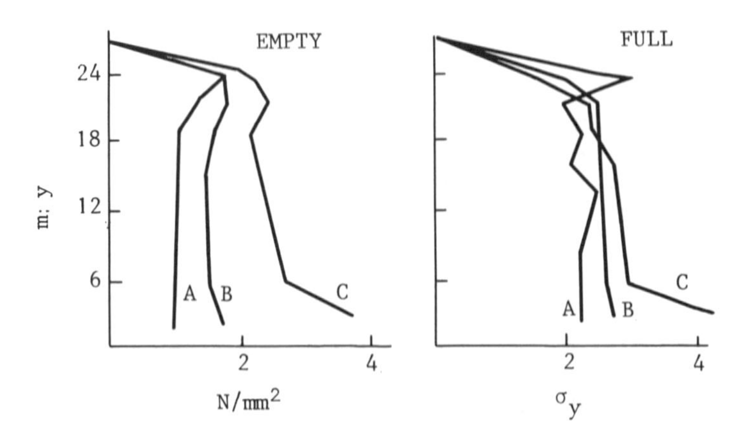

Fig. 3. Owen Falls dam δy on upstream face. 2D

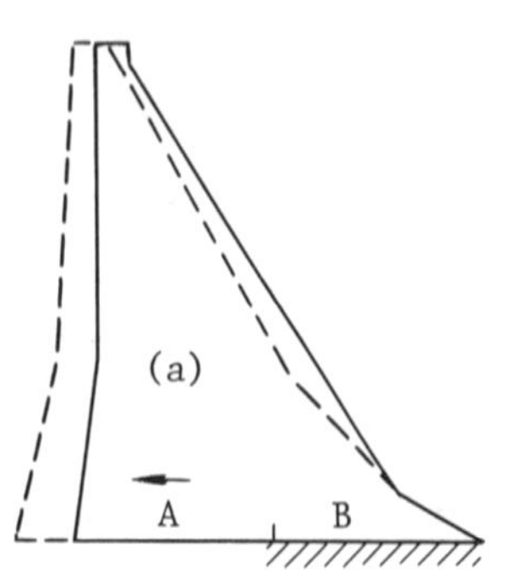

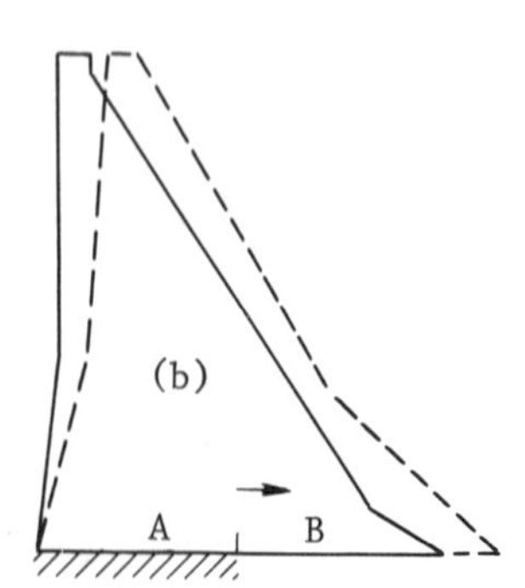

Fig. 4. Baitings dam. Ground displacement shape vectors

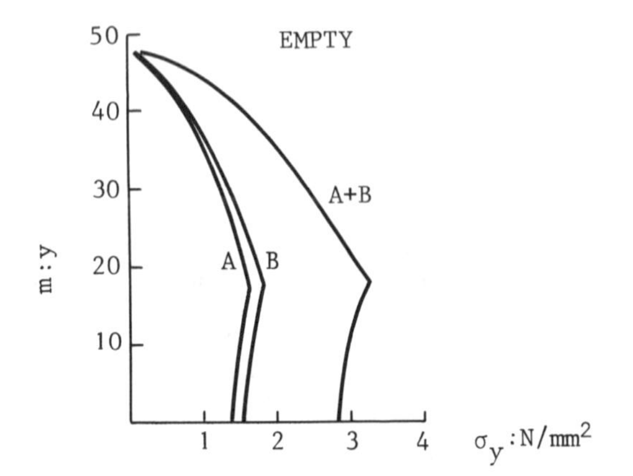

Fig. 5. Baitings dam. Upstream face σy for acceleration of 2 halves

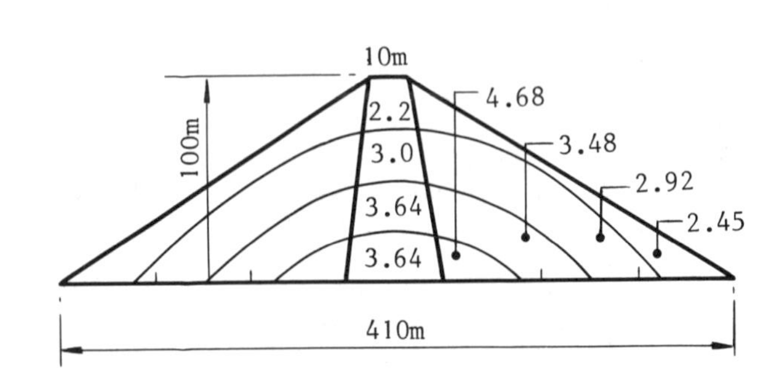

Fig. 6. Rockfill dam with zoned E-values (Kn/mm^2) c.f. Table 4

	S16E San Fernando			Port Hueneme		
No. of Base Divs.	4	7	7	4	7	7
Speed of Pulse (m/s)	1000	1000	1500	1000	1000	1500
Uniform Acceln.	0.865	0.865	0.865	0.851	0.851	0.851
Varying Acceln.	0.647	0.550	0.560	0.485	0.515	0.430

Table 4. Dynamic magnification factors at crest for uniform and varying acceleration at base of rockfill dam.

Inertia forces + Damping Forces + Stiffness Forces = 0, (1)

the inertia forces depend on total displacements, whereas damping and stiffness forces depend on relative displacements. These last refer to what are called the 'response degrees of freedom' (RDOF) relative to the common displacement of all the 'ground degrees of freedom' (GDOF). To clarify this point, and leading towards the analysis of multiple-support excitation, the single-support case will be restated in general form.

Analysis for a single-support excitation

19. Because matrix formulation will later be a necessity, it is used here on the vertical cantilever of Fig A1. The three node-points are each restricted to a single DOF, and total displacement from a fixed axis is denoted by v^t. The equations of motion are then

$$\begin{bmatrix} M \end{bmatrix} \begin{pmatrix} \ddot{v}_1^t \\ \ddot{v}_2^t \\ \ddot{v}_3^t \end{pmatrix} + \begin{bmatrix} C \end{bmatrix} \begin{pmatrix} \dot{v}_1^t \\ \dot{v}_2^t \\ \dot{v}_3^t \end{pmatrix} + \begin{bmatrix} K \end{bmatrix} \begin{pmatrix} v_1^t \\ v_2^t \\ v_3^t \end{pmatrix} = 0 \quad (2)$$

for viscous damping. M, C and K are 3 x 3 mass, damping and stiffness matrices. In general, not all terms of these matrices are non-zero. If the cantilever were rigid, total displacement of any point would be the same as the ground displacement. But if elasticity is introduced, an additional displacement of any point becomes possible. Writing this for the 3DOF,

$$v^t(t) = v(t) + v^s(t) = v(t) + rv_g(t) \quad (3)$$

where $v^t(t)$is a time-dependent vector (3 x 1) of total displacement, v(t) is the vector of dynamic displacement (i.e. due to elasticity of the cantilever), and $v^s(t)$ is termed the 'pseudo-static' vector, which depends on the ground motion. This last term is also written as $rv_g(t)$ in (3), where $v_g(t)$ is the vector of ground acceleration - 1 x 1 in this illustration; r is therefore a 3 x 1 vector, to be referred to as 'ground displacement shape vector' (GDSV), because it will later be seen to determine the deformed shape of the structure when any one of the GDOF is given unit displacement, with all other GDOF held fixed. Clearly the r-vector is generally different for each GDOF. In this cantilever example a unit displacement of the GDOF v_1 results in a unit displacement of both v_2 and v_3 because no other point is held fixed; the r-vector is then a column of 1's. Using (3) in (2)

$$\begin{bmatrix} M \end{bmatrix} \begin{pmatrix} 0 + \ddot{v}_g \\ \ddot{v}_2 + r_2\ddot{v}_g \\ \ddot{v}_3 + r_3\ddot{v}_g \end{pmatrix} + \begin{bmatrix} C \end{bmatrix} \begin{pmatrix} 0 + \dot{v}_g \\ \dot{v}_2 + r_2\dot{v}_g \\ \dot{v}_3 + r_3\dot{v}_g \end{pmatrix} + \begin{bmatrix} K \end{bmatrix} \begin{pmatrix} 0 + v_g \\ v_2 + r_2 v_g \\ v_3 + r_3 v_g \end{pmatrix} = 0$$

which can be rewritten

$$\begin{bmatrix} M \end{bmatrix} \begin{pmatrix} 0 \\ \ddot{v}_2 \\ \ddot{v}_3 \end{pmatrix} + \begin{bmatrix} C \end{bmatrix} \begin{pmatrix} 0 \\ \dot{v}_2 \\ \dot{v}_3 \end{pmatrix} + \begin{bmatrix} K \end{bmatrix} \begin{pmatrix} 0 \\ v_2 \\ v_3 \end{pmatrix} = - \begin{bmatrix} M \end{bmatrix} \begin{pmatrix} \ddot{v}_g \\ r_2\ddot{v}_g \\ r_3\ddot{v}_g \end{pmatrix} - \begin{bmatrix} C \end{bmatrix} \begin{pmatrix} \dot{v}_g \\ r_2\dot{v}_g \\ r_3\dot{v}_g \end{pmatrix} - \begin{bmatrix} K \end{bmatrix} \begin{pmatrix} v_g \\ r_2 v_g \\ r_3 v_g \end{pmatrix}$$

and abstracting the RDOF

$$\bar{M}\ddot{v} + \bar{C}\dot{v} + \bar{K}v = -(\bar{M}r + M_g)\ddot{v}_g - (\bar{C}r + Cg)\dot{v}_g - (\bar{K}r + K_g)v_g \quad (4)$$

where

$$\bar{M} = \begin{bmatrix} m_{22} & m_{23} \\ m_{32} & m_{33} \end{bmatrix}; \quad \bar{K} = \begin{bmatrix} k_{22} & k_{23} \\ k_{32} & k_{33} \end{bmatrix}; \quad \bar{C} = \begin{bmatrix} c_{22} & c_{23} \\ c_{32} & c_{33} \end{bmatrix} \quad (5)$$

and

$$M_g = \begin{pmatrix} m_{21} \\ m_{31} \end{pmatrix}; \quad K_g = \begin{pmatrix} k_{21} \\ k_{31} \end{pmatrix}; \quad C_g = \begin{pmatrix} c_{21} \\ c_{31} \end{pmatrix} \quad (6)$$

In words, the matrices in (5) relate the RDOF with each other; those in (6) are vectors which relate RDOF to the GDOF. For the single GDOF these matrices can be extended for any number of RDOF. If in (4) all dynamic terms are made zero,

$$(\bar{K}r + K_g)v_g = 0, \text{ which gives } r = \bar{K}^{-1}K_g \quad (7)$$

and indicates why r was previously described as the 'ground displacement shape vector'. For the shear-type cantilever of Fig. A1

$$K = \frac{12EI}{h^3}\begin{bmatrix} 1 & -1 & 0 \\ -1 & 2 & -1 \\ 0 & -1 & 1 \end{bmatrix}; \quad \bar{K} = \frac{12EI}{h^3}\begin{bmatrix} 2 & -1 \\ -1 & 1 \end{bmatrix};$$

$$K_g = \frac{12EI}{h^3}\begin{Bmatrix} -1 \\ 0 \end{Bmatrix}$$

from which, using (7) the transpose of r is (1,1), as was obtained intuitively. Returning to (4), the final term is zero by virtue of (7). If damping is proportional to stiffness, the penultimate term is also zero; but not if it is proportional to mass. However, even in this case, its contribution to the right-hand side of (4) is small enough to be ignored in comparison with the first term. Thus

$$\bar{M}\ddot{v} + \bar{C}\dot{v} + \bar{K}v = -(\bar{M}r + M_g)\ddot{v}_g \quad (8)$$

For lumped-masses the mass matrix is diagonal, giving $M_g = 0$. The well-known form of (8) is then obtained. Even when a consistent mass matrix is employed M_g can often be ignored in comparison with $\bar{M}r$.

Analysis for multiple-support excitation

20. To extend the above analysis to multiple-support excitation, consider the dam-like structure of Fig. A2, in which each node is allowed only one degree of freedom, v, in the upstream/downstream direction. Nodes 1,3, 5

are GDOF, whereas nodes 2 and 4 are RDOF. The deformation of this 'dam' to static load is expressed by the matrix statement, using standard notation

$$\begin{bmatrix} k_{11} & & & & \text{SYM.} \\ k_{21} & k_{22} & & & \\ k_{31} & k_{32} & k_{33} & & \\ k_{41} & k_{42} & k_{43} & k_{44} & \\ k_{51} & k_{52} & k_{53} & k_{54} & k_{55} \end{bmatrix} \begin{pmatrix} v_1 \\ v_2 \\ v_3 \\ v_4 \\ v_5 \end{pmatrix} = \begin{pmatrix} f_1 \\ f_2 \\ f_3 \\ f_4 \\ f_5 \end{pmatrix} \quad (9)$$

Not all stiffness terms will be non-zero, but all have been retained for completeness. If $v_3 = 1$, and $v_3 = v_5 = 0$, and noting that $f_2 = f_4 = 0$, from (9) may be abstracted the following statement for the RDOF

$$\begin{bmatrix} k_{22} & k_{24} \\ k_{42} & k_{44} \end{bmatrix} \begin{pmatrix} v_2 \\ v_4 \end{pmatrix} = - \begin{pmatrix} k_{23} \\ k_{43} \end{pmatrix} . 1 \quad (10)$$

It is seen that the square matrix relates RDOF with each other, previously referred to as $\bar{K}$, whilst the right-hand side vector relates the non-zero GDOF, v_3, to the RDOF; this was earlier called K_g. In matrix form, (10) rewrites as $v = \bar{K}^{-1}K_g = r$; that is, r is the vector of displacements of the RDOF resulting from unit displacement of one support point, with all other support points fixed. If there are m GDOF it is necessary to calculate m different r-vectors in general. In these calculations, $\bar{K}$ will be the same, but K_g will be different for each vector. For example, if node 1 is accelerated, the transpose of K_g will be $(k_{21}\ k_{41})$. If these m r-vectors are assembled into an R-matrix, such that each vector is a column of R, the equations of motion for the dynamic part of the n RDOF are

$$\underset{(n.n)(n.1)}{\bar{M}\ \ddot{v}} + \underset{(n.n)(n.1)}{\bar{C}\ \dot{v}} + \underset{(n.n)(n.1)}{\bar{K}\ v} = \underset{(n.n)(n.m)(m.1)}{-\bar{M}\ R\ \ddot{v}_g} \quad (11)$$

where vector $\ddot{v}_g$ specifies the m accelerations of the GDOF. These equations may be solved either by forward integration, or by the modal method, which is used here. Looking at the R matrix more closely, if all components of the $\ddot{v}_g$ vector are the same, those rows in (11) corresponding to RDOF which are in the same direction as $\ddot{v}_g$ will sum to unity, whereas all other rows will sum to zero. The R-matrix then becomes a 'direction-vector', , composed of 0's and 1's, which picks out from the n nodal displacements those which are in the same direction as v_g. More precisely, it picks out elements of the mass matrix associated with these degrees of freedom.

Calculation of total displacement

21. To dynamic displacements calculated from (11), pseudo-static displacements must be added, to give total displacement. Thus

$$v^t(t) = v(t) + R\, v_g(t) \quad (12)$$

Calculation of dynamic forces

22. Equation (9) is also a valid relationship between dynamic forces and total displacements. To avoid excessive use of subscripts, dynamic actions are assumed in what follows. Con-

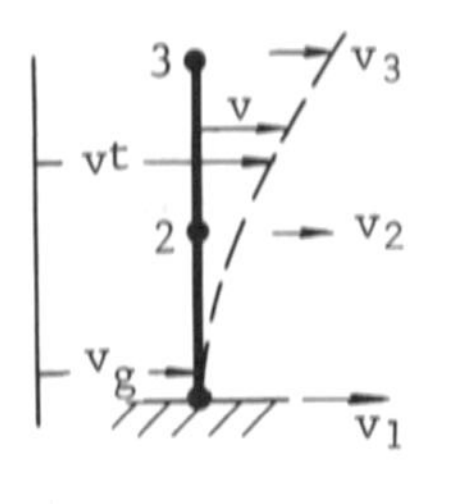

Fig. A1

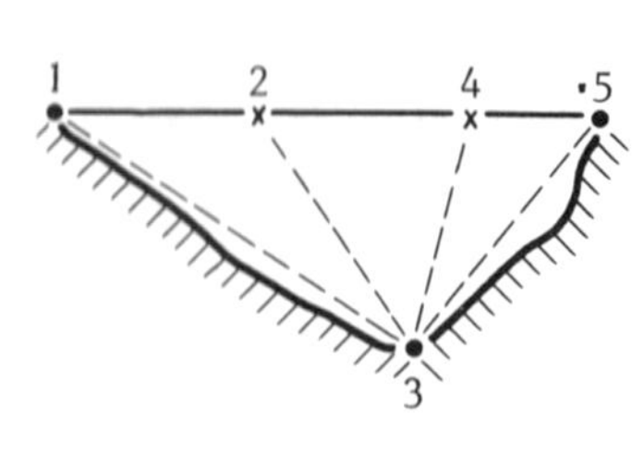

Fig. A2

sidering the RDOF first, and taking $v_3{}^t = v_g$ whilst $v_1{}^t = v_5{}^t = 0$, (9) gives

$$f_2 = k_{22}\, v_2{}^t + k_{22}\, v_g + k_{24}\, v_4{}^t$$
$$f_4 = k_{42}\, v_2{}^t + k_{43}\, v_g + k_{44}\, v_4{}^t$$

which in matrix form is

$$f = \bar{K}v^t + K_g\, v_g$$
$$= \bar{K}\,(v + v^s) + K_g\, v_g = \bar{K}v + (\bar{K}r + K_g)\, v_g \quad (13)$$

By definition of r, the second term is zero, so that dynamic forces corresponding to RDOF are calculated from dynamic displacements only. Now consider the dynamic forces corresponding to GDOF. From (9), with $v_3{}^t = v_g$ and $v_1{}^t = v_5{}^t = 0$,

$$f_1 = k_{12}\,(v + v^s)_2 + k_{13}\, v_g + k_{14}(v + v^s)_4 \quad (14)$$
$$f_3 = k_{32}\,(v + v^s)_2 + k_{33}\, v_g + k_{34}(v + v^s)_4$$
$$f_5 = k_{52}\,(v + v^s)_2 + k_{53}\, v_g + k_{54}(v + v^s)_4$$

which may be recast as

$$\begin{pmatrix} f_1 \\ f_3 \\ f_5 \end{pmatrix} = \begin{bmatrix} k_{12} & k_{14} \\ k_{32} & k_{34} \\ k_{52} & k_{54} \end{bmatrix} \begin{pmatrix} v_2 \\ v_4 \end{pmatrix} + \begin{bmatrix} k_{12} & k_{14} \\ k_{32} & k_{34} \\ k_{52} & k_{54} \end{bmatrix} \begin{pmatrix} r_2 \\ r_4 \end{pmatrix} v_g + \begin{pmatrix} k_{13} \\ k_{33} \\ k_{53} \end{pmatrix} v_g \quad (15)$$

Considering the rectangular matrix, each vector links a single RDOF with each of the GDOF; it is the transpose of K_g when this is the matrix of the K_g vectors for all RDOF; it will therefore be called $K_g{}^T$. The final vector in (15) gives the contribution to forces at the GDOF due to unit displacement of the chosen GDOF, in this case node 3. Following ref. 9, this will be described as K_{gg} - indicating the cross linkage between one GDOF and its companions. Finally, in matrix form, for dynamic forces at the GDOF,

$$f_g = K_g{}^T\, v + \left[K_{gg} - K_g{}^T\,(\bar{K}^{-1}K_g)\right] v_g \quad (16)$$

since $r = -\bar{K}^{-1}K_g$. The second term in (16) represents forces due to the differential movement of the supports. The foregoing analysis of forces is for movement of a single GDOF. The adaptation for movement of several, or all, GDOF is obvious.

REFERENCES

1. PRISCU R. The Behaviour of Romanian Dams during the Vrancea Earthquake of March 4, 1977. I.C.O.L.D. 13th Congress, New Delhi, 1979, Q51 R12.

2. CHOPRA A.K. and CORNS C.F. Dynamic Method For Earthquake Resistant Design and Safety Evaluation of Concrete Gravity Dams. Ibid.Q51,R6
3. CALCIATI F., CASTOLDI A., CIACCI R. and FANELLI M. Experience Gained During In Situ Artificial And Natural Dynamic Excitation Of Large Concrete Dams In Italy; Analytic Interpretation of Results. Ibid. Q 51, R 32.
4. ZHAO-XUAN, L and LIANG-SHEN W. Safety Evaluation In Earthquake Resistant Design Of Gravity Dams. Ibid. Q 51, R 26.
5. ACADEMIA SINICA. Earthquake Loads For Hydraulic Structures. Ibid. Q51, R27.
6. CHEN, H, TANG, J, QIAN, W, WANG, L. and SUN C. Dynamic Behaviour and Earthquake Response Of The Dam Section With Inserted Power Station Of Fengshuba Dam. Ibid. Q51. R28.
7. WULFF J.G. and VAN ORDEN R.C. Evaluation of the Earthquake Stability of Lower Crystal Springs Dam, California, U.S.A. Ibid.Q51, R1.
8. FU Z.X., CHEN H.C., and WANG D.X. Earthquake Resistant Design of Luijiaxia Dam, Ibid. Q 51, R 29.
9. CLOUGH, R.W. and PENZIEN J., Dynamics of Structures. McGraw Hill Kogakusha Ltd. 1975.
10. SEVERN, R.T., JEARY A.P. and ELLIS, B.R. Forced Vibration Tests and Theoretical Studies on Dams. Proc. I.C.E.(Pt.2)Sept 1980,Paper 8362.
11. ALTINISIK D. Aseismic Design of Concrete Dams. Ph.D. thesis, Univ. of Bristol, Sept.1980.

24

Aseismic design of arch dams: particularly the contribution from the reservoir, and multiple-support excitation of the base

D. ALTINISIK, P. A. A. BACK, S. R. LEDBETTER, R. T. SEVERN and C. A. TAYLOR, (Bristol University/Sir Alexander Gibb and Partners, Reading)

The first part is concerned with frequency, mode-shape and response measurements on a thin, cylindrical model dam. Tests include steady-state vibration and transient input from a shaking table, with measurement of hydrodynamic pressure. The second part presents similar measurements and calculations on a 1 : 250 model of the Victoria arch dam. Consideration is given to excitation of the base of the dam by different inputs, generally leading to a reduction of response.

INTRODUCTION

1. From papers presented at the 1979 ICOLD it is clear that the aseismic design of arch dams shares its problems with other concrete dams. In particular, the rôle of the reservoir, the contribution from foundation elasticity, and the possibility of dispensing with computer calculations by using a simplified approach based on experience. Only one paper (Ref. 2) allows that all parts of the base are not subjected to the same acceleration simultaneously although Ref. 6 refers to it as a topic for research.

2. For the added-mass of the reservoir, there appears to be general acceptance of Westergaard's approach (Ref. 11), presented in 1931, and applying strictly only to rigid dams with vertical upstream face in an infinitely wide valley. Perhaps the reason for this is that direct evidence of hydrodynamic pressure in a reservoir is hard to come by, and one of the few successful measurements (Ref. 12), at the North Fork dam U.S.A., concludes that Westergaard's approach provides reasonable results if rationally applied. Whilst the measured curves presented there are different from Westergaard's curves, there is some similarity in the crest region of the crown-cantilever for the first two modes, and it is here that the added-mass is of greatest significance. Another possible reason for acceptance of Westergaard's method is that added-mass is then obtainable from a simple formula which can be added to the structural mass in such finite element packages as ADAP and SAP, whereas the proper treatment of the reservoir involves a coupled finite element analysis of structure and reservoir, which is not available in these packages.

3. Study of various 1979 ICOLD papers on arch dams (Refs. 2, 3, 4, 5, 6) illustrates the difficulties of drawing out any general rules about, for instance, the value of the fundamental antisymmetric and symmetric resonances in terms of height, or span. Nearly always, however, these resonances are close together, with antisymmetric first. Despite these difficulties, as well as the fact that as many as a dozen modes can often be calculated within the significant frequency range of typical spectral curves, an attempt has been made (Ref. 6) to produce a 'design-office' approach, eschewing computers, and using only one mode in the upstream/downstream direction and another for cross-valley excitation. Detailed validation of this approach for a wide variety of arch dams would be a major forward step. Meanwhile, it is noted that this is the only paper in which an alternative to Westergaard's reservoir-effect approach is used. Regrettably, the basis of this alternative is not presented, although a formula involving tabulated coefficients is given, showing the same features as those produced by the coupled structure/reservoir analysis used in this present paper.

4. A disappointing feature of the 1979 ICOLD papers on arch dams was the absence of a comprehensive study which included computer analysis of foundation elasticity, Westergaard's

Hz	RESERVOIR	f_1 sym	f_2 asym	f_3 sym	f_4 asym	f_5 sym
Finite	Empty	190.5	204.5	329.8	393.8	403.5
Element	Full	89.8	96.4	155.9	185.7	190.3
Shaker	Empty	132	146	200	280	295
	Full	65	112	130	182	207
M.A.M.A.	Empty	138.7	145.8	195.1	294.9	334.7
	Full	69.4	88.1	111.0	148.1	264.3

Table 1. Ferrocement cylindrical model. Comparison of calculated and measured frequencies.

approach versus others for the reservoir effect, model studies of various kinds (both steady-state and transient), prototype steady-state tests, and, finally, the performance of the dam in an actual earthquake. Ref. 2 is the nearest approach to this ideal. After recording results from an idealised cylindrical model, this present paper gives the first results from such a comprehensive study, although it is doubtful if the dam involved will ever experience a major earthquake in Sri Lanka where it is to be built.

FERROCEMENT CYLINDRICAL MODEL

5. With studies of structure-fluid interaction in mind, particularly under transient acceleration input, an idealised cylindrical, constant thickness arch dam was made in ferrocement. Its height, radius and thickness were 495, 563 and 8.5 mm respectively. The sides and base were cast into stiff steel channels, which in turn were bolted to the sides and base of a stiffened steel box. Fig. 1 shows the arrangement. The difficulties of attempting to produce such encastré boundaries are well appreciated, and allowance must be made for this when comparing calculations and measurements. The thinness of the model is also a critical parameter, and the 8.5 mm figure given is the most representative value obtained from thickness measurements on the finished model.

Calculated and measured hydrodynamic pressure

6. Fig. 2 gives the results for hydrodynamic pressure obtained in three ways. First, by calculation, using an improved version of the fluid-structure coupling process initially described by Dungar (Ref. 10), with 32 3D finite elements in the structure and 96 in the reservoir. These calculated values are shown by full-line curves in Fig. 2, where the upper set of diagrams are values on the crown-cantilever, and the lower set gives values at a level 12 cm below the crest. Symmetrical modes are in (a), (b), (c) and (f), antisymmetrical modes in (d), (e) and (g), and the values have been normalised to a maximum acceleration value of g. For the experiments, this meant that it was necessary to measure the response acceleration at the points where hydrodynamic pressure was measured. The dashed curves in Fig. 2 were obtained using the MAMA multipoint, steady-state shaking system (Ref. 7), and the dash-dot curves by shaking the complete dam-reservoir system in its box, Fig. 1. The shaking-table used is driven by a Ling Dynamics V825 vibrator, controlled through a Varian computer. During the pressure measurements the system is shaken at one of the dam resonance frequencies, with Druck PRDC 10F pressure transducers (0 - 2.5 p.s.i. range) placed close to the dam wall, and moved from point to point to map the whole pressure field. Nine sockets for the pressure transducers were also built into the dam wall; these were used in the beginning to check the values given by the 'wandering' transducers. The reservoir length was three times the dam height, and pressure measured along this length, indicated a fall to zero at about half-way. In Fig. 2, the resonance frequency values given are those obtained using MAMA; as will be seen in Table 1, there is some spread in the values obtained by the three methods used, and the MAMA values are probably the most reliable.

7. Comparison with Westergaard (Ref 11). Values obtained from Westergaard's formula are given for the upper part of the model by the dotted curve in Fig. 2(d). It continues parabolically to a value of 436 mm at the base, or nearly 3 times the maximum measured value. It is wrong to think of this overestimate of hydrodynamic pressure, and hence added-mass, as a conservative design feature of Westergaard's approach, since exaggerated reduction of resonance frequency will not be advantageous if it results in a higher value on spectral curves.

Calculated and measured resonance frequencies and mode-shapes

8. A comparison of values for the first 5 resonance frequencies appears in Table 1, for reservoir empty, and full. The two sets of experimental values are in good agreement, but both are appreciably lower than those calculated. This was expected, because of assumed encastré boundary conditions in the calculations which cannot be realised in practice. Another major contribution arises because bending stiffness is proportional to (thickness)^3, and small departures from nominal thickness (8.5 mm) has a large effect. Microcracking in the dam also has the effect of reducing the nominal thickness. Shaking-table tests referred to in Table 1 were carried out by incrementing frequency of vibration of the table in steps of 5 Hz away from resonances, and 2 Hz when close to them, measuring displacement and acceleration response at various points, including a reference accelerometer on the table itself. Fig. 3 shows the response obtained with an accelerometer on the crest crown-cantilever, so that the resonances are symmetrical ones. With water in the reservoir, Fig. 4, the resonance frequencies are reduced and, particularly with reservoir full, the peaks are less well defined. It will be observed that the horizontal scale in Fig. 3 is 20% larger than in Fig. 4, so that the real difference in 'peakiness' is more pronounced than a cursory glance indicates. From Table 1, the reduction in frequency due to a full reservoir is about 50% from all three assessments.

9. As to mode-shapes, Fig. 5, which gives the first three symmetrical modes, indicates that the first mode-shape is hardly changed by the presence of water (as is usually assumed), whereas the second mode-shape changes significantly, and the third mode-shape is entirely different when plotted on the crown-cantilever. Whereas Fig. 5 gives calculated mode-shapes, Fig. 6 shows the surface water pattern of the first symmetrical mode produced during excitation by MAMA with the force level increased for the purpose of the photograph.

10. A comparison of the two experimental methods, between Fig. 3, Fig. 4(b) and Table 1

Fig. 1

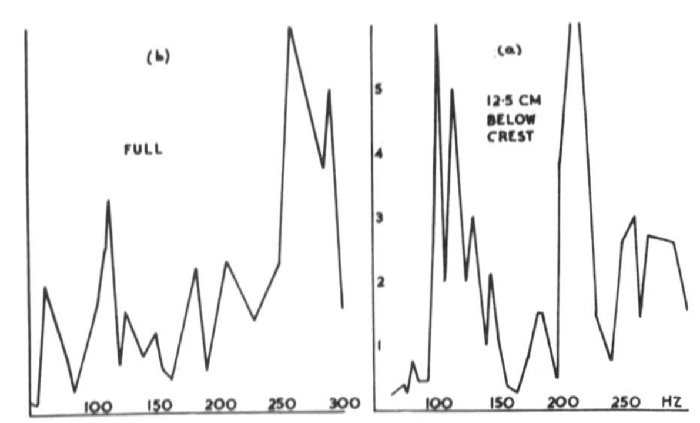

Fig. 4. Cylinder model on shaking table. Acceleration response at crest crown cantilever. (a) Reservoir full (b) water level 12.5 cm below crest ($\frac{3}{4}$ full)

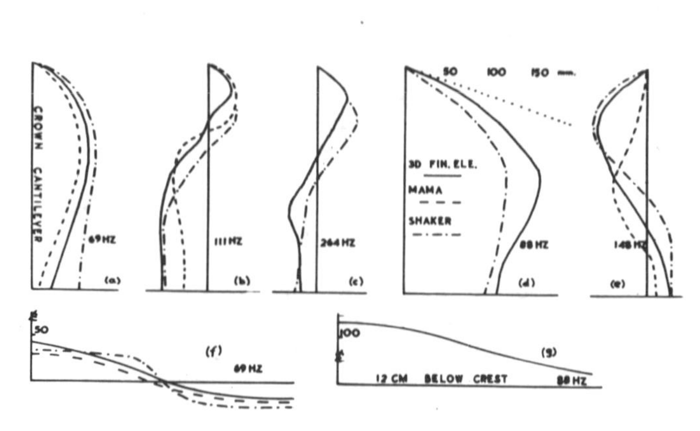

Fig. 2. Cylinder model. Calculations and measurements of hydrodynamic pressure

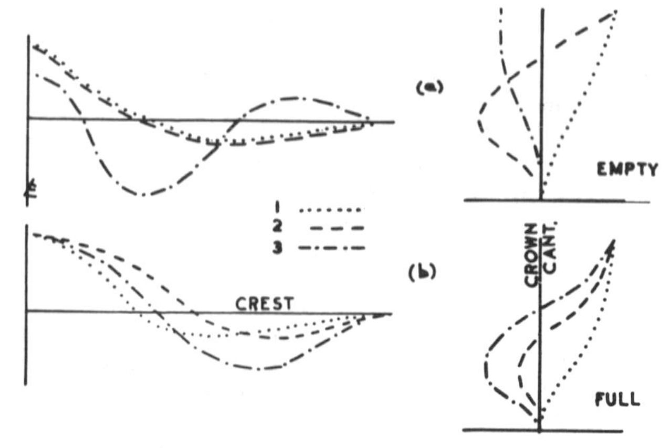

Fig. 5. Cylinder model. Calculated symmetric mode shapes (a) reservoir empty (b) reservoir full

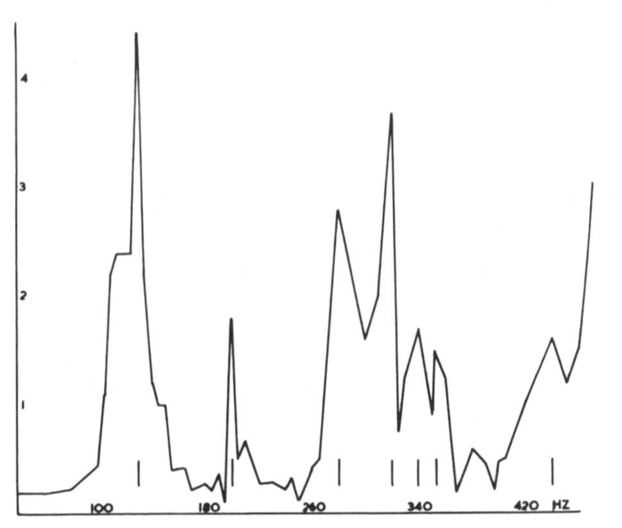

Fig. 3. Cylinder model on shaking table. Acceleration response at crest crown-cantilever. Reservoir empty

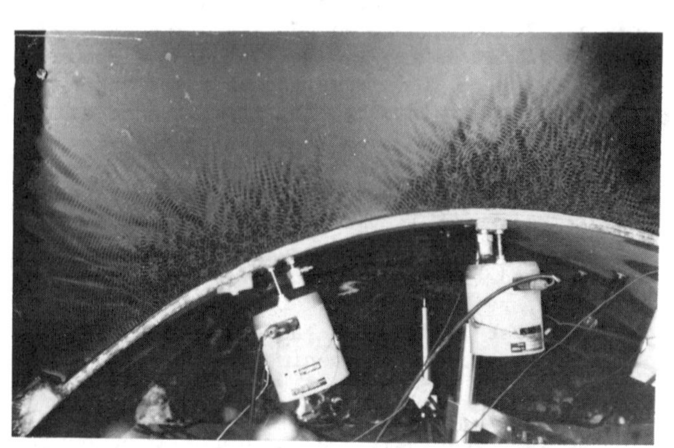

Fig. 6

shows good correlation for the two lowest symmetrical modes, but less so at higher frequencies. This was also true for the antisymmetrical modes (Ref. 14), and the probable reason is interference from modes of the rest of the system. MAMA is designed to produce pure modes, and is a sophisticated approach compared to the peak-amplitude method of Figs. 3 and 4. Such data needs to be processed in a more revealing way, and the required transfer-function approach will be discussed later for the Victoria dam.

Transient response

11. The ferrocement model was made thin so that response could be easily measured, with reservoir full, and with shaking facility available (26 kN max force), under the transient input of earthquake records. It has been seen that this caused difficulties in comparing calculated and measured frequency values, so much so that a similar, but thicker (50 mm) model is now under construction in hard urethane rubber, having an E-value of 25 MPa, which bonds chemically to the surrounding steel. The ferrocement model was, however, satisfactory for the transient acceleration input tests, as Fig. 7 indicates. Here, (a) is the acceleration input measured at the base of the dam. The actual acceleration fed from the computer to the shaker was a factored S16E San Fernando 1971 record, and a comparison with the published record will show the fidelity of the various processes involved in getting from this record to that given in (a). However, comparison by eye is not sufficient as far as structural response is concerned, and it was therefore the recording (a) that was used for the comparison of measured and calculated response; it was recorded in digital form by the Varian computer and processed to obtain spectral acceleration curves. These were similar to the published S16E curves, but not identical. Fig. 7 (b) gives the acceleration measured at crest crown-cantilever in response to (a). Noting that the acceleration scale in (b) is to be multiplied by 10 to compare with (a) a maximum acceleration magnification factor at the crest of about 3.5 occurs; the corresponding calculated value, using 1 mode and 2 % damping, was 5.05. This damping value was taken as the average value for the first two modes obtained by both MAMA and the shaking-table tests. It should also be said that the response calculations were made with an E-value of 26 GPa, which is the value deduced from Table 1 as that required to bring the calculated and measured resonance frequencies into coincidence, rather than with the E-value used for the first two rows of Table 1, which was obtained from tests on specimens.

12. Fig. 7 (c) gives the hydrodynamic pressure (a) for full reservoir, measured 125 mm below crest level on crown cantilever, whilst (d) is measured 325 mm below crest level. The pressure scale (mm water) in (d) is to be multiplied by 10 to compare with (c), then showing a ratio of about 11.5/9.5 of maximum values. In general, taking a few peaks in (d) and (c), the ratio is consistent with the Westergaard's parabolic shape, but as previously noted, the Westergaard value is too big. The correspondence between peaks of acceleration, (b), and peaks of pressure, (c) and (d), should also be noted.

VICTORIA ARCH DAM

13. The Victoria arch dam is under construction in Sri Lanka as a project sponsored by the Overseas Development Administration of the U.K. Government, with Sir Alexander Gibb and Partners as Consulting Engineers. It is of logarithmic spiral type, 120 m high and almost symmetrical about the crown cantilver. Its span is 450 m with developed crest length of 510 m, crest thickness of 9.3 m and base thickness 31 m. The geometry can be appreciated from Fig. 8 which is a plot of the 3D, 20-node finite element mesh, with 66 elements in the dam and 95 in the foundation block, giving a total of 795 nodes. The nodes at the edge of the foundation block have zero displacement.

14. The study of the aseismic design of Victoria dam comprises the following features, only part of which has been accomplished at this time.

1. Calculation of resonance frequencies and mode shapes for different foundation conditions, and different approaches to the assessment of the added-mass of the reservoir.
2. Calculation of response to different inputs, particularly the study of multiple-support and asynchronous excitation of the base.
3. Resonance testing of 1 : 250 scale models in microconcrete using the MAMA multipoint equipment, as well as the abstraction of resonance frequencies and mode-shapes from transient testing of models, using transfer function analysis.
4. Response measurement in models on a shaking table, with and without water, subjected to various earthquake records. As well as microconcrete, a hard urethane rubber is being used as model material, having a dynamic E-value of approximately 25 MPa.
5. Prototype tests on the completed dam using a system of synchronised, eccentric-mass exciters (Ref. 15).

Model testing: frequencies, mode-shapes, hydrodynamic pressure

15. The 1 : 250 model of Victoria dam and reservoir is shown in Fig. 9. The foundation block extends 25 cm upstream, downstream and below the base of the dam. The 5 shakers of the MAMA system can be seen. At the present stage of the study, the model is continuous across the spillway section, and this is to be remembered in making comparisons in Table 2. During the next stage the spillway will be cut out, as indicated in Fig. 8. The testing programme consists of measurement of resonance frequencies and mode shapes using MAMA for empty and full reservoir conditions; measurement of hydrodrodynamic pressure on the dam face and in the reservoir whilst MAMA is exciting the dam in one of its resonances; and measurement of resonance frequencies by excitation of a crest point with random noise through

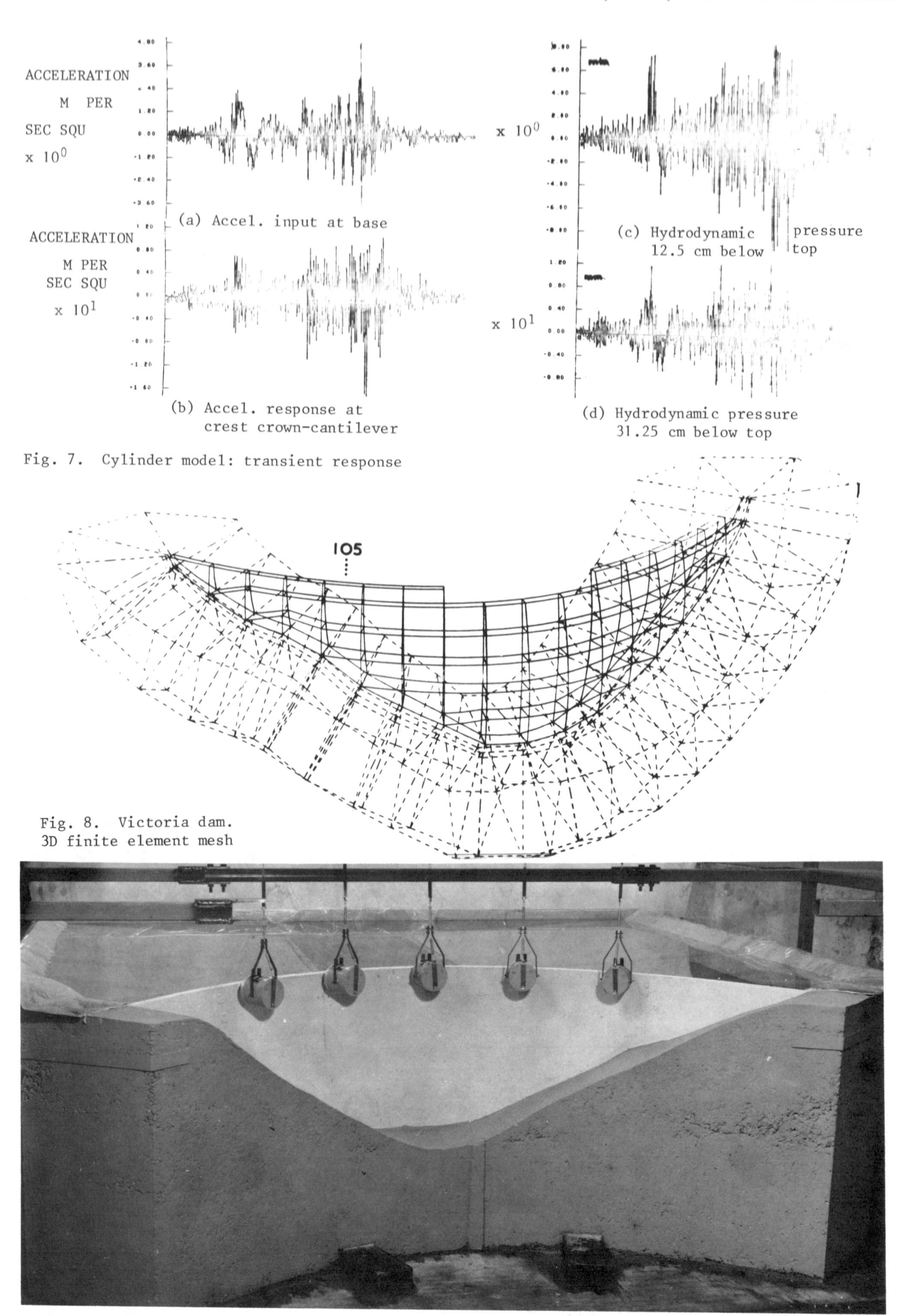

Fig. 7. Cylinder model: transient response

Fig. 8. Victoria dam.
3D finite element mesh

a MAMA vibrator, followed by processing of the acceleration response at other points using a Hewlett Packard Spectrum Analyser.

Theoretical studies : frequencies, mode-shapes, hydrodynamic pressure

16. Using the mesh of Fig. 8, 12 resonances were calculated below 6 Hz in the reservoir-empty condition for E = 30 GPa in the dam, and either 10 or 9000 GPa in the foundation. 10 GPa is the most probable value, whereas 9000 approaches the rigid-base condition. The S.A.P. program, which was used for these calculations, has no structure-fluid coupling facility; Westergaard's method was therefore used with the mesh of Fig. 8 for full reservoir. The Bristol finite element package does have this facility, and was used with a mesh having only 18 20-node elements in the dam and no foundation elements. This coarser mesh, which ignores the spillway cutout, was designed for the multiple-support excitation studies described later, but it has served to calculate hydrodynamic pressure.

Comparison of measured frequencies and mode shapes

17. The first 3 rows of Table 2 give 6 frequencies using Fig. 8 mesh. The first mode is antisymmetric, the second symmetric, as is usual for arch dams. An elastic foundation decreases the frequencies by about 15%; Westergaard's added-mass approach decreases them still further, by about 40% in the first two modes. The mode shapes are plotted in Fig. 10; (1) is for the crown-cantilever, where modes 1, 4 and 6 are indistinguishable from zero, whilst 2, 3 and 5 are single lobe in the vertical direction. For clarity, (2) presents only the first 4 modes, plotted on the crest. The slight lack of symmetry in the dam is reflected in the mode-shapes.

18. The next two rows of Table 2 are for the coarser mesh with rigid foundation, but with rigorous fluid-structure coupling. Direct comparison is appropriate only between rows 1 and 4, from which very good agreement is seen for the first two modes, but the coarse mesh cannot produce the higher modes accurately. An interesting indirect comparison comes from looking at the reduction in frequency caused by the reservoir; the coarse mesh, with fluid-structure coupling, gives 40 and 33% reduction of the reservoir-empty frequencies in the first and second modes, respectively. This accords well with the 38 and 40% given by Westergaard's formula.

19. The remainder of Table 2 gives experimental results, which have been scaled to prototype values. A minor difficulty here was that the ratio of dam/foundation E-value was 3 for the prototype and 0.7 for the model. Rows 6 and 7 were obtained using MAMA. Row 6, for empty reservoir is in satisfactory agreement with row 2 for the first 3 modes, and row 7, for full reservoir, remarkably close to row 3. From these results it would appear that Westergaard's approach is valid. Perhaps this is not surprising, because measurements of hydrodynamic pressure, Fig. 12 and elsewhere, show that Westergaard's formula is close to the truth in the important upper regions of a dam. Its lack of validity in the stiffer base regions appears to be of little consequence. A comparison of measured and calculated (coarse mesh) mode-shapes on the crest is given in Fig. 11 for the first two modes, with reservoir empty.

20. In rows 8 and 9 resonance frequencies from the transient testing method are given, which are in very good agreement with the MAMA values. A hammer, and a random noise generator were both used for the transient input (Ref. 16).

Calculated and measured hydrodynamic pressure

21. From the frame surrounding the model, attachments containing pressure transducers were lowered into the reservoir at any desired position whilst the dam was vibrating at a resonance. Fig. 12 presents results for the first two modes, down the face of the dam at the stations indicated. Calculated values have only been obtained at this stage for the coarse mesh.

Response calculations for Victoria dam

22. Using the S.A.P. programme, the S16E San

RESERVOIR	FOUNDATION	f_1 asym	f_2 sym	f_3 sym	f_4 asym	f_5 sym	f_6 asym	
Empty	9000 GPa	2.605	2.845	3.145	3.939	4.652	5.365	Fine Mesh
Empty	10	2.210	2.516	2.777	3.499	4.144	4.784	
Full	10	1.372	1.510	1.771	2.247	2.697	3.126	
Empty	∞	2.713	2.876	4.315	5.[illegible]91	6.141		Coarse Mesh
Full	∞	1.638	1.936	2.825	4.[illegible]15	5.102		
Empty	see	2.12	2.31	2.618	2.[illegible]6	3.45	3.70	MAMA
Full	text	1.36	1.61	1.804	2.[illegible]9	2.74	3.00	
Empty	see	2.11	2.30	2.62	2.[illegible]5	3.42	3.78	Transient
Full	text	1.37	1.59	1.82	2.[illegible]0	2.74	3.00	

Table 2. Victoria Dam. Comparison of calculated and measured frequencies. Prototype scale.

Fernando (1971) record factored by 1/3 was applied in the upstream/downstream direction and cross-valley, and the same record factored by 1/6 applied in the vertical direction; all 3 components simultaneously. Six modes have been included in the response spectrum approach, with 2% damping. The results for the crown-cantilever are given in Fig. 13. Curves a and b are hoop stresses on downstream and upstream faces, respectively; curves c and d are vertical stresses on the same faces, and e is the deflection.

23. In view of the proposal (Ref. 6) that response is contained essentially in the first symmetric mode for upstream/downstream input, and in the first antisymmetrical mode for cross-valley input, the 1/3 S16E record was applied to Victoria dam upstream/downstream only, with stresses calculated for an increasing number of modes, starting with one, and ending with the sixth. Fig 14(a) gives vertical stress on upstream crown cantilever, showing that the first mode contributes very little, as expected, since it is asymmetric. The symmetric second mode makes the major contribution, and the addition of higher modes produces little increase in stress. Fig. 14(b) gives hoop stress on upstream crown-cantilever, resulting in similar conclusions. Whilst these two results might be sufficient vindication of the Ref. 6 proposal, Fig. 14(c) giving vertical stress on the upstream face of a cantilever at chainage 105 (see Fig. 8), shows that stress is reduced when modes 3 and 4 are added to 1 and 2, but increases again appreciably on addition of modes 5 and 6.

<u>Multiple-support excitation of Victoria dam</u>

24. A distinction is drawn between asynchronous excitation and multiple-support excitation. The former term is used when the same record affects every part of the base, but with finite travel time, so that a phase-difference occurs between points. Such evidence as exists, for example at the Ambiesta dam (Ref. 2) where records were obtained on the two abutments, indicates that the record is not the same at different points of the base, and true asynchronous excitation is unlikely. Nevertheless, it probably gives a better indication of the correct response than would the more usual constant phase analysis. The term multiple-support excitation on the other hand, refers to different records applied simultaneously to different parts of the base, and it is the analysis used here on the Victoria dam, with the S16E component of San Fernando 1971 applied to the whole of the left-half of the base and the S74W component applied to the right-half, together with various combinations of these.

25. Ref. 1 gives an outline of the required analysis, and Ref. 13 a more detailed explanation. Basically what is required, additional to standard response analysis, is calculation of the pseudo-static shape vector of the dam when a single base point is given unit displacement, with all other base points fixed.

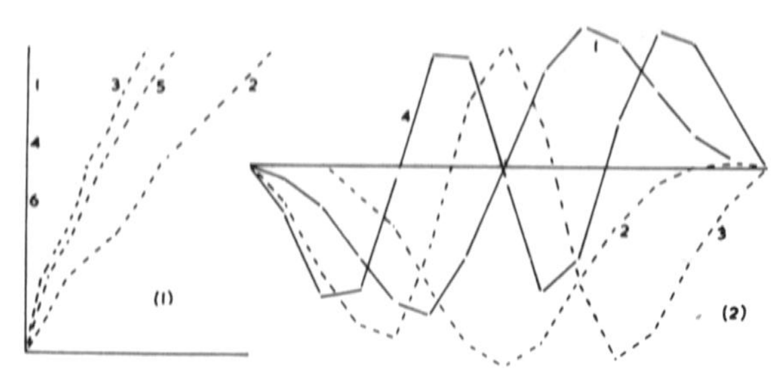

Fig. 10. Victoria Dam Mode Shapes. (1) Crown Cantilever, (2) Spillway Level. Fig. 8 mesh

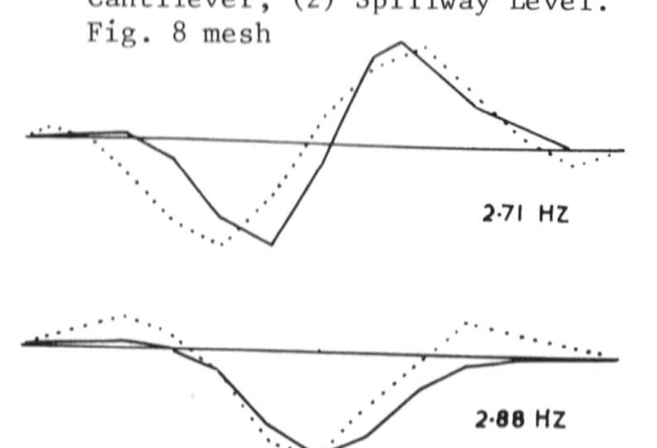

Fig. 11. Victoria dam: mode-shapes: calculated (coarse mesh) ———; measured .....
Reservoir empty

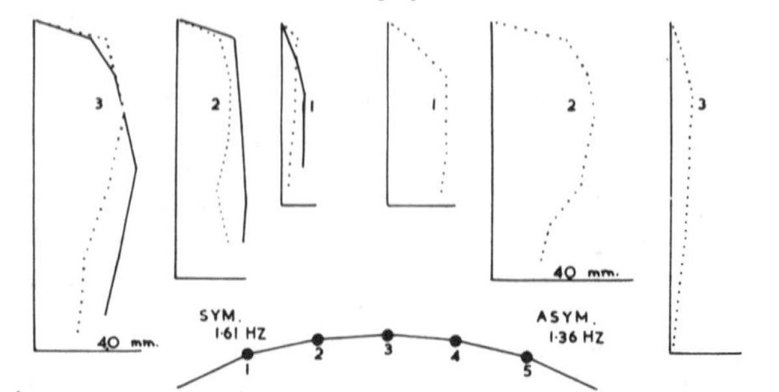

Fig. 12. Victoria dam model: hydrodynamic for first two modes· measured ...., ——— calculated

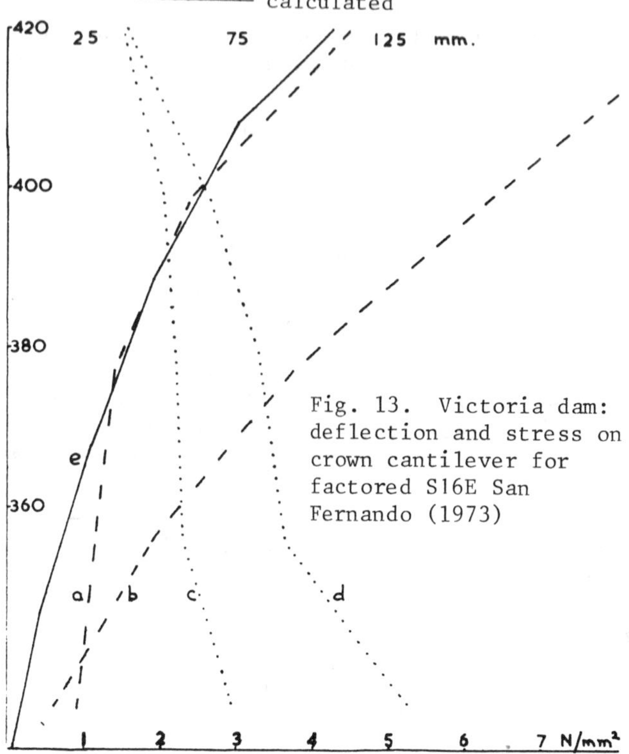

Fig. 13. Victoria dam: deflection and stress on crown cantilever for factored S16E San Fernando (1973)

When a number of base points are to be given the same acceleration, as here for the Victoria dam, a shape vector is obtained for a unit displacement given to all these base points in the direction of the acceleration which is to be applied. Two different shape vectors have therefore been obtained for the Victoria dam, one each for the left and right halves of the base. The implementation of this multiple-support excitation analysis in a computer program requires familiarity with the structure of the program, and it has therefore been carried through only on the Bristol finite element package (Ref. 14), although implementation in S.A.P. is intended.

26. At the present stage, the multiple-support excitation work is exploratory, with many solutions required; it has therefore been carried out only on the coarse mesh, having 18 20-node elements and rigid base. Fig. 15 gives maximum radial deflections of the crown-cantilever in (a), and of crest-level in (b), for the first symmetrical mode of vibration only and with 1% damping. Curves A result from application of S16E in the upstream/downstream direction to the left-half of the base, with the right-half fixed. Curves B result when S74W San Fernando is applied to the right-half, with the left-half fixed. As a check, S16E was applied to the left-half simultaneously with S74W to the right half; this gave curves C, which are seen to be the sum of A and B. Omitting curves D for a moment, curves E are for S16E applied to the whole base, using this multiple-support excitation technique, and the results are appreciably larger than curves C. The maximum value in E is 270 mm, whilst curve Fig. 13e gives 107 mm for the same base input, but to a dam with (massless) foundation, and 2% damping, using the SAP program. This result is considered to indicate the correctness of the multiple-support program, and the vital importance of damping. Finally, returning to curves D, these are for S16E applied to the whole base, but using the simple 'direction-vector' rather than the pseudo-static shape vector. The former consists simply of 1's associated with masses in the earthquake direction only, and 0's elsewhere, needing no calculation, whereas the latter generally has non-zero components associated with each degree of freedom, which are calculated within the program. This accounts for the difference between D and E.

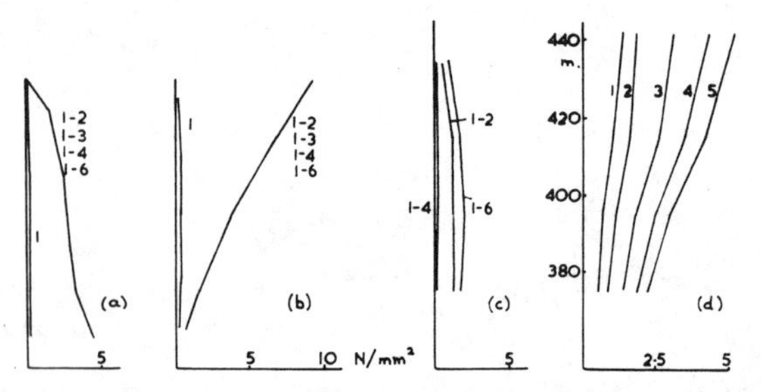

Fig. 14. Victoria dam: (a), (b) and (c), stress stress due to modal addition; (d) multiple support excitation stress

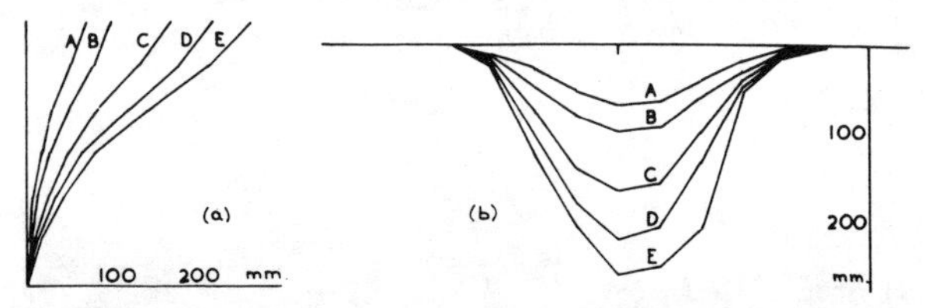

Fig. 15. Victoria dam. Displacements, (a) crown-cantilever, and (b) crest due to multiple-support excitation (see text)

REFERENCES

1. CLOUGH R.W. and PENZIEN J. Dynamics of Structures. McGraw Hill 1975
2. CALCIATI F., CASTOLDI A., CIACCI R., and FANELLI M.A. Experience Gained During In Situ Artificial And Natural Dynamic Excitation Of Large Concrete Dams In Italy: Analytic Interpretation of Results XIII ICOLD, Q 51, R 32, New Delhi, Oct. 1979
3. TARBOX G.S., DREHER K., CARPENTER L. Seismic Analysis of Concrete Dams, Ibid R.11.
4. WIDMAN R. The Dynamic Behaviour of Arch Dams Investigations By Means of Calculation and Measurement. Ibid R.9.
5. SWANSON A.A. and SHARMA R.P. Effects of the 1971 San Fernando Earthquake On Pacoima Arch Dam. Ibid R.3.
6. ACADEMIA SINICA (China) Earthquake Loads for Hydraulic Structures. Ibid R.27.
7. TAYLOR G.A., GAUKROGER D.R., and SKINGLE C.W. MAMA - A Semi-Automatic Technique For Exciting The Principle Modes of Vibration Of Complex Structures. Roy. Aircraft Estab., Tech. Rep. 67211, Aug. 1967.
8. SKINGLE C.W., HERON K.H., GAUKROGER D.R., Numerical Analysis of Vector Response Plots. Roy. Aircraft Estab. Tech. Rep. 73001, Feb. 1973.
9. GAUKROGER D.R. and COPLEY J.C. Methods For Determining Undamped Normal Modes and Transfer Functions From Receptance Measurements. Roy. Aircraft Estab. Tech. Rep. 79071, June 1979.
10. DUNGAR R. An Efficient Method of Fluid-Structure Coupling In the Dynamic Analysis Of Structures. Int. J. Num. Meth. in Eng. Vol. 13, 93 - 107 (1978).
11. WESTERGAARD H.M. Water Pressure On Dams During Earthquakes. Trans. A.S.C.E. Paper No. 1835, Nov. 1931.
12. CROWSON R.D. and NORMAN C.D. Comparison of Vibration Test Results For A Model Arch Prototype Arch Dam. U.S. Army, Waterways Expth. Stn. Report N-77-1, March 1977.
13. ALTINISIK D. and SEVERN R.T. Natural Frequencies And Response Characteristics of Gravity Dams. Conf. On Design of Dams to Resist Earthquake, I.C.E. London, Oct. '80, Paper No.
14. ALTINISIK D. Aseismic Design of Concrete Dams. Ph.D. thesis, Univ. of Bristol, 1980.
15. SEVERN R.T., JEARY A.P., and ELLIS B.R. Forced Vibration Tests and Theoretical Studies on Dams. Proc. I.C.E. (Pt. 2) Sept. 1980, Paper 8362.
16. SEVERN R.T. and TAYLOR C.A. Transient input tests on a model arch dam. To be published.

25 The use of models in assessing the behaviour of concrete dams

G. OBERTI, PROF DR ENG, President, ISMES; and A. CASTOLDI, DR ENG, Director, ISMES, Bergamo

The present "state of the art" of the modelling techniques as to the determination of the earthquake response of dams is described. After having discussed the requirements posed by the similitude theory as far as both the modelling materials and the reproduction of the static and dynamic loads are concerned, the two different techniques adopted at ISMES are presented:

a) the first is based on the use of large scale models, excited by actuators and allows an accurate determination of the natural frequencies and the modal shapes, as also of the modal strains and the hydrodynamic pressure distribution. The interaction of the dam with the foundation and the reservoir can be correctly taken into account;

b) the second is based on the use of small scale models, tested on the shaking table. These models are made up of special materials so as to reproduce faithfully - together with the static and dynamic loads - the structural behaviour up to failure.

Finally, the first results of a new testing technique now under study, by means of which the model is excited, through its foundation, by "travelling" waves, are described.

1. INTRODUCTION

The lack of sufficient direct information on the behaviour of dams subjected to high intensity quakes and the complexity of the phenomena set up in the system: dam, foundation, reservoir, presents the seismic verification of a dam as a problem that is still open and one that, notwithstanding recent progress, certainly calls for further study.

In fact, in recent years the availability of calculation techniques like the Finite Element Method have offered an approach more in consonance with the reality of the phenomenon, which allows a seismic verification based on the employment of modal analysis rather than on the use of equivalent static forces. These techniques are, however, burdensome and insufficient when simultaneous account has to be taken of the interaction with the reservoir and the foundation or when requiring information on the behaviour of the structure beyond the elastic field. From these points of view experimentation on models has been found to be an extremely powerful method of investigation, it being able to provide a highly reliable and detailed information on the dynamic behaviour of the structure.

What follows is meant to illustrate the characteristics and the possibilities of experimentation, as also the different techniques employed, in particular those based on the use of elastic and failure models. In addition a brief review is provided of a number of current experience at ISMES with a view to putting together a new technique capable of facing, in a comprehensive manner, the problems of the seismic verification of a dam.

2. ELASTIC MODELS

2.1. Design criteria of the model

The elastic model is meant to study the behaviour of a dam on the basis of the hypothesis that, for extreme loading conditions, the maximum stress level should always be below the elastic limit of the materials for both the dam as also the foundation, and that non-linear phenomena - caused for example by the opening of joints, do not take place. In this case, the stress state provoked by the quake may be determined independently of that caused by static loads - hydrostatic thrust, weight, temperature, etc.- to which it must be added in order to obtain the maximum stresses.

The criteria of designing the model and of the tests are therefore based on the following hypotheses:

- The system under examination - and therefore reproduced on the model - is made up of the dam and of a substantial extension of the foundation and the reservoir.

- The dam is considered to be a homogeneous and monolithic body - thus not calling for the reproduction of the vertical joints, nor the possible perimetral joint.

- The quake is a transient motion applied simultaneously along the entire boundary of the foundation reproduced.

- The effect of the reservoir during the quake is that of generating hydrodynamic pressures that, in the first approximation, depend solely on the density, the cubic compressibility module of the water and on the elastic characteristics of the dam - the effect of the surface waves being thus neglected.

On the basis of the foregoing hypotheses, the determination of the response to a quake may be effected in a simple manner by analytical means, starting from the knowledge of the modal parameters of the system under study, which therefore become the main aim of the tests.

In the case of a model being based on the aforesaid criteria the ties imposed by the laws of similitude are reduced substantially. In fact, it being no longer necessary to reproduce the weight and the hydrostatic thrust, the density scale S_ρ of the materials is found to be independent of the scale of elastic modulus S_E. Table 1 shows the relevant scales.

Further, it needs to be underlined that a system of external forces is generally used for the determination of the modal parameters. As such it is possible to set aside the use of the shake table, thus gaining greater freedom for what concerns the choice of the geometric scale and the utilization of larger-sized models.

2.2. Construction technique

The model is built inside a large concrete tank that, in its turn, is sunk in the ground. This allows the adoption of geometric scales in the order of $S_L \simeq 100$, as well as enabling the simultaneous reproduction of a substantial extension of the foundation and the reservoir (see Fig. 1).

The model is built up in the following manner: at the outstart the foundation is cast and shaped using a material - or several materials if the reproduction of zones with different mechanical characteristics is called for - made up of concrete with heavy aggregates having such elastic modulus (in the order of 130,000-270,000 Kg/cm^2) and specific weight (in the order of 2.5-2.7 g/cm^3) that $S_E \simeq 1$ and $S_\rho \simeq 1$.

Formworks are then deployed for the casting of the dam, which is carried out with a material composed of resin and heavy aggregates - iron grains, sand, quartzite. The utilization of these materials is recommended by the guarantee of uniformity of the mechanical properties of the casting - even when the model is built up in two or more layers, as also by the ease with which strain gauges may be applied on the faces.

Following the removal of the formworks, the successive operation consists of the application of tri-directional estensimetric rosettes on the two faces for the measurement of the principal stresses relative to each mode, as well as of the deployment of the hydrodynamic pressures transducers. The pins for attaching the exciters and for the application of the accelerometers are instead set in the dam body - during the casting. The pins facilitate deployment of the accelerometers both in the radial and tangential directions and, if necessary, in the vertical one. Finally, the upstream face and the reservoir are water-proofed by means of an epoxy based paint; in this way the test can be carried out in full reservoir condition without influencing the mechanical properties of the materials.

With the scale adopted for the elastic modulus and the density of the materials, full reservoir conditions may be reproduced with water (see Fig. 2).

2.3. Test criteria and data processing

In carrying out the tests the simplifying hypothesis is usually adopted that the behaviour is such as to exibit normal vibration modes. This hypothesis, which is fairly well verified and as such acceptable in the case of elastic models, is, however, often contradicted by the results of forced vibration tests on prototypes, for which, at times, it is necessary - because of the nature of the damping forces - to have recourse to the concept of complex modes (ref.1).

The determination of modal parameters is made by exciting the dam by means of small electrodynamic shakers, deployed on the downstream face in a radial direction.

The test technique, first of all, determines the most suitable combination of exciting sinusoidal forces in order to eliminate or reduce the contribution, to the response, of the modes in the neighbourhood of that under examination, and successively, utilizes the response obtained with this system of forces, for the final estimate of modal parameters.

In practice, following the choice of a certain number M of excitation positions (normally M = 4), for each of these, a series of transfer functions $h_{ji}(\omega)$ is determined, these latter linking the acceleration response a_j in the j-th point (with j = 1, ..., N) with the i-th excitation force F_i (with i = 1, ..., M). The analysis of these transfer functions makes it possible to work out an in-

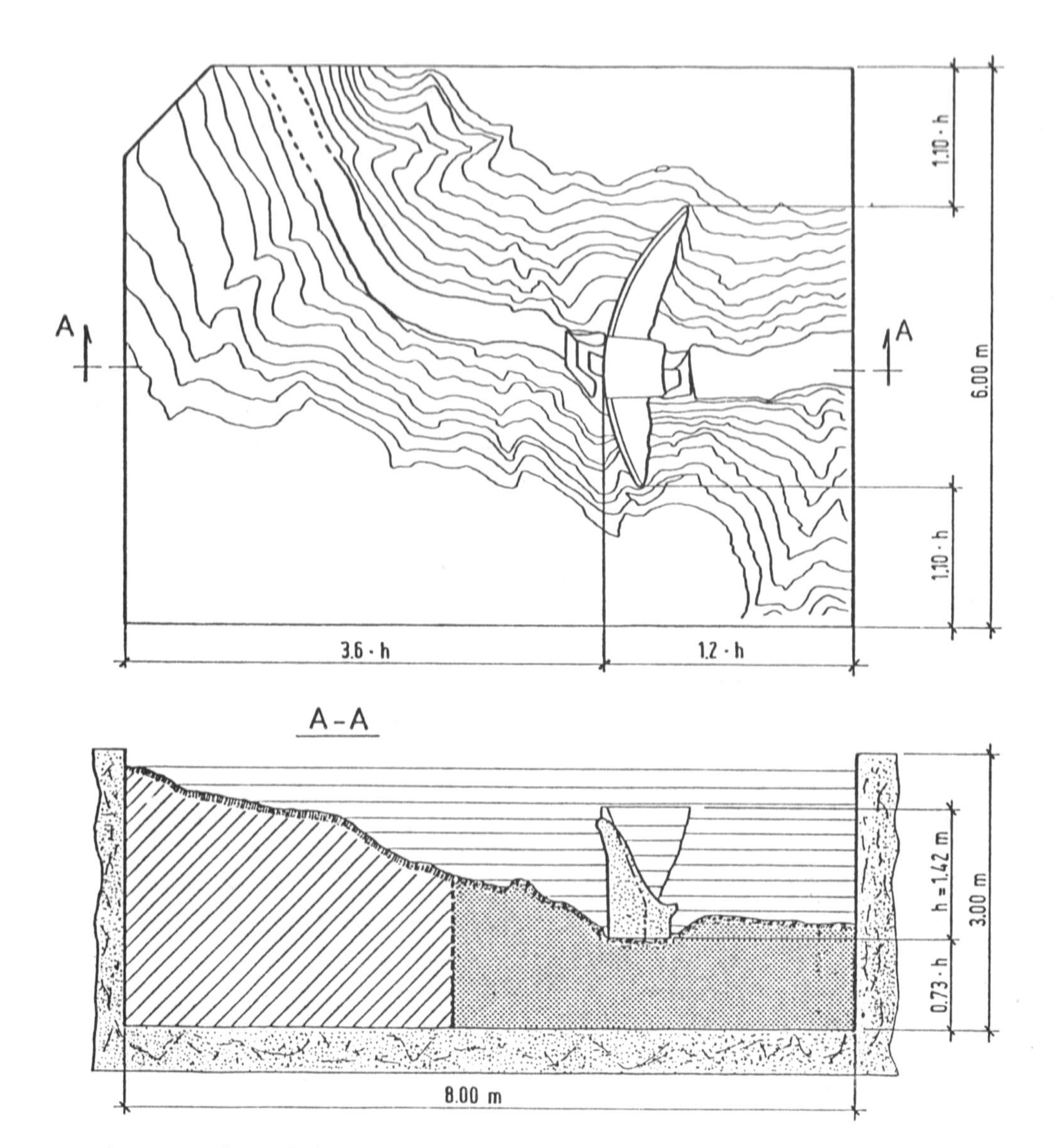

Fig. 1. Disposition of the model in the testing tank. Amaluza arch-gravity Dam (Ecuador).

Fig. 2. Llauset arch-dam (Spain). Elastic model tests with full reservoir.

itial estimate of the natural frequencies and the modal shapes.

An iterative process is begun on the basis of these preliminary data and consists in:

- calculating the forces F_i (with $i = 1,M$) capable of annulling the generalized forces relative to the M-1 modes in the neighbourhood of the mode k under examination. These forces are obtained solving the expressions:

$$\sum_{i=1}^{M} F_i^* \phi_i^k = 1$$

$$\sum_{i=1}^{M} F_i^* \phi_i^s = 0 \qquad \text{with } s \neq k$$

where ϕ_i^s is the modal shape at the position i-th for the mode s;

- determining analytically, via the transfer functions, of the response of the dam to such a system of forces:

$$\{a_j\} = \sum_{i=1}^{M} h_{ji} F_i^*$$

- determining a new and more precise estimate of the modal shapes, starting from the response curves.

At the end of the process, the structure is excited with the distribution of the forces so obtained, exploring the frequency range around the natural frequency of the mode being looked for. The response that is obtained is used for the definitive estimate of the modal parameters: natural frequencies, dampings, modal shapes, distribution of stresses and of the hydrodynamic pressure. This estimate is achieved by employing a programme based on the analytical fitting of the response curves themselves. The verification of the orthogonality of the vibration modes is carried out to check the accuracy of the estimates effected.

It is important to emphasize that all the test phases, from the preliminary determination of the transfer functions up to the identification of the modal parameters, are effected in a highly automatized manner, with the aid of a minicomputer and a series of interactive programmes.

By way of example of the quality of the results that may be obtained with this technique, Fig. 3 carries the modal shapes - in terms of displacements and of principal stresses - for the first mode of the Amaluza Dam (ref. 2).

2.4. Seismic verification

Having effected the determination of the modal parameters, the response of the dam is calculated with the method of modal analysis. In empty reservoir conditions the modal shapes are associated with a mass matrix $[M]$, obtained lumping the mass of the dam in the nodes of the gage point network; in the case of full reservoir, a matrix $[M_{add}]$ is added to the matrix $[M]$, the former taking into account - according to a well-known concept - the hydrodynamic forces. The matrix $[M_{add}]$ is obtained starting from the measurement of the hydrodynamic pressures.

Generally, four vibration modes are considered in calculation, the first two symmetrical and the first two antisymmetrical. The design quake may be defined either in terms of time-history or of response spectra; in the latter case the maximum probable value of the response is calculated (ref. 3, 4).

3. FAILURE MODEL

3.1. Design criteria and construction method

In the models put forward for the study of the behaviour of a dam beyond the elastic field, up to failure, the principle of the superimposition of effects being no longer valid, it is necessary to reproduce simultaneously all the acting forces and the discontinuities present (such as joints and faults). In particular the following forces need to be reproduced:

- the weight force: as is known, this is obtained by selecting a model material for which the condition $S_E = S_\rho \cdot S_\ell$ is respected;

- the hydrostatic thrust: the fluid needs to have a density in the scale S_ρ preselected for the model materials;

- the hydrodynamic pressure: while neglecting the effect of surface waves, it is necessary to reproduce not only the density but also the compressibility of the fluid. Further the dimensions of the reservoir reproduced - in particular the length - is of substantial importance;

- elastic forces: the stress-strain characteristics of the model material need to reproduce in similitude those of concrete;

- inertial forces: these are fundamentally determined by the quake applied to the base, which therefore must reproduce in similitude the design time history;

- damping forces: in this case as well a correct simulation of the rheological characteristics of the material is called for, but it is also necessary to reproduce the phenomena of dissipation via radiation along the boundary.

The dam model is cast inside a rigid metal tank, in which a limited extent of both the foundation and the reservoir are reproduced. The tank is then fixed to a shake table (Fig. 4) capable of generating a given time history.

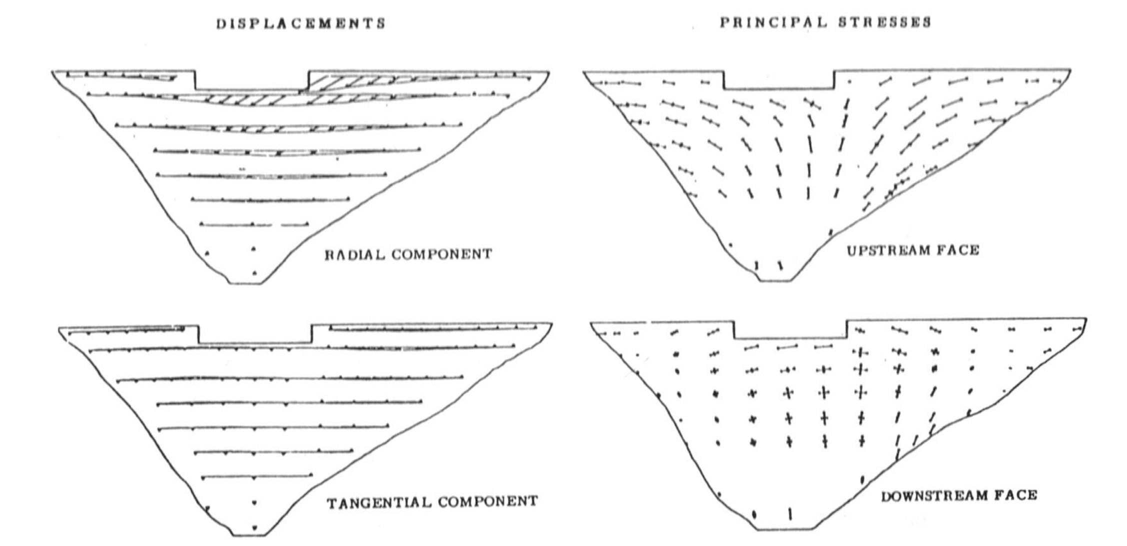

Fig. 3. Example of the results obtained from the elastic model. Amaluza ar-gravity Dam, first symmetric vibration mode.

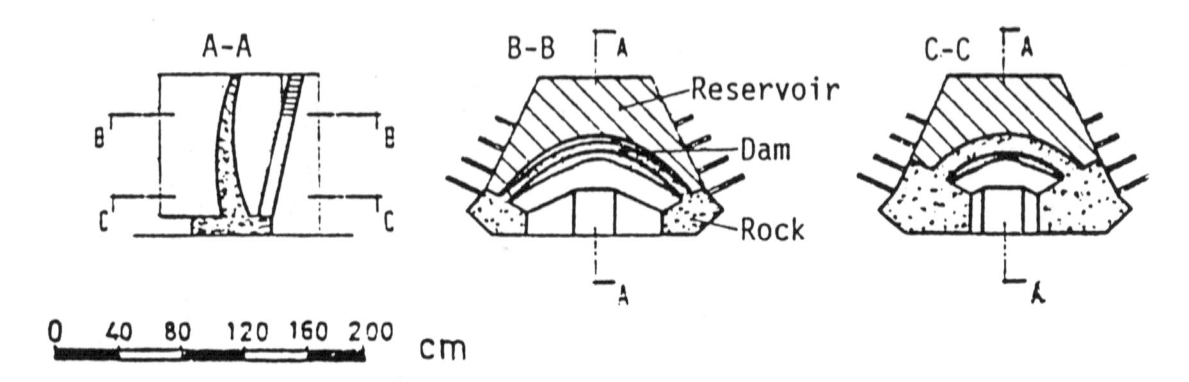

Fig. 4. Failure model. Setting up of the model in the testing tank.

Table 1. Similitude ratios

PARAMETERS		ELASTIC MODEL	FAILURE MODEL
Length:	ℓ	S_ℓ	S_ℓ
Elastic modulus:	E	$S_E = E_r/E_m = 1$	$S_E = E_r/E_m$
Specific gravity:	ρ	$S_\rho = 1$	$S_\rho = S_E \cdot S_\ell^{-1}$
Time:	t	$S_t = S_\ell \cdot (S_\rho/S_E)^{\frac{1}{2}} = S_\ell$	$S_t = S_\ell^{\frac{1}{2}}$
Velocity:	v	$S_v = 1$	$S_v = S_\ell^{\frac{1}{2}}$
Acceleration:	a	$S_a = S_\ell^{-1}$	$S_a = 1$
Sound wave velocity:	c	$S_c = 1$	$S_c = 1$
Strength:	σ		$S_\sigma = S_E$
Strain:	ε	$S_\varepsilon = 1$	$S_\varepsilon = 1$
Pressure:	p	$S_p = S_E$	

The material used at ISMES has a matrix made up of plaster, while the artificial aggregates are composed of lead beads covered with a vinyl resin film. The average characteristics of the materials are shown in Table 2, according to the diameter of the aggregates employed.

With these materials the relationship $S_E = S_\rho \cdot S_\ell$ is found to be satisfied for the values of $S_\ell \leq 100$. On this scale the dimensions of the model turn out to be conspicuous, in particular in relationship to the dimensions and weight which the shake table is capable of carrying. For this reason it is necessary to limit as much as possible the foundation area and the reservoir extention that are reproduced, which, in some cases, may lead to an incorrect simulation of the interaction of the dam, with the reservoir and the foundation. The fluid employed is a zinc chloride saturate solution with baryte suspension ($\gamma \leq 2.2$ g/cm^3), that facilitate a correct simulation of the density, while the compressibility and viscosity values are found to be altered. Fig. 5 shows the deployment of a model on the shake table.

3.2. Test execution

The tests are generally carried out in three stages:

- at the outstart tests are effected with a sinusoidal motion in order to determine the natural frequencies;
- the second stage consists in generating a series of time-histories corresponding to the assigned spectrum, which are applied to the base of the model via the shake table;
- in the third stage only one of the time-histories is utilized - that which has provoked the highest response; its intensity is increased step-wise until failure of the model is achieved.

Usually the tests are carried out on two separate models excited respectively in upstream-downstream and transversal direction.

4. CONSIDERATIONS ON THE RESULTS

Numerous tests, on models as well as on prototype, have been carried out at ISMES with the forced excitation techniques described above (ref. 5). The Table 3 shows the main characteristics of some of the dams that have been studied and carried the experimental values relative to the first four modes. Besides making a valid contribution towards clarifying the various aspects connected with the seismic behaviour of a dam, the data have also facilitated in calibrating the accuracy of the illustrated modelling techniques. In this context, an analysis was made of the results obtained both on models and on prototypes, taking into consideration the two fundamental phenomena of the interaction between dam and foundation and between dam and reservoir, and their influence on the modal parameters.

4.1. Dam-foundation interaction phenomenon

In order to show up the validity of the modelling criteria adopted for the foundation, the determination of the natural frequencies and of the damping coefficients was effected for the Amaluza arch-gravity Dam, in two totally different boundary conditions of the reproduced foundation area: in the first case, the foundation was free along the boundary, in the second it was in direct contact with the walls of the container tank - in its turn sunk in the ground.

The results (see Table 4) show that the reproduction of the foundation for an extension that is of the order of 1-2 times the height of the dam - both depthwise as also laterally - suffices to achieve a correct determination of the natural frequencies. On the other hand, owing to the absence of energy radiation, the damping offers lower values on the first case.

A further confirmation of the accuracy of this type of modelling was obtained when carrying out a comparative study for the Talvacchia Dam between prototype and elastic model, in which the foundation was reproduced with the criteria put forward above; as may be seen from Table 5, the comparison shows satisfactory agreement.

4.2. Dam-reservoir interaction phenomenon

The analysis of the available data was carried out with the intention of determining:

- the variation of the natural frequencies with the water level in the reservoir;
- the variation of the natural frequencies and of the modal shapes with the varying of the length of the simulated reservoir;
- the added mass relative to the first modes.

For what concerns the variation of the natural frequencies with the varying of the water level in the reservoir, the diagrams of Fig. 6 bring together the data relative to arch and arch-gravity dams. Although the cases studied are relative to dams having different geometric characteristics, nevertheless the following general conclusion may be drawn:

- the experimental values show similar behaviour for the two types of dams that have been considered - arch and arch-gravity; in both cases only the highest part of the water level - over 70% of the maximum - turns out to be significant; the maximum variations are in the order of 25%;
- the frequency variations are not identical

Table 2

COMPARISON BETWEEN EXPERIMENTAL MEAN VALUES (S) AND THEORETICAL VALUES (T) OBTAINED FROM THE RELATIONS:(*)

STRENGTH	STRAIN TO FAILURE	
$\sigma_{R_c} \simeq 1.2 \cdot E \cdot 10^{-3}$ (Kg/cm^2)	$\varepsilon_{R_c} \simeq 250 \cdot 10^{-5}$	Compressive monoaxial tests
$\sigma_{R_t} \simeq 0.1 \cdot \sigma_{R_c}$	$\varepsilon_{R_t} \simeq \sigma_{R_t}/E$	Tensile tests
$\sigma_{R_g} \simeq 1.5 \div 2.0 \cdot \sigma_{R_t}$	$\varepsilon_{R_g} \simeq \sigma_{R_g}/E$	Flexural tests

Ø aggregate (mm)	Elastic modulus E (Kg/cm^2)	S_E	Strength σ (Kg·cm^{-2}) Compression S	Compression T	Tension S	Tension T	Flexure S	Flexure T	Strain to failure $\varepsilon \cdot 10^5$ Compression S	Compression T	Tension S	Tension T	Flexure S	Flexure T
3	12,435	24	11.2	14.9	1.3	1.5	2.6	2.4	301	250	13.5	12.0	21.0	19.2
4	8,640	35	9.3	10.4	1.4	1.0	1.7	1.7	283	250	17.8	12.0	23.3	19.2
4.9	8,328	36	8.3	10.0	1.1	1.0	1.6	1.6	237	250	15.7	12.0	23.0	19.2
7.4	6,675	45	7.0	8.0	0.8	0.8	1.3	1.3	251	250	13.0	12.0	21.0	19.2

J. J. Woddel: "Concrete Construction Handbook" – Ed. Mc Graw – Hill

Table 3

VIBRATION TESTS ON CONCRETE DAMS

	DAM	TEST ON	h (m)	R(m)	C (m)	l (m)	t (m)	1° f(hz)	1° ξ(%)	2° f(hz)	2° ξ(%)	3° f(hz)	3° ξ(%)	4° f(hz)	4° ξ(%)	Reservoir
GRAVITY AND ARCH GRAVITY	Alpe Gera	prototype	174	–	–	528	5÷127	3,47^{S}	4,40	4,72^{A}	4,50	6,16^{S}	4,50	7,43^{A}	3,43	EMPTY
								3,25	5,40	4,56	5,12	–	–	–	–	0,78·H
	Amaluza	model	170	200 ÷ 289	345	410	6,5÷48,6	2,55^{S}	3,60	3,07^{A}	3,50	4,03^{S}	3,34	5,35^{A}	3,25	EMPTY
								1,96	4,00	2,38	3,69	3,18	3,50	4,86	3,35	FULL
	Fiastra	prototype	87	157	–	254	3,4÷31	4,72^{S}	3,27	5,97^{A}	2,46	7,87^{S}	2,38	9,72^{A}	2,50	0,88·H
								4,29	3,30	5,53	2,56	7,34	2,80	9,16	2,50	0,72·H
	Ridracoli	model	100	120	341	432	8,00÷30	2,19^{A}	3,00	2,75^{S}	4,00	3,74^{A}	3,00	4,01^{S}	3,00	EMPTY
								2,02	3,20	2,41	4,30	-	-	-	-	FULL
ARCH	Place Moulin	prototype	155	287	490	678	6,5÷42	2,032^{A}	1,18	2,033^{S}	1,15	2,96^{A}	1,22	3,63^{S}	1,20	0,95·H
	Llauset	model	77,5	87 ÷ 142,0	214	280	7,1÷24,4	3,42^{A}	1,60	4,65^{S}	2,64	5,06^{S}	2,60	6,70^{A}	2,16	EMPTY
								2,67	1,62	3,76	3,00	4,10	2,60	5,91	2,16	FULL
	Baserca	model	87,5	127,5 ÷ 235,5	260	320	7,2÷21,2	4,38^{S}	2,90	4,27^{A}	2,80	[illegible],83^{S}	1,70	7,83^{A}	1,70	EMPTY
								3,30	4,60	3,71	3,20	5,37	2,50	7,11	2,40	FULL
	Talvacchia	prototype	78	119	173	216	4,7÷16	3,68^{A}	3÷5	3,80^{S}	3÷5	5,35^{S}	3÷5	6,70^{A}	3÷5	0,9·H
		model						4,05	2,00	4,49	2,20	6,12	1,90	7,28	2,00	EMPTY
	Barcis	prototype	50	36	–	71	2,0÷4,5	7,60^{A}	7,00	10,10^{S}	4,00	15,30^{S}	4,00	16,30^{A}	3,50	FULL
	Ambiesta	prototype	59	74	117	145	2,1÷7,8	3,90^{A}	2,15	4,27^{S}	3,02	6,61^{A}	4,12	7,30^{S}	6,65	FULL
								4,11	2,00	4,70	1,90	7,12	4,00	-	-	0,94·H

h: maximum height — H: maximum water level — R: upstream crest radius — C: crest width — l: crest developped length — t: thickness — A: antisymmetric mode — S: symmetric mode

for all the vibration modes; the tendency of the higher modes to undergo minor variation - around 10% - would seem to be clear, although the available data do not suffice for the establishment of empirical laws;

- although the data are dispersed there is no disagreement the values obtained from tests on models and on prototypes.

The influence of the length of the reservoir was studied on the elastic model of the Talvacchia Dam by reproducing two different lengths of the basin: the first equal to 5 x H (with H = dam height), the second equal to 2 x H. In both cases, at the opposite ends of the dam the reservoir was closed by means of a wall which is "rigid" in the relevant frequency range. The natural frequencies, the modal shapes and the distribution of pressure do not show significant variations for the first four vibration modes. The damping determined turns out to be of the same order of magnitude in both test conditions.

Usually the effect of the reservoir is considered by having recourse to the concept of "added" mass, which deserves an in-depth look in the light of some experimental results. Recent experiences with models, carried out with the employment of multiple sinusoidal excitation which facilitated in making the structure vibrate according to an almost pure mode, have made it possible to measure the hydrodynamic pressure for a number of modes - amplitude and phase shift of the pressure relative to the acceleration of the point under examination. From the analysis of such data it is found that only in the case of the first modes may the hydrodynamic pressure be treated as an "inertial force" associated with an added mass; in fact, in this case the pressure and acceleration of the points considered are essentially in phase. As against this, for the higher modes there is a greater influence of the quadrature component of the pressure. Valued by utilizing the real components of the pressures measured, the added mass normally turns out to be different from mode to mode, as confirmed by the different frequency variations.

The foregoing leads to the conclusion that the use of the equivalent mass in the study of a dam in full reservoir conditions is a valid approach only in first approximation and is at any rate acceptable, if the added mass is obtained starting from the effective distribution of the hydrodynamic pressures.

5. FUTURE DEVELOPMENT OF THE EXPERIMENTAL TECHNIQUES

In view of what has been stated above, it would seem to be clear that, while responding in a quite satisfactory manner to the numerous problems posed by seismic verification, the modelling techniques in use suffer from two main limitations:

- the hypothesis, in the case of elastic models, that the stress level in the heaviest load conditions should remain below the elastic limit of the materials;

- the impossibility, where failure models are concerned, of reproducing in complete similitude the acting loads, the behaviour of the materials

Table 4

TEST CONDITION	MODE							
	1° S		1° A		2° S		2° A	
	f (Hz)	ζ %	f (Hz)	ζ %	f (Hz)	ζ %	f (Hz)	ζ %
1	2.52	2.05	3.00	1.56	3.95	1.89	5.35	1.95
2	2.55	3.60	3.07	3.50	4.03	3.34	5.35	3.25

Table 5

TALVACCHIA DAM	RESERVOIR	NATURAL FREQUENCIES						
		1° A	1° S	2° S	2° A	3° S	3° A	4° S
Prototype	Near full	3.68	3.80	5.35	6.70	8.35	-	-
Model	Full (L = 5 H)	3.53	3.69	5.27	6.56	8.66	10.07	12.97
	Full (L = 2 H)	3.57	3.72	5.28	6.60	-	-	-
	Empty	4.05	4.49	6.12	7.28	9.57	11.61	14.43

and the discontinuities present in the dam as also in the foundation - joints and fault surfaces.

In order to overcome these limitations, research is currently in progress at ISMES with the aim of putting together a more accurate modelling technique, that would facilitate the total and simultaneous simulation of all the relevant phenomena.

The goal aimed at is that of abandoning the shake table as an exciting system and of simulating the quake as a travelling wave in the base of the model. It is thought that this would open the way for operating on a geometric scale in the order of 80-100: this scale would make it possible to repeat on the model, in a more faithful manner, the construction techniques of the prototype - joints, grouting conditions etc. - and to reproduce a substantial extension of the foundation; further, the materials already available for failure model, may be considered adequate, although there is room for improvement for what concerns the constancy and the reproducibility of the mechanical characteristics, by replacing the plaster with a concrete.

The fundamental problem becomes therefore that of the "excitation". The research programme in hand is looking into various solutions, in particular:

- Employment of electrohydraulic and electrodynamic exciters

These exciters simplify and give accuracy to the control of a wave with definite characteristics, but they are unable to supply sufficient energy to lead to the failure of the model.

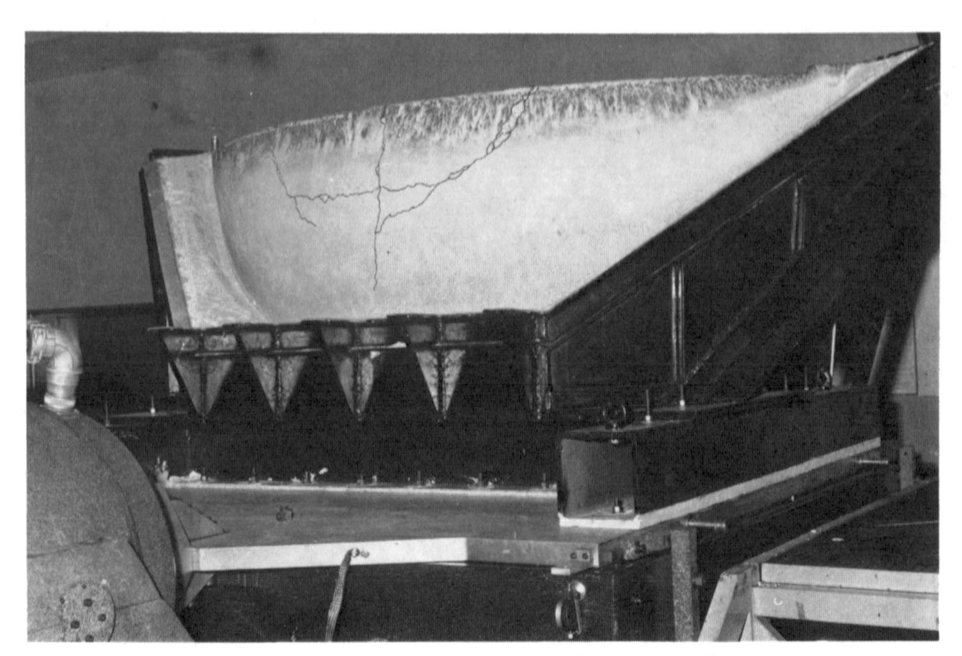

Fig. 5. Ridracoli Dam. Failure model.

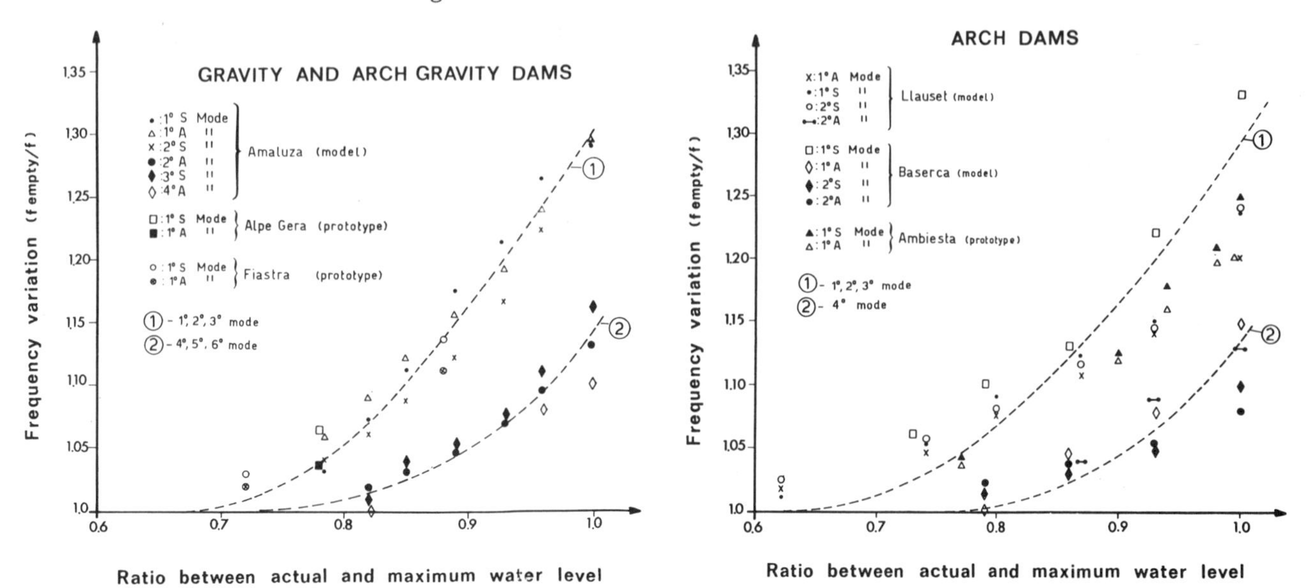

Fig. 6. Ratios between natural frequencies at empty reservoir and natural frequencies at different water levels.

Use of explosive charges

In this way it is possible to obtain the desired intensity of the motion and its energy distribution in the frequency range. Currently, the possibility of using several micro-delayed charges is being researched, the aim being to control, at least in first approximation, the duration of the "quake".

On the basis of theoretical considerations as well as drawing from experience, the reflection of the waves in the boundary area of the model represents the greatest obstacle in the setting up of the new technique. It is thought that one possible solution to the problem - for which an experimental verification is foreseen - consists in sinking the model and the foundation in ground with suitable elastic properties so as to make the energy radiation efficient and to include a strong dissipation via internal damping of the materials.

Although the setting up of the new dynamic modelling technique presents numerous difficulties it is thought possible on the basis of the experience in progress to reach a satisfactory solution. The authors believe that in the near future a failure model will be available, which is capable of better representing the complex phenomena met with in the dam-foundation-reservoir system during a quake.

REFERENCES

1. ISMES Report no. 1370. Dynamic tests on Passante Dam. July 1978.

2. ISMES Report no. 1384. Dynamic tests on Amaluza Dam model. October 1978.

3. OBERTI G., CASTOLDI A., MAZZIERI C. Analysis by means of physical models of the seismic behaviour of concrete arch dams. 5th WCEE, Istanbul, 1975.

4. SEVERN R.T., TAYLOR P.R. Earthquake effects on arch dam by response spectra methods. Symp. on Vibs. in Civil Eng. - I.C.E., April 1965.

5. CALCIATI F., CASTOLDI A., CIACCI R., FANELLI M. Experience gained during in situ artificial and natural dynamic excitation of large concrete dams in Italy: Analytical interpretation of results. 13th International Congress on Large Dams. Q.51 - R.32.

26 Theoretical assessment of the behaviour of arch dams for seismic loading

G. L. HUTCHINSON, DipCE, MEngSc, DPhil, MICE, and T. G. TSICNIAS, MSc (NTUA), King's College, London

This paper considers the main problems encountered in the design of arch dams against earthquakes and places an emphasis on the selection of appropriate design earthquake inputs. Results from earthquake dynamic analyses of two configurations are presented. It is concluded that it is necessary to use a particular spectral level accounting for vertical ground motion.

INTRODUCTION

1. Past behaviour of arch dams under earthquake loading has been seen to be generally satisfactory. For example at Lower Crystal Springs, a curved gravity dam at height 154 ft located 0.25 miles from the San Andrean fault resisted with no apparent damage the San Francisco earthquake of 1906 (magnitude 8.3). Similarly, Kariba arch dam with a height of 420 ft resisted, with no apparent damage an earthquake of magnitude 6.1 in 1963.

2. In 1971 the San Fernando earthquake of magnitude 6.6 caused no apparent damage to the Santa Anita and Big Tujunga arch dams, both of height 249 ft and located 17 and 20 miles respectively from the fault. The Pacoima arch dam, located only three miles from the fault suffered no visible damage to the arch structure itself but the joint between the arch and the thrust block opened.

3. Three arch dams in Italy, with heights varying from 164 to 446 ft experienced with no damage, horizontal ground accelerations of 0.33g in an earthquake in 1976 which had a magnitude of 6.5. Ref.7 contains further details about the behaviour of a number of actual arch dams subject to various earthquakes.

4. Although the record of the behaviour of arch dams in resisting earthquakes around the world has been satisfactory the actual safety factor of such dams against dynamic loading remains uncertain. Unlike in pseudo-static earthquake analyses (ref.2) a set of design criteria do not exist for the evaluation of the dynamic stability of arch dams.

5. This paper briefly considers the main problems encountered in the design of arch dams with an emphasis on the selection of appropriate design earthquake inputs. Particular consideration is given to the effect on the earthquake induced stresses in the concrete structure induced by applying either a common input for the three components of the earthquake ground motion or a common input for the two horizontal components and a distinct input for the vertical component. A comparative study of three different structural types of high arch dams has been undertaken to demonstrate the significance of this difference.

The interactive subsystems

6. The response of an arch dam to an earthquake ground motion results from an interaction between three sub-systems; namely the arch structure, the reservoir and the underlying rock medium. (fig.1).

7.
Well known finite element analyses are available to evaluate the interaction between the underlying rock medium and the arch structure. These analyses consider a large portion of the foundation rock to be elastic.

8. Compared with gravity dams there has been a little progress on the evaluation of the hydrodynamic pressures (ref.5) resulting from the arch dam/reservoir interaction. The reason for this appears to be the three dimensional form of the interaction as well as the complexity of the contact surface (actual arch dams usually have a double curvature upstream face). Theoretical studies (ref.12) have considered simple arch dam models of cylindrical shape in a valley with vertical walls. Results have indicated that the dynamic response of the arch dam is increased two to five times with a full reservoir. However, because these models do not accurately simulate actual arch dams the conclusions of these studies cannot be incoporated in earthquake resistant design. Moreover, there is not yet wide agreement (ref.13) about the controlling parameters of this interaction (e.g. water compressibility). Furthermore, the effect of the vertical ground motion, on the hydrodynamic pressures,seen to be significant in gravity dams (ref.3), and the significance of the interaction between the reservoir and the underlying rock remain uncertain.

Methods of analysis against earthquake

9. Two methods are available for the analysis of arch dams: (a) the trial load method,
(b) the finite element method.

(a) Trial load method (ref.2).

10. Using this method the arch dam is idealized as a set of horizontal arches and vertical cantilevers. The distribution of the loads between the arches and cantilevers is such that compatability at the intersection points is satified.

The inertia forces and the hydrodynamic pressures resulting from the earthquake ground motion are accounted for by pseudo-static horizontal forces applied at the intersection points. The inertia forces are estimated by multiplying the nodal masses by the expected maximum ground acceleration which is considered to be uniform throughout the height of the dam. i.e. the dam is assumed to respond to the ground motion as a rigid body.

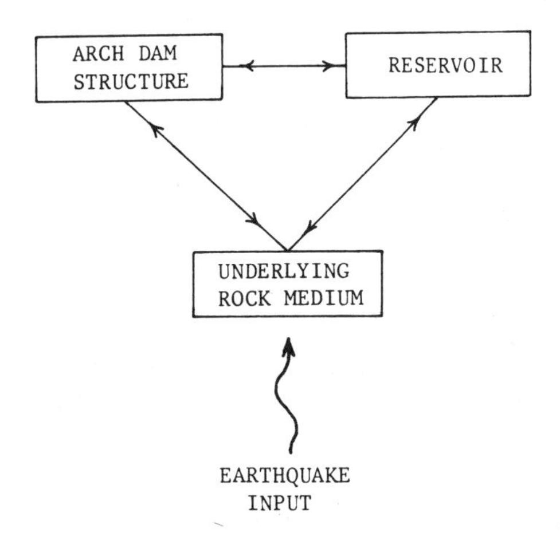

Fig.1. Interactive Sub-systems

An estimate of the hydrodynamic pressures is obtained by using Westergaard's (ref.14), "added mass" approach. This involves considering a volume of water moving with the dam. The equivalent static force resulting from this motion (i.e. the product of the added mass and the seismic coefficient) is applied at the upstream face of the dam. This procedure can only take into account the hydrodynamic pressures due to a ground motion in the upstream/ downstream direction and to allow for the curvature of the dam the added volume of the water for each cantilever section is multiplied by the cosine of the angle to the reference plane (Fig.2). However, Westergaard's assumptions of an infinitely long, rigid structure with a vertical upstream face are hardly satisfied in the case of an arch dam.

(b) Finite element method (ref.6).

11. The arch dam structure and a portion of the foundation are discretised by a finite element mesh. The earthquake input is applied as a three dimensional rigid base translation at the boundary of the elastic portion of the foundation. The input may be in the form of a ground acceleration time history or a response spectrum. The hydrodynamic pressures are accounted for by distributing the Westergaard "added mass" on the nodal points of the upstream face of the dam.

12. The difference between the pseudo-static and the dynamic earthquake analysis of a high arch dam (660 ft) was presented in ref.9. It is made clear from the comparison of the response stresses that the assumption of a rigid arch structure results in a totally unrealistic stress distribution not only with respect to the magnitude but also with respect to the sign of the stresses. The difference appeared to be most marked in the arch stresses at the top elevation.

13. For a system with n degrees of freedom the equation of motion relative to the rigid base is

$$[M]\{\ddot{V}\} + [C]\{\dot{V}\} + [K]\{V\} = -[M][r]\{\ddot{V}_g\} \quad (1)$$

where

$[M]$: the (nxn) matrix of nodal masses
$[C]$: the (nxn) matrix of damping coefficients
$[K]$: the (nxn) stiffness matrix
$\{\ddot{V}\},\{\dot{V}\},\{V\}$: the (nx1) nodal acceleration, velocity and displacement vectors of the system respectively
$\{\ddot{V}_g\}$: the (3x1) vector of the ground motion components $\ddot{V}_{xg}$, $\ddot{V}_{yg}$, $\ddot{V}_{zg}$
$[r]$: the (nx3) matrix of pseudo-static influence coefficients where columns 1, 2 and 3 represent the nodal displacements due to unit rigid base displacement in the x, y and z directions respectively.

14. The solution of equation (1) may be obtained by a direct integration scheme, however, the large number of degrees of freedom in an arch dam/foundation system may decrease the efficiency of this method.

15. In the earthquake response of an arch dam/ foundation system the major contribution is due to the lowest few normal modes and so in the case of damping orthogonality the normal mode technique is more efficient. In a normal mode analysis the major computational effort is in the calculation of the natural frequencies and associated mode shapes. The undamped free vibration solution of equation (1) yields the natural frequencies and mode shapes of the system and powerful eigen-solution techniques (ref.1) may be used to obtain the first k required modes. The eigenvalue problem is defined by

$$[K][\Phi] = [M][\Phi][\Omega] \quad (2)$$

where

$[\Phi]$: the (nxn) matrix of mode shapes
$[\Omega]$: the nxn) diagonal matrix of squared natural frequencies.

Using the normal node transformation

$$\{V\} = [\Phi]\ \{Y\} \quad (3)$$

yields the uncoupled equation of motion

$$\ddot{Y}_m + 2\xi_m \omega_m \dot{Y}_m + \omega_m^2 Y_m = P_m \quad \text{for } m = 1,2..k \quad (4)$$

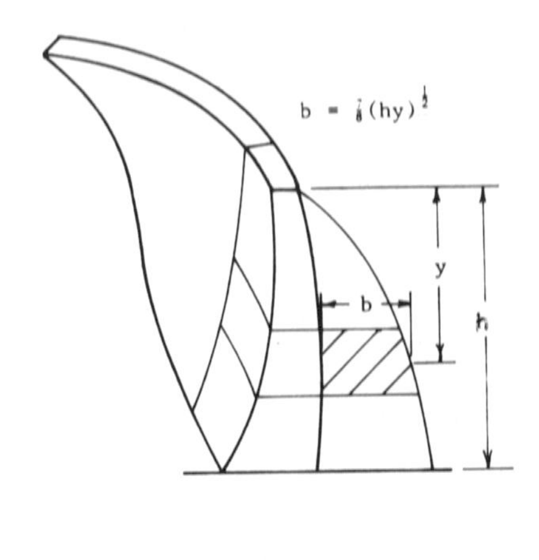

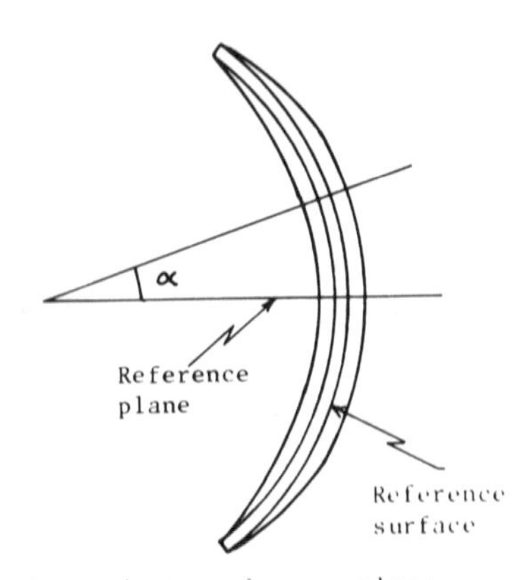

α = angle to reference plane

Plan view of typical symmetric arch dam.

Fig. 2. Schematic diagram of water mass assumed accelerated with dam

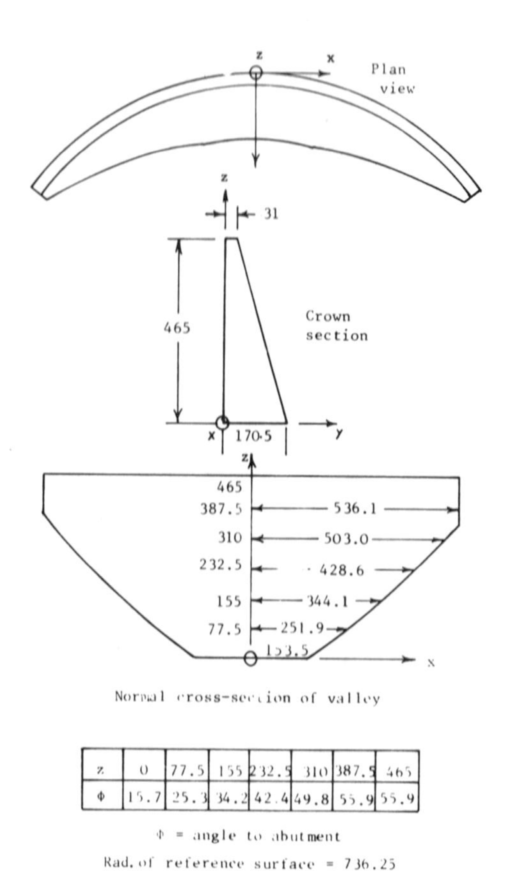

z	0	77.5	155	232.5	310	387.5	465
Φ	15.7	25.3	34.2	42.4	49.8	55.9	55.9

Φ = angle to abutment

Rad. of reference surface = 736.25

Fig. 3. Dam Case (a)

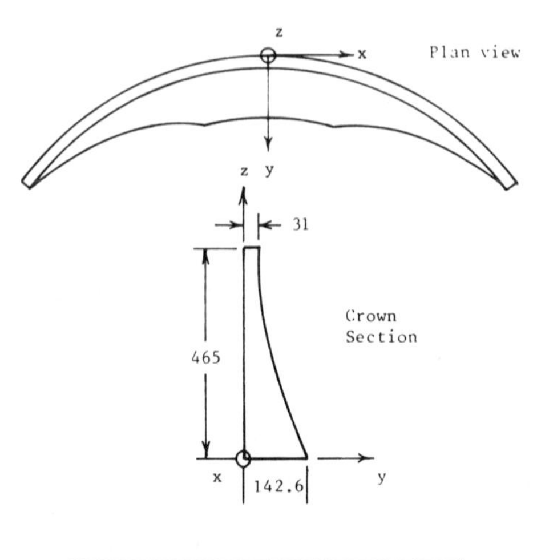

z	0	77.5	155	232.5	310	387.5	465
R.I	565.8	599.8	627.7	649.4	664.9	674.2	677.3

R.I. = radius of intrados

Rad. of reference surface = 708.35

Valley cross-section as in Fig.3.

Fig. 4. Dam Case (b)

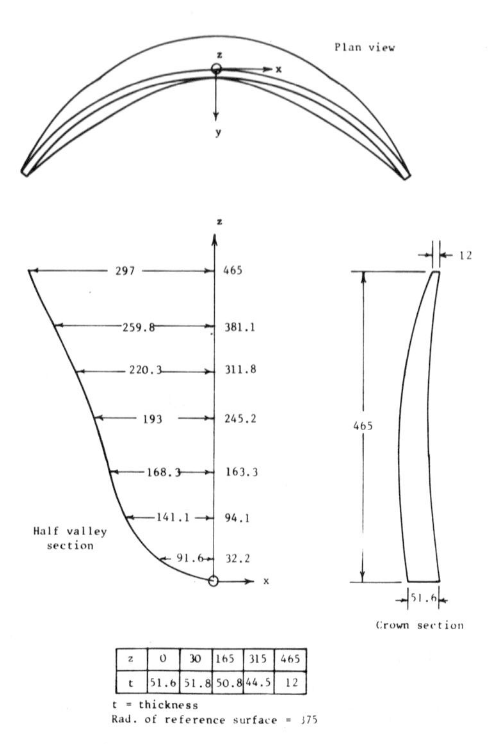

z	0	30	165	315	465
t	51.6	51.8	50.8	44.5	12

t = thickness

Rad. of reference surface = 375

Fig. 5. Dam Case (c)

where

Y_m : the generalised co-ordinate for mode m

ω_m : natural frequency of mode m

ξ_m : percentage of critical damping for mode m

P_m : modal loading defined by

$$P_m = \{\Phi_m\}^T [M][r]\{\ddot{V}_g\} \quad (5)$$

16. For the case of a ground acceleration time-history input the solution of equation (4) is obtained from the Duhamel integral and substitution in equation (3) yields the response displacements.

17. When the input is represented by acceleration spectra, the maximum value of the generalised co-ordinate due to each component of ground motion is calculated seperately.

$$\max Y_m^x = \frac{L_x S_a^x(\omega_m, \xi_m)}{\omega_m^2} \quad (6)$$

$$\max Y_m^y = \frac{L_y S_a^y(\omega_m, \xi_m)}{\omega_m^2} \quad (7)$$

$$\max Y_m^z = \frac{L_z S_a^z(\omega_m, \xi_m)}{\omega_m^2} \quad (8)$$

$$\text{where } L_x = -\{\Phi_m\}^T [M][r]\begin{Bmatrix}\ddot{V}_{xg}\\0\\0\end{Bmatrix} \quad (9)$$

$$L_y = -\{\Phi_m\}^T [M][r]\begin{Bmatrix}0\\\ddot{V}_{yg}\\0\end{Bmatrix} \quad (10)$$

$$L_z = -\{\Phi_m\}^T [M][r]\begin{Bmatrix}0\\0\\\ddot{V}_{zg}\end{Bmatrix} \quad (11)$$

S_a^x, S_a^y, S_a^z : the acceleration spectra for the x, y and z components of ground motion.

The maximum modal displacements are obtained from

$$\max \{V_n\}^{x,y,z} = \{\Phi_m\} \max Y_m^{x,y,z} \quad (12)$$

The maximum response values corresponding to any degree of freedom are estimated by taking the square root of the sum of the squares of the maximum modal responses due to all three components of ground motion (the square root of 3k values).

Selection of the design earthquake

18. The design earthquake is defined as the vibratory ground motion to be used for the aseismic design of the dam structure. It is often considered (ref.10) in accordance with the effect it may have on the dam and consists of two intensity levels.

(a) the dam must survive the largest earthquake which reasonably could occur at the site (maximum credible earthquake) without collapse although considerable structural damage may occur,

(b) the dam should suffer no structural damage for the earthquake which might be expected to occur at the site during the lifetime of the dam (maximum expectable earthquake).

It is generally recognised that the design ground shaking for a particular site depends on:

(i) the magnitude of the earthquake
(ii) the source mechanism
(iii) the paths of the seismic waves
(iv) the distance from the earthquake source
(v) the local geology and the site soil characteristics.

19. After estimates of these five factors have been made a number of approaches may be used for the generation of artificial acceleration time-histories,

(i) generation of accelerograms having a given response spectrum.
(ii) combination or modification of actual recorded accelerograms.
(iii) generation of random noise inputs.

20. It should be noted that the maximum value of the generalised co-ordinate in equation (3) is very sensitive to small variations in the natural frequency. This is clearly demonstrated by the abrupt changes in the response spectrum of any historic earthquake. Since the natural frequencies of arch dam/foundation systems cannot be computed accurately (because of the lack of knowledge of the hydrodynamic effect and the uncertainty of the foundation properties) it is necessary to assume a range of values of frequencies in the vicinity of the estimated frequency and this substantially complicates the design procedure.

21. An attractive alternative to the use of an acceleration time-history input is the smoothed elastic response spectrum. This provides a quantitative description of both the intensity and the frequency content of a ground motion. A significant deficiency of the response spectrum is that it does not provide a clear indication of the duration of the ground motion which is a parameter that can affect the elastic behaviour of the structure. The selected spectral curves should correspond to soil conditions similar to those at the arch dam site.

22. Since arch dams, in contrast to concrete gravity dams, behave as three dimensional structures, two spectral levels should be used: one for the two horizontal ground motion components and one for the vertical component. The use of a common design spectrum curve for both of the horizontal ground motion components is to account for their equal probability of occurance.

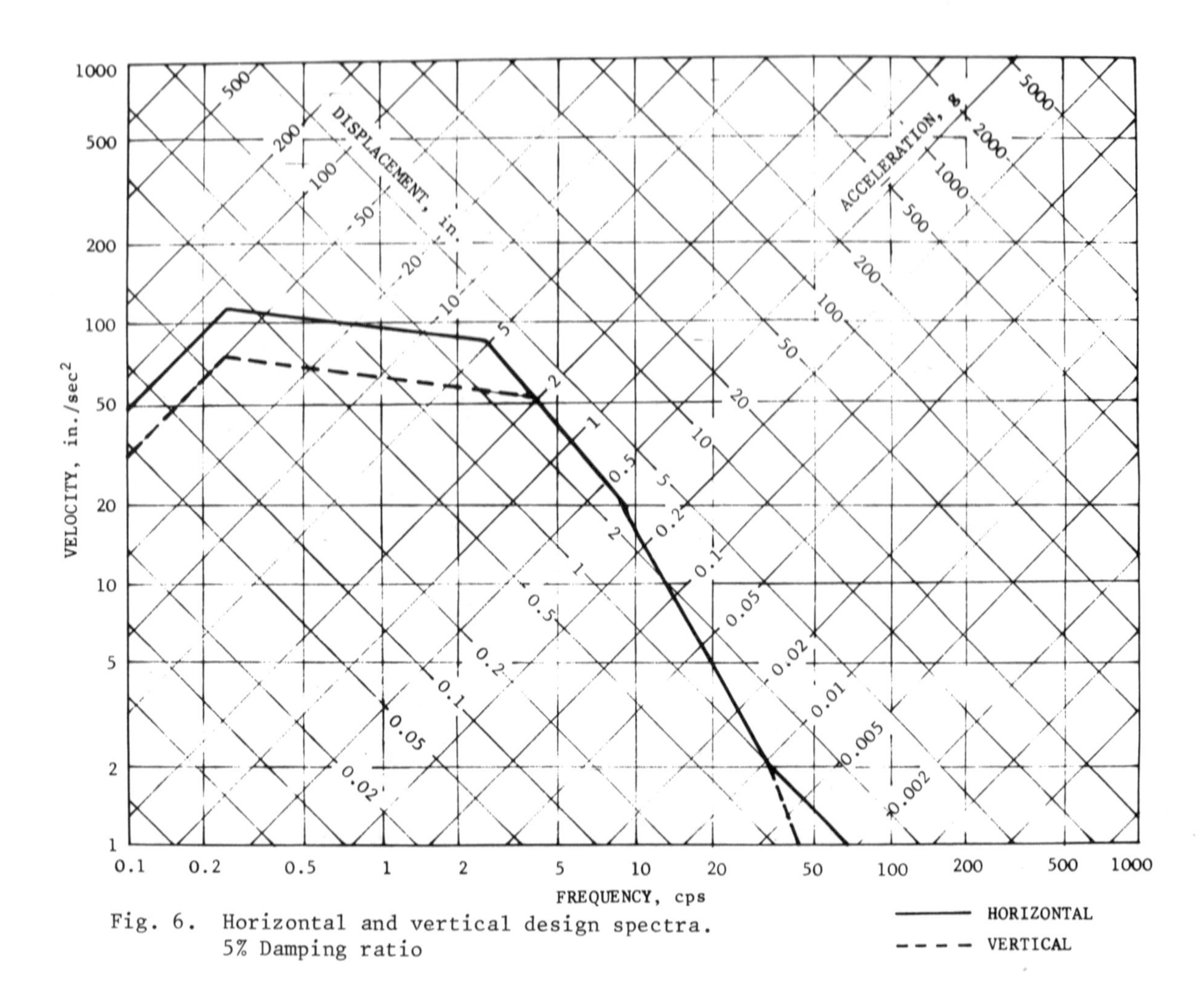

Fig. 6. Horizontal and vertical design spectra.
5% Damping ratio

TABLE 1 Stresses in right hand portion - Case (a)

UPSTREAM FACE

Level	β	Cantilever Stress		Arch Stress		Shear Stress		Max. Stress		Min. Stress		Angle α	
		InA	InB	InA	InB	InA	InB	InA	InB	InA	InB	InA	InB
406.875	22.79	64	62	292	253	36	33	298	258	59	57	81.21	80.50
	36.12	54	50	327	281	60	56	340	294	41	37	78.01	77.07
	47.71	56	51	276	243	32	31	281	248	52	47	81.86	80.92
	54.89	59	52	201	178	127	112	276	244	-15	-13	59.61	59.65
290.625	22.79	116	111	189	164	50	45	215	190	91	85	62.81	59.96
	36.12	126	115	155	132	69	63	211	187	70	60	50.86	48.87
	45.06	156	137	150	130	128	117	281	251	25	16	44.29	44.24
174.375	22.79	262	237	94	85	37	24	270	244	85	75	12.09	12.02
	32.98	217	190	196	175	157	107	326	290	85	74	42.45	43.02
58.125	19.26	324	285	294	263	253	223	562	498	56	50	43.27	43.63
DOWNSTREAM FACE													
406.875	22.79	59	53	277	244	59	54	292	259	43	39	75.75	75.38
	36.12	39	38	168	144	110	99	232	204	-24	-21	60.10	59.00
	47.71	35	31	309	268	114	99	230	304	- 6	- 5	70.10	69.98
	54.89	70	61	309	269	21	62	329	287	50	43	74.63	74.55
290.625	22.79	69	65[5]	257	231	77	70	285	257	42	39	70.24	70.00
	36.12	99	89	82	69	157	139	248	218	-67	-60	43.46	42.94
	45.06	147	128	327	283	118	103	385	334	89	77	63.66	63.46
174.375	22.79	109	94	253	225	103	90	307	271	56	48	62.47	62.86
	32.98	162	146	207	177	175	152	361	314	8	8	48.63	47.91
58.125	19.26	168	143	154	133	146	121	308	266	14	10	43.67	43.88

TABLE 2 Stresses in right hand portion - Case (b)

UPSTREAM FACE

Level	β	Cantilever Stress		Arch Stress		Shear Stress		Max. Stress		Min. Stress		Angle α	
		InA	InB	InA	InB	InA	InB	InA	InB	InA	InB	InA	InB
406.875	22.41	68	66	337	290	75	74	357	313	48	44	75.40	73.29
	75.19	66	60	460	423	104	101	486	449	40	34	76.01	75.79
	47.15	98	94	336	327	36	34	341	332	93	84	81.58	81.87
	56.01	114	103	157	129	111	108	249	225	22	6	50.51	48.42
290.625	22.41	106	103	204	177	76	74	246	223	64	57	61.32	58.17
	35.19	172	164	194	182	89	86	273	261	93	86	48.57	48.01
	44.22	244	224	108	94	160	156	350	328	2	-10	33.50	33.69
174.375	22.41	243	220	89	79	48	47	257	234	75	64	16.17	16.99
	32.14	245	220	192	183	93	87	316	291	122	112	37.12	39.15
58.125	19.07	290	225	241	216	202	179	469	416	62	55	41.66	41.89
DOWNSREAM FACE													
406.875	22.41	110	104	433	421	95	93	459	447	84	79	74.85	74.78
	35.19	56	49	188	160	151	144	287	258	-42	-49	56.80	55.51
	47.15	35	34	356	294	107	93	388	323	3	4	73.19	72.14
	56.01	75	64	333	304	76	60	354	319	52	50	74.79	76.61
290.625	22.41	107	101	260	254	105	102	314	305	53	49	62.99	63.38
	35.19	104	102	125	112	161	147	276	255	-46	-39	46.95	46.03
	44.22	155	137	325	281	151	145	421	371	69	47	60.42	58.15
174.375	22.41	90	77	234	215	105	94	289	263	34	29	62.19	63.03
	32.14	131	120	222	192	169	146	351	307	2	5	52.44	51.88
58.125	19.02	124	104	143	129	159	137	294	255	-25	22	46.69	47.61

TABLE 3 Stresses in right hand portion - Case (c)

UPSTREAM FACE

Level	β	Cantilever Stress		Arch Stress		Shear Stress		Max. Stress		Min. Stress		Angle α	
		InA	InB	InA	InB	InA	InB	InA	InB	InA	InB	InA	InB
390	8.31	53	43	454	399	54	48	461	406	46	37	82.43	82.32
	24.42	72	59	210	186	107	97	269	239	13	6	61.26	61.53
	36.52	125	104	232	218	40	36	246	229	112	93	71.64	73.55
	44.31	142	113	210	191	76	68	260	231	92	24	56.90	59.79
240	8.31	82	45	351	303	53	47	360	312	43	36	80.21	29.90
	24.42	84	67	115	105	87	76	188	165	11	8	50.04	51.99
	34.46	167	137	360	335	52	27	373	346	154	126	75.81	77.10
97	8.31	167	135	128	110	32	27	185	153	110	92	29.65	33.02
	20.52	167	134	182	166	94	74	269	226	80	73	47.20	51.02
15	4.15	177	144	508	414	181	144	588	477	97	81	66.21	66.49
DOWNSTREAM FACE													
390	8.31	38	30	268	241	55	49	280	252	25	19	77.20	77.42
	24.42	77	72	266	240	124	110	328	294	15	17	63.61	63.71
	36.50	81	73	560	504	120	101	589	527	53	51	76.69	77.42
	44.31	54	44	335	298	84	72	358	317	31	25	74.53	75.24
240	8.31	87	68	383	340	75	66	401	350	69	53	76.51	77.02
	24.42	69	63	151	135	192	167	307	270	-85	-71	50.98	51.08
	34.46	109	85	481	430	157	132	539	475	51	40	69.84	71.24
97	8.31	105	82	323	286	86	76	351	312	74	57	70.71	71.63
	20.52	158	123	133	120	152	135	298	257	6	-13	42.70	44.66
15	4.15	267	230	180	147	121	114	353	310	94	67	35.16	34.96

23. The recorded vertical ground accelerations of past earthquakes (ref.8) compared with the associated horizontal ground accelerations have a smaller amplitude (50 to 70%) and a relatively increased number of zero crossings. Chopra, (ref.4) investigating the response characteristics of vertical ground motions, concluded that the spectrum intensity of the vertical components for 20% critical damping is about 20 or 30 percent of that of the horizontal components. Also the spectra for vertical components are relatively accentuated in the shorter periods and reduced in the longer periods. A dependence of the spectrum intensity of the vertical component on the direction of slip on the fault was noted.

24. Design spectra for vertical ground motions may be developed using two procedures:
(a) by an appropriate scaling of the design spectrum of the horizontal components,
(b) by processing the response spectra of vertical ground motion records in a manner similar to that used for the horizontal ground motions.
This latter process involves appropriately normalising, averaging and smoothing the historic response spectra.

Structural significance of different horizontal and vertical design spectra.

25. In order to establish design criteria concerning the seismic stability of arch dams it is valuable to study the effect of the earthquake loading on the stresses in the dam. The significance, from a structural point of view, of using two spectral levels for the three dimensional earthquake input will be considered in this section.

26. For the purposes of this study two structural configurations of arch dams of height 465 ft. were analysed. The characteristics of these dams are given in the appropriate figures.
Configuration 1. Case (a) Fig.3.
Case (b) Fig.4.
Configuration 2. Case (c) Fig.5.

27. The design spectra used for the analyses are those presented by Newmark, Blume and Kapur (ref.11). They consist of two spectral levels; one for the horizontal ground motion components and one for the vertical component (see fig.6). The maximum vertical ground acceleration is taken to be approximately two thirds the horizontal ground acceleration and the viscous damping ratio was taken as 5 percent of the critical. Equal modal damping was assumed for all the modes.

28. For each arch dam case two inputs were considered,
1. the horizontal spectral level was taken for all three directions (Input A),
2. the horizontal spectral level was applied in the horizontal direction and the vertical spectral level in the vertical direction (Input B).

29. The analyses were conducted using the Arch Dam Analysis Program (A.D.A.P.) (ref.6) which is a finite element program for arch dam/foundation analysis

Results and Conclusions

30. The results of the analyses are presented in Tables 1, 2 and 3 which refer to dam cases (a), (b) and (c) respectively. Stresses are given in lbf/in^2, compression positive and In.A, In.B refer to ground motion Input A and Input B respectively. A study of the results indicates that:
(i) as expected in all cases, using a particular design spectrum for vertical ground motion results in lower earthquake stresses at all points in the dam structure.

(ii) In case (c) the difference in earthquake stresses produced by the two inputs is significantly greater than for the other two cases. This is a consequence of the vertical curvature of the dam structure.

(iii) For case (c) the difference is most significant around the mid-height elevation of the crown cantilever.

31. When the combined static and dynamic stresses are considered, the percentage difference resulting from the use of Input A and Input B is obviously reduced. However, for the evaluation of the dynamic stability of a particular arch dam it is clearly necessary to use a spectral level consistent with the actual characteristics of the vertical ground motion.

Unit Conversion Table

1 ft	=	0.3048m
1 mile	=	1609.34m
1 lbf/in^2	=	6.894×10^3 N/m^2
1 lb/ft^2	=	16.0185 kg/m^3

REFERENCES

1. BATHE K.J. Solution methods for large generalized eigenvalue problems in structural engineering. Structural Engineering Laboratory Report No.UCSEM 71-20, University of California Berkeley, 1971.
2. BUREAU OF RECLAMATION. Design of Arch Dams 1976.
3. CHAKRABATI P. and CHOPRA A.K. Hydrodynamic pressures and response of gravity dams to vertical earthquake component. Int.J.of Earthquake Engg. and Struct. Dynamics, No.4, 1973, 1, 315-335.
4. CHOPRA A.K. The importance of the vertical component of earthquake motions. Bulletin of the Seismological Society of America, No.5, 1966, 56, 1163-1175.
5. CLOUGH R.W. and CHOPRA A.K. Earthquake response analysis of concrete dams. Structural and Geotechnical Mechanics, Hall W.J.(Ed.) Prentice-Hall Inc., Englewood Cliffs, N.J. 1977.
6. CLOUGH R.W., RAPHAEL J.M. and MOJTAHEDI S.

ADAP - A computer program for static and dynamic analysis of arch dams. Earthquake Engineering Research Centre, Report No.EERC 73-14, Univ. of California, Berkeley, June 1973.
7. HANSEN K.D. and ROEHM L.H. The response of concrete dams to earthquakes. Water Power and Dam Construction, 1979, April, 27-31.
8. HUDSON E.D. ET AL. Strong motion earthquake accelerograms, Volume II. Earthquake Engineering Research Laboratory. Report No. EERL 71-50, California Institute of Technology, Pasadena, 1971.
9. KOLLGAARD E.B. and SHARMA R.P. Seismic stability evaluation of existing concrete dams. The evaluation of dam safety, American Society of Civil Engineers, 1977, 343-376.
10. LINDVALL G.E. Developing design earthquakes The evaluation of dam safety, American Society of Civil Engineers, 1977, 329-338.
11. NEWMARK N.M., BLUME J.A. and KAPUR K.K. Seismic design spectra for nuclear power plants. Journal of the Power Divison, ASCE, PO2, 1973, 99, November 287-303.
12. PERUMALSWAMI P.R. and KAR L. Earthquake behaviour of arch dam reservoir systems. Proc. of the Fifth World Conference on Earthquake Engineering, 1, 1973, Rome, 977-993.
13. SEVERN R.T. The aseismic design of concrete dams. Water Power and Dam Construction, No.2, 1976, 28, February, 41-46.
14. WESTERGAARD H.M. Water pressures on dams during earthquakes. Transactions, American Society of Civil Engineers, 1933, 98.

27

Criteria for the earthquake resistant design of concrete dams

J. LAGINHA SERAFIM, FICE, FASCE, University of Coimbra, and President of COBA, International Consultants, Lisbon

Foreword: When alerted to this meeting, the writer felt he could write a thought provoking paper. Now he must apologize for his incomplete knowledge given the great importance of dams to mankind of today.

1. Statistical Notes on Effects of Earthquakes on Dams

Dams, in general, when compared to other structures, resist well to earthquakes. However, while references indicate (Ref.1,2) that a few earth dams, most of them old or poorly built, were destroyed and many (around one hundred) significantly damaged by earthquakes, there was no known total failure of a concrete or masonry dam, old or new, well or poorly built, due to this cause. The only concrete dam significantly cracked in a localised zone but not destroyed by such an event was Koyna, in India (1967). This gravity structure, with a very inconvenient profile, was subject to a nearby strong earthquake (M=6.5, maximum horizontal components of the accelerations at the site 0.63g and 0.49g, and 0.34g in vertical). Minor incidents (one or few cracks or opening of joints) took place in some concrete or masonry dams, namely Hsinfenkiang, of buttress type in China (1962), Honen-ike, multiple arch in Japan (1964), Pacoima, arch in California (1971), Ponteba, old gravity masonry in Algeria (1955), and Blackbrook, old gravity masonry in England (1957) (Ref.3).

The Preliminary Work of October 1979 of the ICOLD Committee on Deterioration of Dams and Reservoirs has indicated 15 cases of deterioration of earth and rockfill dams due to earthquakes and vibrations and only 4 cases of concrete and masonry dams.

A very large number of references could be presented on concrete dams that have been subject to earthquakes (some of these very strong) and suffered no damage at all: Lower Cristal Springs, very close to Sant Andreas fault, Gibraltar, Hoover, Big Tujunga and Santa Anita, arches in the U.S.A.; Kamishiba, Ayakita, Kurobe, arches, Ohashi and forty others, gravity in Japan; Makta, Grankarevo, arch and others in Yugoslavia; Monteynard, arch, and Grandval, multiple arch, in France; Ambiesta, Mauna de Sauris, Barcis, Lumiei, Piave di Cadore, ValGallina, Vajont, arches in Italy; Canelles, arch, Camarillas and El Grado, gravity in Spain; Kariba, arch in Rhodesia; Vidraru, Negovanu, Tarnita and Paltinul, arches, Isvorul and Iron Gates, gravity, Poiana and Strimtory, buttresses, in Rumania; all Swiss dams in the Valais (1945); 4 concrete gravity dams in New Zealand; others in China and various elsewhere. The last international meetings on the subject (ICOLD Congresses included) dealt with various observations of harmless responses. But 2 cases of exceptional shocks against arch dams seem worthwhile to refer. The one of Pacoima, built in 1929, which was subject to the most severe accelerations recorded (1.2g horizontal and 0.7g vertical) during the 1971 San Fernando earthquake, only suffering the opening of a joint between the arch and the left thrust block. Some references also indicate that a slight crack appeared downstream in the arch in the same zone. The other case is that of Vajont arch dam that suffered the tremendous impact of an asymmetrical wave of water, soil and rock above 200m high in the right bank without losing its structural integrity.

Past studies in Japan indicated that "before modern construction technique was developed, earth dams were damaged by earthquakes of an intensity of over IV" ... "As for concrete gravity dam construction in the same era, the maximum earthquake intensity experienced was VI and no serious damage occurred" (Ref.4).

Although it must be recognized the very good endurance of concrete dams in relation to seismic action, due to the fact that their safety is a basic consideration, every dam must be designed for the geophysically possible earthquakes at its site.

2. Recurrence Periods of Earthquakes

There are two methods for finding what earthquakes are possible at a site: the seismological method (seismic data) and the geological data (tectonic evidence). The first is statistical and probabilistic, the second being correlative and pretensely deterministic.

More or less frequent earthquakes over a certain capacity of destruction have been felt almost everywhere. However, as it was shown above, significant shakings capable of introducing a risk or impairing the safety of dams can only take place in certain zones of the crust, the most active areas being those close to the boundaries of the plates. For those zones, very destructive earthquakes take place (say, magnitudes above 7.5) within relatively short periods of recurrence. They produce significant dynamic forces until some hundred kilometers from the source. In other areas, where active faults exist, also important earthquakes (say magnitude above 6) are felt until moderate distances from the epicenters, but their periods of recurrence can be rather large.

Once the recurrence period, or the number of earthquakes in an area per year or century (also called excedance rate) of a magnitude exceeding a certain value has been found statistically and theoretically (probabilistically) related (a logarithmic law was proposed by Richter), the criteria have been used of extrapolating this law for the economic life of the dam (100 years). But also it has been extended for a very large period of time, say 50.000 or even 100.000 years, to find the largest earthquake for which the dam must still be safe against release of water, or will not fail (Ref.5). This last practice can lead to unreasonable high values of the dynamic action, since there is a physical limit at each site, as it was seen, and infinite time can not bring infinite magnitudes. In any case, it is possible to define in this way a maximum credible earthquake (MCE) for each site together with a design basis earthquakes (DBE).

In cases analysed by the writer, it was found that the various points representing, in a logarithmic scale, the number of earthquakes above a certain magnitude per year (excedance rate) was not well adaptable to a straight line, while the logarithm of the logarithm of the return period, adapts better to the statistical data. Obviously such a function leads to lower values of the magnitude for a very high number of years. Others have suggested limiting functions for high recurrence periods to account for the physical limits (Ref.6).

Here lie two of the most important points of criteria: the law to be used for extrapolation of the excedance rate and the time limit set for the MCE. The fact that, for the large time limits previously mentioned, it is possible to find geological indicators for the fault activity and for the magnitude of the past highest earthquakes in the zone (shearing movement of the youngest formations, length of fault movement, etc.) gave support for such large time limits. On one hand, there is a strong tendency to take those indicators (regardless of how meager they are) with a very conservative view and on the other hand, it must be recognised that so long periods of time (50 or 100 thousand years) make no sense in terms of the useful life and risk caused by a dam. There we have two other causes that usually introduce over-conservatism. But viewing from another side, this is support for action in relation to permanent analysis of safety and eventually replacement of dams.

3. Active Faults

A maximum credible earthquake, MCE, was defined as the largest one that can take place for given geologic conditions. Its finding leads then to the exaustive study of such conditions until distances from the dam site depending on the possible magnitudes.

The earth's crust presents, near to its surface, especially in highly cohesive or rocky formations, many planes of rupture due to very high tectonic forces. In many cases those planes are discontinuities of the mass, have a certain thickness and show offsets of the formations, indicating movement between the two sides, and sometimes slickensides in the milonitic material that fills the discontinuity. They are called faults and are usually very old. In some cases (very few indeed), evidence can be found of very recent movements, in terms of the geological time scale (much less than the age of the Pleistocene). Since earthquakes are originated by the release of energy during the rupture of the crust at certain faults (causative faults), there is a tendency to call active faults to all those that show signs of recent movement. This can be wrong since the movements can be sympathetic. Also it can be wrong because experts have tendency to accept as evidence doubtful indications, sometimes calling the faults "indeterminate active" instead of "indeterminate inactive". Such attitudes put a large amount of unnecessary and unjustified conservatism in the evaluation of the seismic risk of an area.

An active fault, from a seismologic point of view, is a fault capable of originating earthquakes, and the best indicator of activity is the location of epicenters of observed earthquakes in the fault surface. It is true that earthquakes above a certain magnitude ($M>4.5$) are always associated with fault surface rupture, also, there are strong correlations between the surface displacement or the length of ruptured crust, and the magnitude of the earthquake. But, because the foci of the earthquakes are sometimes very deep, because in many cases they are under the water and because the surface evidence is usually so meager and debatable, it is very questionable if the possible activity of a known or a new found fault (and the seismic risk of a dam) can be established with confidence by such procedures. Also, in many places it does not seem possible to locate or detect all faults existing in the crust until a reasonable distance from the dam.

For all these reasons it seems to the writer that the geological method for establishing the seismic risk for a dam site can lead to an even greater conservatism than the seismological method.

4. Characteristics of the Design Earthquakes

As it is known today, a single parameter, like the maximum acceleration (that was considered in the past for the equivalent static method) can not define an earthquake from the point of view of its effect on a high risk structure like a concrete dam. The necessary design data are the response spectrum and the accelerogram of the chosen earthquake. However, Housner's spectrum intensity, or the "area" under the velocity response spectrum curve between the periods of 0.1 and 2.5 sec is, alone, a good indicator of the importance of the structural action that a given earthquake can have (Ref.5). The duration of a strong earthquake is very important for its destructiveness. All those factors can be derived from earthquakes similar to those predicted for the site, taking into consideration the site condition and distance from the causative fault. The damping value to be used in calculation is sometimes taken from observations but is basically a matter of judgement.

The magnitude, maximum accelerations response spectra, spectral intensity, time history (accelerogram), duration, etc., of the DBE must be established in a reasonable basis from all the accumulated data, and not with over-conservatism, taking into consideration that the safeguard of the structures will be the safety factor and safety criteria. While the DBE must be a dynamic load ressembling the observed earthquakes at the site or at sites with similar geological and tectonic environments, and its effects can be analysed by current refined methods (Ref.2,5) considering usual elastic and continuous structures, the MCE, necessarily a stronger load, must not be assumed to have characteristics different from the above. What is important to say here is that dam engineers must, for the sake of safety, consider safe loads and criteria, but for the sake of economy can not exagerate in them.

This is why we can not accept establishing the site response spectra and syntetic accelerograms by summing data from various inconvenient earthquakes with different frequencies and energy distributions although all of them with the same magnitude. Neither do we see any good justification to use the 84% percentile amplification factors for rock sites. Also it is not supported the use of a low damping coefficient of the structure in calculations for a low probability load.

5. The Acceptable Effects of the MCE

Once the maximum credible earthquake at a site is, already in itself, the limit of the possible shaking it has no sense to use a safety factor to "protect" the structure. Usually stresses are computed elastically. While in compression they are not accepted above the dynamic compressive strength, in tension cracking is admitted but not in the two sides of the dam. Usually the dam is considered completely loaded at the time of its occurrence. However, it seems more appropriate to compute the dam for the most usual water level and for the maximum height of the water and judge safety and endurance from the results for both situations. Cracks upstream and downstream in the dam above the water level for the MCE have no significance from the point of view of total risk. Even life and property will not be endangered downstream of a concrete dam if the portion above water falls down with the MCE.

An effort must be made to make non linear analysis in a discontinuous structure for the MCE given their low probability of occurrence. Appropriate model tests up to rupture can be justified. It is obvious that near rupture, even in tension, the non elastic behaviour of the material will increase safety once the stresses will be lower than those found by linear analysis. Computations of several dams that suffered strong earthquakes (Ref.5,7) indicated that this is a reasonable statement. Also the dumping in the structure increases near rupture, as observed in real tests.

Another important point that needs research and must be deeply investigated is the amount of stored dynamic energy near failure and the energy necessary for causing total failure or the destruction of the dam by pieces to fall down. Certainly when cracks are produced there is a large loss of accumulated dynamic energy (equal to the energy necessary for rupture) causing a sharp reduction of the vibration if not stopping it. This situation has been observed during dynamic tests on models for sustained motion. A preliminary calculation of an arch dam indicates that only the rupture in tension by bending of the cantilevers (horizontal cracking) can absorb a very significant amount of the maximum dynamic elastic energy stored in the dam. A good shape for the dam will be very helpful to avoid cracking.

REFERENCES

1. H.B. SEED, F.I. MAKDISI and P. DE ALBA, "Performance of Earth Dams During Earthquakes", Journal of Geotechnical Division, Proc. ASCE, Vol. 104, GT7, July 1978, pp 967-944.
2. R. PRISCU, A. POPOVICI, D. STEMATIU, L.ILIE and C. STERE, "Ingeneria Seismica a Marilor Baraje", Edito. Academici, Bucuresti, 1980.
3. K.D. HANSEN and L.H. ROHEM, "The Response of Concrete Dams to Earthquakes", Water Power and Dam Construction, London, April, 1979, pp. 27-31.
4. ICOLD Committee on Earthquakes, "A Review of Earthquake Design of Dams", March, 1974, Bull. 27, pp. 71.
5. G. TARBOX, K. DREHER, L. CARPENTER, "Seismic Analysis of Concrete Dams, Proc. 13th ICOLD Congress, New Hehli, 1979, Vol. II, pp. 964-994.
6. L. ESTEVA, 1976, "Seismicity" in "Seismic Risk and Engineering Decision", Editor C. Lomnitz and E. Rosenbluth, Elzevier, Amsterdam.
7. Papers on Question 51. Proc. 13th Intern. Congress on Large Dams, New Dehli, 1979, Vol. II.

SYNOPSIS
The paper presents considerations on the effects of earthquakes on dams particularly in the case of concrete dams and briefly discusses the methods for finding the Design Basis Earthquake (DBE) and Maximum Credible Earthquake (MCE) . The periods of recurrence of these earthquakes are referred to as well as searches for active faults and characteristics of the earthquakes to be considered in the design. The possibility of the design of a dam at the point of complete failure for MCE is indicated. The comparison of the maximum energy stored in a dam during an earthquake with the energy necessary for rupture is envisaged as well as the possible non-elastic behaviour of concrete dams during very strong earthquakes. The importance of shapes in the response is indicated.

STRUCTURE BEHAVIOUR WHEN SUBJECTED TO EARTHQUAKE

Professor H. Bolton Seed (Session Chairman; University of California, Berkeley). In reviewing the ten Papers included in this Session, I note that one of them deals with the design criteria for Alicura Dam in Argentina, two of them deal with design and analysis of earth dams, and seven with analysis techniques for concrete gravity and arch dams.

The Papers provide a comprehensive review of many aspects of earthquake-resistant design problems. However, in reading them one cannot help but note that our analytical abilities at the present time seem to be much further advanced than our ability to develop representative models of material characteristics to incorporate in the analyses.

Thus it appears that our primary limitation lies not in the development of improved analytical procedures, important as this might be, but in the improved characterization of material properties. This necessarily involves the development of better models for representing concrete behaviour in the inelastic range and under cyclic loading conditions, as well as better methods for obtaining representative samples of soil through improvements in undisturbed sampling and sample handling techniques as well as soil testing procedures.

In view of the wide range of topics covered by this Session, I would like to suggest that we try to direct our considerations particularly towards the following topics.

(1) Observations of the field performance of dams during earthquakes (past observation, possible future efforts, and required instrumentation)

(2) Comparison of field performance with predicted performance

(3) Types of behaviour requiring evaluation

(4) The degree of sophistication required in analytical techniques for predicting performance - recognizing current limitations in techniques for determining material properties

(5) Criteria for determining acceptable levels of stress in seismic analysis of concrete dams

(6) The role of model testing in evaluating the seismic stability of dams

(7) Determination of the appropriate level of conservatism to be used in evaluations of seismic stability.

Professor J.L. Serafim (Paper 27). Mr Dreher (Paper 19) presented us with the first approach to the solution of a problem that has always given concern to dam designers - the offsetting of faults under dams. Of course dam builders have faced this problem many times. We all know that in the foundations of certain dams there are possibilities of an offset. What the engineers have done in the past is to excavate the fault, fill it with concrete and embed steel rods, or even rails, across the path of the fault. Concerning the effect of foundation movements on dam behaviour we had the opportunity to model this problem in 1964/65 at MIT. The original dam in our model had three separate sections or foundation blocks which could move in different directions. Unfortunately we only remained at that Institute for one year, but some tests had been done which tended to prove that differential displacements in the foundation block were not critical for safety. I think that Mr Dreher is now finding the first scientific approach to this problem.

Mr P. Londe (President, ICOLD). It seems to me that in such a case the most important factor for stability will be the change in pore pressure under the dam rather than the distribution of tensile stresses, and I would like to have a discussion of this point.

Mr K.J. Dreher (Paper 19). Mr Londe's point is correct. The change in pore pressure in a dam's foundation is critical to safe performance following fault displacement beneath the dam. The tensile and compressive stresses themselves are not particularly significant, which is exactly the point of my Paper. The particular structure for which these results have been developed is a gravity-arch dam, although it has been idealized in my Paper as a simple two-dimensional gravity dam. While the fact that the dam is a gravity-arch does not entirely eliminate the concern for changes in pore pressure, the capability of the dam to develop arch action does eliminate any concern for overturning, which is a significant advantage over a straight gravity dam. The remaining effects of pore pressure in the foundation that require consideration for fault displacement conditions are the changes in driving forces from pore pressure acting on potentially unstable wedges of rock and any cahnge in pressure gradients which could indicate piping of foundation materials. Although not presented in my Paper, we have attempted to bound potential worst case pore pressure conditions. Unfortunately, we have not developed a pore pressure model comparable to the analytical model developed for the dam structure.

Mr W.A. Wahler (W.A. Wahler Inc., U.S.A.). The analysis of the Lower Crystal Springs Dam, often referred to in this Conference, was made for the City of San Francisco by my firm. The axis of this dam is approximately parallel to and about 1000ft downstream from, the San Andreas fault. The reservoir is formed by the fault valley and is just upstream from a densely populated city. This dam was carefully designed and built in the 1890s using current technology. It survived the infamous 1906 earthquake with only minor fracturing.

Because the failure of this dam could be very hazardous to a large number of people, the City of San Francisco had us analyse the dam using current finite element technology. We analysed the dam considering various foundation condition assumptions including one that the foundation was pervasively and intimately fractured (which was the case). Our analyses were based on extensive foundation and abutment testing results. We obtained very close correlation between calculated and actual behaviour of the dam due to a simulated earthquake of the 1906 variety. This, however, in my opinion, did not verify the suitability of the original 1890 design which did not correctly take into account the actual foundation and abutments condition as much as it reflected the influence of a non-rigid interface between the abutments and foundation and the dam. Had the abutments and foundation been 'ideally solid and rigid' the actual and calculated behaviour may not have been so favourable. This challenges very well established foundation and abutment criterion - that they should be consolidated (consolidation grouting) when fractured.

In general a lot of money is spent to grout a fractured foundation. Obviously a finite fault or fracture plane may not be tolerable, but is pervasive non-planar fracturing always bad? This can be an argument against consolidation grouting downstream of the cut-off where in some cases drainage may produce more reliable stability.

Did the fractures save the dam? Should we blast rather than grout concrete dam foundations?

Mr K.J. Dreher. I do not agree entirely that the theory for dynamic non-linear analysis of concrete dams has been developed adequately. The main problem in adequately modelling the non-linear dynamic response of concrete arch dams and some gravity dams is the inclusion of contraction joint effects. I believe that any cracking and subsequent non-linear behaviour will be largely influenced by contraction joints. If contraction joints are ignored and the concrete modelled as a homogeneous, isotropic continuum, analyses for large excitations might indicate cracking at orientations other than nearly perpendicular to the direction of contraction joints. Cracks relatively parallel to contraction joints are not likely to occur physically. Unfortunately, we have not yet developed, nor are we aware, of an acceptable finite element model of contraction joint behaviour.

Professor R.T.Severn (introducing Papers 23 and 24). There are two points that I want to make relating to Paper 23. We were testing a gravity dam at the same time that the ICOLD papers were produced for the thirteenth Congress, and it occurred to me to try to pull together the various results that people had obtained. It struck me that the proposed simplified method of Professor Chopra was not really adequate. I was surprised that he had not taken the foundation into account.

Let me refer to the results in Paper 23. If one is to design a gravity dam by a spectral approach then it is absolutely vital to get the frequencies right. The 2-dimensional approach for gravity dams is, I think, quite insufficient, either by any finite element method or by the simplified approach of Chopra. The Crystal Springs Dam, admittedly slightly curved, and therefore a gravity-arch dam, was a very interesting case. The great value here was that the frequencies had been measured by analysis of ambient response and these were 6.1Hz and 8.40Hz. The frequencies calculated by those authors were very near to the measured values. They were 3-dimensional calculations however. The 2-dimensional calculations give values almost twice as large.

If one looks at Paper 25, there were some dams for which foundation was not taken into account and some dams for which water was not taken into account. Here again, frequencies were measured and compared with 3-dimensional calculations. These are fixed base calculations and so the 4.22Hz obtained is higher than the 3.47Hz measured. If a slightly flexible base was used instead of infinity this frequency would come down. The 2-dimensional solution gives something quite different, and what worries me here is that in a 2-dimensional solution such as is proposed in the simplified approach a single value of the height is necessary and few gravity dams are built in wide trapezoidal valleys where plane strain conditions apply. I think that is equally valid for rockfill dams. Professor Seed mentioned the value of the vertical component of acceleration, as did Mr Lane. Professor Seed also suggested that it might be possible, because the vertical component occurs before the horizontal, to take them separately, and I am sure he had the 2-dimensional solution in mind. But it seems to me that this has to be considered with care because a vertical component occurring before the horizontal component effectively changes the gravity; we are effectively hitting something when it is in mid-air - but that is much easier to do than when it is on the ground. So I am not sure that one can take vertical and horizontal components independently. As I showed in rockfill dams, there is a vertical mode at the low frequency end.

The main point I want to get across is that with the highly sophisticated state we have reached in 2-dimensional analysis we should be considering whether 2-dimensional analyses really do answer all the questions.

Now the second point is the asynchronous motion case. Professor Seed suggested that the effect of asynchronous motion was not quite so large as our Paper would suggest. I agree with that, and in fact if you look at the Paper the values given are for acceleration at the crest of the dam. The asynchronous stresses are not affected in the same way. The point is that with the asychronous motion the different parts of the base are being moved relatively and therefore there are stresses produced by that differential motion, which are added to the dynamic stresses. But in total the asynchronous stresses are only 10 or 20% less than the synchronous ones for the dam that we were studying.

For Paper 24 I shall just refer to model testing. The method that most of us have used hitherto has been multi-point excitation. The multi-point excitation system is what we might call steady state and is a mechanically oriented system requiring some considerable skill to operate it sensibly. As in other branches of science our mechanical systems are being replaced by electronic systems and the approach which we, and Dr Castoldi, are attempting is the computer analysis of transient data. This approach, although the manufacturers of the various pieces of equipment will tell you otherwise, does have its disadvantages. For the Victoria Dam, our steady state test (in which we have great confidence) and our 3-dimensional finite element calculations, agree very well with fundamental frequency (about 500Hz), and with the second frequency. The peaks of the curve show the frequency in this transfer function approach, which is basically a computer approach, with very little skill required by the man doing the tests. The skill is in the computation. But the spectral approach, which is causing us concern at the moment, appears to give a much lower fundamental frequency, both with and without water. My colleague Mr Jeary suggests that this is some fault of the recording system, and he has done a lot of work on this method for tall buildings. So the new transfer function approach still has a lot of faults.

Dr A. Castoldi (introducing Paper 25). Following a well established approach in many fields of engineering, surveillance is now a complex process consisting of three different stages

(a) creation of a mathematical model validated on the basis of experimental results

(b) installation of a monitoring system capable of giving information on the actual behaviour of the structure

(c) correction and updating of this model in order to take into account any possible changes in the behaviour of the structure

This complex process is not feasible in its entirety at least in the light of current experience.

Creation of a mathematical model. The accuracy of present practice in building mathematical models of dam dynamic behaviour is considered satisfactory for design purposes, assuming a suitable choice of material properties (both for dam and rock). However, a greater accuracy is needed if the model has to account for any possible change in the dynamic behaviour of the dam. Unfortunately, better results cannot be expected at present, because there are still too many uncertainties in the schematization of the problem. On the contrary, experimental accuracy in determining modal parameters (natural frequencies, dampings and modal shapes) is already high and can be increased by using new excitation techniques and data processing methods.

From this point of view, a research programme, now in progress at ISMES and sponsored by ENEL, concerns

(a) the use of a multi-point excitation which allows the adjustment of the energy transmitted to the structure, so as to make the response of one mode more evident than the response of the other modes. The excitation system consists of six mechanical vibrators generating sinusoidal forces, whose frequency, amplitude and phase can be adjusted by means of a minicomputer

(b) the use of explosive charges. This kind of excitation is a better simulation of the mechanism of an earthquake and, in many cases, makes it possible to obtain more easily information concerning the amplification of the abutments, the time delays and therefore the phase lag of the motion at various points of the structure and of the foundation, and in general a better comprehension of the interaction between the dam and its foundation

Monitoring. Monitoring is aimed at determining the modal parameters (natural frequencies and modal shapes) of the dam. This kind of information can be obtained by analysing the natural micro-activity of the site and the response of the structure to this activity. A check on the feasibility of this monitoring has been made by comparing the estimate of the model parameters obtained through this process and the same parameters obtained through forced vibration tests for the Barcis dam in Northern Italy. An instrumentation capable of automatically and continuously performing this kind of analysis is being designed at ISMES.

Updating of the reference model. At present we are just at the beginning, and additional work is needed to make the method practicable. The main problems are

(a) the sensitivity of the proposed method, that is an analysis of the expected change in the natural frequencies due to a certain amount of damage (to be done analytically and by simplified physical models)

(b) how to separate changes in natural frequencies of a real dam, due to a change in water level and temperature, and to the opening of joints, etc., from the changes resulting from possible damage. An answer to this problem may be obtained by installing on the dam a suitable monitoring system.

Professor J.L. Serafim (introducing Paper 27). In 1969 in Lisbon and the south of the Iberian peninsula we had one of those strong earthquakes which come at intervals of about 200 years. This provided the singular opportunity of taking measurements of pore pressures and water flow before and after the earthquake, in the foundations of a dam near Barcelona and of another dam near a village close to Sagres, where lives were lost. We were observing the former dam during the first filling; the latter had been under observation for eight years. In both of them there was an important increase in flow from the foundation, and an increase in the uplift pressures in some zones of the foundations. I would say that this problem of increase in flow in rock foundations after earthquakes is an indication that the joints do open. There is no doubt that there was some movement in the rock mass, but this is also an indication that we must be careful when designing the drains. I would like to indicate to Professor Zienkiewicz that all the dams must have drains; concrete dams must have drains as they are the best solution for abnormal pore pressure.

It is also very important to find out what earthquakes are possible at a site. If we imagine that every kind of earthquake is possible, we do not know where to stop. I have seen cases where, besides finding that the magnitude of the maximum credible earthquake at a site was 6.5, all the inconvenient aspects of all the 6.5 magnitude earthquakes were added in. I cannot support such an attitude. Neither can I support the approach of taking finally the 85% earthquake. Of course if we are building a dam somewhere we must assess safety, but we must not add conservatism to conservatism. I agree that dams are very high risk structures, but we must not be over-conservative in our design parameters.

In addition we are considering some new safety criteria for dam building. After discussing the meaning of traditional safety factors all of us agreed to leave that aside as a thing of the past. In the future we will have to look at the reliability of dams in probabilistic terms. For instance, when looking at the probability of the maximum credible earthquake (if we take 10 000 years as the interval) we have immediately a load probability of 10^{-4} per year. This is equal to an indicated value, from the historical record of the frequency of failures per dam per year, thus we will not need for this load a safety coefficient above one. I am convinced that most of the dams have a reliability much above that value (10^{-4}) and also that it increases with automatic observation and warning systems.

Finally, experience shows that reliability of arch dams is in fact very high. One important example is that of Vaiont. This dam was overflooded by a wave which was completely asymmetrical. The height of the wave on the right bank was over 200m and on the left bank not more than 100m. A second case was the very good behaviour of Pacoima Dam during the San Fernando earthquake. However, the very high intensity of the shock at Pacoima has been questioned. What can not be doubted is that the foundation rock in the left bank was completely crushed by the earthquake, and that the dam did not suffer.

Mr K.J. Dreher. I would like to comment concerning the use of peak velocity as the intensity measure for ground motions. Peak velocity may be an appropriate measure of ground shaking for earth dams since the peak ground velocity is normally determined by the longer period components of the ground motion. However, for concrete dams, peak velocity may not adequately account for higher frequency components which have a pronounced effect on their response. Unfortunately, neither does peak ground acceleration. Consequently, the parameters we use to represent ground motion intensity, and they are not necessarily the only parameters that can be used, include peak velocity, spectrum intensity, and the root mean square values of acceleration and velocity in the strong motion part of the earthquake. To various extents, these parameters can be tailored to the frequency characteristics of the particular structure that we are analysing.

Dr S.K. Sarma (Paper 18). I agree generally with Dr Dreher. The reason for choosing peak velocity is that it is an energy related parameter. However, in the analyses, acceleration records are used which are normalized to the chosen peak velocity, maintaining their frequency characteristics.

Professor H. Bolton Seed. The cyclic pore pressure parameter, A_n, is likely to be influenced not only by sampling, testing and handling, but also by the initial stresses in the dam. It is sometimes measured by testing an isotropically consolidated sample, but in fact one obtains quite different values of A_n on non-isotropically consolidated samples; accordingly one would have to be careful in selecting the right value of A_n to put in the calculations; I assume however, that you intend that to be done anyway?

Dr S.K. Sarma. Yes.

Dr J.A. Veltrop (Harza Engineering Company, USA). The title of this Conference is Design of Dams to Resist Earthquake. Speakers' comments have been concerned with research for earthquake design in the future rather than with earthquake design for the present. While perhaps valuable in future years information provided is of limited use to the practising engineer at this moment. Let me cite two examples. There is a 700ft high arch dam 25 miles from Mount St Helen's. Is it safe? Should the reservoir be drawn down or not?

This is not a scientific matter. It is a matter of convincing journalists from the newspapers. This is a very different matter compared with trying to determine frequencies, and how close we can get with our analyses. None of that will convince any journalist or the public. There will have to be many more fundamental considerations of the kind that we heard from Professor Serafim, such as how safe is an arch dam, and is it really the arch dam that we are concerned with, or is it the foundation that supports the arch dam? I cannot be concerned at all about arch dam design against earthquake. There is no need for this, in my opinion, no matter how elaborate the analysis is.

My second example is the design for a fill dam in Venezuela which has to have the ability to absorb an offset in the foundation of potentially more than 2ft. We do not make analyses of such an offset. We make provisions for making sure that the proper material is in the neighbourhood of this offset, so that on the basis of judgment this material will fill the gaps should displacement occur during the earthquake; but no analyses are necessary.

Mr D.M. Finlayson (W.S. Atkins and Partners, Surrey). My remarks are presented from the viewpoint of a designer of one of the large number of embankment dams now planned or being constructed in the 25 to 75m range, which are not an element of a large or prestigious project, and therefore the amount of effort to be expended on design is limited. Paradoxically, these structures are, if anything, subject to greater damage in an earthquake than their taller cousins because the fundamental period of vibration is higher and coincides more nearly with the frequency of the earthquake.

It is worth briefly looking at the difficulties which pervade aseismic design. First of all, we do not understand the earthquake mechanism. We do have some models for strike slips which we can use but they in no way cover those kind of earthquakes which are not associated with a particular tectonic structure, and in any case are very simplistic. Secondly, despite what has already been said, we are not able to model the propagation of seismic waves from the source to the dam through any reasonable kind of geology which might be experienced in a distance anything between 10 and 100 km. Thirdly, although this subject has been much addressed at this Conference and our knowledge is clearly improving, we are not able at the moment to model the response of a dam on foundations adequately. Lastly, as Professor Seed has pointed out, our operating experience is extremely limited. This differentiates dynamic from static analysis totally, where we have a very large body of data, concerning both dams and natural slopes, which we can use for back analysis.

One might think, on the basis of all this, that meaningful analysis is precluded because of so many uncertainties. Clearly, whatever calculations are performed, however complex, they deal with only a part of the phenomenon and the results are largely untested. There is a tendency to increase one's confidence in line with the amount of calculation one does on a topic which is in itself rational, and I think that this is slightly dangerous. The remedy is to seek information on the performance and modes of damage of embankments which have suffered earthquakes, as Professors Seed (Paper 12) and Ambraseys (Paper 18) have done, and also Mr Howells (Paper 15).
The list of possible effects given by Sherrard some 17 years ago is still comprehensive. Firstly, there may be a fault actually passing through the body of the dam which may displace. Secondly, there may be slope failures induced in the body of the dam by the shaking of the dam and the amplification of the ground motions which result. Thirdly, the dam may slide on its foundations, and this is an interactive phenomenon with the stability of the dam itself. Fourthly, there may be cracking induced by the earthquake. This, as other speakers have remarked, commonly occurs parallel to the crest, but particularly where one is talking of a rock valley with steeply sloping abutments it is possible to get horizontal cracks at depth, following compaction due to earthquake shaking. Any cracking that does occur in an embankment may then lead to a failure following from piping. A fifth hazard is that one might actually get vertical movement on a fault within the reservoir basin, which can itself induce over-topping of the reservoir just because of the difference in levels and the downthrow of the dam. Sixth, one may get seiches in the reservoir itself, although I think that in recent years this has been shown to be not of great importance. Seventh, there may be overtopping of the dam due to slides or rock falls into the reservoir as at Vaiont. Almost any earth fill embankment that one could postulate would not survive that. Lastly, there can be over-topping due to the damage to the spillway or the outlet works.

We have therefore eight possible effects from an earthquake, of which only two are dealt with by stability analysis, that is slope failures and sliding of the dam on the foundation. In addition it is possible to model the problem of cracking by finite element analysis, but only partially. It seems to me that the message is fairly clear, and that is that analysis is not the primary tool of the designer, particularly within the framework that I have described.

If we return then to the vantage point of the designer, his task is to provide a robust structure which is insensitive to the severity of the extreme events which may occur; that is what I would call an intrinsically safe design. If we now restrict ourselves to the embankment only, the designer for the embankment has to consider three principal factors, one of which is possible disruption on a fault, the second of which is loss of free-board, whether this is due to a general settlement or a discrete slip, and the last factor is the extent of likely cracking and the control of any consequential seepage. If movement on a fault in the dam foundation is considered possible, the embankment must at least approximate to the concepts

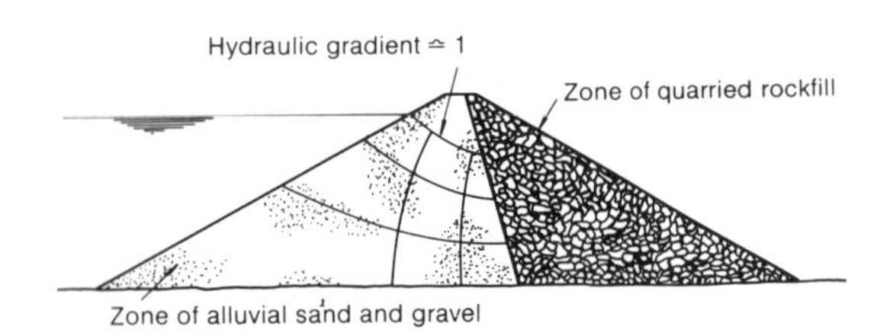

Fig.1. Fundamentally safe embankment dam

of the fundamentally safe embankment elucidated by Professor Cluff (Fig. 1); that is, principally, that the stability of the embankment does not depend on the amount of flow through it, and that the element of the embankment which prevents very large flows is not capable of sustaining cracks. Calculation of the loss of the freeboard requires some assumptions about design earthquakes, be they maximum expectable or maximum credible, and hence requires a method of analysis which contravenes the previous conclusion that analysis is of dubious value. My solution is to provide special details at the top of the dam which allow over-topping of limited height, thus avoiding having to place absolute confidence in the analysis. This safeguard is also required against some of the other effects which I enumerated previously which may involve over-topping. With this kind of safeguard there is no reason not to use the limit equilibrium methods proposed by Dr Sarma (Paper 18), which I prefer to call pseudo-dynamic, as they are different in kind from pseudo-static analyses used ten years ago. This leaves the question of cracking, and here I believe that the designer best directs his attention to the large body of literature on the settlement and cracking of dams under normal static conditions, since the principles of sound detailing, such as the shaping of abutments, are equally applicable to general settlement of a bank in an earthquake. The problem is how to avoid low minor principal stresses, and the top of the dam is especially vulnerable being at a low level of ambient stress and more severely shaken than lower levels. Earthquake designs should assume that the core is cracked to the full height sustainable according to the amount of cohesion, piping being prevented by substantial transitions and filters. These are the main elements of embankment design under the restraints which I have outlined, which I feel present a pragmatic approach.

Finally, I have two comments arising from previous Papers. The first concerns Dr Sarma's work The concept of limit equilibrium on a single failure surface is clearly incorrect if the material behaviour is not markedly brittle, since the critical surface at each instant in time would differ according to the history of accelerations at various levels in the dam. I think this was demonstrated quite well by Mr Kutter's figures from his work in the centrifuge at Cambridge (Discussion, Session 2) where he showed a zone of deformation rather than a discrete slip surface. I do not feel that this is a fundamental objection to the use of the method as, for example, static analyses which postulate circular surfaces have been used for many years in conditions where it is known that the failures are not going to be properly circular. The problem is the lack of correlation with any field observation, and in the absence of true prototype observations I feel that Mr Kutter's experiments in the centrifuge at Cambridge will provide important data.

Secondly, there has been some disagreement about the importance of vertical accelerations. Dr Sarma has shown elsewhere that for a given level of horizontal acceleration the angle at which it acts to the horizontal which produces the worst displacement in the sliding block model is slight. Professor Severn, on the other hand, has shown us interesting mode shapes involving vertical deformations but has not indicated the consequences on embankment stability which, in the absence of pore pressure rise, one would expect to be slight. This needs further elucidation, and I hope that somebody will attempt it.

Professor Bolton Seed. What Mr Finlayson is really advocating is the use of defensive design measures, which most earth dam designers regard as the first approach to earthquake resistant design. Many of the manuals on design of earth dams start by outlining defensive design measures which one should take before starting any calculations. That is the way that most of the dams that I have been involved with have been approached, and I suspect most other dams have been handled in a similar way.

Mr W.A. Wahler. There are many reasons why designs must be analysed and reported. One very important one not mentioned by the previous speaker is that most designs will be analysed at some time or another by other people without the presence of the designer. This could happen after distress or failure but now, more commonly, regulatory reviews are being made of dams and other critical structures before construction is approved, and properly reported analysis could be critical to project acceptance. The project must be 'mathematically' defensible, and that defence must be made available to reviewers to assure timely approval. Perhaps the speaker was meaning to say that too much emphasis is put on slope analysis. I agree. Unfortunately, design analysis for embankment dams usually consists only of slope stability analysis. I believe that if we were to count the papers on slope stability at any conference, and papers on other equally important considerations, we could come to the conclusion that slope stability was the only, or at least the most critical, question to be answered.

This should not be the case. Consider for a moment some of the site development problems that are often ignored such as the possible transverse cracking of an embankment in a deep, steep-walled canyon where there is a sharp break in one or both of the abutment slopes. Transverse cracking can develop with or even without earthquake-induced consolidation. Can this type of problem not be even more critical than slope stability especially since the cracking problem is often ignored, and the slope problem is never ignored even if it is not always analysed ad infinitum? This is just one example of a non-slope type analysis that can and should be made for every dam where it is applicable. In this type of analysis, strains are more critical than stresses. In this case the design defence is not a flatter slope but rather transition, filter and/or drainage zonation. There are many non-slope types of analysis that can be critical and must be analysed.

The design of a dam must consider all of the site and material factors that can affect its operation and performance. Perhaps excessive conservatism (whatever that is) generally permits success without specific design for every applicable factor, but this is not a systematic, honest or reliable way to design a dam. We cannot reasonably rely on general conservatism to protect us for every specific potential mode of failure.

Professor A.N. Schofield (Cambridge University). The numerical solution and constitutive relations for non-linear damage analysis of earth dams in earthquakes need more checks against good data from geotechnical centrifuge model tests before too great confidence is placed in results of analysis. At present any constitutive relations involving critical states make assumptions about principal axes of the tensors of stress, stress increment, and plastic strain increment, such as assuming coincidence of axes and isotropy. It is important to check the sensitivity of numerical solutions to such assumptions, and one use of a geotechnical centrifuge could be that soils which tend to develop transverse isotropy could be studied in direct analogue tests. It is my opinion that an understanding of the problem of liquefaction requires us to think also in terms of failure mechanisms, and the geotechnical centrifuge is important in that connection.

In continuum calculations for metals the stress tensor at each point cannot exceed the limiting stresses for Mises or Tresca material. Soil material in some states differs from the Mises or Tresca material in that mean normal effective pressure can change the strength; for Mohr-Coulomb material strength increases as effective pressures increase. This increase ends when the effective pressure reaches a critical state value. In states of higher effective pressure the Cam-Clay material is appropriate, because the elastic limit stress falls as mean normal pressures increase. By combining these materials: soil at low stresses exhibits the mechanism of failure of Coulomb rupture, and at the same density but at higher stresses its strength at first yield is reduced and it distorts with increase of pore pressure or plastic collapse of volume, exhibiting the plastic compression commonly associated with Terzaghi's consolidation theory.

The Cam-Clay model defines a limit to the effective pressure that can be applied to a granular aggregate at any particular density. It was supposed originally that in soil in states of even higher pressure, liquefaction occurs; early studies of liquefaction were based on the assumption that the soil structure spontaneously collapsed under relatively high effective pressure. However, more recent studies of liquefaction under cyclically induced pore pressures show that the soil's effective stress state moves far below critical state pressures towards states of very greatly reduced effective pressure, or even zero effective stress, as pore pressure increases. What happens when this cyclically induced effective stress relaxation brings soil to regions of near zero effective stress?

On moving through states of successively lower mean effective stress, first the Coulomb failure limits take over from the Cam-Clay yield locus, and then the soil reaches the end of the range of stresses in which plastic yielding and continuum behaviour occur. One class of failure in this low effective pressure range is that fracture or fissuring can appear. In typical compression tests of cylindrical specimens one can observe tensile strain cracking in axial compression with longitudinal tensile strain cracks, or in radial compression there may be transverse cracks: in both cases as one approaches low effective stresses the soil will cease to be a continuum. The same class of material behaviour also occurs in the index tests as soil approaches the plastic limit state. An important feature of such behaviour is the development of a secondary permeability through the crumbling, cracking soil: such behaviour has been observed in geotechnical centrifuge tests. Another class of behaviour also observed near zero effective stress in centrifuge tests is upward migration of voids and channelling and piping through soil. In general the more dense the soil the higher the effective stress needed to ensure that soil behaves continuously and does not exhibit this low effective stress discontinuous behaviour.

At a certain relative density soil under low effective stress is a material liable to have openings; under low to medium effective stress it is continuous but liable to rupture; and under high effect stress it will yield homogeneously and show plastic compression. With increase of relative density the pressures at which such changes of behaviour occur will increase. It is generally supposed that increased relative density in a design will lead to the benefit of increased strength. It is not sufficiently appreciated that with the strength increase also comes the hazard of a different mechanism of failure when soil changes into a material liable to have open gaps or fissures, with a rapid increase of permeability. In water-retaining structures it seems preferable to have soil in states nearer to critical pressures where it generally behaves more like a continuum, either at effective pressures a little below critical pressure where soil acts as a material that ruptures and develops negative pore pressures, or at higher pressure where it becomes a material exhibiting plastic compression. In a centrifuge model each element will be at the correct stress and will exhibit the correct material behaviour.

In a centrifuge test, when each element of soil has had the correct stress history and is at the correct mean normal effective stress, the mechanisms of failure of the model as a whole will include discontinuous behaviour in appropriate regions. So far there are no continuum calculations which can generate discontinuous behaviour in the appropriate regions of a soil body under analysis. In the few models of liquefaction failure which we have tested up to now on the Cambridge Geotechnical Centrifuge an essential feature of each failure has proved to be a region of crumbling, cracking, piping or channelling which when it is induced by stress or strain allows rapid transmission of pore fluid and leads to rapid transformation of some part of an initially solid body into a flowing or fluidized state. Further tests are in hand including studies of earthquake loading, and as data from these tests are analysed it is hoped to make advances in finite element analyses, and thus increase their value to designers.

Professor O.C. Zienkiewicz (Paper 20). I just wanted to assure Professor Schofield that the symbolic tangent matrix $D_{ij}KL$ includes all plastic, elastic and creep strains. Such symbolic writing is used throughout the calculations and indeed a very extensive range of plastic models are used in the calculations presented. The model which is used for concrete and for soil is the critical state model and includes the zone of cracking. What has not yet been done as I mentioned yesterday - and I think that is for the future - is the inclusion of changes of permeability due to straining, but that obviously involves additional calculaions and a better knowledge of the materials. That is the next stage of work which one hopes very much will be accomplished soon. What we hope has been clearly presented is that with the present state of the art in numerical computation, practical problems of soil liquefaction, etc., can be generally dealt with.

Mr D.A. Howells (Chairman, SECED). I hoped that we would have at least one paper at this Conference dealing with the structures directly associated with dams: valve towers, bellmouth spillways, spillway gates and tunnels. The problem of spillway gates in particular is one on which we should welcome discussion. I should also like to make some comments on tunnels.

Firstly the wall of a cavity is a closed boundary in which static waves or oscillations may conceivably be set up thus forming an energy trap and amplifying the ground motion. However, the frequency of oscillation of even a very large power station cavern would be high (in the region of about 100Hz) and would be unlikely to be excited by earthquake. Also, such surface waves normally travel over plane surfaces; if they are forced to follow a curve their energy is transformed into body wave energy, transmitted away at a tangent; the radiation damping is high. For practical purposes one may say, 'You can shake a lump but you can't shake a hole'. Reports of damage to tunnels by shaking are so scarce that as a hazard it corresponds to the risk of intersection of the tunnel line by a fault displacement. There are a number of classic reports of this and the one most completely investigated was one of the effects of the Izu-Oshima earthquake of 14 January 1978, in which the fault plane cut through a modern metre-gauge railway tunnel. This is described in a paper by Tsuneishi and his colleagues published in english in the Bulletin of the Japanese Earthquake Research Institute, in 1978. I think that people tend to assume that the point where the fault plane will cut the tunnel can be predicted well enough to justify the use of sleeves and devices like that, but I doubt it.

Rigid tunnels in soft ground cannot be treated as holes. There have been lately quite a number of developments of the basic theory given by Newmark. The problems are not so much those of inertia forces but those of compliance with the imposed displacements of the ground. The propping of power station plant from the walls of a cavern greatly increases the earthquake resistance, and in areas of natural earthquake activity or possible induced activity this makes underground power stations more attractive. In deciding the earthquake instrumentation of a dam, the tunnel provides an attractive location.

Professor H. Bolton Seed. I should like to bring this Session to a close by emphasizing some of the points that have been made during the course of the discussion.

Mr Finlayson noted a number of design problems which are not amenable to analysis but are resolvable through the use of practical design measures. Most of these problems are illustrated by the performance of Hebgen Dam in the Hebgen Lake (Montana) earthquake of 1959. In this particular earthquake the causative fault was located about 1000ft from the left abutment of the dam and the earthquake, with magnitude 7.6, caused

(1) an overall lowering of the dam and reservoir area by about 10ft due to tectonic movements of the underlying rock

(2) settlement of the crest of the dam varying from 1 to 3ft due to consolidation of the foundation alluvium

(3) slumping of the crest on the upstream side of the dam due to lateral movements of the upstream shell

(4) severe cracking of the embankment, with cracks varying from 3 to 12in in width on the upstream and downstream sides of the crest

(5) several cracks, the largest being about 3in wide, in the concrete core wall

(6) some leakage near the abutment contact in the general vicinity of the core wall cracks

(7) a number of waves in the reservoir which caused the dam to be overtopped at least four times for periods of about 10 minutes; the depth of water flowing over the crest was estimated to be about 3ft during the second of these overflows

(8) a number of landslides around the edges of the reservoir

(9) a massive rock slide, estimated at about 50 million tons of rock which dammed up the river about 7 miles downstream and created a lake which backed up almost to the downstream toe of the dam.

Clearly not all of these effects, many of which could possibly lead to failure of a dam, are amenable to analysis. However, most of them can be rendered harmless by the use of defensive design measures such as the provision of ample free-board, the use of wide transition sections of filter materials not vulnerable to cracking, provision of crest details to minimize erosion in the event of overtopping and other practical design measures of this type.

At the same time, we must recognize that there are also problems, such as the failure of a tailings dam in Japan in 1978 some 24 hours after the earthquake which triggered the failure, which can only be examined by analytical means. It is only through the development of new analytical procedures that we can gain insights into the conditions where such events might occur.

In my own view, therefore, there is great scope for improved analytical procedures to provide the insight we need to attack problems we do not fully understand or anticipate, but there is also a need for the use of pragmatic design measures which can help to minimize the problems which could occur. Used appropriately, in conjunction with each other, they will lead to better, more economical and safer designs than we have produced in the past.

28 Dams, natural and induced earthquakes and the environment

E. T. HAWS, MA FICE FNZIE, Rendel Palmer & Tritton and N. REILLY,
BSc(Eng) MICE, Rendel-Parkman

Environmental aspects requiring assessment when considering earthquake effects on a dam are set out, with investigational methods and parameter ranges. Case histories are summarised including reservoir induced seismicity, and environmental impacts caused by seismic events at dams and reservoirs are listed. In conclusion, a brief qualitative assessment of the state of the art is given.

Introduction

Dams represent substantial modifications to the environment in which they are constructed, and these modifications should be the subject of study before commencement of any new project. Earthquake activity is a natural environmental factor of which due account must be taken in the early project studies. This paper indicates environmental aspects requiring assessment when considering the susceptibility of a new or existing dam to earthquake effects. The modifying effects of the completed dam and its reservoir upon the environment of the region as affected by subsequent earthquake activity will also be considered. An aspect of seismicity related to dam projects alone concerns the phenomenon of seismic events now commonly recognised as being induced by the presence of the dams and reservoirs themselves.

Environmental Aspects to be Considered (Table 1)

A very important aspect in considering susceptibility of a dam project to earthquake risk is the degree of seismicity of the area in which the project is to be built. Earthquake maps are available for the world, a good example being that due to Bullard. (ref. 1). This example shows earthquakes recorded over a period of 10 years plotted at their place of occurrence on a map of the world. The probability of earthquake at any particular location is thus approximately proportional to the density of points plotted in that area but account must be taken of how representative were the 10 years recorded, with due cognisance of longer term earthquake trends. If the area of the project has been the subject of good seismic records for a reasonable period of time, then a statistical extrapolation is possible for the intended life of the project.

A thorough assessment should be made of the geological fault systems within the area of the project using ground reconnaissance, aerial photography and satellite imagery, involving the disciplines of geology and geomorphology and taking special account of the quaternary geology for evidence of movement on faults over substantial periods of time, though recent in geological terms. The type, inclination and rate of strain accumulation on the faults are of great importance, particularly if induced seismicity is a possible risk, as with the larger reservoirs.

From the historical records and the fault analysis, the maximum credible earthquake on each relevant fault can be assessed and relationships exist which permit peak accelerations at the dam site to be estimated from the other data. Using these peaks, accelograms can be synthesised for the dam site based on recorded earthquakes in similar exposures, taking account of solid geology and the soil mechanics properties of overlying deposits. The peak accelerations may be taken as equal for the two horizontal components and a simple assumption often adopted is that the vertical acceleration is equal to half the horizontal. For less important dams and simpler design approaches using quasi static analysis, these factors are sufficient. As an alternative in such cases, or where significant data is lacking, a simple empirical approach can be made on the basis of the various earthquake intensity scales and observed effects in the project area.

For more important structures where dynamic analysis is considered essential, a further factor in the seismicity assessment concerns the response of the dam to earthquake stimulus and relevant damping factors. Resonance measurements have now been made by coupled shaking machines for a variety of dam types and the many different modes of vibration are available in the literature.

Another important environmental aspect is that related to soil and rock mechanics properties of the foundations and abutments. In the earthquake analyses, knowledge of all the normal parameters obtained from a properly conducted site investigation plays an important part. However, additional emphasis must be placed on susceptibility of the foundation and abutment materials to significant abnormal deformation under earthquake conditions. Soft foundations

TABLE 1 ENVIRONMENTAL ASPECTS TO BE CONSIDERED

* N-Natural, I-Induced

ASPECT	PARAMETERS	N,I*	INVESTIGATION OR INSTRUMENTATION	RANGE OF PARAMETERS & REMARKS
Seismicity	Seismic risk.		Seismic zoning maps, first and second level. Statistics if good records.	
	1. Magnitude $M = \log_{10}$ (CA) where amplitude is recorded by standard seismograph at 100 km. $= \log_{10} A/A_o$ where A_o = .001 mm.	N & I	6. Assessment of maximum credible earthquake on each relevant fault considering all of 1-5. Regional seismograph network for long term records of all events. At least one, and preferably not less than 3 seismographs to record shocks of M=1.5 to 3.5 during filling of major reservoirs. Strong motion accelographs on dams and abutments.	$M \ngtr 8.6$ $M > 6.5$, ground surface displacement $M < 5$ unlikely to be damaging
	2. Duration (increases with magnitude)			(important for potential earth slides)
	3. Location of epicentre.			
	4. Depth to focus.			Focus in upper 20km for destructive effects.
	5. Number and magnitude range of aftershocks.			
	7. Peak acceleration at damsite.		Assessment from distance relationship for all earthquakes of 6.	
	Time dependent variations during design earthquakes, including amplitude, acceleration, duration of strong shaking, frequency.	N & I	Computer simulated accelograms from well recorded earthquakes in similar exposures, modified as appropriate for 7.	Two horizontal components similar, vertical less. (Common assumption $a_v = 0.5a_h$). Vertical component also has higher frequency.
	Dam response.		Resonance measurements by coupled shaking machines for dams higher than 60m. (But amplitudes very small). Observations of installed instrumentation during earthquake events (seismographs; strong motion accelographs; pore pressure, deformation & stress measuring instruments).	Many different modes of vibration participate in motion of dam.
	Intensity. Scales according to effects. Modified Mercalli I - XII	N		
	USSR scale of degrees 0 - 10°.	N		Seismic factors used for 7°-9° and special design measures for stations >1000 MW in 9° areas.
	Japanese met agency 0 - VII	N		
		N	It has been suggested that some earthquake resistance should be provided for dams outside recognised seismic zones. It has also been suggested that any type of dam, well built, should withstand a peak acceleration of 0.2g. This assumption should be treated with extreme caution.	
		I	It is commonly accepted that induced earthquakes will not exceed maximum credible natural earthquake for the area involved. Also that induced earthquakes are of the strike-slip or normal type.	

TABLE 1 continued

ASPECT	PARAMETERS	N,I	INVESTIGATION OR INSTRUMENTATION	RANGE OF PARAMETERS & REMARKS
Geology	Tectonic history	N,I	Assessment by geologists and geomorphologists.	
	Faults and their activity (over thousands of years). Inclination. Length involved in slippage.	N,I	Assessment by geologists and geomorphologists. Use of satellite imagery; trenching across faults.	Locate and assess all major fault zones within 130 km of dam site.
	In situ stresses. Absolute values, h & v. Ratio, horizontal to vertical.	I	From tectonic assessment. Also, in special cases, by overcoring strain gauge clusters or by jacking tests or by hydrofracture.	Combined effect of increased vertical load and pore pressure most likely to induce earthquake where max compressive stress is vert. (normal faulting)
	Rate of strain accumulation	I	In areas of high strain accumulation and seismicity, reservoir induced stress changes small compared with natural, and induced effect on seismicity small. In areas of moderate strain accumulation (low natural seismicity or adjacent high natural seismicity) stress changes from reservoir significant and induced effects on seismicity more obvious. In areas of low strain accumulation (stable seismically) general stress levels well below failure and induced earthquake unlikely.	
Foundations and abutments	Shear strength Poisson's ratio Elastic modulus	N & I	Standard static site investigation techniques; undisturbed sampling and laboratory tests. Vane tests In situ penetration, jacking or plate load tests.	
	Damping factor Elastic wave velocity	N & I	In situ seismic geophysical tests.	
	Susceptibility to deformation (and increased loading)	N & I	Soil or rock mechanics analysis from above parameters.	Soft foundations prone to large earthquake displacements (problem with blankets which should preferably be avoided); note possibility of (a) phase shift from one abutment to other, particularly for large arch dams. (b) loss of freeboard due to differential tectonic ground movements or soil compaction. (c) leakage and piping through earthquake formed cracks (avoid rigid impervious cores). USSR practice allows for seismic soil pressure on retaining walls.
	Susceptibility to liquefaction	N & I	Determination of in situ density and simulated laboratory liquefaction by cyclic loading. Determination of range of possible phreatic surfaces.	Very important for saturated silts and sands and their mixtures. Determination from laboratory tests of densification, surcharge load or high static factor of safety necessary to avoid susceptibility.
	Time dependent variations of above	N & I	Accelerated ageing, slaking, or changed moisture content tests.	It is important to check for deterioration of resistance due to raised water table and time variant effects.

TABLE 1 continued

ASPECT	PARAMETERS	N,I	INVESTIGATION OR INSTRUMENTATION	RANGE OF PARAMETERS & REMARKS
Reservoir water	Dynamic pressure	N & I	Estimation by formulae of Westergaard, Zangar, US Bureau of Reclamation, or USSR Hydraulics Division of Standards. Dynamic water pressure gauges.	
	Waves		Major waves can be caused by reservoir bank slides, see under Reservoir area. See under geology and seismicity for assessment of risk of relative movements.	Combination of some or all, of subsidence; fault displacement; tilting of reservoir bed; landslides; and seiches; can cause overtopping.
			Analyse earthquake case also for reservoir empty, perhaps with lower design earthquake.	
Pore Water	New water tables	I	From geology & borings assess aquifers, aquicludes, permeabilities. By analogue or flow nets assess variant and max. new water tables & any likely anomalies between loading & pore pressure changes.	
Reservoir area, large earth or rock slides	All normal soil and pore water parameters required for static & dynamic stability analysis.		Survey by photogrammetry etc. Site and laboratory investigation of soils, rocks & groundwater levels. Installation of piezometers, inclinometers and deformation markers with regular monitoring. Static & dynamic stability analysis.	Note that reservoir drawdown to alleviate another post-earthquake condition may reduce stability of reservoir or dam bank. Very important in areas of narrow valleys, high seismicity, tectonic or geotechnically unstable slopes.
Path & Depth of escaping water	Mode of dam overtopping and maximum discharge for emergency drawdown.	N,I	Assess overtopping in maximum credible earthquake & volume of water released. Flow & duration for emergency drawdown.	Overtopping not such as to cause failure.
	Downstream channel hydraulics.	N,I	Survey by photogrammetry etc. Estimate of friction coefficient for flood channel from measurements of natural floods. Flood surge wave calculations and available freeboard for downstream habitation and installations.	
Downstream Consequences	Human habitation, occupation, & installations. Infrastructure.	N,I	Demographic, social & economic survey.	Note probability of developments downstream with time.
	Downstream dams and reservoirs.	N,I	Effect of flood wave from affected upstream dam & reservoir, considering flood routing & local earthquake effects at downstream reservoir.	Spare reservoir now sometimes being provided downstream of vulnerable dams in high hazard areas.
Dam	All necessary static and dynamic parameters.	N,I	Design may permit some damage & deformation but no failure at extreme event maximum credible earthquake; no significant structural damage for normal design earthquake.	

may be subject to substantial deformation both due to additional loading during the period of shaking and due to accentuation of movement on faults. In such circumstances, designs incorporating sealing blankets should be avoided as they are liable to rupture or fail by piping. The designer should be alert to possibilities of differential movements over the length of a dam through phase shifts such that movements at one part of the dam or abutment are not in synchronisation with those at another. Again, deformation due to compaction, differential ground movements or partial failure under earthquake conditions may cause loss of freeboard. Lower San Fernando Dam in California was an extreme case of such a problem where approximately only one metre of freeboard remained after earthquake damage and 80,000 people had to be evacuated temporarily from downstream areas (ref. 2).

A significant special case concerns the susceptibility of foundations or the dam itself to liquefaction under earthquake shaking. Saturated sandy and silty sandy soils are a particular risk which may be extreme if in situ densities are low (ref. 2). Where these conditions are known to exist a specialised investigation programme including in situ density measurements and laboratory cyclic loading tests should be undertaken to assess suitable design criteria. Measures which might result include densification of in situ deposits, application of defined surcharge, and adoption of unusually high static factors of safety.

If the dam design is to be subjected to dynamic analysis then the site investigation should include measurement of dynamic properties of natural and construction materials. In situ seismic geophysical test methods are available as are such specialist laboratory techniques as cyclic load tests but further development work is required concerning dynamic design parameters and failure criteria.

With all materials, assessment should be made of time dependent variations of properties as there could be long term deterioration due to new conditions of wetting and loading. Estimates should be made of phreatic surfaces and, in the laboratory, accelerated ageing and slaking tests should be undertaken.

Apart from possibilities of induced earthquake, the reservoir reacts with its surrounding environment under earthquake conditions. Shockwaves exert an additional dynamic pressure on the dam. This pressure can be estimated by several available formulae and observed by gauges installed on new or existing dams.

Waves can be established on a reservoir surface during earthquake and these may be very serious, as was the case at Vaijont. In certain circumstances, low stability areas of the reservoir bank can finally be rendered unstable by the additional dynamic effects of earthquake. Extremely large volumes of material may slide into the reservoir, and the displaced water may overtop the dam with serious consequences. A similar effect can be caused by fault displacement or tilting of the reservoir bed. These possibilities require investigation during planning and design, including determination of soil and rock mechanics properties and pore water conditions for static and dynamic stability analyses for potential slip areas. Often, the areas involved are substantial and the use of photogrammetry and satellite imagery may be an economic first approach.

Another factor which may influence reservoir bank as well as dam stability, is the possible desirability of lowering the reservoir rapidly after earthquake damage, and the analyses should include such drawdown conditions.

Dam overtopping should be only a remote possibility or else of very limited extent. However, if it remains a possibility within the design life of the dam, then the overtopping mode and the water flow downstream should be assessed. The latter would be on the basis of a flood surge, using a best assessment of the friction coefficient appropriate for conditions of flow in excess of bank full. Any overtopping should take place without dam failure and designs may need to be modified accordingly.

In assessing the performance of a dam and reservoir under earthquake conditions and the impact on the surrounding environment, the effect on human life, habitation and installations downstream, is of the utmost importance. Dam projects often lead to additional human activity and, in assessing risk and exposure, an assessment must be made of demographic, social and economic conditions, as likely to develop over the design life. A special case concerns any cascade of dams and reservoirs. In such an instance, releases of excess quantities of water from an upstream reservoir must be passed safely through the lower reservoirs. This situation is being met in some instances in California by providing downstream of the main operating dam an additional dam normally kept empty and intended to act as a catch pond for any unusual flows.

For the dam itself, as part of the total environment, design criteria may permit damage under extreme maximum credible earthquake conditions. However, this should never imply total failure or major release of stored water and for the defined normal design earthquake, damage should be minimal.

Performance of Dams Under Natural Earthquake

A study of the literature reveals that although a substantial number of dams have been subjected to earthquakes, relatively few have collapsed completely. Table 2 lists some dams affected by natural earthquakes. In the first section there are named 5 dams which suffered significant or serious damage and 3 which failed completely. In the second section are listed 17 dams which suffered little or no damage in earthquakes of varying intensity. In addition to those named, Seed (ref. 2) reports that a

TABLE 2 SOME DAMS SUBJECTED TO NATURAL EARTHQUAKES (see also ref. 8)

Dam	Country	Description	Ht. m	Built	E'quake Date	E'quake M(or I)	Damage
FAILURE OR SERIOUS DAMAGE IN NATURAL EARTHQUAKES							
Sheffield	U.S.A.	Embankment: silty sand	8	1917	1925	6.3	Collapse
Eklutna	Alaska	Embankment:	6	1929		8.5	Severe
Lower San Fernando	U.S.A.	Embankment: sand shells	42	1912/30	1971	6.5	Very severe
Upper San Fernando	U.S.A.	Embankment: sand shells	25	1921	1971	6.5	Serious
Hebgen Dam	U.S.A.	Embankment	37	1915	1959		Serious
El Soldado	Chile	Tailings Dam			1965	7.1	Collapse
(On Izu Peninsula)	Japan	Tailings Dam			1978	5.7	Failure
Baihe	China	Embankment	60		1976	7.8	Serious

In addition Seed (ref 2) reports failure of Volcano Lake, Coleman, Rogers & 2 private dams in USA, Dobry & Alvarez report failure of 10 Chilean dams during a mag. 7 earthquake (1965). (ref. 3)

LITTLE OR NO DAMAGE IN NATURAL EARTHQUAKES

The following dams suffered little or no damage in earthquakes of magnitudes between 5.8 & 8.2 recorded since 1906: Oroville, Dry Canyon, Santa Felicia, Lower Crystal Springs, Pacoima, USA; Poteba, Algeria; Madan, Bayano, Panama; Cogoti, Recoleta, Chile; Ono, Upper Murayama, Lower Murayama, Otani-ike, Honen-ike, Japan; Douhe, China; Corfino, Italy. Their heights are between 22 & 235m and the types covered include embankment, concrete gravity, and concrete arch.

TABLE 3 SOME CASES OF RESERVOIR INDUCED EARTHQUAKES (see also ref. 9)

Dam	Location	Description	Height m	Res.Vol. $10^6 m^3$	Impounding Date	Largest E'quake M	Largest E'quake Year
MAJOR EARTHQUAKES							
Marathon	Greece	Gravity	63	41	1930	5.0	1938
Hoover	U.S.A.	Arch-gravity	221	36,703	1936	5.0	1939
Kariba	Zimbabwe/Zambia	Arch	128	160,368	1959	5.8	1963
Hsinfengkiang	China	Concrete buttress	105	10,500	1959	6.1	1962
Koyna	India	Concrete gravity	103	2,708	1964	6.5	1967
Kremasta	Greece	Embankment	165	4,750	1965	6.3	1966
Roi Constantine	Greece	Embankment	96	1,000	1969	6.3	

MINOR EARTHQUAKES

At the following dams, minor induced earthquakes of magnitudes between 3.2 & 5.0 were recorded: Clark Hill, Jocassee, USA; Grandval, Monteynard, France; Canalles, Spain; Vajont, Italy; Benmore, N.Z.; Kurobe, Japan; Bajina-Basta, Yugoslavia; Nurek, USSR; Talbingo, Australia; Manic 3, Canada. Heights ranged between 67 & 317m & reservoir volumes between 61 & 10,400 x 10^6 m^3. Delay between impounding & maximum induced earthquake was less than 4 years except for a single case of 22 years.

TABLE 4 IMPACT ON ENVIRONMENT CAUSED BY SEISMIC EFFECTS AT DAMS & RESERVOIRS

PHENOMENON	IMMEDIATE EFFECT	PRECAUTIONS OR ACTIONS
Ground surface displacement (M>6.5)	Cracking of superimposed structures. Contribution to waves & seiches.	Mainly avoidance of building across known active fault lines. Self sealing materials, plastic concretes, bitumens, sands, may have a place.
Water waves & seiches due to ground motions, landslides. Failure of spillway and/or outlets in earthquake.	Possibly overtopping, washout of crest, spilling on power-house at toe & loss of power to gates etc; flood surge in river downstream.	Analyse & stabilise reservoir slopes; provide large freeboard. Reinforce crest. Protect downstream areas by levee banks or relief channels if hydraulic analysis shows necessary.
Dynamic water pressures.	Increase in water load.	Design for effect.
Emergency emptying of reservoir following damage to dam.	Flooding in downstream area. Possible drawdown condition failures in dam & reservoir banks.	Protect downstream areas from flooding. Design dam & stabilise reservoir banks against most rapid drawdown.
Foundation & dam cracks during earthquake.	Possible piping or erosion problems or failure.	Provide wide filters & self sealing materials. Do not use brittle materials in construction. Use rubber or plastic water stops, not metal. Controlled drawdown if necessary.
Major e'quake damage.	Low factor of safety, lack of adequate water control because of partial failures, as shown by immediate inspection.	Evacuate downstream population.

large number of embankment dams in Japan were affected by the 1939 earthquake, of which 12 failed completely, and 40 had slope failures. The majority of the failed and damaged dams were built of sandy soils and there were no complete failures in banks of clay soils. A further 33 embankments in California were affected by the 1906 San Francisco earthquake without significant damage. They were mostly constructed of clayey soils. A large number of small irrigation dams in Japan were affected by the 1968 Tokachi earthquake. The dams were reported to be constructed of loose volcanic sand and some 93 were damaged.

It will be noticed that a large number of dams have been subjected to severe shaking without incurring sufficient damage to release the reservoir and many have suffered only slight damage. Dams constructed of fine granular materials are the most vulnerable and all the complete collapses have involved this type of material. Nevertheless not all dams constructed of such material present an earthquake hazard. This is demonstrated by the rather surprising result of the study of five hydraulic fill dams in California reported by Leps et al (ref. 4). Leps concluded that although some dams constructed by this method constituted a hazard, many, if not most, had a satisfactory margin against failure during moderate earthquakes.

Comparatively few earthquake incidents involving concrete dams have been reported. Limited evidence suggests that concrete dams have a good resistance to shaking. Clearly arch dams will tolerate relatively little ground movement within the structure and gravity dams become sensitive if the crest mass is excessive.

Induced Seismicity

A discussion on seismicity and dams would be incomplete without consideration of induced seismicity. This was first noticed as a result of Magnitude 5 events apparently associated with Lake Mead (USA) in 1939 and Marathon (Greece) in 1938. For two decades these were thought to result from special conditions at these locations until the spate of construction of large reservoirs in the late 1950s and 60s. Then within a very few years events of magnitude in the range 5.0 to 6.5 occurred at Koyna (India 1967), Kremasta (Greece 1966), Hsinfengkiang (China 1962), Kariba (Rhodesia 1963) and Benmore (New Zealand 1966). (Table 3) Significant events below magnitude 5 occurred at about 11 other reservoirs in various countries. That these events were induced is evidenced by the short period between impounding and the largest earthquake (usually, though not always, less than 4 years) and their coincidence in many cases with the achievement of maximum water level.

Except for two, virtually all the reservoirs associated with induced earthquakes are large (i.e. greater than 200 x $10^6 m^3$ and many greater than 1000 x $10^6 m^3$) and it was initially assumed that the great weight of the reservoir was the major factor in triggering the earthquake. A study of alignments and attitudes of the earthquake generating faults in various cases has led to the view, now gaining greater acceptance, that the increase of pore pressure at depth is also a major contributing factor to the occurrence of induced seismicity. Evidence for this is provided by the seismic effects resulting from high pressure fluid injections reported by Healy and others (ref. 5).

Understanding of the mechanism of induced seismicity has hitherto been inhibited by lack of detailed information concerning the seismic regime prior to construction of the reservoir. Nurek reservoir in USSR (ref. 6) is an exception to this and much information is available concerning the seismic history of the area prior to impounding. This is likely to be the case for many future large reservoirs.

It must be borne in mind that the great majority even of the largest and deepest reservoirs do not stimulate seismic effects. For example, of the 25 largest reservoirs quoted in the ICOLD World Register of Dams, only 4 have shown induced seismicity. The propensity for induced seismicity is a function not only of the size parameters of the reservoir but also of the tectonic state of the earth's crust at the reservoir location prior to impounding.

Impact on Environment Caused by Seismic Effects at Dams and Reservoirs (Table 4)

Many of the potential impacts on the environment associated with seismic events at dams and reservoirs have been referred to earlier. However, some deserve further mention and emphasis.

Dams actually built across active faults which suffer displacement in earthquake are clearly prone to deformation and damage and only in cases of extreme necessity should such known sites be developed. Without proper precautions, the dam and its ancillary structures can be subject to cracking and possibly failure. Even if cracking does not lead to an immediate structural problem, it could well be followed in the case of an embankment dam or the foundations by progressive piping and erosion with serious consequences. Some safeguard can be provided by incorporating within the design elements of self-sealing materials such as plastic concretes, bitumens and sands. Filter layers in embankments should be wide and also have self-sealing properties. For water stops, rubber or plastic based materials should be used instead of sheet metal. In the event of cracking and relative movements, early inspection of the dam could lead to a decision on the desirability of drawdown which should be controlled in order to avoid rapid drawdown type instability.

The reservoir water can be the subject of major waves and seiches due to a variety of ground motions or reservoir landslides as well as due to the dynamic forces within the water body. Such phenomena can lead to overtopping of the dam and without adequate protection, this could involve crest washout and loss of freeboard. Overtopping spillage could lead to damage and failure of any power house at the toe of the dam which could be particularly serious if it were the sole source of supply for power to gates. Overtopping would also lead to a flood surge in the river downstream. The spillway outlets may be made inoperable by earthquake movements, requiring immediate action for emergency water level control. These problems can be guarded against by such measures as analysis and stabilisation of vulnerable reservoir slopes; provision of extra freeboard and crest reinforcement; and the protection of downstream river channel areas by embankments, training works or relief channels.

In rare cases a large reservoir may result in induced seismicity. Present understanding suggests that this happens only in areas where tectonic conditions exist for natural seismicity. If this is so then the induced earthquake is an earlier occurrence of an event likely to have happened naturally. It has been argued from this that, because the build up of strain (being a function of time) is less, the induced event must be smaller in magnitude. This theory ignores the possibility that areas of reduced effective stress consequent on a rise in pore pressure may result in a greater extent of the fault being involved in the movement and consequently an increase in seismic magnitude.

Older dams in seismic areas represent a special problem and many are being investigated and reanalysed by modern techniques. Strengthening by such methods as ground anchors and drainage may be undertaken, and in extreme cases, replacement can be required.

Finally, there remains the problem of major damage due to an earthquake far greater than that predicted or allowed for in the designs. If there is no immediate failure of the dam structure but considerable damage and perhaps lack of adequate water control, then an immediate and thorough inspection is necessary by suitably qualified engineers. This may well lead to an emergency decision for evacuation of population from vulnerable areas downstream and the chain of decision making and communications should be adequate for the necessary prompt action.

Conclusions

A dam interacts with its environment, including seismicity, on a global and local scale. Globally, the world can be divided into seismic zones, but seismologists advise that no areas are completely aseismic. Locally, the dam's immediate environment may include foundations close to a potentially active fault or on susceptible material such as low cohesion soils. The reservoir environment is also vital and vulnerable slopes should be investigated and stabilised to prevent massive displacement water waves.

The dam itself is susceptible to earthquake effects. Experience shows that many dams have reasonable resistance to quite severe shaking but some types are vulnerable, particularly certain embankments with shells of silts and sands. Potential major environmental risk relates mainly to sudden substantial releases of water and their effect on downstream population, habitation and installations. The effects would be serious and much attention is being directed to dynamic techniques in the field, laboratory and design office to ensure adequate earthquake resistance. These techniques are being applied increasingly but little prototype verification has yet been possible because of the rarity of major events and lack of instruments.

Old dams and reservoirs may require re-analysis and modification to meet modern criteria.

A factor relevant to the safety margins to be built into a dam, is the degree of development downstream. This can be a major unknown considering the life of a dam and the pace of development in many countries.

An aspect brought to light by larger reservoirs is that of induced seismicity. Understanding is in its infancy and accurate prediction is probably a long way off, but theories are being developed. The literature (Table 1, ref.7 et al) indicates geological and tectonic conditions of particular significance.

In the present state of the art, added precautions such as extra freeboard, self healing materials, strengthening and even catchpond dams downstream are being adopted to protect the environment where seismicity, foundation and reservoir conditions, and downstream hazard conjoin adversely.

REFERENCES

1. BULLARD Sir Edward. The origin of the oceans, Scientific American Inc, 1977.
2. SEED H.B. Considerations in the earthquake-resistant design of earth and rockfill dams. Geotechnique 29, No.3, 1979, 215-263.
3. DOBRY D., ALVAREZ L. Seismic failures of Chilean tailings dams. ASCE Soil Mechanics and Foundations Division, Nov. 1967.
4. LEPS T.M., STRASSBURGER A.G., MEEHAN R.L. Seismic stability of hydraulic fill dams. Water Power & Dam Construction, Oct. & Nov. 1978.
5. HEALY J.H., RUBEY W.W., GRIGGS D.T., & RALEIGH C.B. The Denver earthquakes, Science 161, 1968.
6. SIMPSON D.W. Reservoir induced seismicity at Nurek & Toktogul Reservoirs, Soviet Central Asia. Proc.Conf. on Intra-Continental Earthquake, Ohrid, Yugoslavia, Sept. 1979.
7. SIMPSON D.W. Seismicity changes associated with reservoir loading, Engineering Geology 10, 1976, 123-150.
8 & 9. HAWS E.T. Tabulation & data of dams affected by natural & induced earthquakes. Unpublished, available upon request, 1980.
10. ICOLD Committee on Earthquakes. A review of earthquake resistant design of dams. International Commission on Large Dams, Bulletin 27, March 1974.
11. ICOLD Q51 Seismicity & aseismic design of dams. Proc 13 Int.Congr.Large Dams New Delhi'79

29 Hazards of natural and induced seismicity in the vicinity of large dams

P. L. WILLMORE, MA PhD, Institute of Geological Sciences, Edinburgh

The seismicity problem for large dams is compared and contrasted with that of nuclear installations, in which one seeks to determine the size and character of the Operational Base Earthquake (OBE) and the Safe Shutdown Earthquake (SSE) from the study of natural seismicity. In the case of large dams, seismicity induced by the impoundment of water is important, and may dominate the statistics. The methods available for the assessment of hazard on a particular site, and the returns from a proper programme of seismic observations during and after construction, will then be discussed.

INTRODUCTION

1. Our approach will be first to consider the problem of natural regional seismicity around the proposed dam-site to see the extent to which we can set limits on the nature of the ground motion which is most likely to produce disturbance of any given level of intensity.

2. Our statistical expectations may be greatly modified by the possibility of the inducement of seismicity, now thought to be mainly due to the increase in water pressure in pores and cracks, and will consider the ways in which inducement can affect our expectations in highly seismic areas, as contrasted with those in areas normally thought to be aseismic. This will lead us to consider the possibility of using static strain measurements to assess the condition of the bedrock in the site area.

3. Next, we shall consider the various types of instrumentation available for use on site and in surrounding regions proceeding then to the types of seismic observations, and to the role which the observational programme can play in the periods before and after impoundment.

4. In drawing final conclusions, we will summarise the extent to which the various classes of data may be relevant to the agency responsible for the dam construction, as compared with those which will contribute to the general body of knowledge relevant to the construction of other dams. This will provide some indications of the sources of funds and the nature of the organisations most appropriate to work of this type.

5. It is emphasised that the opinions expressed in this paper are those of the author, and should not be considered to represent official policy of the Institute of Geological Sciences or of the Natural Environment Research Council.

PRELIMINARY SITE ASSESSMENT

6. It is assumed that the Geological Site Report will include mention of any faults visible at the surface, and major earthquake zones at plate boundaries are, of course, typically marked by extensive belts of multiple faulting. Individual faults in other regions, however, exhibit comparatively little correlation with seismicity, as a fairly high proportion are left over from extinct orogenies, and many substantial intra-plate earthquakes occur in regions devoid of related surface faulting.

7. Given that the major data on seismic hazard came from the historical pattern of earthquake occurrence, one needs the longest possible historical time-span to smooth out the effects of short-term "bunching" and to provide evidence of any longer-term migration of centres of activity. The most directly relevant data would reside in the detailed histories of seismic intensity derived from motions actually felt in the vicinity of the construction site, but these, because of the tendency for dam sites to be in fairly inhospitable regions, are seldom available.

8. The main body of data is therefore derived by the application of magnitude-intensity-distance relations to known locations, and this process can be applied by computer to every point of a raster covering the site area. Thus, after allowing for the often gradual lowering of the detection threshold as regional coverage develops, we derive an estimate of the maximum intensity experienced in a given historical time-span (which can be projected into the future if the historical data-set is assumed to be a good random sample of statistically stationary population), or an estimate of the recurrence interval for any given level of intensity.

9. The most elementary statistical relationship between frequency of occurrence

(N(T) events in time T) and the severity of an event (Intensity I at any given point or magnitude M within a given area) is of the form

$$\mathrm{Log}_e\, N(T) \doteq a(T) - bI \text{ (or M)} \qquad (1)$$

$$= a + \log T - bM$$

if the events described are earthquakes of magnitude M and a is taken to characterise the population of events within unit area and time, (Fig. 1). The Intensity population will be zoned (as in Fig. 2) if the local variations within the historical time-span are assumed to be stable over longer time intervals i.e. historically active regions continue to be active, quiet ones remain quiet). This entirely probabilistic approach (as in Milne and Davenport, 2) can be modified (as by Cornell, 3) if a causative link between earthquake occurrence and tectonic features can be identified. In this approach, for example, "gaps" in linear features can be expected to fill up in the course of time, although the lineaments as a whole maintain long-term stability.

10. If M(T) is taken to be the magnitude of the largest earthquake to be expected in a time interval T, log N = 0 and the solution to equation (1) becomes

$$M(T) = a/b + 1/b \log T \qquad (2)$$

Equation (2) can be extended to cover a variable time-span comprising S intervals of time T by writing

$$M(ST) = a/b + 1/b (\log S + \log T) \qquad (3)$$

which is plotted as the straight line in Fig. 3. This approach has the advantage that we need only collect evidence of the largest event in each time interval T, thereby avoiding many of the problems of lack of completeness, and of the tendency for earthquakes to occur in "swarms".

11. If the individual samples within our available time-span are ranked in descending order of magnitude, the probability that the jth sample will represent the largest event in any future time interval ST is

$$P = j/(1 + S) \qquad (4)$$

this being the form of presentation to which Yugulalp and Kuo (4) applied Gumbel's first and third probability distributions (5). Of these two distributions, the first, corresponding to events of unlimited magnitude, corresponds to the straight line of Fig. 3, whereas the third (magnitude tending asymptotically to an upper limit) corresponds to the curved line.

12. Now let us consider the annual number of events occurring within the radius r of a given site, in an extended region of uniform seismicity. From equation (1) we now have

$$N(r) = 2\pi r^2 \exp (a - bM) \qquad (5)$$

and let us also assume an intensity-distance relationship of the form

$$I = \alpha M - \beta \log r \qquad (6)$$

Eliminating r and consolidating the constant terms, we find

$$\mathrm{Log}\, N = 2/\beta\ (\alpha - b\beta/2)\ M - I \quad + \text{consts} \qquad (7)$$

so that the expectation of a given value of I will be dominated either by large, distant earthquakes or small, near ones, according as to whether $(\alpha - b\beta/2)$ is positive or negative.

13. In the middle range of magnitude for most earthquake regions, $(\alpha - b\beta/2)$ is positive, so that fairly large earthquakes contribute most of the hazard. If, however, an upper limit of magnitude exists, b will tend to infinity as the asymptotic limit is approached, and the maximum expectation of given intensity will correspond to the value of M for which $(\alpha - b\beta/2) = 0$, (see Fig. 4, references 6 and 7). In Britain, the body-wave magnitude of the "most perceptible" earthquake is about 5.25, whereas larger values are common in regions of higher seismicity. This approach gives us a criterion for selecting or synthesising a "design eathquake" appropriate for any given site. This "most perceptible" event seems the appropriate one to consider for the "Operational Base Earthquake" of a nuclear power station, or for its counterpart (the threshold of minor damage) in dam design.

INDUCED SEISMICITY

14. It is now well known that a substantial proportion of large reservoirs have been the site of earthquake swarms within a few years of impoundment, the figure having been set at some 7% of all cases for which the dam height exceeds 100 m and the volume of water exceeds 10^9 m^3. The incidence of this phenomenon bears little relation to the natural seismicity of the dam site region. Weakening of the rock formations by the penetration of water into pores or cracks is the most probable mechanism.

15. One unifying hypothesis which can be applied at the very elementary level of the above remarks is that the whole of the earth's crust is subjected to some strain input from crustal movements. Earthquake zones are those in which relative motions between plates are sufficiently rapid to require release by the "stick-slip" process. Regions of little or no natural seismicity are those in which strain accumulations are relieved by creep. As creep rates increase rapidly when stress differences approach the level of fracture there should be a tendency to produce equilibrium stresses at levels for which a sudden, slight weakening of the rock can lead to fracture. The following conclusions emerge from this hypothesis:

i) In earthquake zones, an increase in pore

pressure can advance the phase of the stick-slip cycle but will, if anything, reduce its severity by reducing the stress-drop. The characterisation of the "most perceptible" earthquake is shifted from the large and rather distant combination towards a higher expectation of a nearer event of lesser magnitude.

ii) In areas of little or no natural seismicity, the pore-pressure increase can generate events which might otherwise never have occurred, and can increase, perhaps by a very large factor, the expectation of an earthquake occurring very near to the dam. The maximum magnitude for such an event should be considered to be the maximum expected, on statistical or other grounds, for the whole of the seismic province which includes the dam site, and therefore demands consideration at a level comparable to that of the "Safe Shutdown Earthquake" of a nuclear power station.

iii) Contrary to popular belief, the immediate aftermath of an earthquake which fits in to the long-term seismicity pattern of the region is a comparatively good time for design and construction to proceed, as the prior expectation of such an event should, in any case, have been included in the site assessment. The fact that the event actually has occurred will modify (usually downwards) the stress-field of the region, and will improve the data base for assessing the spatial and temporal distribution of natural earthquakes.

iv) It has been strongly urged (see, for example, 8) that in-situ stress should be measured on reservoir sites in advance of impoundment. In view of comment (ii) above, such measurements are particularly relevant in regions of low natural seismicity.

v) Given the very limited extent of the space-time window within which much of the hazard is concentrated, dam-site investigation should be recognised as a potentially very fruitful area for earthquake prediction studies. The simplest of monitoring networks will yield seismicity data within which one could look for the characteristic patterns of foreshock activity. More ambitious observational programmes could be aimed at monitoring changes in the VP/VS ratio, or the development of anisotropic wave-propagation characteristics during the period of impoundment.

EQUIPMENT AND METHOD

16. The basic requirement for the location of a seismic source is to record the times of the first onset of vibration at at least four suitably located seismograph stations, or to determine the times of arrival of the "P" and "S" phases for at least 3 stations. As the waves in question have propagation velocities between about 3.5 and 8 km/sec, relative times of arrival on the different channels should be determined at least to within 1/10th of a second to yield locations to the nearest kilometre, and the best instrumentation provides accuracy of the order of 1/100th of a second. The stations should be sufficiently remote from the main centres of activity to avoid the worst of the effects of locally-generated vibration and should enclose the entire area within which local earthquakes could be induced by, or relevant to, the reservoir. Typically, a network radius of 10-20 km would seem to be appropriate. The following three types of network are worth consideration:

i) Separate stations, independently timed, recording on rotating drums with pen and ink or on smoked paper. Capital cost is in the vicinity of £4k per station, operational costs include daily service visits, but staff can be relatively unskilled. Record interpretation is simple and rapid at an elementary level, but limited in accuracy of location, and in the extent to which it can be developed to higher levels of sophistication.

ii) Telemetry on to continuously running magnetic tape. Requires a good central viewpoint for inexpensive line-of-sight radio linking and a good electronics technician for maintenance. Capital cost around £20k for the central recording station, £20k for the playback system and about £2.0k per outstation channel. Consummables for a 3-component base and 8 outstations would amount to about £6k per annum. The comparatively high cost of the playback equipment could be shared between more than one record system if there were other dam sites in the area for which similar monitoring requirements existed. If both sites were to be instrumented, a longer range of telemetry or a multiplex link could lead to economies of recording and processing, and in this situation the cost-effectiveness of the tape system would be enhanced in relation to the all-drum approach.

iii) Triggered digital systems. Competitive in capital cost with continuously-running magnetic tape, and offering rather better options for the most sophisticated interpretation. Centralised recording, however, involves all of the constaints which arise from telemetry requirements, separate recording suffers from the problem of doubtful uniformity of triggering across the network, and the clearest requirement for this type of instrumentation is to cover the strong-motion range of intensities. Strong-motion instrumentation should certainly be installed in the completed dam before impoundment commences, and it is likely that digital recorders and their playback systems will soon have established a clear margin of superiority over the small film recorders which provide most of the existing strong-motion recording capacity around the world.

17. The high quality of the ultimate record, and the fact that the telemetry brings together the output of all the recording stations to a common point for assessment, makes it highly desirable for such a system to be in

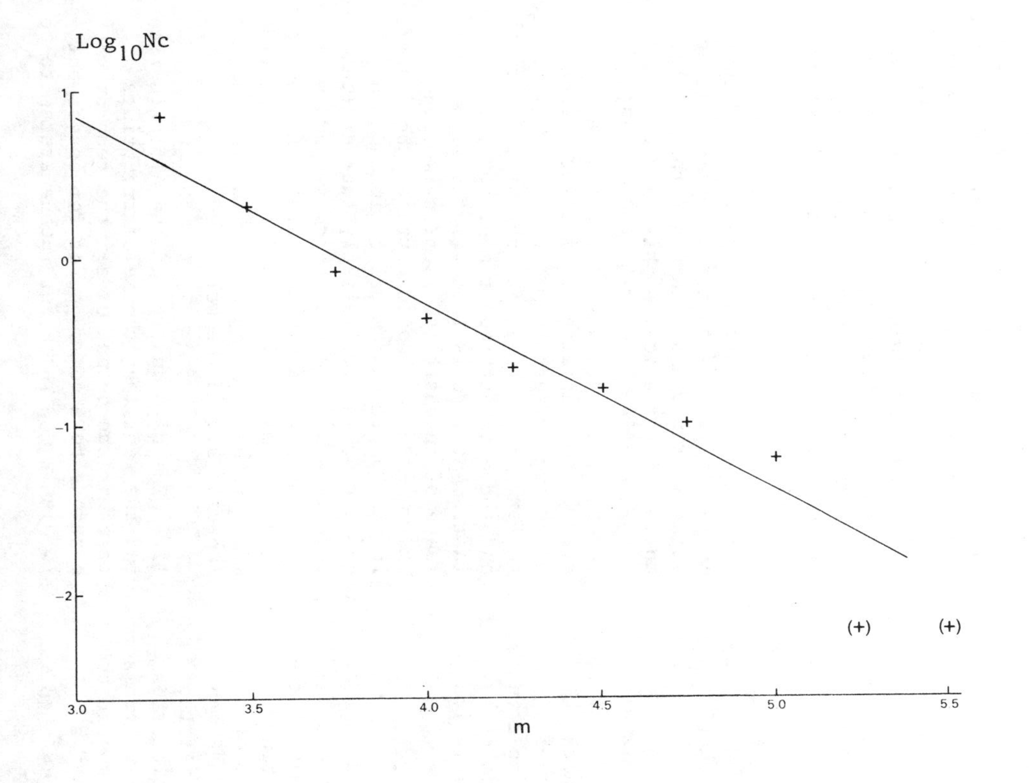

Fig 1 (above). Number of events per annum for British events on land, as a function of body-wave magnitude m (adapted from ref 4).

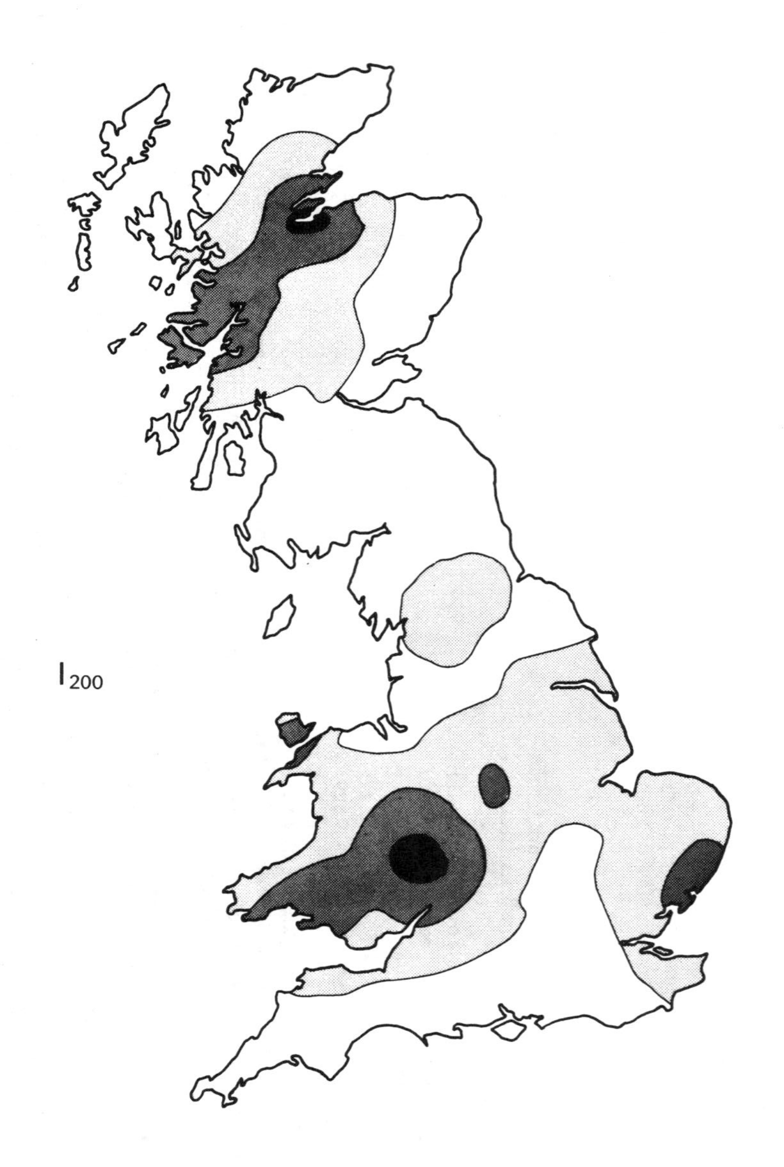

Fig 2 (right). Intensities in Great Britain with return period of 200 years (ref 4).

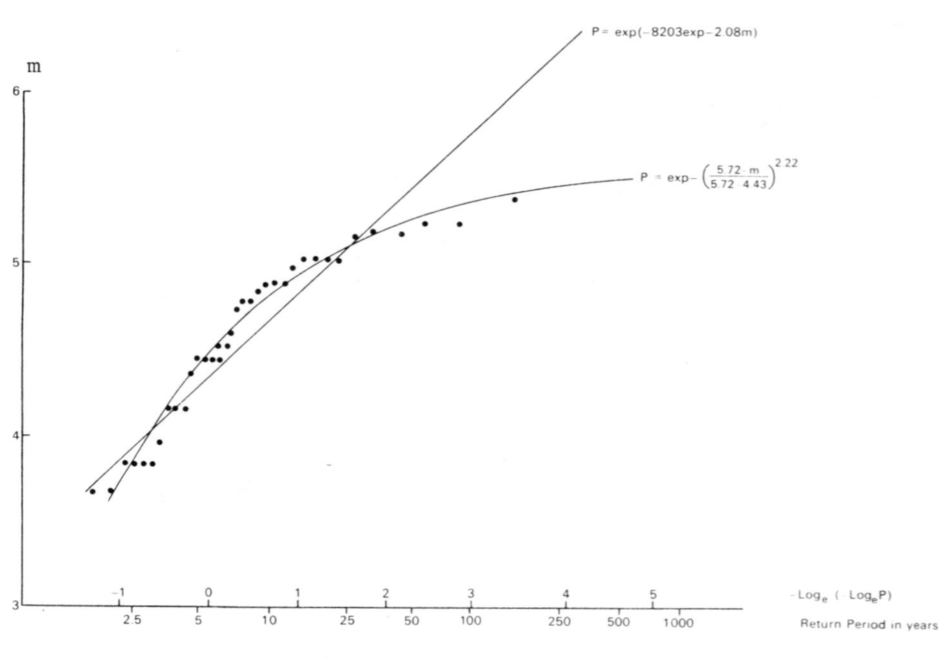

Fig 3 (above). Five-year extreme magnitudes plotted against return period for m as an annual extreme (ref 4).

Fig 4 (right). Annual expectation of Intensity ⩾ I in Great Britain, generated by earthquakes of magnitude ⩾ m. Solid lines, surface point focus as equation (7). Dotted lines, modified hypocentral distance corresponding to finite focal radius and depth (ref 6).

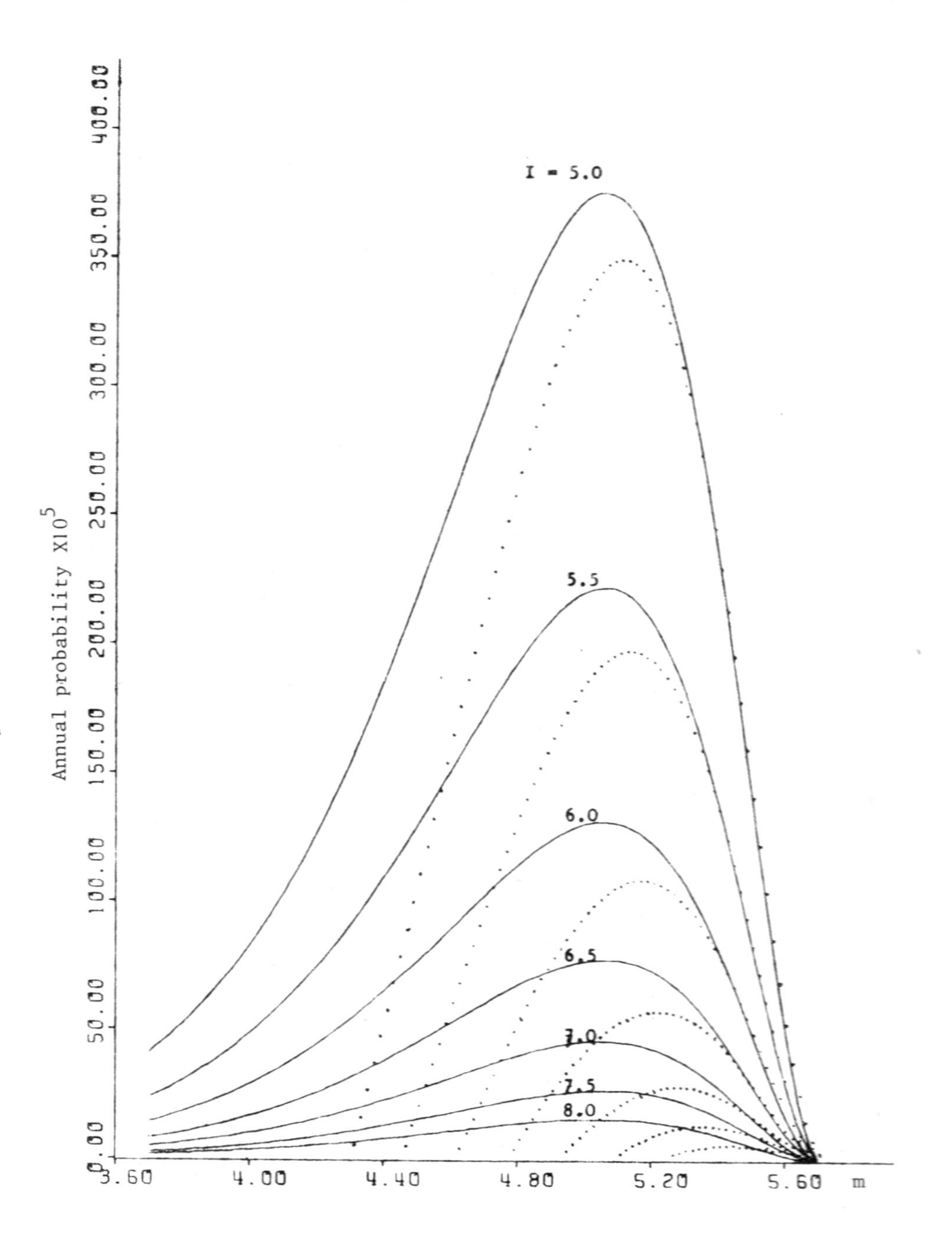

operation well in advance of the commencement of impoundment. Taking all of these considerations into account, the recommended approach would be:

i) Acquire about four small drum-recording systems for installation as soon as possible, perhaps coincidentally with the assessment of the possibilities for a more powerful system. Operations could be sustained by locally engaged part-time staff, guided by a small amount of expert consultant's time.

ii) Acquire and deploy the main body of sensitive equipment, needing at least one full-time technician for operation and one scientist for interpretation. This phase should be under way during the construction phase of the dam.

iii) Instal the strong-motion equipment towards the end of construction, in time to monitor the impoundment stage.

18. By these means, we have the opportunity to locate and indentify natural and artificial seismic sources in the region, thereby providing the necessary understanding of the background activity against which one will have to distinguish important events during the impoundment stage and its aftermath. Note, however, that stage ii could yield side benefits to the dam constructors, by providing a record of the seismic side-effects of any untoward incidents during the construction period.

CONCLUSIONS AND RECOMMENDATIONS

19. The main points made above may be summarised as follows:

i) Historical studies of regional seismicity can give useful guidance on the characteristics and occurrence probability of a "design earthquake", but the inducement potential of large dams can greatly enhance the expectation of events at short range.

ii) Low or unobserved levels of historical seismicity do not imply immunity from induced effects, and a shortage of historical data might even require enhanced factors of safety to allow for inducement. Inclusion of stress measurements during the site-assessment stage seems to be the most desirable approach to this problem.

iii) Site monitoring (which can initially be undertaken with quite simple equipment) should start as early as possible, in order to identify natural and artificial sources in the region, and to provide date on the velocity structure for subsequent earthquake location

iv) Full-scale monitoring, starting before impoundment and continuing for several years afterwards, requires telemetry into high-quality recording systems, to permit the application of modern methods of earthquake prediction in relation to possible induced effects.

v) Triggered strong-motion recorders should be deployed at a late stage in construction, and maintained in operation throughout the lifetime of the dam.

vi) The cost of a monitoring service to the construction agency can, at least to some extent, be offset as an "insurance" against side-effects of untoward incidents during construction.

20. Having drawn the above conclusions, notably point (vi), from the foregoing argument, we should point out that the major long-term benefit of each experiment will arise in connection with the design and construction of dams other than the one on which any particular experiment has been carried out. It is therefore suggested that some international body, such as ICOLD, should undertake the initiative of planning a coherent experimental programme, acquiring a body of equipment with experienced operators who could move from one construction project to another and finally acting as a clearing house for the results of the successive research projects.

ACKNOWLEDGEMENTS

21. The author acknowledges the financial support of the Natural Environment Research Council, and the work is published by permission of the Director of the Institute of Geological Sciences.

REFERENCES

1. LILWALL R.C. Seismicity and seismic hazard in Britain. Institute of Geological Sciences, Seismological Bulletin No. 4, 1976.
2. MILNE W.G. and DAVENPORT A.G. Distribution of earthquake risk in Canada. Bull. Seismol. Soc. Am. Vol 59 part 2, 1969, 729-754.
3. CORNELL C.A. Engineering Seismic risk analysis. Bull. Seismol. Soc. Am. Vol 68 part 2, 1968, 1583 - 1606.
4. YUGULALP T.M. and KUO J.T. Statistical prediction of the occurrence of maximum magnitude earthquakes. Bull. Seismol. Soc. Am. Vol 64 part 2, 1974, 393 - 414.
5. GUMBEL E.J. Statistics of extremes. Columbia University Press, New York and London, 1958.
6. WILLMORE P.L. and BURTON P.W. The U.K. approach to hazard assessment. Proceedings of the E.S.C. symposium on earthquake risk for nuclear power plants, in Luxembourg, 20 - 22 October 1975. Publication No. 153 of K.N.M.I., de Bilt, 1976, 35 - 37.
7. BURTON P.W. Perceptible earthquakes in the United Kingdom. Geophys. J.R. astr Soc. Vol 54 No 2, 1978, 475 - 479.
8. UNESCO Final report of the intergovernmental conference on the assessment and mitigation of seismic risk. Paris 1976, 11 and 13.

30 Computed and observed deformation of two embankment dams under seismic loading

M. P. ROMO, PhD and D. RESENDIZ, DrEng, Instituto de Ingeniería, UNAM, Mexico

On March 14, 1979, El Infiernillo and La Villita dams were shaken by an earthquake of 7.6 Richter magnitude. Both dams underwent measurable settlements. This provided an unusual opportunity to assess the accuracy of available methods to determine embankment deformation under seismic loading.

Given the state-of-the-art and available information, three simplified methods were selected to estimate the settlements induced in both dams. The results are compared to the observed settlements.

INTRODUCTION

1. It is broadly acknowledged that embankment dams subjected to seismic shaking may undergo a number of failure or damage modes (refs. 1 and 2). Cracking, sliding, settlement and slumping are some of them and their extent depends strongly on the characteristics of the embankment foundation and materials. Loose to medium dense sands in the embankment or its foundation are likely to undergo large slides under earthquake loading, whereas denser and coarser materials tend to result in lesser embankment damages.

2. Important dams are currently constructed under stringent specifications and control, so as to attain uniform, dilatant embankments with low deformability. Likewise, loose or weak deposits in the foundation are either eliminated or appropriately treated, particularly if the site is in a seismic area. Thus, the most severe damage modes are generally avoided.

3. However, high embankment dams are susceptible to undergo at least moderate permanent deformations (i.e. slumping and settlement), irrespective of its foundation and material properties, when hit by a strong earthquake. Hence, this damage mode should be explicitly analyzed in dams located in seismic zones.

4. Permanent deformations induced by earthquake loading depend on a number of factors such as the intensity and duration of the seismic event, the static stress condition acting in the dam before the earthquake and the response to cyclic loading of the materials that constitute the dam.

5. A number of elaborate procedures based on the finite element method have been developed recently to compute both the initial static stresses (prior to earthquake loading) and the dynamic stresses induced by the seismic event (refs. 3 to 5). Simpler, approximate procedures to compute embankment deformations exist as well (refs. 1, 6 to 10). However, neither elaborate nor simplified procedures have been sufficiently calibrated against actual prototype behaviour.

6. This paper contributes to such a calibration. Given the state-of-the-art and the available information, three simplified methods are selected, namely Newmark's (ref. 6), Makdisi-Seed's (ref. 10) and Reséndiz-Romo's (ref. 1) to estimate the permanent deformations induced in La Villita and El Infiernillo dams by the earthquake of March 14, 1979. Estimated and observed deformations are compared in order to assess the accuracy of these methods.

7. The results presented in this paper show that Newmark's and Makdisi-Seed's methods give values far on the unsafe side. The method developed by Reséndiz and Romo gives results on the safe side and somewhat closer to the observed behaviour.

GENERAL CHARACTERISTICS OF THE DAMS

8. El Infiernillo dam is located some 70 km from the mouth of the Balsas River. It has a reservoir with a total capacity of 12 x 10^9 m^3 and is a rockfill embankment with a thin, central clay core. The embankment is 146 m high over the rock foundation and has a volume of 5.5 x 10^6 m^3, of which 5.0 x 10^6 m^3 are granular materials. The core has slopes of 0.089:1 and its thickness is only 20% of the hydraulic head. Well graded, 2.5 m thick, sand filters were envisaged both upstream and downstream, together with transition zones having slopes of 0.15:1. The remaining zones of the section comprised rockfill with outer slopes of 1.75:1 and two berms downstream (see Fig. 1).

9. La Villita dam is about 55 km downstream of El Infiernillo dam and some 13 km from the mouth of the Balsas River. La Villita is a 60 m high earth and rockfill dam whose vertical core has slopes of 0.2:1. At both sides of the core there are filters, transition zones and rockfill shells; the outer slopes are 2.5:1. In plan the dam axis is slightly concave downstream, and is founded on a deep alluvial deposit (see Fig. 2). To prevent water seepage a 0.6 m thick

and 74 m deep concrete cutoff wall was provided. In order to reduce the core settlements and to improve the confinement and strength of the cutoff wall, the alluvial deposits underlying the impervious core were grouted down to 26 m. A more detailed description of both dams has been presented elsewhere (ref. 11).

THE EARTHQUAKE OF MARCH 14, 1979

10. This seismic event had a 7.6 Richter magnitude and its epicenter was located in the Pacific Ocean, some 40 km from the City of Zihuatanejo (Fig. 3). The maximum ground accelerations recorded at SICARTSA steel mill, which is some 8 km downstream of La Villita dam, were as follows:

Component	Acceleration(cm/s^2)
NS	263
Vert.	83
EW	309

The recording accelerometer is founded on an alluvial deposit 40 m deep.

11. The earthquake was also recorded at both El Infiernillo and La Villita dams which are located about 87 km and 108 km from the epicenter, respectively (Fig. 3). The main characteristics of the records at the dams sites are summarized in Table 1.

12. The damage induced to modern buildings in the Mexico City area, some 355 km from the epicenter, ranged from slight to severe and one building collapsed. In the epicentral area, high buildings suffered damage ranging from slight to moderate, and masonry constructions suffered severe damage. The earthquake induced liquefaction in a poorly graded, fine sand deposit in an island at the mouth of the Balsas River.

METHODS USED TO COMPUTE THE EARTHQUAKE-INDUCED DEFORMATIONS

13. Most of the simplified procedures in current use, are based on the model proposed by Newmark (ref. 6), where the sliding mass of soil is assumed to behave as a rigid block sliding along a well defined failure surface. The permanent deformations induced to an embankment by a seismic event, are calculated from the acceleration time history of the base. Whenever the ground acceleration exceeds the yield acceleration (an acceleration at which a potential sliding surface attains a factor of safety of unity), sliding will occur along the failure plane and the magnitude of the displacement is computed by double integration of the acceleration time history. Newmark (ref. 6) proposed the following equation to compute the average earthquake-induced horizontal permanent displacement of the block:

$$D = \frac{2V^2}{N}\left(1 - \frac{N}{A}\right)^2 \qquad (1)$$

where V is the maximum ground velocity; A is the maximum ground acceleration and N is the yield acceleration to be computed from stability analyses. Furthermore, the freeboard loss, L, can be calculated using the following relationship:

$$L = D \tan \alpha \qquad (2)$$

Table 1. Principal characteristics of the accelerograms recorded at El Infiernillo and La Villita dams on March 14, 1979

Station	Componet	Maximum acceleration* (cm/s^2)	Maximum velocity* (cm/s)	Maximum displacement* (cm)	Digitized length (s)
El Infiernillo Elev. 180 Crest	Trasversal (TRAN)[a]	355	43.5	7.58	68
	Vertical (VERT)	334	22.5	2.84	68
	Longitudinal (LONG)[b]	327	45.2	7.14	65
El Infiernillo Powerhouse[c]	TRAN	105	9.1	1.46	9
	VERT	69	5.5	0.92	16
	LONG	120	12.0	1.76	19
La Villita Crest	TRAN	311	14.5	1.54	31
	VERT	184	7.3	0.62	27
	LONG	155	11.0	1.41	31
La Villita Right bank[c]	TRAN	17	0.7	0.06	5
	VERT	15	0.7	0.05	4
	LONG	18	0.7	0.06	5
La Villita Berm elev. 13.5	TRAN	133	8.2	1.09	14
	VERT	60	5.1	0.57	12
	LONG	124	8.7	1.28	14

* After base-line correction
a Normal to crest
b Parallel to crest
c Considered as maximum ground acceleration

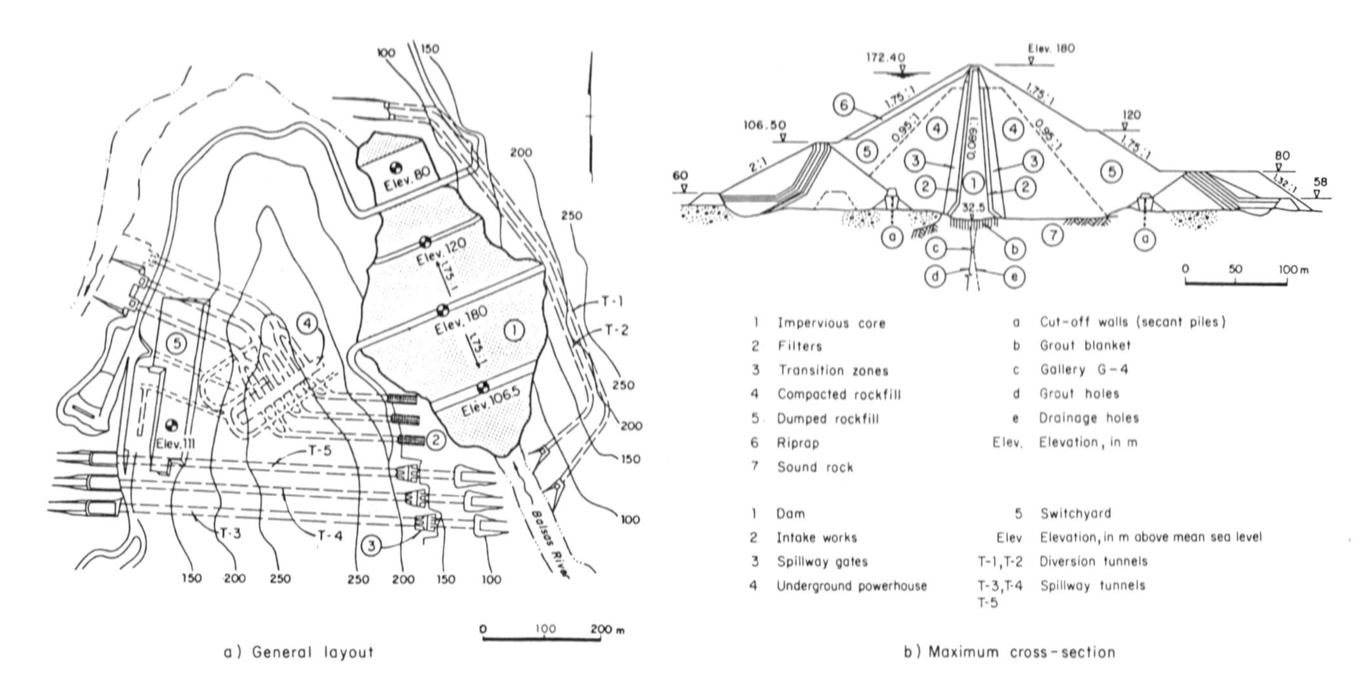

Fig. 1. El Infiernillo Dam: plan view and main section

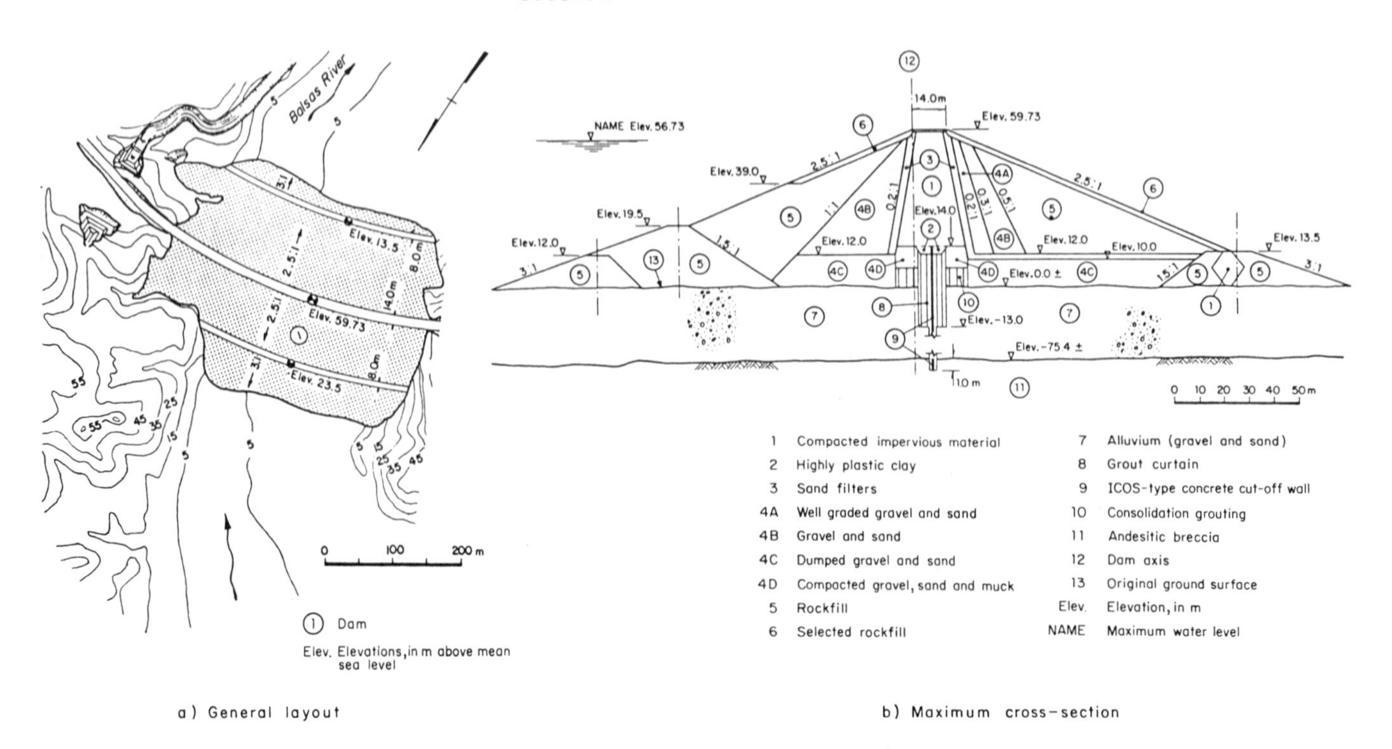

Fig. 2. La Villita Dam: plan view and main section

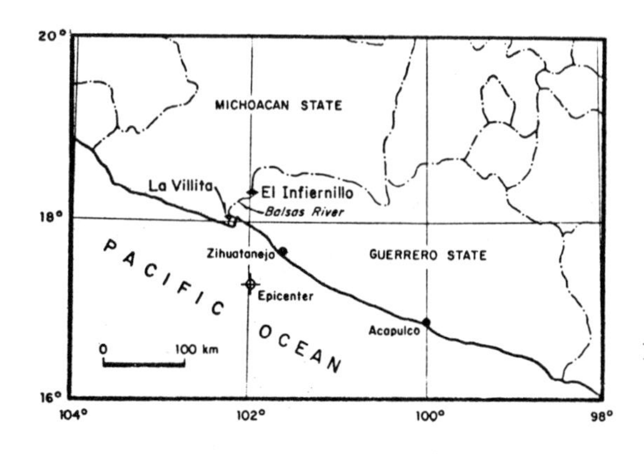

Fig. 3. Location of the epicenter of the March 14, 1973 earthquake

where α is the angle of the sliding plane with the horizontal.

14. Makdisi and Seed (ref. 10), proposed an alternate approach which is equivalent to Newmark's except that the earthquake excitation is obtained from the dynamic response of the embankment using either shear-beam or finite element models. They presented the results of a number of available studies in normalized charts as shown in Figs. 4 and 5. The method assumes perfectly elasto-plastic soil behaviour, and the yield acceleration is computed from slope stability analyses, as previously indicated. The numerical application of this method can be carried out using Figs. 4 and 5 together with Fig. 6, where the yield accelerations obtained from stability analyses for both dams, are shown. The basic steps required in the calculations are:

a) Obtain the maximum acceleration ratio $k_{max}/\ddot{u}_{max}$ from Fig. 4, entering with the depth of the sliding mass calculated from the stability analysis. In this ratio k_{max} is the average maximum acceleration of the sliding mass and $\ddot{u}_{max}$ is the maximum crest acceleration (both given as a fraction of the acceleration of gravity, g)

b) Obtain the corresponding yield acceleration, k_y, from Fig. 6

c) For a given earthquake magnitude and the ratio k_y/k_{max}, obtain the normalized displacement $U/(k_{max}\ g\ T_o)$ from Fig. 5. Here T_o is the first natural period of the embankment and g is the acceleration of gravity.

15. Both, Newmark's and Makdisi-Seed's methods assume that block displacements occur along well-defined sliding surfaces. Yet, in most cases deformation is the result of small strain increments throughout the whole embankment. In these cases, earthquake-induced deformations should be calculated with a procedure that permits the integration of local strains to obtain the overall deformation of the embankment. With this purpose, a number of sophisticated methods were recently developed (refs. 3 to 5). However, considering the uncertainties involved in the dynamic properties of the soils and other variables, elaborate methods of analysis are seldom justified for practical purposes, since their complexity does not necessarily lead to a better estimation of the relevant features of embankment behaviour.

16. Simplified yet rational procedures are preferable as guiding tools in the seismic design of embankments. In order for these procedures to be reliable they should at least account for:

a) The stress-strain behaviour of the embankment materials

b) The mean shear strain of soil elements contributing to the overall embankment deformation

c) A relationship between a measure of that mean shear strain and the corresponding embankment distorsion

17. A simplified deformation method which takes into account all these three conditions is available for computing the freeboard loss of embankments under seismic loading (ref. 1). The method is an extension of a procedure developed by Reséndiz and Romo (ref. 12) to calculate permanent deformations occurring in embankments during their construction. This method has been shown to predict accurately the measured outward slope displacements in a number of dams (refs. 13 and 14).

18. With reference to Fig. 7, the settlement of the embankment crest due to earthquake-induced plastic deformations can be calculated by integrating the horizontal outward movements, δ, given by the expression,

$$\frac{\delta}{\delta_{max}} = \frac{1-\cos\left[2\pi\left(\frac{Y}{H}\right)^{1.85}\right]}{5.5\left(\frac{Y}{H}\right)^{2.20}} \tag{3}$$

assuming that both the embankment volume and the crest width remain unchanged during the deformation process. Therefrom it is easy to demonstrate that the freeboard loss, L, is given by the following approximate relation,

$$\frac{L}{H^2} = \frac{1}{(B+b)}\left[\left(\frac{\delta_{max}}{H}\right)_u + \left(\frac{\delta_{max}}{H}\right)_d\right] \tag{4}$$

where H is the embankment height, B is the base width, b is the crest width, suscripts u and d denote the upstream and downstream slopes, respectively, and the corresponding values of (δ_{max}/H) are calculated from the following equation,

$$\frac{\delta_{max}}{H} = \frac{1}{93(F-1)} - \frac{1}{535(F-1)^2} + \frac{1}{9310(F-1)^3} \tag{5}$$

where F is the actual safety factor which is related to safety factors calculated with conventional stability methods (i.e. Bishop's modified), by means of the relationship shown in Fig. 8.

19. If the critical failure surface is shallow, the geometric parameters of the embankment are redefined as shown in Fig. 9.

20. It is important to stress that eq 5 was developed for static loading. However, if F is properly defined, it may be extended to dynamic loading. For this purpose the safety factor, F, is defined in terms of an equivalent strength, namely: strength equals the sum of sustained plus pulsating stress necessary to produce the failure strain, ε_f, after a number of constant stress cycles in laboratory samples (ref. 3). A gross approximation of this equivalent strength in many cases is the conventional CU strength (ref. 15).

21. Therefore, in practice this method may be applied simply using pseudostatic slope stability analyses with CU strength parameters. The distribution of the seismic coefficients to be

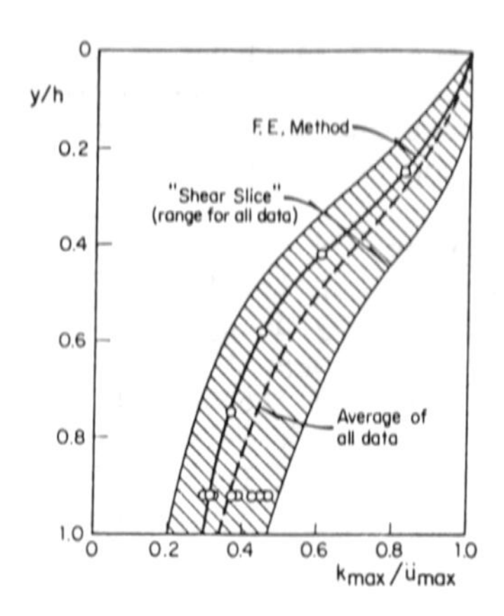

Fig. 4. Variation of "maximum acceleration ratio" with depth of sliding mass (ref. 10)

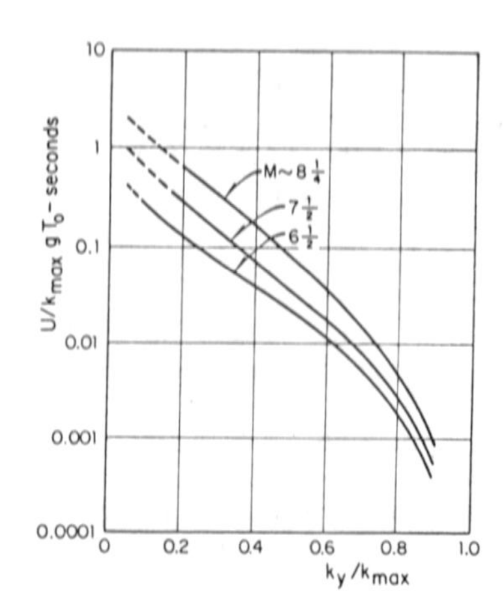

Fig. 5. Variation of average normalized displacement with yield acceleration (ref. 10)

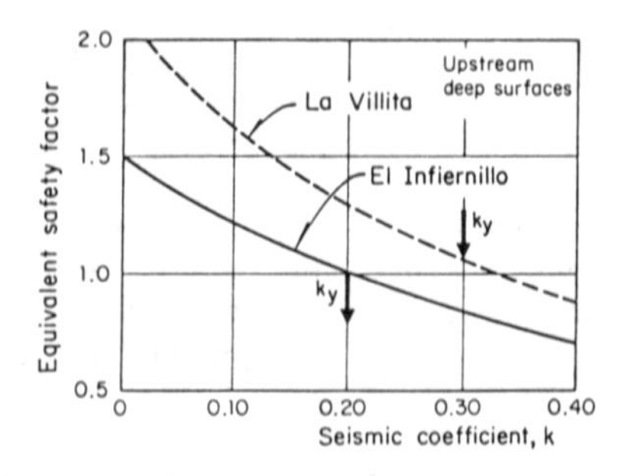

Fig. 6. Correlation between safety factors and seismic coefficients

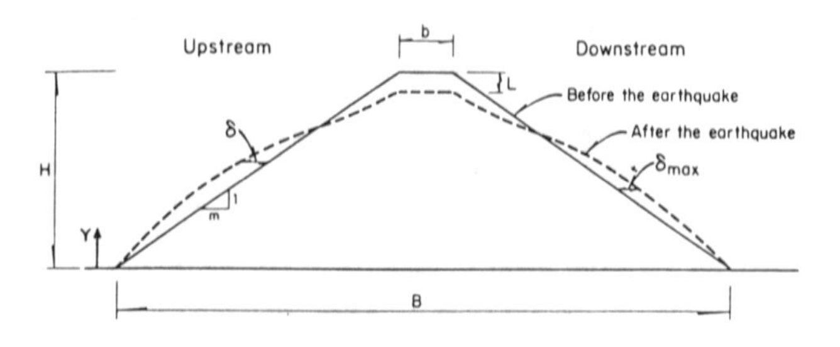

Fig. 7. Freeboard loss due to overall embankment deformations

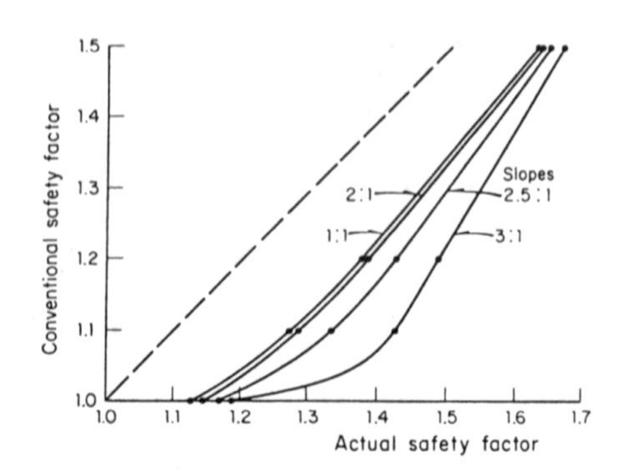

Fig. 8. Relationship between conventional safety factor and actual safety factor (ref. 12)

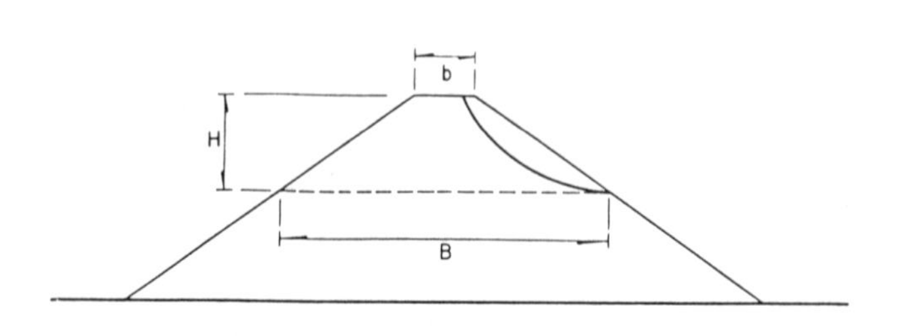

Fig. 9. Geometric parameters for shallow failure surfaces

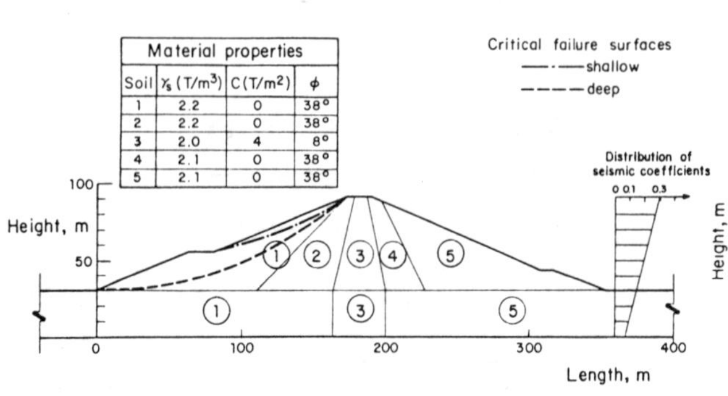

Fig. 10. Geomety, material characteristics and critical failure surfaces for La Villita Dam. Pseudo-static stability analyses

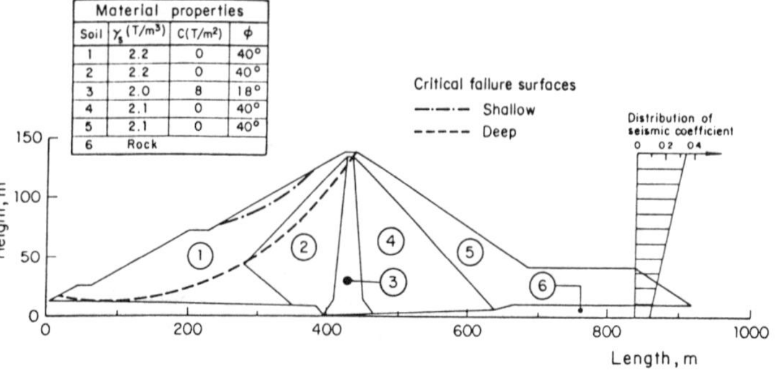

Fig. 11. Geometry, material characteristics and critical failure surfaces for El Infiernillo Dam. Pseudo-static stability analyses

considered along the height of the embankment may be obtained from dynamic response analyses. The resulting safety factor for each slope, is used in eq 5 to compute the maximum outward displacement of the slope, and then, through eq 4, the freeboard loss.

THE CASES OF LA VILLITA AND EL INFIERNILLO DAMS

22. Upon the earthquake of March 14, 1979, the crest settlement measured at the maximum section of La Villita was about 5 cm, and that of El Infiernillo was approximately 13 cm. Furthermore, both dams underwent minor cracking along and across their crests (ref. 16).

23. In this section the settlements of both dams produced by this seismic event are computed with the three described procedures, and the numerical results are compared with the actual vertical displacements measured in both dams in order to gain some information on the adequacy of the different methods to predict earthquake-induced displacements.

24. The geometry and the material parameters used in the slope stability analyses are shown in Figs. 10 and 11. The strength parameters of the clay of the impervious dam cores were obtained from CU tests, whereas those of the rockfill shells were derived from CD tests, on the assumption that free-draining conditions prevail in those dam portions even under earthquake loading.

25. The pseudo-static slope stability analyses were carried out using a linear distribution of seismic coefficients along the height of the dams, as shown in Figs. 10 and 11. These seismic coefficient distributions were obtained from the recorded response of both dams. The critical failure surfaces for shallow and deep slides in La Villita are shown in Fig. 10, and the corresponding ones for El Infiernillo are presented in Fig. 11.

Newmark's method

26. From the available motions recorded nearby the dams, it has been found that the ratio of maximum ground acceleration to maximum ground velocity varies between 8 and 12 $(\text{seconds})^{-1}$. Using the value of 10 $(\text{seconds})^{-1}$ in eq 1, and substituting the result into eq 2, the following relationship is obtained to compute the freeboard loss, L

$$L = \frac{A^2 - 2AN + N^2}{50\ N\ m} \qquad (6)$$

where $m = \dfrac{1}{\tan\alpha}$

27. The numerical values of freeboard loss obtained for both dams, using eq 6, are shown in Table 2 for different values of the maximum ground acceleration. From this table it may be seen that the crest settlements computed with Newmark's method for both La Villita and El Infiernillo dams are practically zero for the maximum ground accelerations recorded during the earthquake of March 14, 1979 (see Table 1). To compute settlements as those measured for El Infiernillo and La Villita, the corresponding ground accelerations would have to be multiplied by 7 and 35, respectively.

Makdisi-Seed's method

28. The numerical application of this procedure may be carried out by using Fig. 6 together with Figs. 4 and 5. The magnitude of the earthquake of March 14, 1979 was 7.6, and the maximum recorded crest accelerations were 0.3 g for La Villita and 0.34 g for El Infiernillo. Using this information, the computed maximum crest settlement is practically zero for La Villita and 0.3 cm for El Infiernillo. The latter value is about 40 times lower than the settlement measured at the crest of El Infiernillo.

Reséndiz-Romo's method

29. The freeboard losses of La Villita and El Infiernillo dams were computed for the shallow and deep critical failure surfaces, obtained from the upstream slope stability analyses. In these calculations, the safety factor of the downstream slope was taken equal to that of the upstream slope. Since the latter is actually smaller than the former, this would lead to overestimate the loss of freeboard in Tables 3 and 4.

30. The crest settlements of La Villita dam, computed by this method ranged between 26 and 38 cm for seismic coefficients varying from 0.20 to 0.35, considering the critical deep failure surface shown in Fig. 10. Similarly, for the critical shallow failure surface shown in Fig. 10, the crest settlements computed ranged between 18 and 23 cm for seismic coefficients varying from 0.20 to 0.30. It is interesting to note that for similar safety factors (see Tables 3a and 3b), larger crest settlements are calculated for the deep failure surface than for the shallow one. This is something that should be expected because the method of analysis takes into account all the deformations occurring in the embankment within the span of the critical sliding surface; hence, as this surface gets deeper the freeboard loss will be larger. The critical seismic coefficient (safety factor = 1) for the earthquake of March 14, 1979 was 0.20. For this value the crest settlements computed for the shallow failure surface was 18 cm, and for the deep failure surface was 26 cm. These

Table 2. Newmark's method results

Maximum ground acceleration $(\text{cm/sec})^2$	Freeboard loss (cm)	
	La Villita Dam	El Infiernillo Dam
300	0.04	0.59
500	2.04	5.09
700	5.50	13.99
1000	15.69	35.60
2000	87.44	179.28

values are to be compared to the measured freeboard loss of about 5 cm.

31. The freeboard loss calculated for El Infiernillo dam ranged between 67 and 102 cm for seismic coefficients varying from 0.20 to 0.40, considering the deep failure surface. Similarly, for the shallow failure surface, the calculated vertical displacement of the crest was 60 cm for a seismic coefficient of 0.20. As in the case of La Villita dam, from the comparison of the measured crest settlement of El Infiernillo dam (13 cm) with its computed value for a seismic coefficient of 0.20 (see Table 4), one can only say that they are of the same order of magnitude.

CONCLUSIONS

32. For rockfill dams, sliding-block methods render grossly underestimated values for the permanent dam deformation under moderate to severe earthquakes.

Table 3. Freeboard loss for La Villita dam

Seismic Coefficient	Conventional safety factors (Bishop's)	Actual safety factor (from fig 8)	δ_{max}/H (from eq 5)	L/H (from eq 4)	Freeboard loss, L (m)	Other parameters
0.20	1.45	1.61	1.308×10^{-2}	4.289×10^{-3}	0.26	
0.22	1.38	1.56	1.385×10^{-2}	4.543×10^{-3}	0.28	B+b=356+16=372 m
0.25	1.30	1.50	1.489×10^{-2}	4.883×10^{-3}	0.30	H = 61 m
0.30	1.18	1.41	1.667×10^{-2}	5.465×10^{-3}	0.33	average slope,
0.35	1.08	1.31	1.884×10^{-2}	6.179×10^{-3}	0.38	2.6:1
0.40	0.99	-	-	-		

a) Deep failure surface

Seismic Coefficient	Conventional safety factors (Bishop's)	Actual safety factor (from fig 8)	δ_{max}/H (from eq 5)	L/H (from eq 4)	Freeboard loss, L (m)	Other parameters
0.20	1.29	1.49	1.507×10^{-2}	5.071×10^{-3}	0.18	B+b=198+16=214
0.25	1.17	1.40	1.688×10^{-2}	5.679×10^{-3}	0.20	H = 36 m
0.30	1.06	1.29	1.926×10^{-2}	6.479×10^{-3}	0.23	average slope,
0.35	0.97	-	-	-	-	2.6:1
0.40	0.88	-	-	-	-	

b) Shallow failure surface

Table 4. Freeboard loss for El Infiernillo dam

Seismic Coefficient	Conventional safety factor (Bishop's)	Actual safety factor (from fig 8)	δ_{max}/H (from eq 5)	L/H (from eq 4)	Freeboard loss, L (m)	Other parameters
0.20	1.71	1.80	1.073×10^{-2}	4.590×10^{-3}	0.67	
0.22	1.65	1.75	1.127×10^{-2}	4.821×10^{-3}	0.70	B+b=929+15=935 m
0.25	1.56	1.68	1.211×10^{-2}	5.182×10^{-3}	0.75	H = 145 m
0.30	1.44	1.58	1.353×10^{-2}	5.790×10^{-3}	0.84	average slope,
0.35	1.33	1.49	1.507×10^{-2}	6.448×10^{-3}	0.98	1.85:1
0.40	1.24	1.42	1.646×10^{-2}	7.040×10^{-3}	1.02	

a) Deep failure surface

Seismic Coefficient	Conventional safety factor (Bishop's)	Actual safety factor (from fig 8)	δ_{max}/H (from eq 5)	L/H (from eq 4)	Freeboard loss, L (m)	Other parameters
0.20	1.01	1.17	2.044×10^{-2}	8.153×10^{-3}	0.60	
0.25	0.92	-	-	-	-	B+b=356+15=371 m
0.30	0.84	-	-	-	-	H = 74 m
0.35	0.77	-	-	-	-	average slope,
0.40	0.70	-	-	-	-	1.85:1

b) Shallow failure surface

33. The deformation method proposed by Reséndiz and Romo gives somewhat conservative results of the right order of magnitude, which is probably as accurate a result as should be expected in this type of problem.

34. However, a more thorough validation of the proposed method is desirable. This would require either an extensive experimental program or additional comparisons of its results to measurements in well documented case histories. It is hoped that this paper prompts further independent calibration.

REFERENCES

1. RESENDIZ D. Optimum seismic design of embankment dams. Pub. N° E18, Instituto de Ingeniería, UNAM, México, 1975
2. SEED H.B. Considerations in the earthquake resistant design of earth and rockfill dams. Nineteenth Rankine Lecture. Geotechnique, 1979, Vol. 29, N° 3, September
3. SEED H.B., LEE K.L., IDRISS I.M. and MAKDISI F.I. Analysis of the slides in the San Fernando dams during the earthquake of Feb. 9, 1971. Report N° EERC 73-2, University of California, Berkeley, June
4. LEE K.L. Seismic permanent deformations in earth dams. Report to the National Science Foundation, School of Engineering and Applied Science, University of California, Los Angeles, 1974
5. SERFF N., SEED H.B., MAKDISI F.I. and CHANG C.Y. Earthquake induced deformations of earth dams. Report N° EERC 76-4, University of California, Berkeley, September
6. NEWMARK N.M. Effects of earthquakes on dams and embankments. Geotechnique, 1965, Vol. 15, N° 2, June, 139-173
7. AMBRASEYS N.N. and SARMA S.K. The response of earth dams to strong earthquakes. Geotechnique, 1967, Vol. 18, N° 3, September, 181-213
8. SARMA S.K. Seismic stability of earth dams and embankments. Geotechnique, 1975, Vol. 25, N° 4, December, 743-761
9. ROMO M.P., AYALA G., RESENDIZ D. and REYES A. Permanent deformations induced to El Infiernillo and La Villita dams by the earthquake of March 14, 1979. Chapter 7 in Performance of El Infiernillo and La Villita dams in Mexico including the earthquake of March 14, 1979, Comisión Federal de Electricidad, México, 1980
10. MAKDISI F.I. and SEED H.B. Simplified procedure for estimating dam and embankment earthquake-induced deformations. Journal of the Geotechnical Division, ASCE, 1978, Vol. 104, GT7, July, 849-867
11. SRH-CFE-UNAM. Behavior of Dams Built in Mexico. Contribution to the XII ICOLD, Mexico 1976, Chapters 5 and 7
12. RESENDIZ D. and ROMO M.P. Analysis of embankment deformations. Proceedings of the Specialty Conference on Performance of Earth and Earth-Supported Structures, 1972, June, 817-836
13. CATANACH R. and McDANIEL T.N. Addendum to lateral deformation of a dam embankment. Proceedings of the Specialty Conference on Perfomance of Earth and Earth-Supported Structures, 1972, Vol. 3, June, 161
14. RESENDIZ D. Discussion on: A.Penman and A. Charles, Constructional deformation in rockfill dams, Journal of the Geotechnical Engineering Division, ASCE, 1974, Vol. 100, N° GT3, March, 370-373
15. SEED H.B. and CHAN C.K. Clay strength under earthquake loading conditions. Journal of the Soil Mechanics and Foundations Division, ASCE, 1966, Vol. 92, N° SM2, March, 53-78
16. MORENO E. Behavior records of El Infiernillo and La Villita dams from construction to March 1979, Chapter 5 in Performance of El Infiernillo and La Villita dams in Mexico including the earthquake of March 14, 1979, Comisión Federal de Electricidad, México, 1980

31. Full-scale forced vibration studies and mathematical model formulation of arch concrete dams

T. A. PASKALOV, BSc, MSc, PhD, J. T. PETROVSKI, BSc, MSc, PhD, and D. V. JURUKOVSKI, BSc, MSc, Institute of Earthquake Engineering and Engineering Seismology, University 'Kiril and Metodij' Skopje

Within the scope of a continuous Yugoslav research project on earthquake resistant design of dams, dynamic forced vibration studies were carried out on six earth and rock-fill dams and two arch concrete dams*. The dynamic tests on one of the arch dams were performed for a reservoir water level, close to the normal, i.e. 8 meters bellow the crest level. This is the first stage of investigation, so there are planned new tests for a minimum reservoir water level. On the second arch dam dynamic forced vibration tests were carried out, for two different reservoir water levels. The first test was performed for a water level slightly less than the maximum, and the second test for a water level corresponding to $^{2}/_{3}$ of the designed reservoir level, or a level 11 meters lower than the first test. Also, there was an attempt to formulate a mathematical model for the second dam,with emphasis to the influence of dam-reservoir interaction on the dynamic properties of the structure. The more significant results of this kid of investigations are presented in this paper.

INTRODUCTION

The earthquake effects on dams and other accompanying structures is of great concern to engineers and scientists mainly for two reasons. First, the dynamic response of these structures to an earthquake motion is rather complex due to the effect of different types of static and dynamic loads, and second, these structures are of special public interest and economic importance. Although the recent extensice studies were carried out on problems related to the dynamic response on these structures to earthquakes, there is a large variety of problems which have not been classified so far, particularly the definition of seismic design criteria to which dams should be analysed and the parameters necessary for accurate formulation of mathematical models.

The investigations being presented in this paper are a part of a continuous Yugoslav project on earthquake resistant design of dams, and part of these was carried out under the YU-US cooperative project titled as "Earthquake Resistant Design of Industrialised Building Systems and Dams", involving four institutions from Yugoslavia and the University of California at Berkeley, from USA. In this report the basic results obtained from experimental and analytical studies on two different in size arch concrete dams have been presented.

OBJECTIVES OF THE INVESTIGATION

The main objectives of the full-scale forced-vibration studies were the evaluation of the dynamic properties, namely, natural frequencies, corresponding mode-shapes and damping characteristics of the two arch concrete dams. These data allow a correlation of experimental and analytically derived data and provide important information for accurate mathematical models formulation. For that reason, the influence of the reservoir water on the dynamic characteristics of the already mentioned two arch dams, under different water levels was included in the scope of these studies. Following these studies, a set of strong motion instruments was installed on each dam body and bedrock, respectively.

EXPERIMENTAL PROGRAMME

Within the scope of the research project several dams have been studied so far, most of them being located in high seismic areas. Under this programme two arch concrete dams are a subject of studying, which, for convenience, are named as Dam No1 and Dam No2. They are both studied under different reservoir water levels, namely, first for a water level close to the spielway level, i.e. for a full reservoir, and second, for a reservoir water level as minimal as possible.

THE ARCH DAM No 1

Description of the Dam. The dam is designed and constructed as a typical arch concrete dam with a peripheral joint. The maximum height of the dam is H=123 meters and the total crest length is L=400 meters. From structural point of view the dam body is shaped as a double-curved thick shell structure, which, having in mind the large dimensions of the dam, gives a very elegant architectural look. The thickness of the shell on the crest level is d=5.0 meters, and the maximum shell thickness at the contact with the foundation rock is D=30.0 meters. The normal reservoir water level is three meters bellow the crest and the maximum water level is 0.65 meters above the normal one. During the test performance the reservoir water level was 8.0 meters bellow the crest level.

Experimental results. Two GSV-101 vibration generators were located at the middle part of the dam crest. The direction of the exciting force was radial as well as tangential. Sixty seven measuring points were selected along the dam crest and the downstream face of the dam body. Fig.1

shows the position of the vibration generators as well as the distribution of the measuring points.

During the test performance, for symmetric and antimmetric excited vibrations, amplitude versus frequency response curves were recorded at their different points of the crest, namely at point "1", "2", "6" and "7" (see Fig.1). The frequency response curves were measured in radial and tangential direction respectively, depending on the way of excitation. During this test five symmetric and three antisymmetric mode shapes were clearly distinguished. The complete eight mode shapes were measured in all 67 measuring points both in radial and tangential directions.

The results of these tests are presented in Table 1, and some of the frequency response curves and mode shapes in Figs 2 through 9.

Table 1 Dynamic Characteristics of the Arch Dam No 1

Type of Vibration	Mode	Exciting Force at Resonance (kg)	Resonant Frequency (cps)	Acceleration Amplitude at Ref. Point (10^{-3} g)	Damping %	
					Half Power Bandwidth	Amplitude Decay Curve
Symmetric	I	3310	2.705	3.329	1.20	1.38
	II	2423	3.725	0.434	1.91	-
	III	4220	5.625	2.750	1.31	2.72
	IV	2608	7.130	1.128	2.07	2.27
	V	3416	8.160	1.551	1.67	2.26
Anti-Symmetric	I	1440	2.510	0.757	1.07	1.03
	II	2148	4.460	1.149	0.99	0.75
	III	2152	6.840	1.697	0.84	1.51

Discussion of the results. Under forced-vibration excitation of the Dam No 1, dynamic characteristics, i.e. resonant frequencies, mode shapes and damping were obtained for both symmetric and antisymmetric shape of excitation. For obtaining of the symmetric modes of vibration the existing forces from both generators were simultaneously applied in phase along the canyon direction, and for antisymmetric modes the forces were applied either perpendicular to the canyon or out of phase for 180° across the canyon.

For symmetric excitation five resonant frequencies (f_1=2.70 cps; f_2=3.72 cps; f_3=5.62 cps; f_4=7.13 cps; f_5=8.16 cps), and for antisymmetric excitation three resonant frequencies (f_1=2.51 cps; f_2=4.45 cps; f_3=6.84 cps) were distinguished.

THE ARCH DAM No 2

Description of the dam. The dam body is rather symmetric. Unsymmetry in the river canyon is avoided by inserting massive concrete abutments on both sides of the canyon. The geometric characteristics of the double-curved dam body are given in the Figs 10 and 11 . The horizontal arches are of uniform thickness while the vertical cantilever members along the height of the dam are of variable thickness. The variation of thickness (in meters) of these members is given by the following expression:

$$d(z) = 1.50 + 0.08\ z$$

where "z" is the distance in meters along the height measured from the crest. As observed from the above expression the minimum thickness is 1.50 meters at the dam crest. The length of the upper arches (from 0 to 12 meters) is 147 meters. Bellow this level the length of the arches is given by the expression:

$$L = 2 \cdot \left[72.0 - 1.625\ (z - 12)\right]$$

As mentioned in Figs.12-13the length of the dam crest is 194 meters, and the total height of the dam from the foundation to the crest is 53 meters.

Experimental results. The two GSV-101 vibration generators were located at the middle of the crest 22.5 meters apart each from the other in order to avoid the bridge overpassing the spilway part of the dam (see Fig. 12). The direction of the existing forces in radial, i.e. along the river canyon, Fig. 12 also shows the seventeen measuring points along the dam crest. The vertical distribution of the measuring points is on the upstream face of the dam body in vertical along the points 8 through 14 on the crest, as it is shown in Fig. 12 . The position of the generators and the distribution of the measuring points was the same in both tests.

During the first test, for symmetrically excited vibrations, frequency-amplitude response curves were recorded in points 8 and 9 (Figs 14 and15). For antisymmetric vibrations, frequency-amplitude response data were obtained in point 10 (Figs 16 and 17). The frequency response amplitudes have been measured in radial direction only. For both symmetric and antisymmetric modes of vibration frequency response data were also recorded in points 11, 12 and 13 at the crest of the dam.The results of the first test are presented in Table 2. During these tests, mode-shapes were only recorded at the dam crest.

Table 2 Dynamic Characteristics of the Arch Dam No 2, First Test

Type of Vibration	Mode	Exciting Force at Resonance (kg)	Resonant Frequency (cps)	Acceleration Amplitude (10^{-3} g)	Damping %	
					Half Power Bandwidth	Amplitude Decay Curve
Symmetric	I	600	3.61	4.73	1.90	-
	II	3100	7.80	16.92	3.00	-
Unsymmetric	I	300	3.29	1.55	2.70	-
	II	680	5.23	7.52	1.80	-

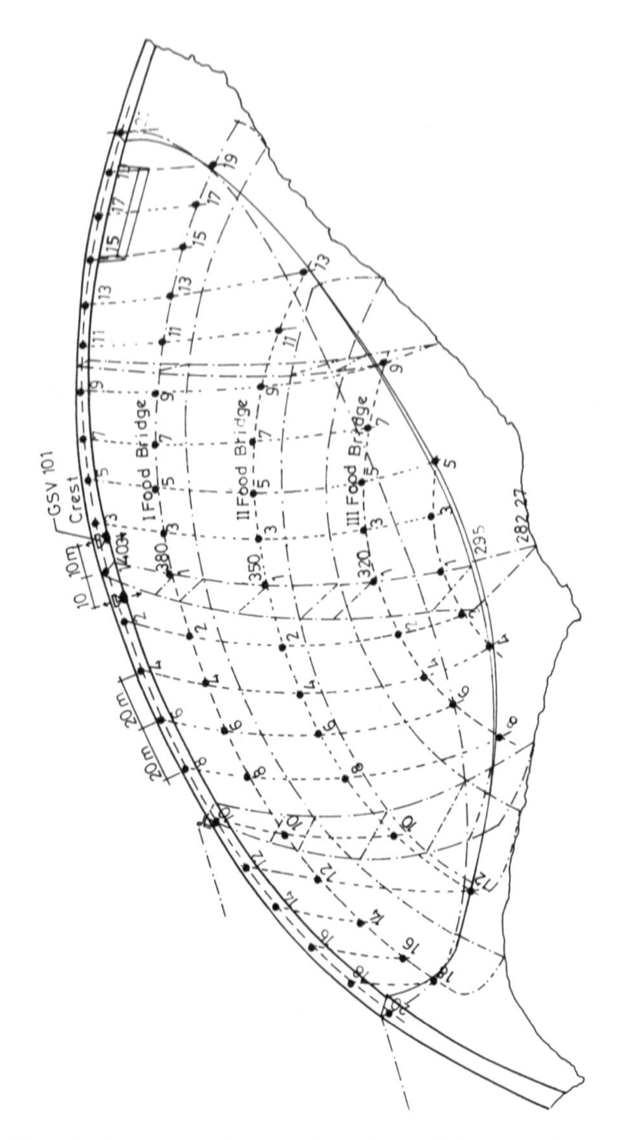

Fig. 1 General view of the dam with position of the shakers measurement points

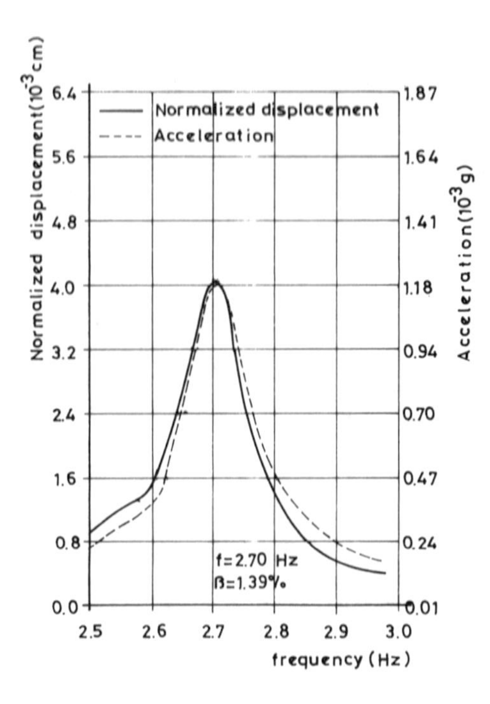

Fig. 3 Comparison between acceleration resonant curves and normalised displacement for radial symmetric vibration in reference point 1 at dam crest and load in generator backets S3+LF

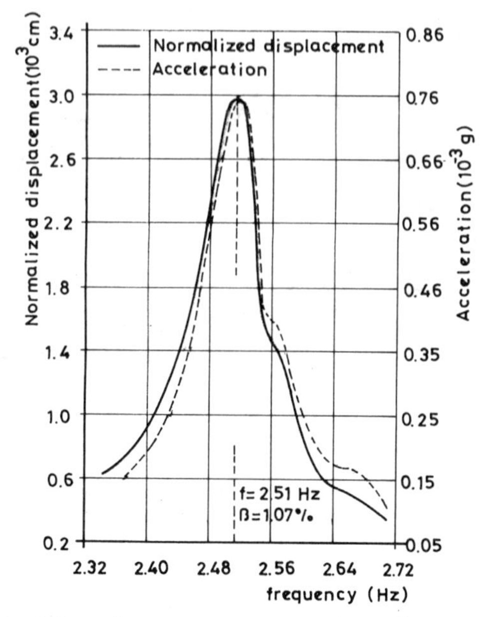

Fig. 2 Comparison between acceleration resonant curves and normalised displacement for radial antisymmetric vibration in reference point 7 at dam crest and load in generator backets SF+L3

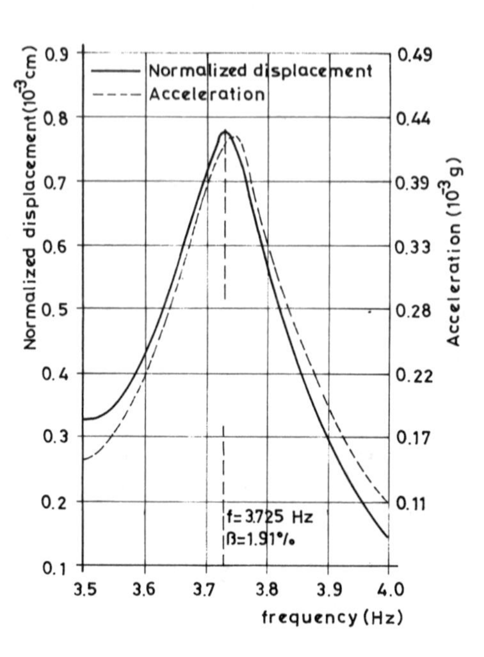

Fig. 4 Comparison between acceleration resonant curves and normalised displacement for radial symmetric vibration in reference point 1 at dam crest and load in generator baskets S3+LF

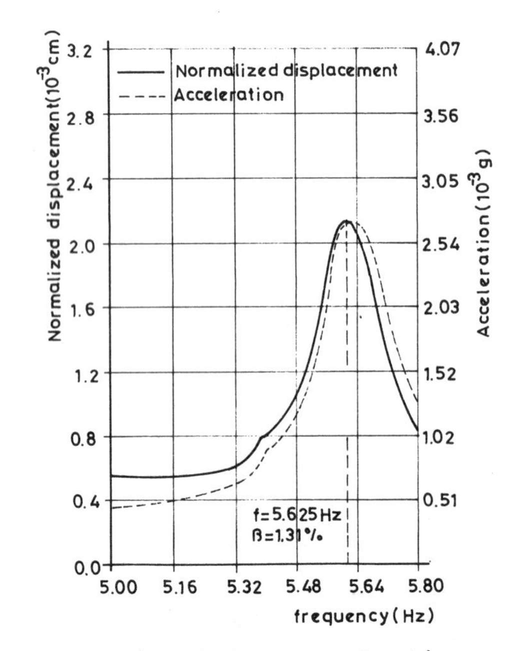

Fig. 5 Comparison between acceleration resonant curves and normalised displacement for radial symmetric vibration in reference point 1 at dam crest and load in generator baskets S2+LF

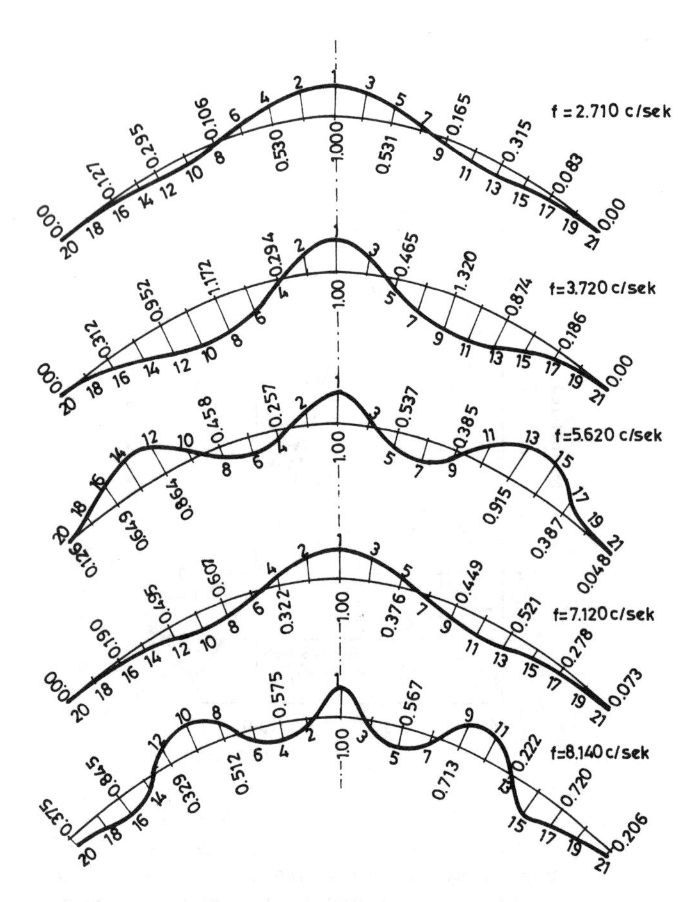

Fig. 6 Horizontal mode shapes at dam crest for radial components under symmetric vibrations

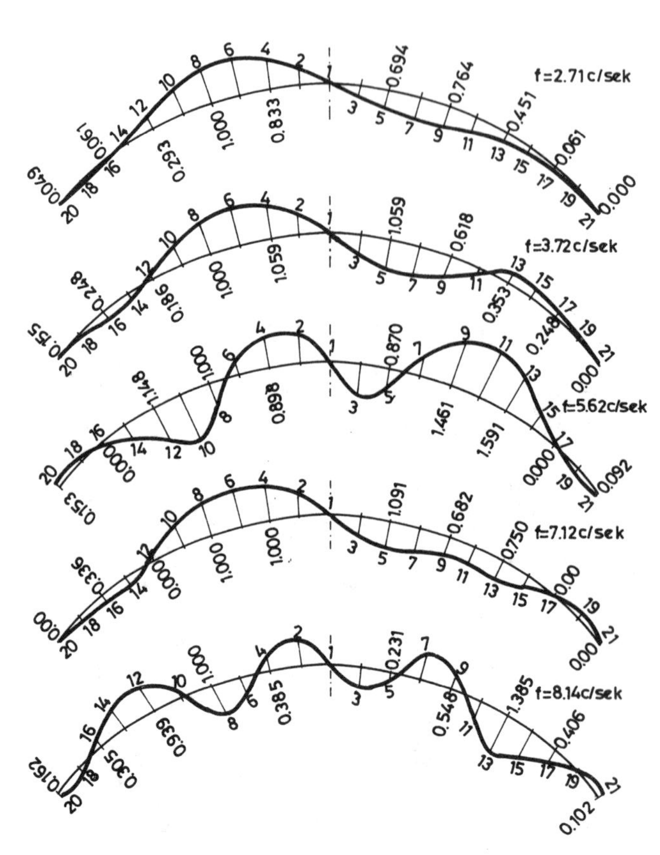

Fig. 7 Horizontal mode shapes at dam crest for tangential components under symmetric vibrations

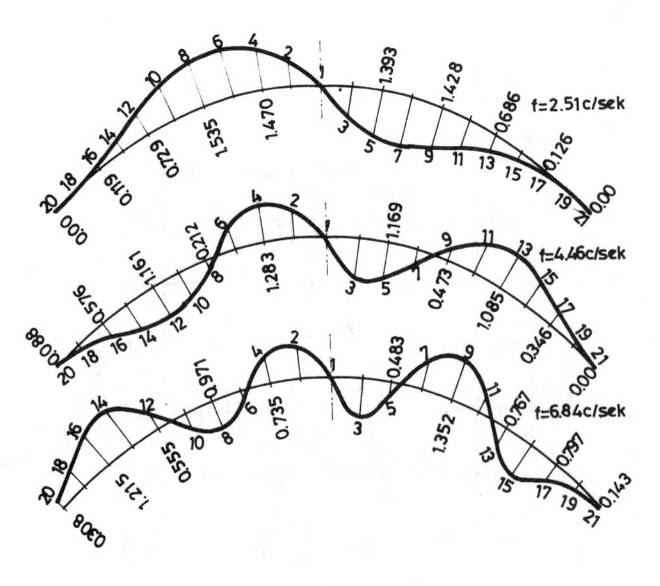

Fig. 8 Horizontal mode shapes at dam crest for radial components under antisymmetric vibrations

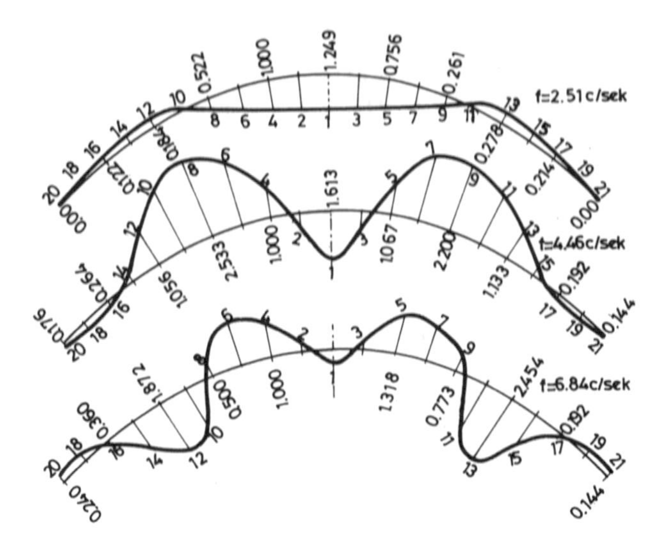

Fig. 9 Horizontal mode shapes at dam crest for tangential components under antisymmetric vibrations

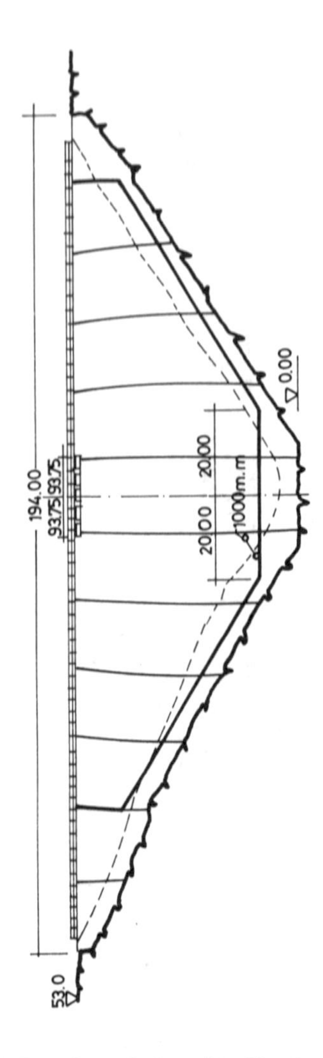

Fig. 11 Developed longitudinal section

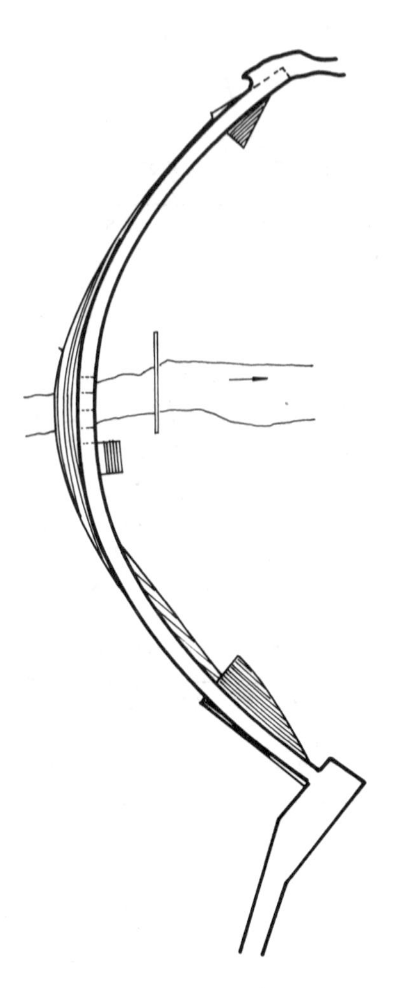

Fig. 10 Plan of the dam

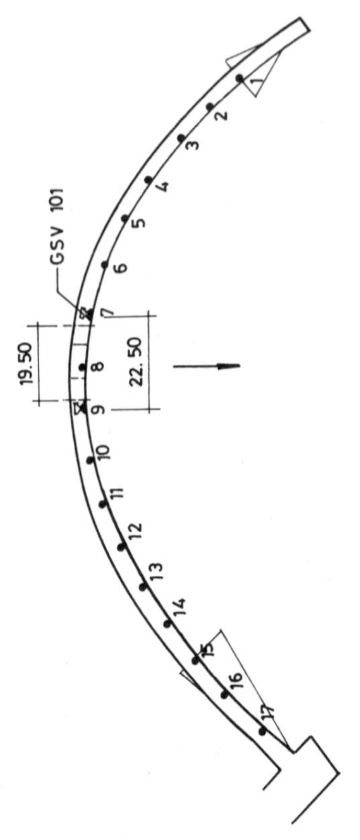

Fig. 12 Position of generators at the dam crest with measuring points

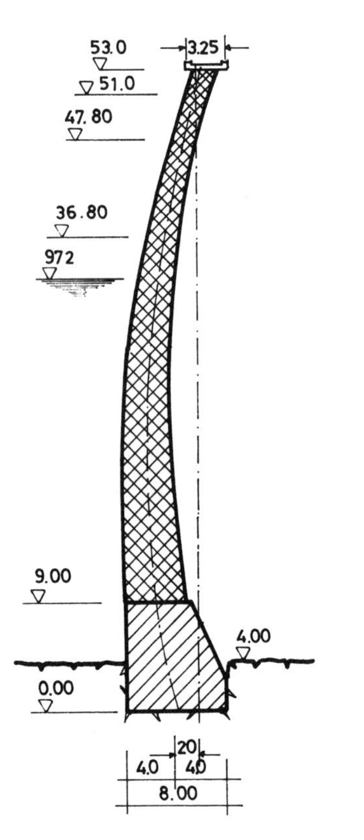

Fig. 13 Cross section

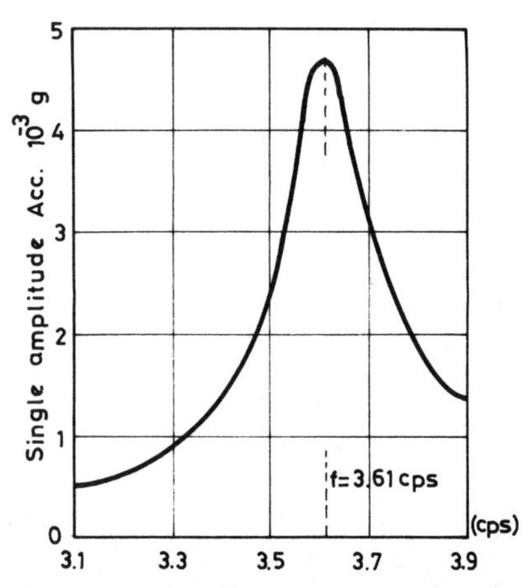

Fig. 14 Frequency response in point 8, first symmetric mode, first test

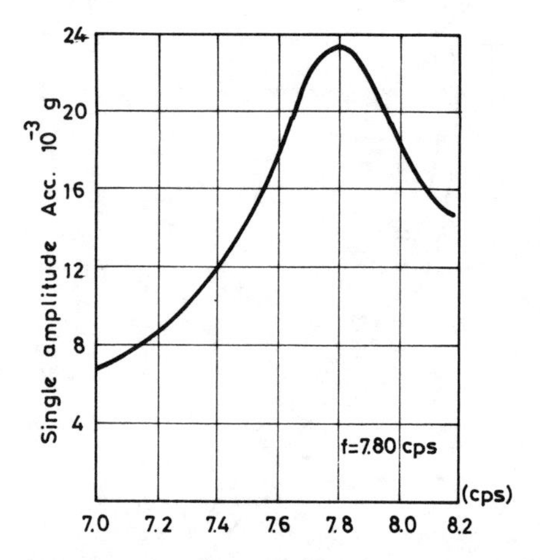

Fig. 15 Frequency response in point 9, second symmetric mode, first test

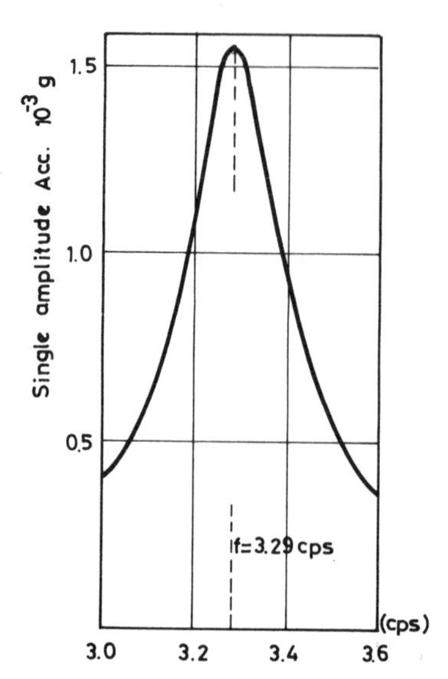

Fig. 16 Frequency response in point 10, first antisymmetric mode, first test

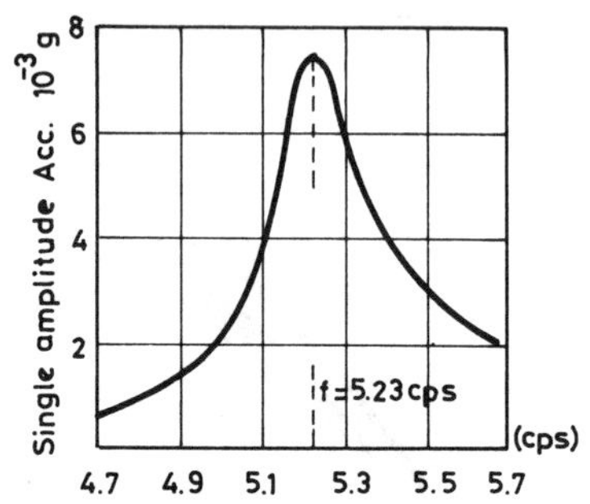

Fig. 17 Frequency response in point 10, second antisymmetric mode, first test

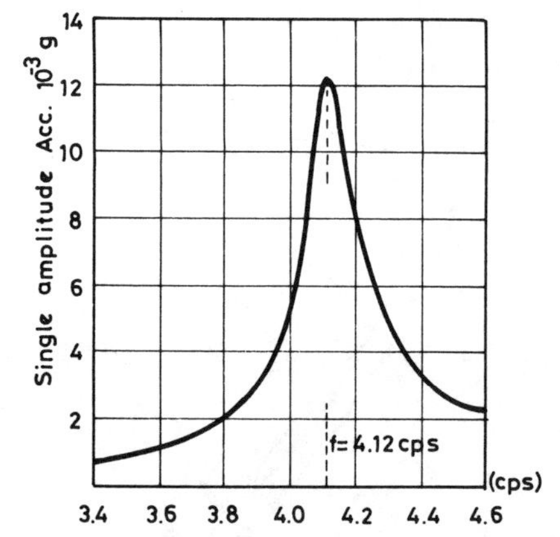

Fig. 18 Frequency response in point 8, first symmetric mode, second test

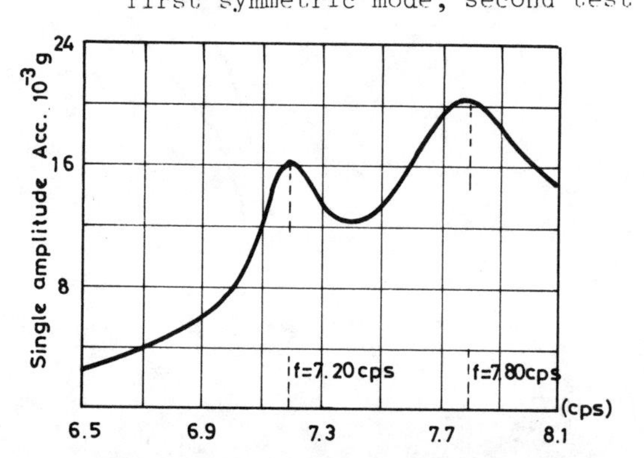

Fig. 19 Frequency response in point 8, second symmetric mode, second test

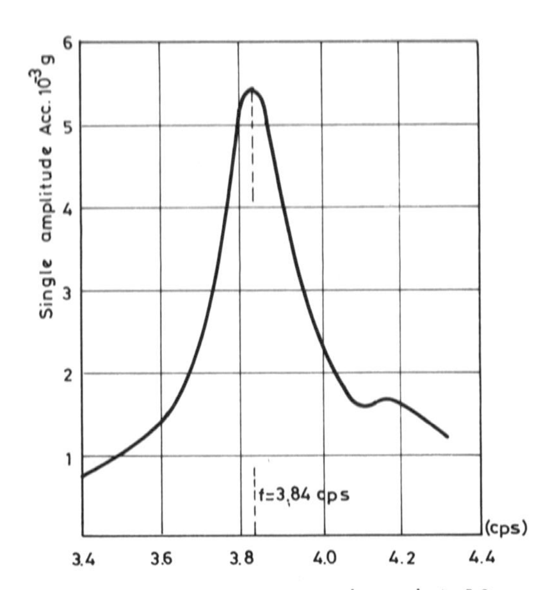

Fig. 20 Frequency response in point 10, first antisymmetric mode, second test

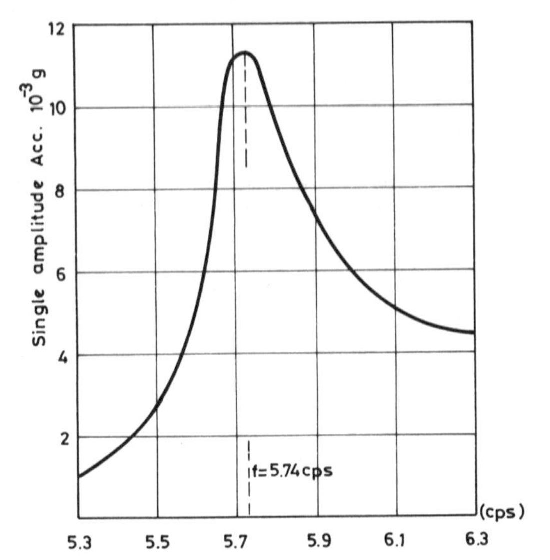

Fig. 21 Frequency response in point 10, second antisymmetric mode, second test

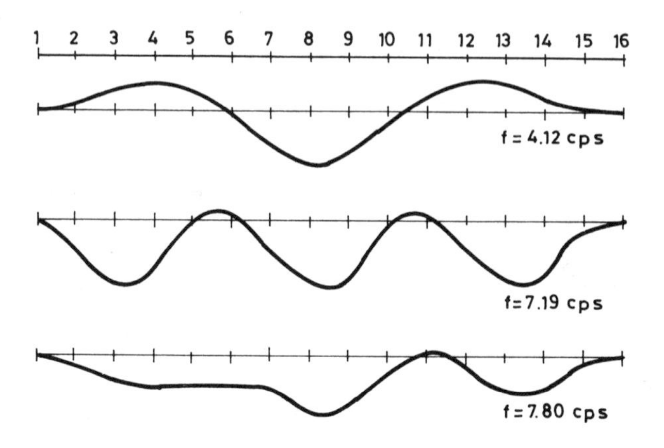

Fig. 22 Symmetric horizontal mode shapes at dam crest in radial direction, second test

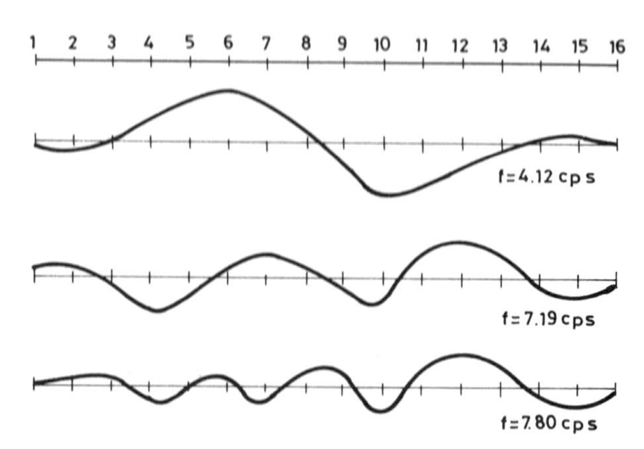

Fig. 23 Symmetric horizontal mode shapes at dam crest in tangential direction,second test

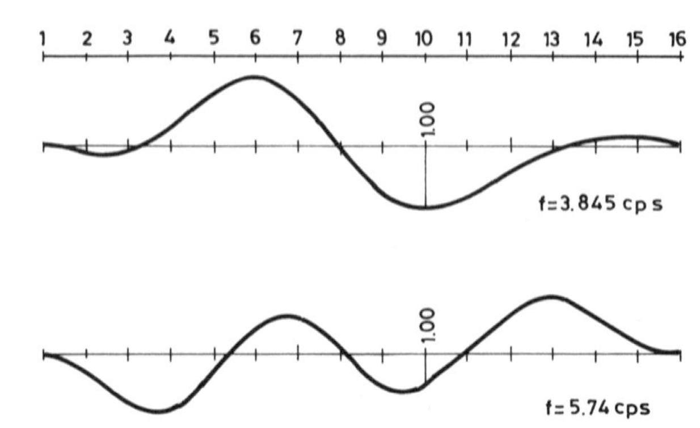

Fig. 24 Antisymmetric horizontal mode shapes at dam crest in a radial direction,second test

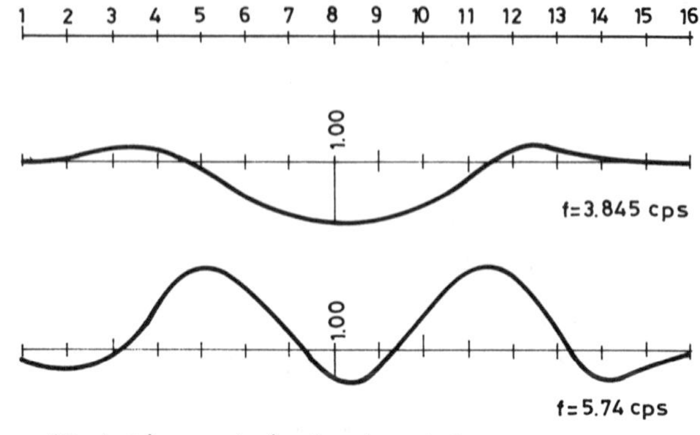

Fig. 25 Antisymmetric horizontal mode shapes at dam crest in a tangential direction,second test

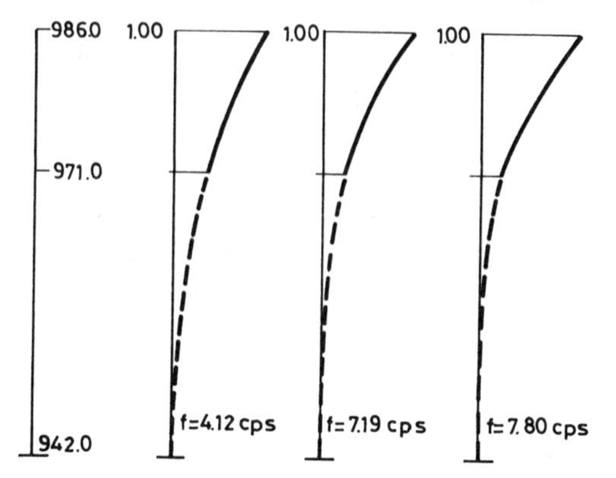

Fig. 26 Vertical symmetric mode shapes in section 8 in a radial direction, second test

The second series of tests were carried out for a reservoir water level about 2/3 of the maximum one. This reflected a difference in the water level between the first and second tests of 11.0 meters. The amplitude versus frequency response curves were also recorded in point 10. The results are presented in Figs. 18 , 19 , 20 , and 21 . Some of the experimental results from the second test are presented in Table 3. The radial and tangential components of the symmetric modes of vibration were recorded at the dam crest and are shown in Figs 22 and 23 respectively, while the components of the antisymmetric modes are presented in Figs 24 and 25 , respectively. The corresponding vertical mode shapes for symmetric and antisymmetric vibrations as recorded in section 8 are presented in Figs. 26 and 27 , respectively.

Table 3 Dynamic Characteristics of the Arch Dam No 2, Second Test

Type of Vibration	Mode	Existing Force of Resonance (kg)	Resonant Frequency (cps)	Acceleration Amplitude (10-3 g)	Damping %	
					Half Power Bandwidth	Amplitude Decay Curve
Symmetric	I	900	4.12	12.35	1.50	-
	II	2550	7.20	16.40	-	-
	III	3150	7.80	20.40	3.90	-
Unsymmetric	I	385	3.84	5.43	1.90	-
	II	840	5.74	11.40	1.90	-

Discussion of the results: Beside determining the dynamic characteristics of the Arch Dam No 2 the most interesting part of the research was to define the influence of the reservoir water level on the dynamic characteristics of the entire structure. When the water level was lowered for 11.0 meters, or approximately 30% of the height of the dam, the resonant frequencies of the first two modes (both symmetric and antisymmetric) increased for about 10-20%. This phenomenon indicates that the reservoir water level influences the inertia (mass) properties of the dam body. Also, the viscous damping coefficients of the modes recorded in the second test tend to decrease (Table 2 and Table 3). If a comparison is made between the maximum unit acceleration amplitudes recorded at the same locations for identical modes of vibration, it appears that the amplitudes of the second test are considerably higher. It is evident that the amplitude level is influenced by the exciting force level, however considerable influence comes from the dynamic amplification factor which in itself depends upon the damping coefficients.

The lowest frequency of the two dynamic tests is the frequency of the first antisymmetric mode. The antisymmetric modes have rather distinguished shapes in both radial and tangential directions (Figs. 24 and 25). However, the mode shapes of the symmetric modes are rather complex as illustrated in by the radial and tangential components of these modes (Figs. 22 and 23). Besides the horisontal mode shapes at the crest, the same mode shapes were also recorded at elevation 971 (Fig. 26). However, the results have not been presented in this paper. In general, comparing only the geometrical shapes of all modes, it appears that almost all modes are regularly distinguished. A small discrepancy of corresponding zero and maximum amplitude points is evident in respect to the radial and tangential mode shape components recorded at the dam crest.

Vertical mode shapes were recorded only in the second dynamic test on the upstream face of the dam, because of the existing vertical curvature. These mode shapes were recorded in profiles 8, 10 and 12 at each five meters along the height of the dam. It should be mentioned, however,that the accelerometers used in this test can record only horizontal components of acceleration. The amplitudes of vibration were normalized according to the referenced point amplitude, so obviously, in the graphical presentation only horizontal components of the radial mode shapes exist. Errors which might have appeared due to unfavourable position of the accelerometers have not been taken into consideration.

Concernig the shapes of the amplitude versus frequency response curves of the two tests, rather good separation is evident exept for the resonant curve of the second symmetric mode. It is evident, from the responses recorded along the dam crest during the first test, that the change of resonant frequencies in all points is negligible, being less than 1%, so it is concluded that the resonant frequencies in any point of the dam for each recorded mode are equal. This is an indication that normal modes of vibration were excited on this dam.

ANALYTICAL RESULTS OF THE DYNAMIC CHARACTERISTICS OF THE ARCH DAM No 2

As it was mentioned above the cross-section of the canyon where the dam is constructed is an irregular trapezoid, which during construction works was turned into a completely symmetrical trapezoid. So, the dam body is shaped completely symmetrical with respect to its longitudinal axis. In such a case the horizontal cross sections of the dam body are arches with constant thickness, while the vertical cross sections of the dam are of cantilever form with variable thickness over the height.

Methodology applied for dynamic analysis of the dam: A computer programme based on finite elements method theory was used for analysis of the dynamic characteristics of the dam. For this case, a corresponding three-dimensional shell element with 16 nodes each and having three degree-of-freedom, has been applied.

In Figs. 28 a, b, and c is given the "parent element" with the notation direction of the nodal points.

The general element obtained on the basis of the presented "parent element" is shown in Fig.29.

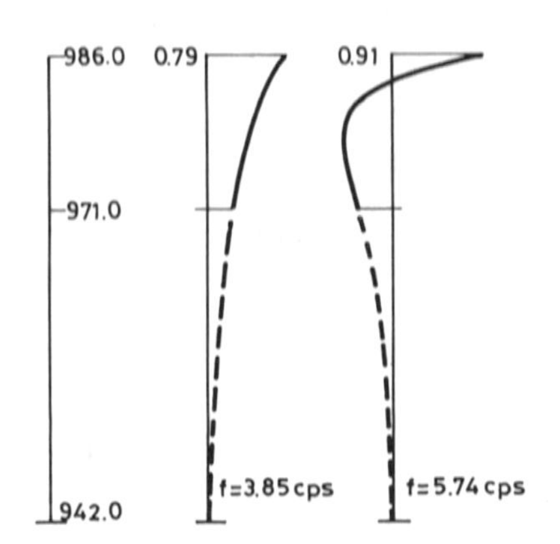

Fig. 27 Vertical antisymmetric mode shapes in section 8 in a tangential derection,second test

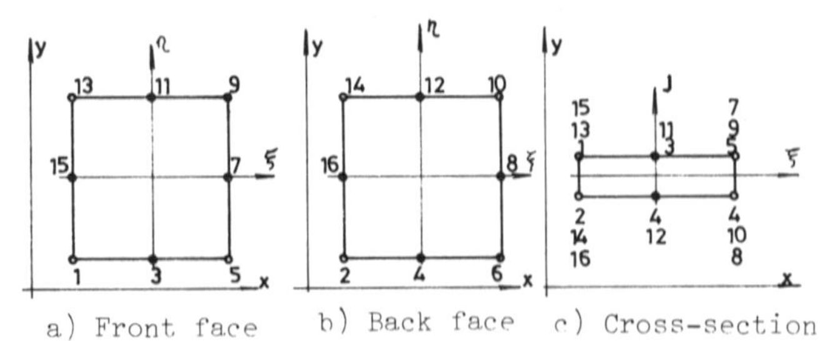

Fig. 28 "Parent Element"

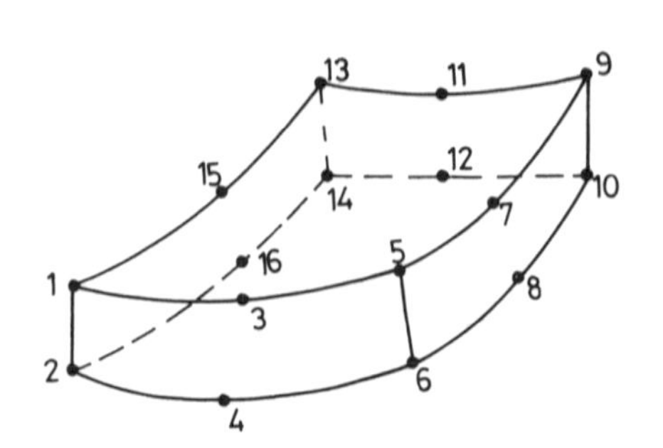

Fig. 29 General 3D-Shell element with 16 nodes

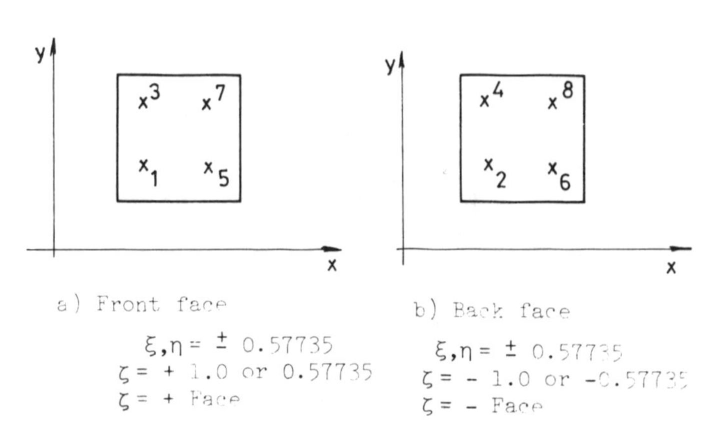

Fig. 30 Integration points within the element

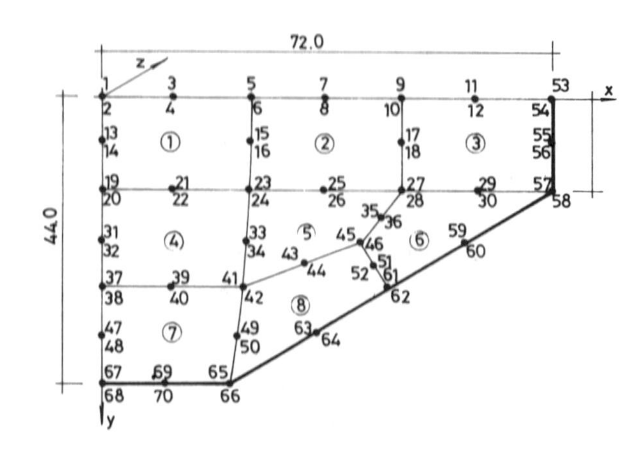

Fig. 31 Mathematical model of the structure

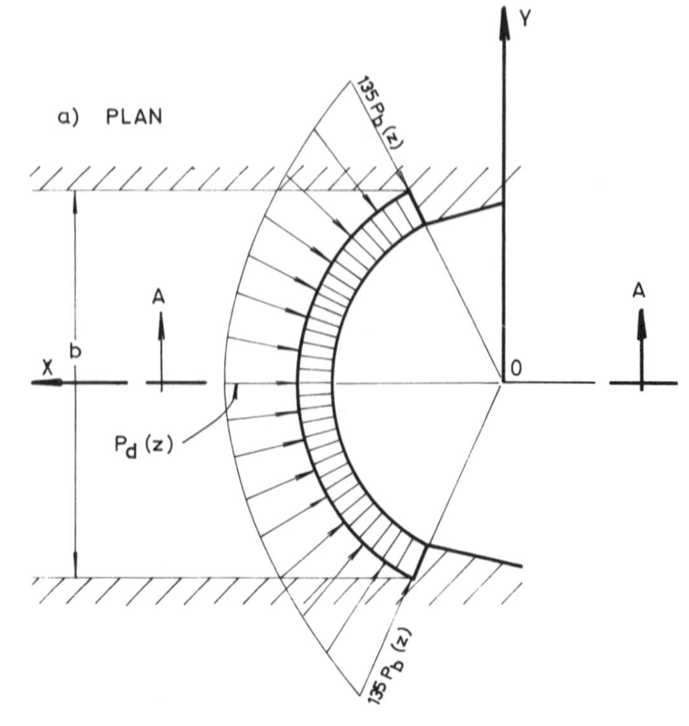

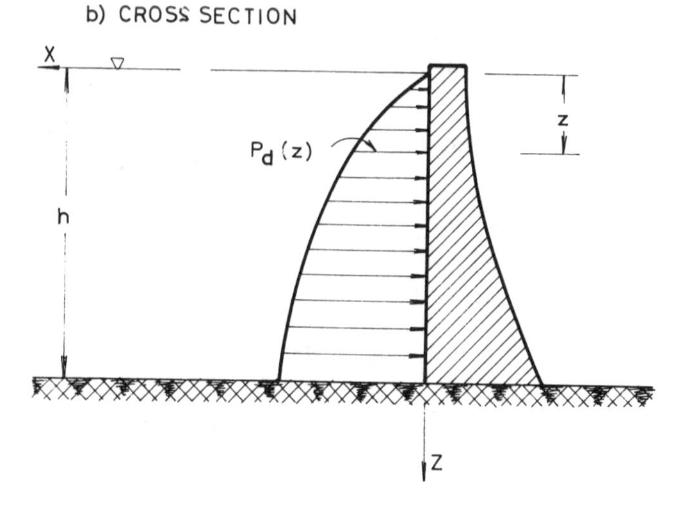

Fig. 32 Dynamic water pressure distribution

The integration points are placed along the coordinate axis within the element (local axis) in the following way:

$$\xi, \eta, \zeta = \pm \frac{1}{\sqrt{3}} = \pm 0.57735$$

which gives totally 8 integration points, Fig.30 illustrates the notation of the integration points.

For the formulation of the element stiffness matrix as well as the mass matrix a standard technique of the finite element method has been applied using shape functions within the element and integration in the already mentioned integration points. For solution of the eigen-value problem the Gramm-Schmidt procedure has been used followed by the inverse integration of the shiffs.

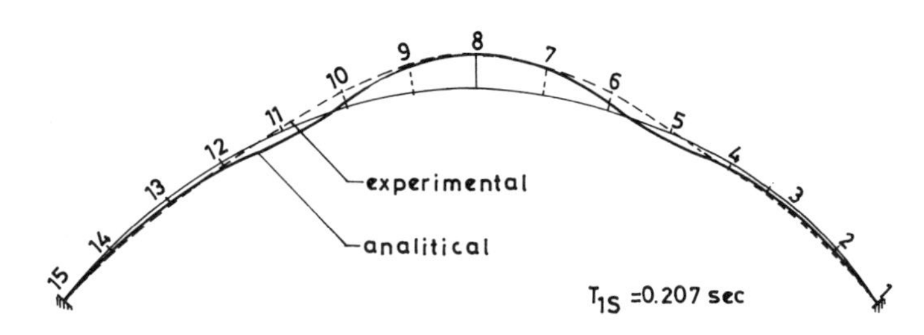

Fig. 33 Symmetric mode shape 1 of the dam Analytical and experimental results

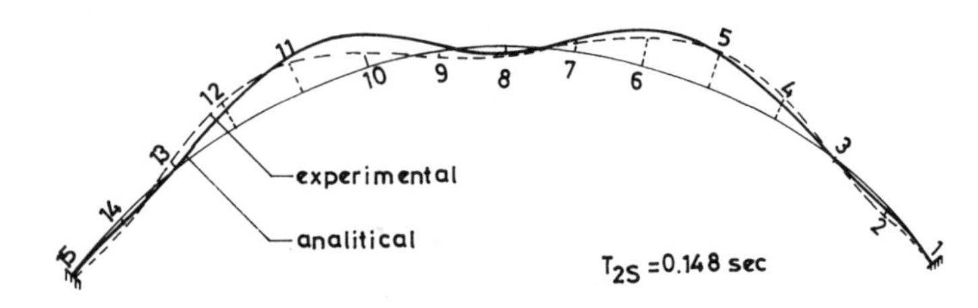

Fig. 34 Symmetric mode shape 2 of the dam Analytical and experimental results

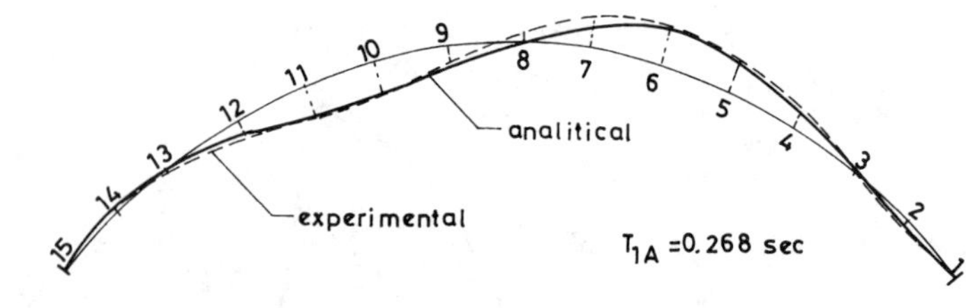

Fig. 35 Antisymmetric mode shape 1 of the dam crest, analytical and experimental results

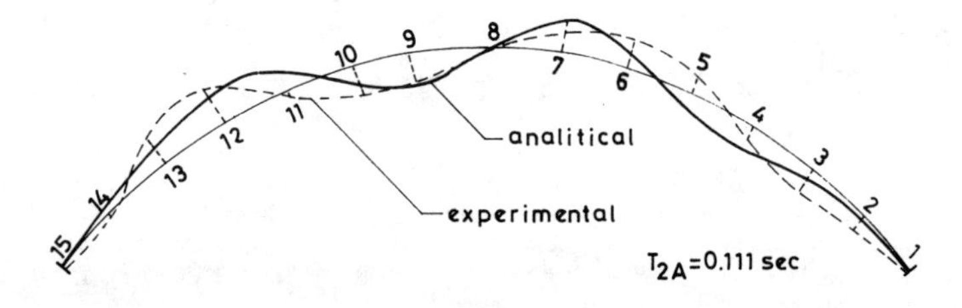

Fig. 36 Antisymmetric mode shape 2 of dam crest, analytical and experimental results

The special set up of the bondary conditions enabled application of the symmetry principle for the specified structure which reduces the amount of calculations.

Mathematical model of the structure: Considering the fact that the structure was shaped symmetrically during construction, a mathematical model for one half of it was formulated. Also, due to foundation abutments supporting the shell applied to achieve complete symmetry of the dam body, a shell model fixed at the abutments has been selected.

The mathematical model is composed of totally 8 elements with 70 nodal points, as shown in Fig.31 in the conventional coordinate system.

The results obtained from the experimental studies of the Arch Dam No 2 show a considerable influence of the reservoir water level on the dynamic characteristics. In order to explain this phenomenon, the interaction between the dam body and reservoir fluid was analytically obtained by means of "added mass" representation of the water, applying modified Westergaard's procedure. For this purpose the following assumptions were made: (i) the upstream face of the dam body was taken as vertical; (ii) the shape function of the dynamic water pressure at the middle part of the arches was assumed as for concrete gravity dams; (iii) this pressure linearly increases by 35% at the arch abutments comparing with the middle part of the arches (See Fig. 32). All these assumptions were applied in the calculation of the "added masses" for two different water levels which correspond to the situation of the experiments performed.

Discussion of the results obtained: By application of the analytical procedure described above, the dynamic characteristics of the Dam no.2 for empty reservoir as well as for the two levels of the water table as 36.80 m and 47.80 m respectively, were defined. The other two water levels are identical to those at which forced-vibration full-scale tests were performed. In the following table are presented the comparable analytical and experimental values of the vibration periods.

Period	Theoretical results			Experim. results	
	Empty Res.	Water level 36.80m	Water level 47.80m	Water leve 36.80	Water level 47.80
1 S	0.207	0.253	0.266	0.243	0.277
2 S	0.148	0.162	0.181	0.139	0.190
3 S	0.106	0.116	0.129	0.128	-
1 A	0.268	0.327	0.344	0.260	0.303
2 A	0.111	-	-	0.174	0.192
3 A	0.100	-	-	-	-

For comparison of interest are only the results which correspond to the reservoir water levels of 36.80 and 47.80 m respectively. Namely, the vibration periods for the first three symmetric modes of the considered cases show satisfactorly coincidence and have similar changing tendency. Foe the antisymmetric modes of vibration the coincidence is rather poor, although the change of the vibration, depending upon the reservoir water level, is of the similar tendency. This means that the increase of the water level causes an increase of the vibration periods of the structure as well. The coincidence of the mode shapes for the considered cases is shown in Figs. 33, 34, 35 and 36, for the first two symmetric and two antisymmetric modes, respectively.

CONCLUSION

In this stage of investigation the influence of the reservoir water level upon the dynamic characteristics of the dam-reservoir-foundation system has been analysed applying rough analytical approximation.in treating the added water mass. However, this treatment showed that the added water mass influences the dynamic characteristics by means of increase of the natural periods of vibration. In the course of further investigation it is obvious that more sophisticated treatment should be applied, as for example representing the reservoir water by means of fluid finite elements. From the above investigations it could be concluded that the interaction between the reservoir water and the dam body increases the inertia properties of the structure by approximately 40%. This considerable influence points out the requirement of further wider experimental and analytical investigations of models and especially of constructed dams.

REFERENCES

1. Petrovski J., Jurukovski D.,Taskov Lj.,Percinkov S., "Dynamic Testing of the Arch Dam 'Rateska reka' in Berovo", Report LDI 77/4 IZIIS, University "Kiril and Metodij", Skopje,Yugoslavia.
2. Petrovski J.,Jurukovski D.,Galboc S.,Percinkov S., "Dynamic Forced Vibration Study of the Rock-Fill Dam 'Kalimanci' in Kocani", Report LDI 77/3 IZIIS, University "Kiril and Metodij" Skopje, Yugoslavia.
3. Hudson D.E., "Dynamic Tests of Full-Scale Structures",Earthq. Eng., John and Wiley 1970.
4. Keightley W.O., "A Dynamic Investigation of Bouqet Canyon Dam " EERL, California Institute of Technology, Pasadena, California, 1964.
5. Napetvaridze S.G., "Research on Earthquake Resistance of Dams in USSR",Proc.3rd World Conf. Earth.Eng., Welington, New Zealand , 1964.
6. Napetvaridze S.G., "Aseismic Design of Hydrotechnical and Civil Engineering Structures", IZIIS, University "Kiril and Metodij",Skopje, Yugoslavia
7. Paskalov, T., "Aseismic Stability of Dams Constructed of Local Materials",IZIIS, University "Kiril and Metodij", Skopje, 1967.
8. Paskalov T.,"Elastic Response of Earth-Fill Dams under Earthquake Effects", IZIIS, University "Kiril and Metodij", Skopje,Yugoslavia, 1970.
9. Rouse G.C., and Roham L.H., "Vibration of Navajo Dam Following a Subsurface Nuclear Blast" US Dept. of Interior Bureau of Reclamation, Denver, Colorado, July 1969.
10. Paskalov T. et al, "Full-Scale Dynamic Study of 'Tikves' Dam", Report OIK 73-9, IZIIS, University "Kiril and Metodij",Skopje, Yugoslavia.
11. Roehm H.L., "Comparison of the Computed and Measured Dynamic Response of Monticello Dam", REC-ERC 71-45, Engineering and Research Center, Bureau of Reclamation, USA, 1971.
12. Rouse C.G. and Roehm H.L., "Vibration of Vega Dam following Subsurface Nuclear Blast", REC-OCE 70-15, April 1970, Division of Design Office of Chief Engineer, Bureau of Reclamation USA.
13. Rouse C.G. and Bouwkamp J.G., "Vibration Studies of Monticello Dam", US Dept. of the Interior, Bureau of Reclamation, Rep. 9, 1967.
14. Cozart W.C., "The Response of an Intake Tower at Hoover Dam to Earthquakes", Engineering and Research Center, REC-ERC 71-50, Bureau of Reclamation, USA, 1971.
15. Paskalov T.,Taskov Lj.,Jurukovski D.,"Evaluation of Seismic Resistance and Stability of Concrete Arch Dam 'Grancarevo'", Rep. IZIIS 79-193, University "Kiril and Metodij", Skopje, Yugoslavia.

32 The potential for induced seismicity—geological approaches

B. O. SKIPP, BSc (Mining) PhD FGS MICE, and M. HIGGINS, BSc (Geol) FGS, Soil Mechanics Limited, Bracknell

The potential for induced seismicity at reservoir and dam sites must now be considered as part of the overall hazard studies, but a rational and economically justifiable procedure is difficult to formulate. Indeed seismicity of significance is rare. The mechanics of induced seismicity is now better understood, but the identification features making a site conducive to such phenomena is not an established or verified undertaking. An approach to the problem in a stable cratonic region is discussed with an example in the West African Craton.

INTRODUCTION

1. Since the earthquakes at Koyna in India and Kremasta in Greece resulting from reservoir impounding seismologists have studied the phenomenon of induced seismicity. The impact of these events upon the engineering world is manifest in various codes and guidelines (UNESCO 1976, IAEA 1979).

2. The early reactions to the recognition that dam and reservoir constructions could in some instances be causally related to earthquakes near the works was alarmist, but more recent studies (USCOLD 1977) have put the issue in better perspective and in broad terms the risk of damage from seismicity is now recognised as being relatively small. The failure of dams is, however, a matter of social and environmental concern. Therefore, in line with the current trend in hazard analysis, the identification of major failure modes and assurances of their very low probability of occurrence based on reasoned arguments demanded in situations which a decade or more ago the judgement of an experienced geologist would have sufficed.

3. It is now recognised that damaging earthquakes can occur in intraplate areas and rarely within old cratons, the so called 'stable' blocks within continents. Plate tectonics, so successful in developing models for seismicity at active plate boundaries, has been less helpful in the formulation of models for such intraplate events. In this paper the problem of developing an understanding of the potential for induced seismicity within a 'stable' continental craton is considered, having particular regard to the temporal and economic constraints.

4. The starting point may be taken as the UNESCO 1976 document in which consideration of induced seismicity was recommended for 'large reservoirs and dams', defined as having either, impounded volumes greater than 10^9 m^3, and/or a depth of water of 100 m or more. Large undertakings on the basis of these criteria showed signs of induced seismicity in one in fourteen cases. More recently (USCOLD 1977) the overall figure is estimated as 0.7% of dams with the World Register. As more dams and reservoirs are instrumented, it is likely that the proportion for which some micro-tremors (M 2)** is observed will increase, but for engineering purposes, concern is for events of M 4.5.

BACKGROUND

1. Most induced shocks have been shown to have focal depth less than 15 km. There seems to be a relationship between magnitude and focal depth in intraplate settings (Bath 1980). Under the mean stress levels of these depths, it is not unreasonable to assume a failure mechanism ranging from brittle fracture to shear rupture under total and effective stresses. Gough D and WI (1970) and Withers and Nyland (1978)have discussed the particular stress states under and adjacent to a water loaded area which may initiate slip on strike slip, dip slip or thrust faults.

2. It is generally agreed that while rapid load and total stress change may play a part in early 'foreshock' symptoms, mainly shocks seem to involve a triggering role in which the intrusion of water (and consequent reduction in effective stress on a fault plane) has the major role. Although there has been a tendency to regard the risks as greater for larger reservoir volumes the acceptance of a 'triggering' role implies no relation between the stress perturbations and the amount of stored energy released.

3. Induced seismicity therefore requires :

i) Stress state with deviatoric stresses approaching the limit state for the existing mean stress.

** M is used for Richter Magnitude, Mb for body wave magnitude.

ii) A stress perturbation leading to slip on old failure surfaces or a generation of new failure surfaces.

4. In a single episode the amount of the energy released will depend upon the strain energy stored and the orientation of principal stresses of an existing failure surface. Generation of fresh fracture through intact rock is not likely for the small magnitudes characteristic of induced phenomena and the apparent prior require -ment of some faulting.

5. It may be argued that the concept of natural earthquakes nearly always occurring on old failts may not be valid for intraplate settings and particularly for cratons where old fractures are now probably 'healed' at depth. Under such circumstances there may be statistically fewer favourably oriented lines of weakness for the release of the accumulated strain energy. Large earthquakes have apparently created fresh faulting in Western Australia. On the other hand there are those who would argue that a shear zone remains a preferred site for further rupture.

6. Assessment of the potential for induced seismicity may be resolved into an appreciation of current stress states and identification of geological features which would allow access of water to possible focal regions. Direct measure -ment of hydraulic fracture stress at depth of 1 to 3 km is feasible, but cost and time preclude its general use. Micro-tremor studies can give information of the current stress direction and potentially active structures, but in commercial practice would be considered only after some preliminary study had established a case for doing so. The nature of that preliminary study is dealt with in this paper.

7. Nikolaev (1973) has argued that induced seismicity is not restricted to rocks of a particular geological age as long as they have been "tectonically" prepared, usually by neotectonic activity and especially along continental/ocean margins, e.g. West Africa, where Pre-Cambrian cratonic areas have been stressed by subsidence along the ocean margin. As may be seen from Table II the latest known faulting at some sites showing induced seismicity may be Mesozoic. It seems reasonable to start a study by tracing, in so far as is possible, the palaeotectonics of a region with a view to understanding the past and present generalised stress states.

SUGGESTED INDICATORS FOR POTENTIAL INDUCED SEISMICITY

1. There is no agreed set of geological criteria which would point to a site as carrying a risk of damaging induced earthquakes, but those that have been suggested relate either to the mechanisms which can give rise to high crustal stress or geological and geophysical manifestations of that high stress. They include :

1. Known tectonic force fields
2. High topographic relief
3. Steep gradients of gravitional anomaly
4. Evidence of meteor impact
5. Residual heating of lower crust as evidenced by hot springs and other recent volcanic features
6. Competent basement rocks capable of storing elastic strain energy and failing in a brittle manner
7. Strong basement inhomogenitees and contrasting rock types
8. Water pathways to deeper levels of faults
9. Rocks susceptible to strength deterioration or volume change under water pressure
10. Evidence of neotectonic features at surface

2. The set of thirty or so dams showing some signs of induced events is not large enough on which to base a pattern recognition programme and as Lomintz (1974) has pointed out that the significance of a reservoir as a trigger has to be judged against the natural background of activity. It would not be unreasonable to then develop a model for greatest risk and compare with it an actual site. A high accumulation of strain energy would be required in a stress field oriented so as to favour release of that energy if the effective stress on a local fault dropped below some threshold after impounding. Recent (Castle et al 1980) work has suggested that a region with recent strike slip or normal faulting is however more susceptible to induced seismicity than one with thrust faulting. It has been suggested that thrust faulting would raise risk to the sides of a reservoir (cf Williams and Myland) and it is generally considered that high horizontal stresses are characteristic of intraplate settings (Sykes, 1978, Ahorner et al 1972, Schafer 1974). A generally high mean stress in the upper crust and a permeable fault zone near the site are probably the main elements in the model. In intraplate cratons one may expect the strong crystalline rocks to accumulate and retain strain energy near the surface but some flexural states near the edges would conceivably increase the risk.

3. Given the as yet statistically unverified geological indicators and the complexity of processes affecting current crustal stress, there is a need to establish some way of carrying out an investigation for a particular site.

METHODOLOGY

1. The first problem in establishing a methodology for geological study is to set target areas in space and time. At a very general level the first target area must be of continental dimensions, or at least large enough to encompass the major features of a current plate tectonic model. This would usually involve a consideration of features of greater distance from a site within several hundred kilometer radius. The point of this may be illustrated by reference to the understanding of stress accumulating in the Deccan (Viz Koyna) the East African Rift system (Kariba), the seismicity of the Eastern Mediterranean (Kremasta). It seems not unreasonable to look at the structural geology, tectonics and seismicity within 300 km radius, 50 km radius and 15 km radius, so increasing the detail of the study in the nearer ranges.

2. The time aspect is controversial. Recent practice (Castle et al 1980), Allen (1976) has been to concentrate on evidence for movement along faults in the Quaternary. Other approaches have included a study of neogene fault (Miocene) movements. The contribution to current stress levels in the crust by older tectonic processes, in particular by episodes in craton formation, is questionable and depends upon the rheological model adopted for crustal materials and its parameters. It is, however, important to examine the paleotectonics of an area in an attempt to identify any major controlling feature or deep seated line of weakness which persists through geologic times, which may influence current stresses. An example of a major line weakness would be the site of the Jurassic/Cretaceous continental separation, which for millions of years prior to break up may have undergone periods of tectonic activity. Many inherent stress states established for millions of years exist in comparatively stable areas, e.g. West Africa, subject to epeirogenic processes, resulting from isostatic adjustment which has been taking place from the end of the Cretaceous.

3. There is an increasing interest in the role played by old continental lines of weakness in controlling oceanic fracture zones and their significance in locating current seismicity (Sykes 1978). Consideration of episodes older than Quaternary is therefore important for general understanding of the current situation within continental plates as regards stress, fracturing and seismicity. Also by having an understanding of the paleotectonic history of an area, comparisons can be with areas of a similar history. The procedures which may be followed can be summarised below:

a) Literature study on paleotectonics and structural geology
b) Landsat and aerial survey
c) Aquisition of any gravity or magnetic survey data
d) Accumulation of instrumental and macroseismic data from world lists and local catalogues
e) Aquisition of any focal plane solutions of earthquakes giving origin or slip stress field, stress drop, etc
f) Aquisition of any data on in situ stress from local mining operations (if any)
g) Field geological studies to establish ground truth for remote sensing to locate and describe faulting near the site
h) Establish much of a local seismograph network

4. In the list items (a) to (e) constitute the desk study. Field work specifically for the purpose of assessing potential for induced seismicity would depend on the outcome of the preliminary study. Direct on-site observations on in situ stress however desirable, would only be undertaken with reluctance.

5. Having obtained information relating to items a to e, it remains to make some preliminary assessment. The first stage is to carry out a seismotectonic interpretation, if possible and identify any faulting close to the dam or reservoir area.

6. General considerations of the structure may permit an estimate to be made of the last date of fault movement. On 'stable' continental areas there may be no evidence to suggest late reactivation unless a datable veneer of recent sediments has been laid down or a disordant suite of fractures can be identified. Statistical study of practice patterns from imagery backed up by ground study may identify the last significant stress state.

7. Association of faults with epicentres may be possible, but again if we are considering continental intraplate sites, this is unlikely as earthquakes of significant magnitude are rare. There may be a basis for a probabilistic evaluation, and this may enable some 'maximum credible' earthquake for the region to be defined. If induced seismicity is a triggering process then the maximum shock which could result, would be not greater (in principle) than the maximum credible for natural seismicity. This reasoning has been adopted recently (Nordstom et al 1979). If there is not sufficient information upon seismicity, such a course would not be reasonable, for example, the maximum earthquake magnitude in the Koyna region known before 1967 was about M = 4.

8. In the absence of sufficient seismicity data to establish such a "maximum event" and until an acceptable algorithm for pattern recognition on geological grounds has been established, the evaluation can only follow a comparative route. This is illustrated in the following example.

EXAMPLE

1. As an example, a site 100 km into the West African Craton (Fig. 1) is studied where a reasoned appriasal of the natural and induced seismic potential was required.

Palaeotectonics of West Africa

The present position of the American and African continents are the result of the opening of the Atlantic which commenced some one hundred and eighty million years ago. Before this opening tectonism took place mainly in the pre-Cambrian with three main events, Libeerian (ending c two thousand five hundred million years BP) Eburnean (c two thousand million years BP) and Pan African (c six hundred million years BP). The deformed and metamorphosal rocks involved in the Libeerian and Eburnean events form the West African Craton which has retained its identity as a geologically stable block to the present day. The Pan African orogenic belt enclosed the craton area and subsequently formed the site for continental separation and the opening of the Atlantic. On the opening of the Atlantic a series of transform faults developed. (Fig. 1) The Chain and Charcot fracture zones are related to the Benin Trough and have controlled the deposional and structural history of the eastern Nigeria since the Mesozoic. Other fracture zones, to the north, can be mapped up to the continental margin and may in some cases be aligned with fractures within the continental crust.

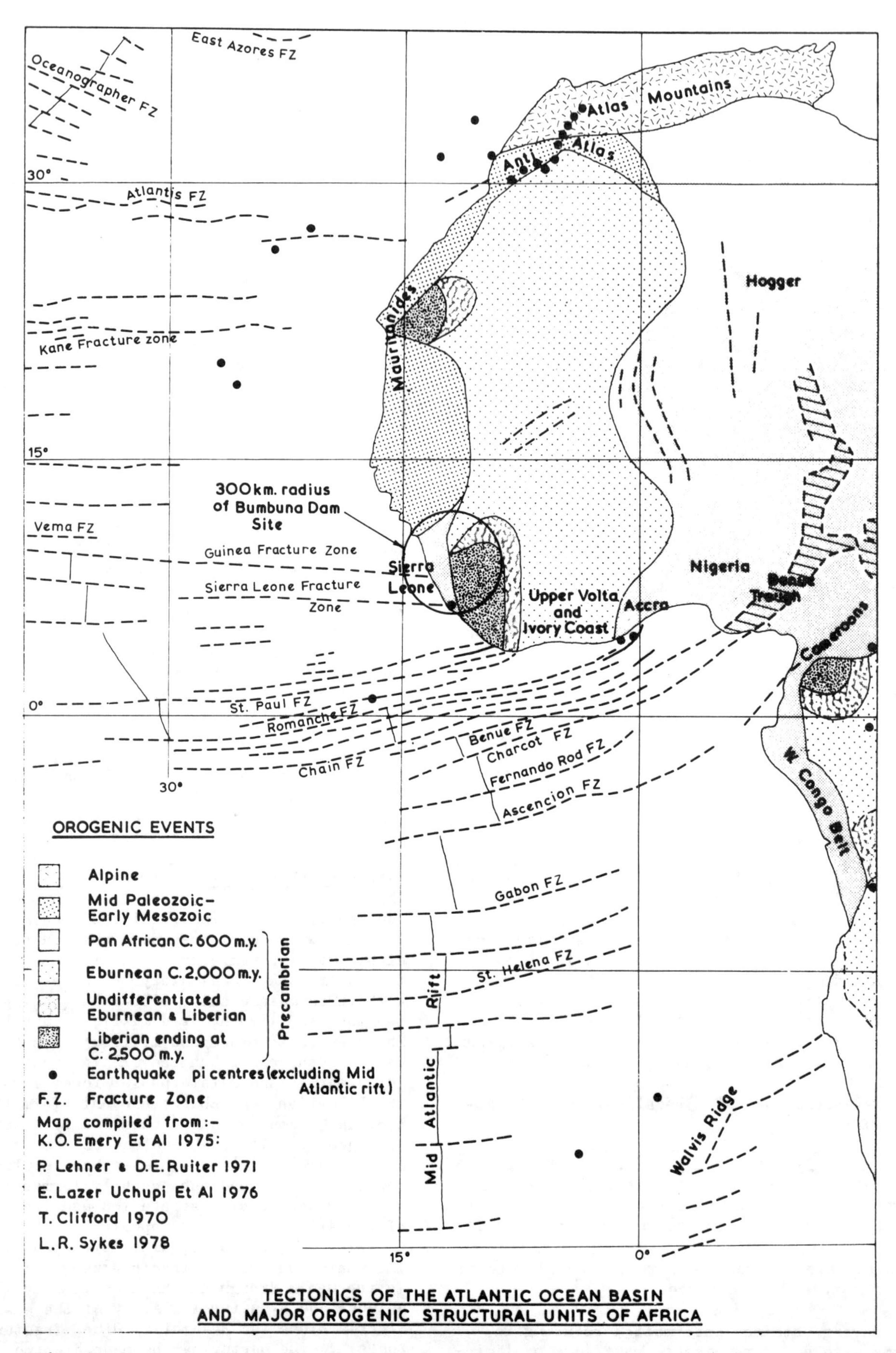
East Azores FZ
Oceanographer FZ
Atlas Mountains
Anti Atlas
30°
Atlantis FZ
Hogger
Mauritanides
Kane Fracture zone
15°
300km. radius of Bumbuna Dam Site
Vema FZ
Guinea Fracture Zone
Sierra Leone Fracture Zone
Sierra Leone
Upper Volta and Ivory Coast
Accra
Nigeria
Benue Trough
Cameroons
0°
St. Paul FZ
Romanche FZ
Chain FZ
Benue FZ
Charcot FZ
Fernando Rod FZ
Ascencion FZ
30°
W. Congo Belt
OROGENIC EVENTS
Alpine
Mid Paleozoic-Early Mesozoic
Pan African C. 600 m.y.
Eburnean C. 2,000 m.y.
Undifferentiated Eburnean & Liberian
Liberian ending at C. 2,500 m.y.
Precambrian
Earthquake pi centres (excluding Mid Atlantic rift)
F.Z. Fracture Zone
Map compiled from:-
K.O. Emery Et Al 1975:
P. Lehner & D.E. Ruiter 1971
E. Lazer Uchupi Et Al 1976
T. Clifford 1970
L.R. Sykes 1978
Gabon FZ
St. Helena FZ
Mid Atlantic Rift
Walvis Ridge
15°
0°
TECTONICS OF THE ATLANTIC OCEAN BASIN AND MAJOR OROGENIC STRUCTURAL UNITS OF AFRICA

Fig. 1

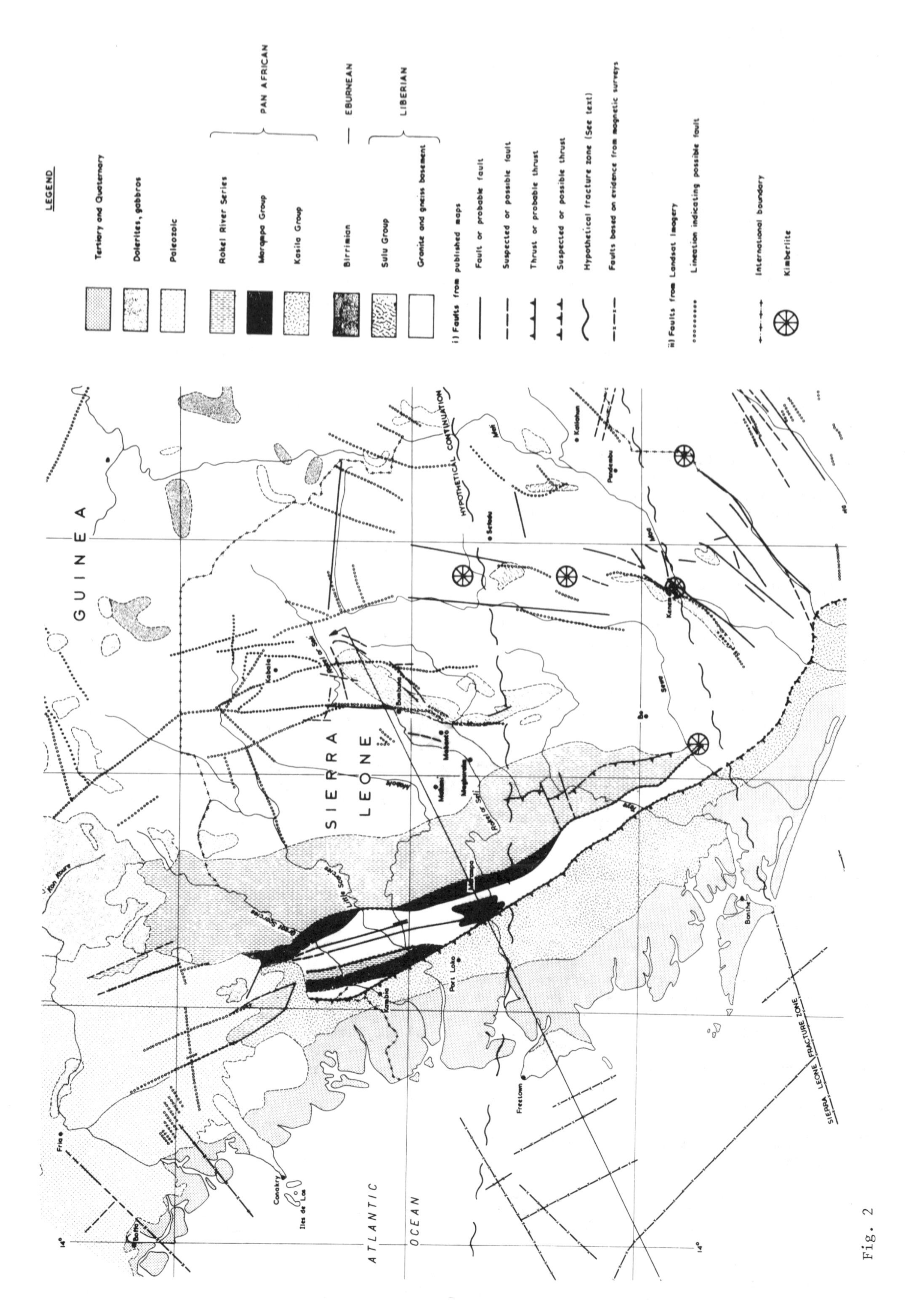
LEGEND
Tertiary and Quaternary
Dolerites, gabbros
Paleozoic
Rokel River Series
Marampa Group
Kasila Group
PAN AFRICAN
Birrimian
EBURNEAN
Sulu Group
Granite and gneiss basement
LIBERIAN
i) Faults from published maps
Fault or probable fault
Suspected or possible fault
Thrust or probable thrust
Suspected or possible thrust
Hypothetical fracture zone (See text)
Faults based on evidence from magnetic surveys
ii) Faults from Landsat imagery
Lineation indicating possible fault
International boundary
Kimberlite
GUINEA
SIERRA LEONE
HYPOTHETICAL CONTINUATION
ATLANTIC OCEAN
SIERRA LEONE FRACTURE ZONE
Freetown
Conakry
Iles de Los
Fria
Port Loko
Kambia
Bo
Bonthe
14°

Fig. 2

Structural Geology within 300 km of a proposed dam site

3. Onshore geology is divided into two units.

1. Pre-Cambrian granitic terrain of the West African Craton with foliation fanning from NNW in the west to ENE in the south-east. Faulting in the Pre-Cambrian consisted of strike slips with some vertical movement and thrusting in the greenstone belts. There has been no evidence of movement since the Cretaceous.
2. The Pan African orogenic belt forming a NW-SE trending zone along the western margin of the craton area from which it is separated by a thrust zone, which may form a crustal dislocation extending to great depth.

4. Offshore geophysics indicate a series of tensional faults running parallel to and normal to the coastline controlling and forming coastal sedimentary basins. These faults developed due to extensional forces, operating during the opening of the Atlantic combined with epeirogenic movements. The fault controlled sedimentary basins which accommodated Cretaceous, Tertiary and Quaternary sediments, are probably still active at the present and are probably the cause of the slight seismicity recorded in the coastal regions.

Transform faulting and the continental margin

5. A relationship has been observed between oceans transform faults and pre-existing zones of weakness within continental margins and it has been suggested that the Guinea fracture zone and the Sierra Leone fracture zone extend several hundred kilometers inland being aligned with ENE-WSW Kimberlite dyke intrusions. These are interpreted as old lines of weakness providing structural control for the transform faults which were reactivated in the Crataceous with the formation of Kimberlite intrusions. Landsat imagery supports this in so far as it reveals a set of lineations reflecting the ENE-WSW tectonic grain of the Liberian craton.

Structural geology within 50 km of the proposed site in Sierra Leone

6. A potential dam site lies in the western edge of a Pre-Cambrian curvilinear greenstone belt surrounded by granitic basement. Laterites overlie Pre-Cambrian rocks, developed on various erosion surfaces formed during uplift and warping of the West African craton since Palaeozoic times. Major faults have been indicated in the area (Wilson and Marmo 1958), one following the course of the River Seli itself. Landsat imagery records several lineations which are consistent with the fault system mapped on the ground, (Fig. 2).

HISTORICAL SEISMICITY AND TECTONIC ASSOCIATIONS IN THE REGION

1. A few small earthquakes have been felt on the coast of Sierra Leone, but the international listing came up with only one event at 0.74°N, 17.52°W on the 4th January, 1951 and Sykes and Landisman 1964 refers to one at 7°N, 11°W. The nearest known active region is in Ghana where events can be associated with the meeting of the Romanche fracture zone and the continental margin.

2. Sykes (1978) suggests that within intraplate regions seismic activity is randomly distributed, but tends to be concentrated either near the ends of major oceanic transform faults along pre-existing zones of deformation or along faults or old fault belts within the thicker lithosphere of the continents. Seismic associations have been suggested where the Atlantic fracture zones meet the Guinea coast (K Burke 1969) and similar arguments have been applied in the Cameroons (Burke 1969). The tracing of such zones into the continent and more specifically into the craton, is still a matter of research. The Sierra Leone fracture zone has been postulated as extending 150 km inland by inference from alignment with the Sava River (Williams and Williams 1977), but extension within the craton has been deduced only by alignment of Kimberlites. This inferred extension is some 50 km south of a proposed dam site.

3. Other potential faults which may be intermittently active in stress relief are in the deep crustal thrust zone separating the craton from the mobile belt and the faults offshore controlling basin formation.

4. These features would be of significance if current crustal stresses are similar to (or conjugate to) these at the time of fault development, or reactivation, but little data on this is available. Sykes (1974) quotes focal mechanism solutions characterised by thrust faulting for earthquakes in Gabon and the Gulf of Guinea. The maximum compressive stress for the latter trending north west-south east agreeing with a near surface strain relief value obtained by Hast (1969). No inference on current state of stress (post Cretaceous) can be made from geological evidence within the craton. The paucity of data does not permit a satisfactory estimate of 'maximum credible' earthquake for the dam site. If however, the maximum epicentral intensity on a possible extension of the Guinea fracture zone is assumed as VI MMI the intensity of site would be around V MMI.

INDUCED SEISMICITY

1. The geological setting of the considered site pertinent to a potential for induced seismicity may be summarised :

a) The site is well within a craton (approximately 100 km)
b) The most recent demonstrable fault movement consists of reactivation of old Pre-Cambrian faults in the Cretaceous
c) The nearest known significant seismically active structural feature is 1000 km to the east-south-east; (Bacon and Bauson 1980)
d) There may be continuing epirogenic processes on the ocean margin
e) A postulated ocean fracture zone extension is 50 km away marked by Kimberlites
f) There is a major but old fault within the Pre-Cambrian basement on the site
g) There is a low topographic relief

Table I. Some characteristics of sites with induced seismicity

Name of Dam	Country	Tectonic Province**	Height	Volume ($x10^6m^3$)	Year of impounding	Year of largest earthquake	Magnitude (Richter) or intensity* (Modified Mercalli)	Prior* Seismicity
			(A) Major induced earthquake					
Koyna	India	Cenozoic volcanics	103	2780	1964	1967	6.5	Low
Kremasta	Greece	Tertiary mountains	165	.4750	1965	1966	6.3	Moderate
Hsinfenkiang	China	Shield or platform	105	10500	1959	1961	6.1	Aseismic
Kariba	Rhodesia	Shield	128	160368	1959	1963	5.8	Low
Hoover	U.S.A.	Tertiary mountains	221	36703	1936	1939	5.0	-
Marathon	Greece	Tertiary mountains	63	41	1930	1938	5.0?	Moderate
Oroville	U.S.A.	Alpine	236	4298		1975	5.7	Moderate
			(B) Minor induced earthquakes					
Benmore	N. Zealand	Tertiary mountains	118	2100	1965	1966	5.0	Moderate
Monteynard	France	Tertiary mountains	155	240	1962	1963	4.9	Low
Kurota	Japan	Alpine or Cenozoic volcanics	186	199	1960	1961	4.9	High
Bajina-Basta	Yugoslavia	Alpine	89	340	1964	1967	4.5 - 5.0	Moderate
Nurek	U.S.S.R.	Alpine	317	10400	1972	1972	4.5	High
Mangalla	Pakistan	Platform	116	7250	1967	1970	4.2	Moderate
Talbingo	Australia	Platform	162	921	1971	1972	3.5	Aseismic
Keban	Turkey	Alpine	207	31000	1973	1974	3.5	-
Vajont	Italy	Alpine	261	61	1963			Low
Pieve de Cadore	Italy	Alpine	112	68	1949	1951		Low
Grandval	France	Platform	88	292	1959		V	Aseismic
Canalles	Spain	Platform	150	678	1960	1962	V	Aseismic
Pukaki	N. Zealand	Alpine	108	10000	1978	1978	4.6	Moderate
Manicougan 3	U.S.A.	Shield	108	10400	1975	1975	4.1	Low
			(C) Changes in micro-earthquake activity					
Grancarevo	Yugoslavia	Alpine	123	1280	1967		1-2	Low
Hendrik Verwoerd	S. Africa	Shield	88	5954	1971	1971	2	Aseismic
Schlegeis	Austria	Alpine	130	129	1971	1971	0	Low
			(D) Transient changes in seismicity					
Oued Fodda	Algeria	Alpine	101	228	1932			Moderate
Camarilles	Spain	Alpine	44	40	1960	1961	3.5	Low
Piasta	Italy	Alpine	93	13	1965	1966	VI-VII	Low
Vouglans	France	Platform or Alpine	130	605	1968	1971	4.5	Low
Contra	Switzerland	Alpine	220	86	1965	1965		Low
			(E) Brazilian induced seismicity					
Carno de Cajura	Brazil	Shield	20			1970	4.6	Aseismic
Paraj buna Paraitinga	Brazil	Platform	80			1977	3.4	Aseismic
Capivara	Brazil	Platform	c40?			1976	4.3	Aseismic

* Tarr
**

h) There is no evidence of local stress raiser (meteor impact)
i) There are no thermal springs or recent volcanic features
j) The local gravity gradients are unknown
k) The general stress field could have maximum horizontal compressive stress acting northwesterly

2. The information suggests that the potential for induced siesmicity is low. It is however, instructive to consider the geological indicators (in so far as they can be assembled) for other sites in or near cratons at which induced earthquakes have been experienced.

Comparative Studies

3. In Table I the general tectonic setting of the sites at which there has been reservoir induced seismicity indicated. Of the thirty-six sites listed, four (Kariba, Verwoerd, Carmo de Cajora, Manic 3) are clearly on a craton, although at its edges. Koyna is on later Decan traps, effusing from a "stable" shield and Talbingo and Hsinfenkiang are on shield/platform boundaries. Of the four reported "shield examples", Kariba may be on a developing rift system and of the remaining examples both Manic and Carmo de Cajura are not far within the craton. Manic 3 is not far removed from the St Lawrence active zone, and the Carmo de Cajura site is not far removed from the junction of an ocean fracture zone with the continental margin.

4. The events at Paraibuna Paraitinga (1977) and Cabivara were on late Pre-Cambrian rocks tectonically similar to the Pan African rocks of Sierra Leone. All those induced earthquakes occur either where old fault zones within the continent meet transform faults offshore or where the junction between a stable craton and a more recent Pre-Cambrian mobile belt coincides with a transform fault offshore.

5. Notwithstanding the Brazilian experience, and in the absence at the site of most of the other diagnostic features in induced seismic potential and despite the probable major fault feature, the site has been judged to carry little potential for induced seismicity. Nevertheless, there always remains a small risk, so it is instructive to consider the character of induced earthquakes.

Strong Ground Motion

6. Instrumental records of strong ground motion associated with induced events are rare. Koyna is still the most quoted instance and from the histogram (Fig. 3) it can be seen to be an extreme case.

7. For a location not having such special features as high topographic relief, strong gravity gradients, or other indications a 'reasonable' approach would be to examine the implications of a small earthquake (M~4.5) 5 to 10 km focal depth at the reservoir/dam site.

8. For an intraplate site, it would then seem appropriate to select from available strong motion records a suite having the characteristics which could be associated with an induced event, i.e. recording within the epicentral region high stress drop and shallow focal depth. A further constraint may be imposed if the extent of faulting within the reservoir area is known. Often however, geological data on faults which would have penetration of 5 to 10 km is obscured. The length of accumulated rupture for a M - 4.5 event may only be approximately 1.0 km so a small existing fault is adequate. In a craton with an exposed crystalline basement old faulting will be in abundance. It will not usually be possible to establish which feature is capable of slipping seismically so such constraints although theoretically attractive are almost impossible to apply.

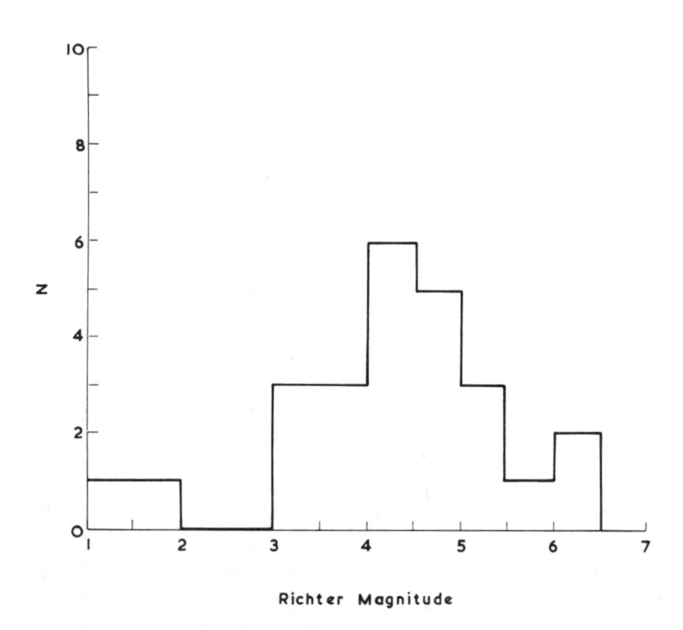

Fig. 3. Histogram of observed reservoir or dam induced earthquakes (25 events)

Table II. Induced seismicity in low aseismic-shield or platforms

Aseismic	Tectonic Province	Maximum age of youngest faulting
Hsinfenkiang	Shield - platform boundary	Tertiary, Quaternary?
Talbingo	Platform	Carboniferous, late Tertiary?
Grandval	Platform	Early Oligocene
Canalles	Platform	Eocene or Oligocene
Henrick Verwoerd	Shield	Mesozoic?
Carmo de Cajura	Shiled	?
Paraibuna Paraitinga	Platform	?
Capivara	Platform	?
Manicougan 3	Shield - Platform boundary	?

CONCLUSIONS

1. So far a satisfactory procedure for assessing site induced seismic potential has not been developed. The general stress states required for induced seismicity are however understood and the strength of faulted crust, more especially in the upper levels is now becoming understood and in principle the shallow crustal stresses can be established. Induced seismicity seems to affect only a few sites and these usually with a low magnitude shocks. In these tech-nical and economic circumstances expensive and elaborate studies are difficult to justify having regard to the confidence in any specific conclusion, and especially at sites well within cratons.

2. All that can then be done is to establish the geological and tectonic setting of a site, sufficiently to enable broad comparisons to be made with the increasing number of instrumented sites a small proportion of which will exhibit induced shocks. Fortunately with the aid of Landsat imagery and the development in palaeotectonic studies this is now economically feasible.

ACKNOWLEDGEMENTS

1. This paper is published with the permission of the Directors of Soil Mechanics Ltd., Foundation House, Bracknell, Berkshire.

REFERENCES

1. UNESCO Intergovernmental conference on the assessment and mitigation of earthquake risk. Final report SC/MD/53, Paris, May 1976
2. I.A.E.A. Earthquakes and Associated Topics in Relation to Nuclear Power Plant siting Safety Guide No. 50-SG-51 1979
3. USCOLD Working Group on Induced Seismicity United States Committee on Large Dams. May 1977
4. BATH M. Fracture risk estimation for Swedish earthquakes. Tectonopysics 61 (1980) T.15-22 1980
5. GOUGH D. and W.I. Stress and deflection in the lithosphere near Lake Kariba I Geophys. J. Rou. Astrn. Soc. 21: p.65-78 1970 Load induced earthquakes at Kariba II ibia. pl.79-101 1970
6. WITHERS R.J. and NYLAND J.B. Time evaluation of stress under an artificial lake and its implications for induced seismicity. Can. G. Earth. Soc. Vol. 15, pp.1520-1534 1978
7. NICOLAEV N.I. Tectonic conditions favourable for causing earthquakes occurring in connection with reservoir filling. Engineering Geology 8, pp.171-189 1974
8. LOMNITZ C. Earthquakes and reservoir impounding - State of the Art. Engineering Geology 8, p.191-198 1974
9. CASTLE R.O., CLARK M.M., GRANTZ A. and SAVAGE J.C. Tectonic state: its significance and characterisation in the assessment of seismic effects associated with reservoir impounding. Engineering Geology 15 (1980) p.53-99
10. SYKES L.R. Intraplate seismicity, reactivation of pre-existing zones of weakness, alkaline magnatism and other tectonism post dating continental fragmentation. Reviews of Geophysics and Space Physics, Vol. 16, No. 4, pp.621-688, 1978
11. AHORNER L., MURAWSKI and SCHNEIDER G. Seismotectonishe traverse von der Nord see bis zum Appennin. Geol. Rundschau 61 p.915-942 1972
12. SCHAFER K. Tectonic evidence of horizontal compressive stresses within European and North American plates. Trans. Am. Geophys. Union V.551443 1974
13. ALLEN C.R. Geological Criteria for estimating seismicity Geol. Sco. Am. Bull. 86: p.1041-1057. 1957
14. NORDSTROM P.A., STEBY B. and TARANDI T. Aseismic design of Mtera dam. CIGB (ICOLD) Q51 R7. 13th Congress New Delhi 1979
15. WILSON N.W. and MARMO V. Geology, geomorphology and mineral resources of the Sula Mountains, Geol. Surv. Sierra Leone, Bull. 1
16. BURKE K. Seismic areas of the Guinea Coast where Atlantic fracture zones reach Africa. Nature Vol.222. May 14th pp.655-657 1969
17. WILLIAMS H.R. and WILLIAMS R.A. Kimberlites and plate tectonics in West Africa. Nature Vol. 8, December 1977, pp.507-508
18. SYKES L.R. and LANDISMAN M. The seismicity of East Africa, Gulf of Aden and the Arabian and Red Seas. Bull. Seism. Soc. Am. Vol. 54, pp.1927-1940 1974
19. HAST N. The state of stress in the upper part of the earth's crust. Tectonophysics Vol. 8. pp.169-211 1969
20. BACON M. and BANSON J.K.A. Seismicity of Southern Ghana 1977-1979 Geology Department, Chelsea College, London, February 1980
21. MENDIGUREM J.A. Personal Communication

33 Probable risk estimation due to reservoir induced seismicity

S. K. GUHA, MSc PhD, J. G. PADALE, MSc, and P. D. GOSAVI, MSc, Central Water and Power Research Station, Pune

Following a survey of reservoirs of the world wherein induced seismicity was reported, it is observed that these earthquakes releasing tectonic strains are triggered primarily through pore pressure effect. Majority of the reported cases occurred in the earthquake magnitude range 3.0 to 5.9. For design purposes, magnitude 7.0 could be taken as the probable maximum earthquake magnitude due to RIS.

Whereas for highly seismic region, the higher design factor deduced on the basis of the regional seismicity is considered adequate, for moderately active regions, due considerations need to be given to nearby seismogenic features and reservoir dimensions for proper evaluation of risk. A methodology for risk evaluation due to RIS incorporating these factors is proposed.

INTRODUCTION

1. Since the classical case history of seismicity following impounding in lake Mead area in USA, over sixty cases of reservoir induced seismicity (Fig.1) (Table) of various levels of intensity (intense I, medium II and minor III) have been detected all over the world (ref.1,2). It is obvious that quite many a case specially those under minor level of intensity (III) have remained undetected. And thus the above number of established cases of RIS could be a fraction of actual case histories. From review of level of activity of RIS world over, it is seen that a large fraction of established cases of RIS falls under the categories of medium (II) and minor (III) levels of intensity while only a few namely Koyna, Kariba, Kremasta and Hsinfengkiang could be classified under the category of intense (I) activity resulting in damaging earthquakes following impounding. Of these cases of intense activity (I), only in the cases of Koyna and Hsinfengkiang, dams have been damaged structurally. The three component accelerogram recorded on the body of the Koyna dam (Fig.2) for the main earthquake (M:7.0) on Dec.10, 1967 (ref.3) following impounding is still the strongest ground motion recorded on dams so far and thus serves as vital design data for dams in the world. In all about 50 accelerograms were recorded in the strong motion net around Koyna dam (ref.3) in the magnitude range 3.5 to 7.0

2. Of the 65 cases of RIS (ref.1) reported so far, 46 cases have been accepted as definite RIS while 12 cases are doubtful RIS and 7 cases are considered not related to RIS. These broad data on RIS have been obtained from 235 cases of deep and very deep reservoirs examined recently. In view of diversity and lack of geological informations on tectonic stress pattern in the reservoir areas, influence of ambient geotectonic field on probability of occurrence of RIS remains obscure. Further, stress in rock due to reservoir impounding is small as such reservoir loading at best could be considered to act only as trigger though such a conclusion could be regarded as premature until in-depth study of role of fluid pressure on geotectonics is available specially with reference to larger number of case histories on RIS. Of the eleven reservoirs where earthquakes of magnitude 5.0 and above occurred following impounding (ref.1) active faulting was present at least in 9 cases, and no fault displacements were observed along the inactive faults. These observations indicate prevalence of triggering phenomena in seismicity associated with reservoir impounding. Therefore, an earthquake which would have occurred later in the normal geologic sequence may be prematurely triggered following impounding. Geographical distributions of RIS cases(Fig.1) would strongly suggest some genetic relationship between RIS and geotectonic phenomena associated with continental margins as most of the RIS cases are situated at or near marginal areas presumably of low tectonic stress.

3. The main hazards to dam from earthquake are due to surface faulting, strong ground movements, water waves in reservoir produced either by seismic waves or by landslides and rock falls and ground deformations associated with faulting. The ground deformations have been observed in number of cases, e.g. the Baldwin Hills reservoir, Buena Vista Hills, Kern County, at locations over the main San Andreas fault, etc.. In case of earth and rockfill dams, the damage occurs

mainly through liquefaction failures. A number of dams was damaged through this process during the Feb.9, 1971 San Fernando earthquake, the significant of which are the Upper and Lower San Fernando dams in which severe cracking and settlement developed. The Pacoima arch dam however sustained severe ground motions with comparatively minor damage (30' clockwise rotation of the axis of the dam and tilting of the dam). Cases of damage to other earth dams are Dry Canyon dam (1952 Kern County earthquake), Sheffield dam (1925 Santa Barbara earthquake); Hebgen dam (1959 Montana earthquake) and the Eklutna dam (1964 Great Alaska earthquake). Of these, the Eklutna dam was very severely damaged and was later reconstructed. The RIS on the other hand poses hazard to dams and pertinent structure through unforeseen rapid movements originating very close to the dam. Intense RIS at Koyna and Hsinfengkiang caused considerable damage during the Dec.10,1967(M:7.0) and Mar.18,1962(M:6.2) earthquakes respectively. Damage was in the form of horizontal cracks due to overstressing.

4. Seismic instrumentation of dams and reservoir area is planned for the following objectives :

a) to assess the level of preimpounding seismicity of the area;
b) to delineate seismogenic crustal weaknesses in the vicinity;
c) to follow up the progress of observed seismicity; and
d) in the event of occurrence of significant earth movements, to record seismic forces (accelerations) experienced by the structure and on ground so as to verify the strength and response of the structure.

An earthquake being a rare event, the instrumentation deployed should preferably be field tested for reliability and prolonged operations and at the same time be sufficiently sensitive (magnification $\sim 10^6$ or so) to record a large number of weak events in as short time as possible. For local seismicity studies a minimum of five seismic stations have been found to be necessary for adequate areal coverage and reliable magnitude estimations. Besides the standard Wood-Anderson Seismographs, highly sensitive(magnification $\sim 10^6$) seismographs with high frequency system response (1-50 Hz) may be installed at each of the stations. Stations of the net should be located to cover the reservoir spread completely and at 10-30 km from one another. Strong motion accelerograph is essential at every dam site. It is relatively inexpensive and operational only during strong earth movements. These instruments should be installed at various levels within the dam and also on ground surface.

CASE HISTORIES OF IMPORTANT RIS

Koyna Dam (INDIA)

5. Following impounding of the reservoir in 1962, about 35000 earth tremors were recorded in the immediate vicinity of reservoir till end of 1979. These tremors (Fig.3) include the main event of magnitude 7.0 and about twenty shocks of magnitude 4.0 and above in the series, and began to occur immediately after impoundment in a very limited area in a highly stable and aseismic Indian peninsula. The earth tremors beginning almost simultaneously with the impoundment, culminated in a major seismic event of M:7.0 after about four years of impoundment. These tremors occurring over a period of 16 years could be easily divided into foreshocks and after-shocks of the main event, (M:7.0) on Dec.10, 1967. Further the main event is accompanied by large increase in fracture volume, very high rate of tilting, large strain and deflection changes and significant lake level changes (yearly moving average), so also the second largest event on Oct.17,1973 in the series (Fig.4). Significant lake level variations (yearly moving average) in triggering largest seismic events on Dec.10, 1967 and Oct.17, 1973 preceded by low b-values are important factors in favour of RIS specially as the area is situated in a highly aseismic region.

Idukki Dam. (INDIA)

6. The dam is situated in very mildly seismic zone with infrequent occurrence of isolated low magnitude earthquakes and earthquake swarms. The reservoir area has six seismic monitoring stations with high capability seismographs having gain upto 10^5 or so and strong motion accelerographs. A geodetic net has been established in the area and has been helpful in assessing tilt of the foundation-rock system. Seismicity pattern (Fig.5) shows that number of earthquakes increased following impoundment, and significant earthquakes occurred following low b-values and anomalous ground tilt. The pattern of activity has been broadly similar to that at Koyna dam excepting that seismic activity was at medium level (II).

Oroville Dam, USA

7. Oroville earthquake (M:5.7) on Aug.1,1975 with epicentre near the reservoir has been one of the recent case histories of RIS following impoundment of the reservoir in 1967 (ref.4). The dam site has been known to be a low seismicity area with an earthquake of magnitude 5.7 in 1940 about 80 km north of Oroville. There has been definite increase in seismicity of the area with number of epicentres around the reservoir following impoundment. Moreover, low b-values prior to the earthquake on Aug.1, 1975 is an indication of accumulation of stress following impoundment.

Kariba Dam, Rhodesia

8. Kariba lake, one of the largest artificial water mass in the world, was first impounded in 1960. The area had no previous record of sizeable seismic activity. The first significant earthquake in the lake area (M:5.7 on Sep.23, 1962) was followed by three more shocks of M:6.1, 5.6 and 5.8 (Fig.6). Two other significant shocks are

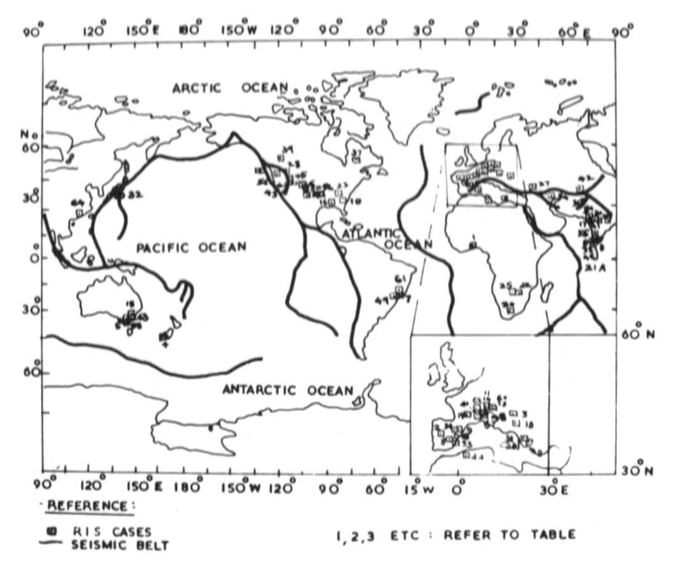

Fig. 1. Reported RIS case histories of the world and seismic belts

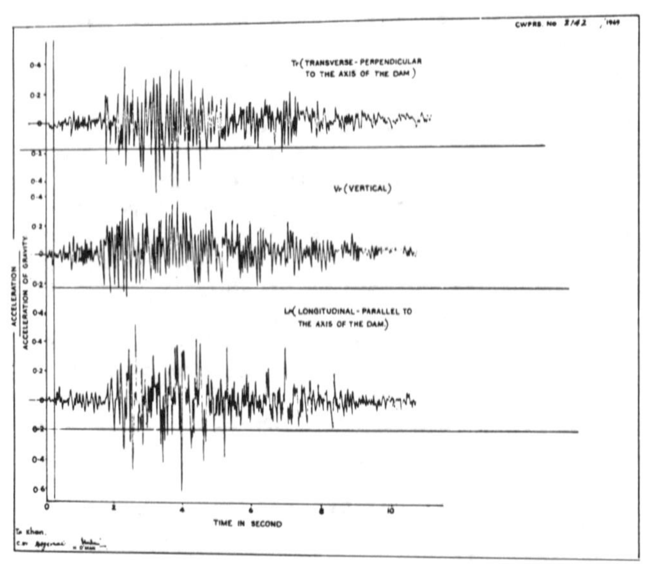

Fig. 2. Accelerogram of the main Koyna earthquake (Dec. 10, 1967; M:7.0, ref. 3)

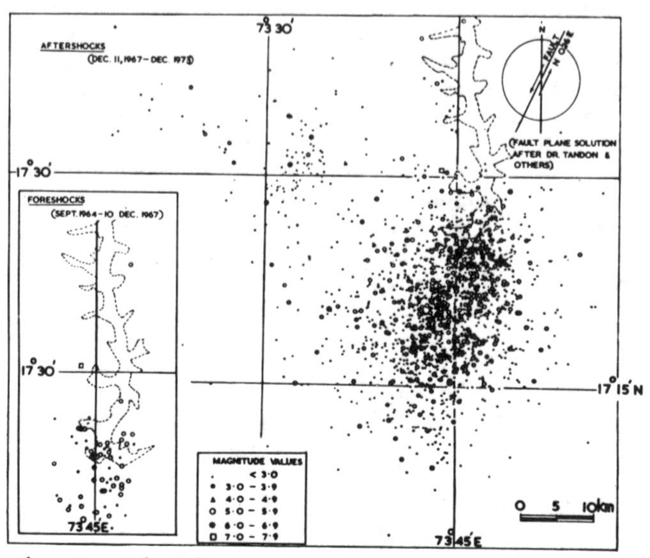

Fig. 3. Distribution of epicentres in Koyna reservoir area (1963-73)

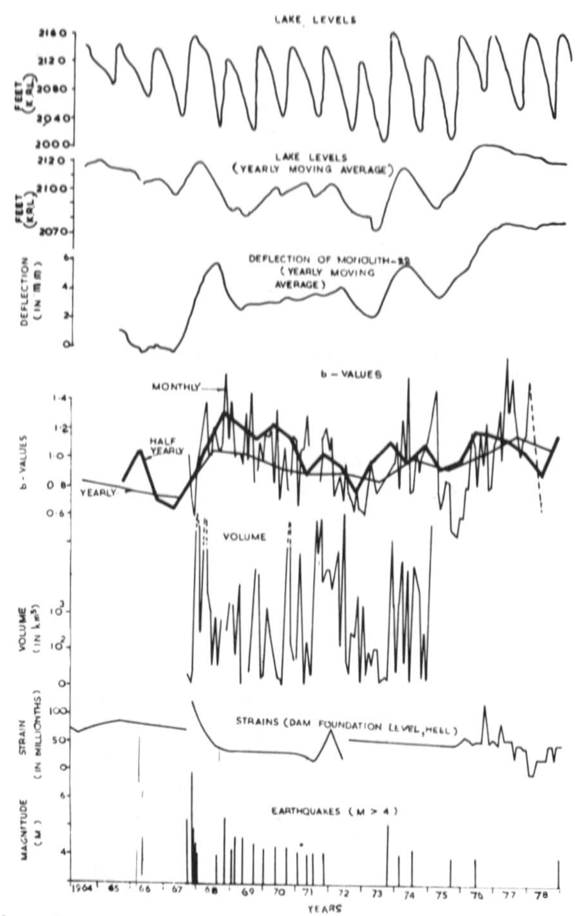

Fig. 4. Lake levels, deflection and strains of the Koyna dam and corresponding earthquake data

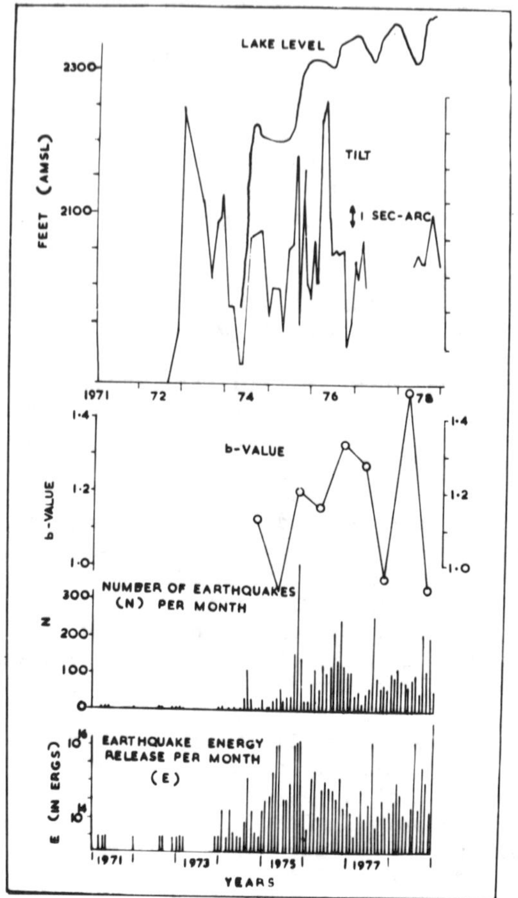

Fig. 5. Seismicity, b-values, ground tilt and lake levels in Idukki reservoir area

M:5.5 (Sep.24, 1962) and M:6.0 (Sep.25, 1962). Since then there has been systematic decline in seismic activity in the lake area though activity on reduced scale is continuing. Seismic activity is spread over longer period and energy is released periodically contrary to experiences at Koyna, Oroville, Monteynard, Hsinfengkiang, etc.. The Itezhitezhi reservoir area in the neighbouring Kafue valley has experienced shocks upto M:4.2 since impoundment.

Kremasta and Marathon Dams, Greece

9. Though the dams were located in tectonically active areas in Greece, occurrences of tremors (Fig.7) soon followed by a moderate earthquake of magnitude 6.3 on Feb.5, 1966 immediately after impoundment of the reservoir, could be attributed to RIS. The lake was full when the moderate earthquake of magnitude 6.3 occurred.

Nurek Dam, USSR

10. The dam is situated in a highly seismic area across the Vakhsh river of Tadjikistan Republic of Soviet Central Asia (USSR). It is presently the highest earth and rockfill dam in the world and has been designed to withstand at least an earthquake of magnitude 6.5. The strongest earthquake of magnitude 4.5 occurred in Nov. 1972, when there was sharp change in lake level, and was preceded by large number of minor tremors.

Manicouagan 3 Dam, Canada

11. Manic 3 dam across the Manicouagan river is situated in Canadian shield (Manicouagan crater). The area has experienced number of low magnitude earthquakes according to seismic map of National Building Code of Canada. The largest earthquake recorded following impoundment was of magnitude 4.1 (Oct.23, 1975) preceded and followed by large number of very shallow (focal depths within 2 km) microearthquakes (ref.5).

Hsinfengkiang Dam, China

12. The dam is situated 160 km northeast of Canton. Construction commenced in 1958 and impoundment began in 1959. An earthquake of magnitude 6.2 struck the dam site on Mar.29, 1962. The earthquake had depth of focus of 5 km and epicentral intensity of VIII. With the increase of reservoir level, seismic activity increased and sixteen earthquakes of magnitude 3-3.5 occurred successively. However, in the first half of 1961 seismic activity decreased with fall in lake level. With the help of a seismic net which began functioning in July 1961, detailed epicentral map of the earthquakes (Fig.8) could be obtained. Within a period of 28 months, 81,719 large and small shocks were recorded in the net (ref.6). Distribution of focal depths shows that maximum number of earthquakes had very shallow focal depths between 5 and 7 km. Thus the seismic history of the reservoir area depicts a typical case of RIS-seismic activity beginning at micro level at low lake level, culminating into a moderate earthquake of magnitude 6.2 at the time of highest water level. Like Koyna dam, the Hsinfengkiang dam also got damaged due to strong ground movements. RIS at Koyna and at Hsinfengkiang has thus great parallelism in all respects. The determining factors could be similarity in geohydrological and tectonic fabric of the two areas.

DISCUSSION AND CONCLUSION

13. The observed phenomenon could be divided in three categories on the basis of the degree of the magnitude range experienced (Table). Intense activity (I) ($M \geqslant 6.0$) has been observed only at four sites so far, viz. Koyna, Kremasta, Kariba and Hsinfengkiang. Of these, Kremasta is in highly seismic zone. Medium RIS(II) (M 3.0 to 5.9) occurs in a large majority of cases (Fig.9). It is possible that the cases wherein the activity was at minor level (III) (microearthquake range $M < 3.0$) are more than have been reported since at this level, earthquakes may be too weak to be felt. In a few cases, the seismicity is triggered or reduced after the depletion or filling up of the reservoir.

14. Annual b-value variations (Fig.4) show consistent decrease over a period of four years prior to Dec.10, 1967 earthquake at Koyna (M:7.0). Both seismicity increase and b-value decrease were simultaneously observed following impounding of the Koyna reservoir. This premonitory period of about 4 years is amply corroborated from the equation of Rikitake (ref.7) $\log T = 0.76 M - 1.83$, as also the pre- and post-earthquake b-values from the relationship obtained by Drakopoulos (ref.8): $b_f = .0.11 + 0.65 b_a$. This change in b-values during pre- (b_f) and post-earthquake (b_a) period is consistent with change in stress level following a major earthquake. Further, monthly b-value variations (Fig.4) show consistent decrease prior to significant earthquakes of magnitude 4.0 and above accompanied by increase in earthquake fracture volume. Similar premonitory variations in b-values have been observed for Mula, Idukki, Manicouagan 3 and Kariba reservoirs. Low b-values and anomalous ground tilts (Fig.5) preceded significant earthquakes in Idukki lake area on Sep.19, 1975 (M:3.0), July 2, 1977 (M:3.5) and on Dec.1, 1978 (M:3.5). Maximum earthquake magnitude recorded during the post impounding period is 3.5. Occurrences of strongest shocks on Sep.23, 1963 (M:5.8), on Apr.20, 1967 (M:5.5) and on Dec.18, 1972 (M:5.4) in Kariba lake region are preceded by low b-values (Fig.6) corresponding to higher stress accumulations. In general, the lower the b-value, the larger the earthquake magnitude depending on tectonic environment. Following Rikitake (ref.7) it could be suggested that whereas b-values could be useful for long term prediction, periodic geodetic surveys and tilt measurements as at Koyna (Fig.10) could be of great help in short term prediction specially in locating the source region of important seismic event. From the present review of

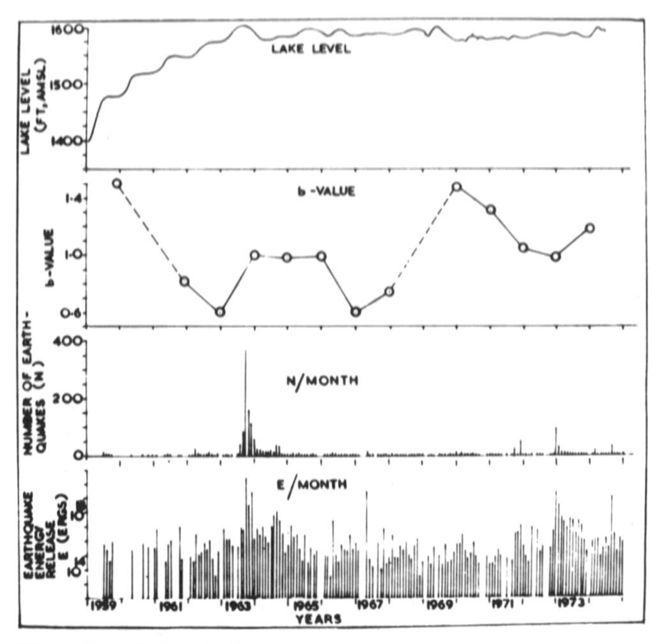

Fig. 6. Seismicity, b-values and lake levels in Kariba reservoir area

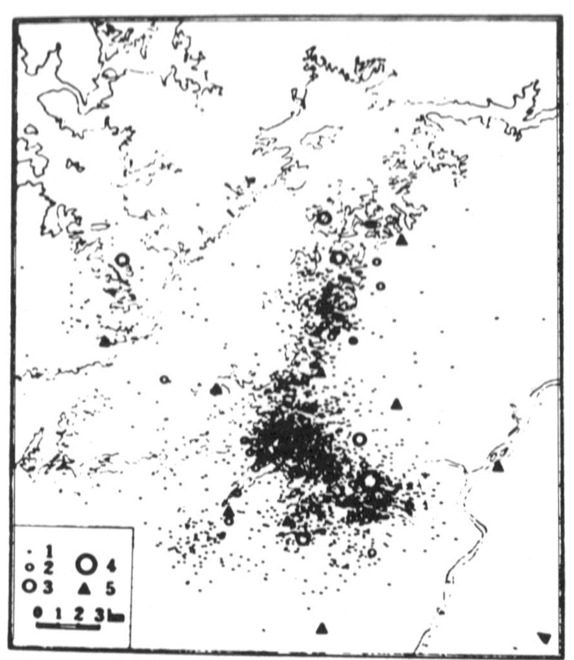
Fig. 8. Distribution of aftershocks in Hsinfengkiang reservoir area (ref. 6) Magnitude M_s : 1. 1.0-2.9; 2. 3.0-3.9; 3. 4.0-4.9; 4. 6.1; 5. seismic station

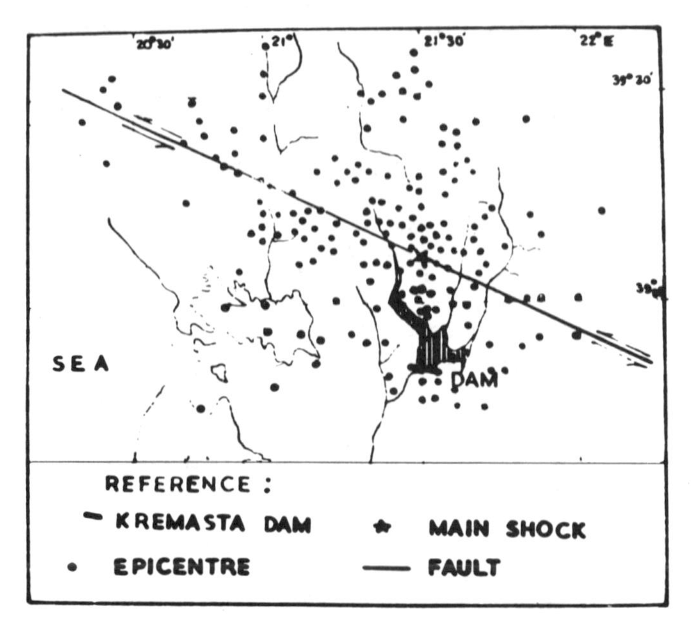

Fig. 7. Distribution of epicentres in Kremasta reservoir area

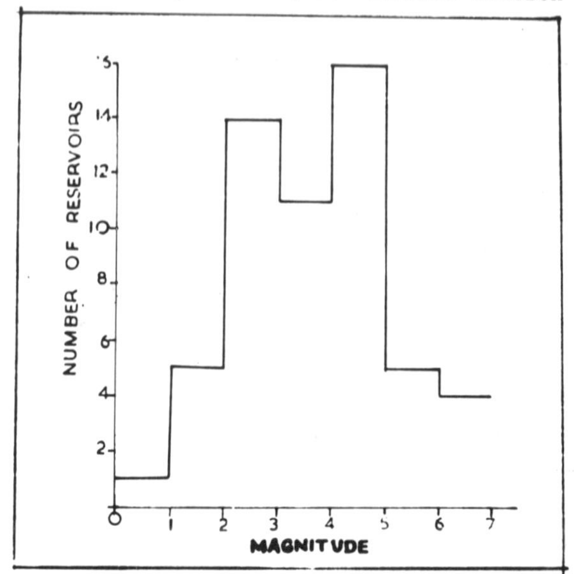

Fig. 9. Magnitudewise distribution of reported RIS cases

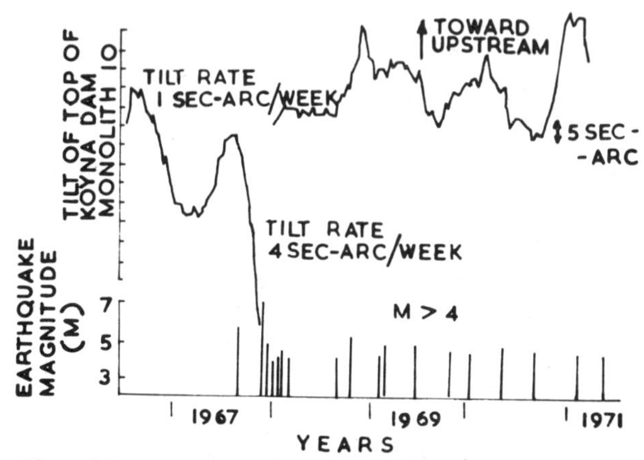

Fig. 10. Ground tilt and significant earthquakes in Koyna reservoir area

reservoir induced seismicity world over, such surveys could profitably be restricted to within twenty km or so of the reservoir periphery.

15. The RIS characteristics are observed to vary from case to case and in general, there is no uniformity. These are as listed below: i) The onset of RIS occurs from 'immediately after' to 'few' years after the first filling; ii) The most probable maximum magnitude is in the range 3.0-5.9 (Fig.9), the maximum earthquake known being that at Koyna of magnitude 7.0; (iii) Enhanced seismicity has been observed following faster loading in a few cases; (iv) No direct relationship between volume or depth of reservoir and the maximum earthquake magnitude could be established; (v) Correlation between reservoir levels/ volume and seismicity has been observed in some cases (not all) at least during the first few years; (vi) As also seen in controlled experiments at Rangely oil fields (USA) and elsewhere, the seismicity depends on the rate of pumping and once triggered, it may continue over some period even after pumping operations are discontinued; (vii) The most favourable tectonic regime seems to be that with moderate tectonic stress (moderate seismicity) and in the vicinity of active normal or strike-slip faulting.
The following mechanisms are essentially based on two evident processes involved, viz. superimposition of reservoir water load on tectonic stress regime and modification in pore pressures in the crustal rock system.
Mechanisms : (i) The reservoir load generated stresses provide the 'last straw' required for the seismogenic movements; (ii) The increase in pore pressure due to reservoir impounding reduces the effective stress necessary for crustal readjustments; (iii) Reservoir load itself causes the earth movements; (iv) Physical interaction between the molecules of water and of the crustal rocks (say quartz as observed in laboratory studies) leads to fatigue and weakening to give rise to seismogenic yield of the material. This could be expected to occur generally in silicate rocks.
The stresses generated by reservoir loading are of very low order ($\sim$ 10 bars) and are thus inadequate for initiating fresh fracturing processes. The pore pressure effect could be expected to be the predominant mechanism in most of the cases especially since the activity is seen to be mostly initiated with time lag after the first impounding. In view of the above observations, it is generally concluded that dam of height over 100 m or large capacity ($\sim 10^9$ m^3) and especially those located in the region of moderate seismicity in the vicinity of active normal or strike-slip faults are more prone to RIS and thus need to be placed under seismic surveillance. Estimation of design seismic coefficient for such cases also deserves special attention particularly if the structure is situated in a region which has been seismically inactive for a long period.

16. From detailed analysis of earthquake data of typical seismic regions of the world, it has been found that Gumbel's approach of extreme statistics and recurrence law of Gutenberg and Richter i.e. $\log N = a - bM$ (N being the number of earthquakes of magnitude M) yield almost similar value of probable maximum earthquake magnitude M_d for period L. As recurrence law of Gutenberg and Richter has been widely applied in earthquake statistics, Campbell's relationship for $M_d = (\log L/L_o + a)/b$ obtained from the same yield a value of 7.0 for L = 50 years for acute seismic regions like Kanto district of Japan and the Himalayas (L_o being the recurrence period). This value of M_d ($\sim$ 7.0) is comparable to the maximum earthquake magnitude (Table) experienced so far as a result of reservoir induced seismicity (RIS). Thus no special consideration needs to be given to added risk due to induced seismicity for reservoirs located in highly seismic areas. Hence maximum risk due to RIS could be covered if structures are designed for ground motions of earthquake $M_d \sim 7.0$ with epicentre in the vicinity of the structure. Two such recent earthquakes are Koyna (M:7.0) earthquake of Dec.10, 1967 (Fig.2) and San Fernando (M:6.6) earthquake of Feb.9, 1971 whose ground motions (accelerograms) have been recorded at close epicentral distances. Hence response spectra of these accelerograms (Figs.11 and 12) could be utilised for design of structures to cover maximum risk due to RIS. Since the reservoirs are found generally to activate faults already present in the vicinity, reservoir impounding seems merely to accelerate the occurrence of a normal shallow tectonic earthquake mainly through pore pressure effect. However, in absence of any methodology to assess accurately the seismic status of reservoir area following impounding, it is suggested that for low seismic regions, the maximum earthquake magnitude (M_{max}) estimated from regional seismotectonics and available earthquake history, may be used as M_d. Quantitatively M_d for low seismic areas following impounding, could be equivalent to an earthquake having stress drop equal to effective stress (σ) given by

$$\sigma = \underset{\text{(tectonic stress)}}{\sigma_o} - \underset{\text{(frictional stress)}}{\sigma_f}$$

This methodology may give upper estimate for M_{max} ($\sim M_d$) but applicability of this approach in fluid induced seismicity is still debatable. Thus, the following methodology is suggested for estimating M_d to cover added risk due to RIS.
1) For highly seismic regions : On the basis of historical and recorded earthquake data, determine Gutenberg -Richter coefficients (a- and b-values) and compute M_d for structural life of L years. For structure of expected life L over 50 years, no added risk due to RIS needs to be considered. (2) For low seismic regions : Find characteristic maximum earthquake magnitude (M_{max}) to have occurred within 100 km radius or so and use it as M_d especially for reservoirs of over

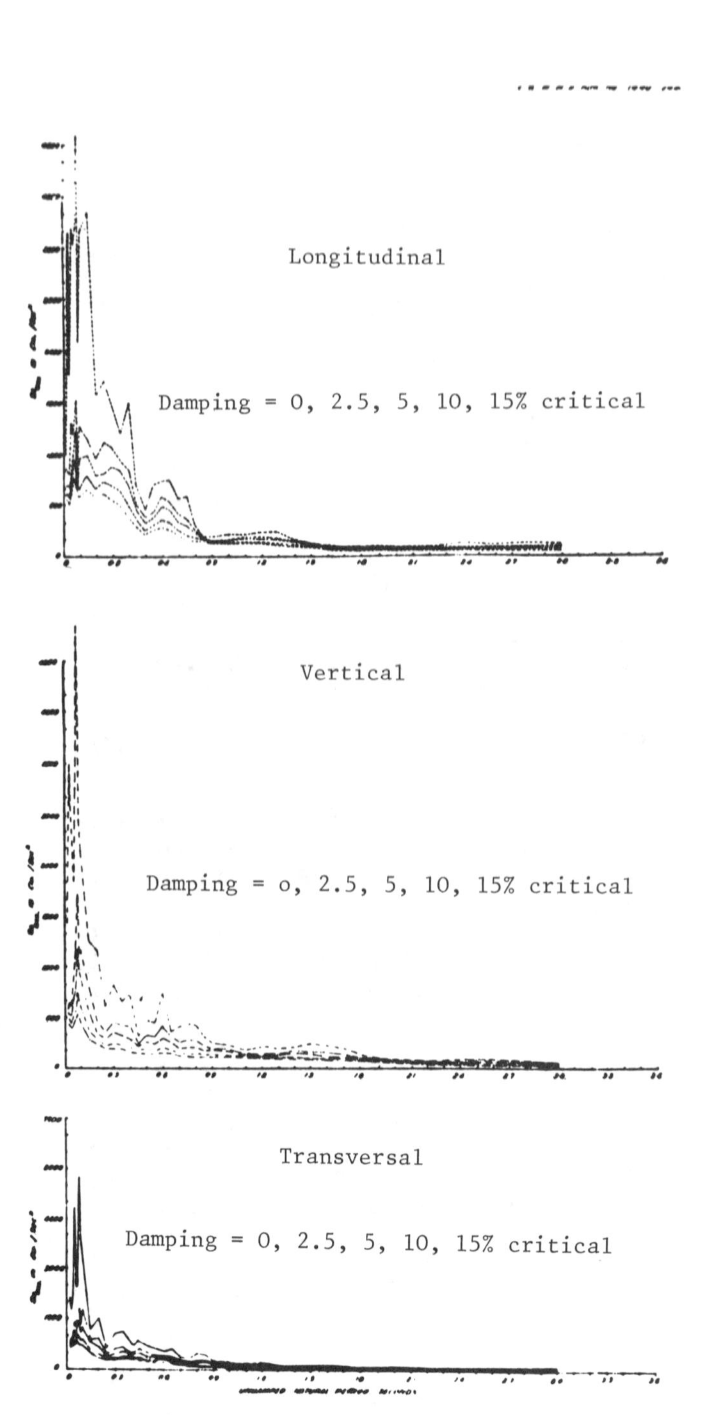

Fig. 11. Maximum acceleration response spectra of the main Koyna earthquake of Dec. 10, 1967 (Fig. 2)

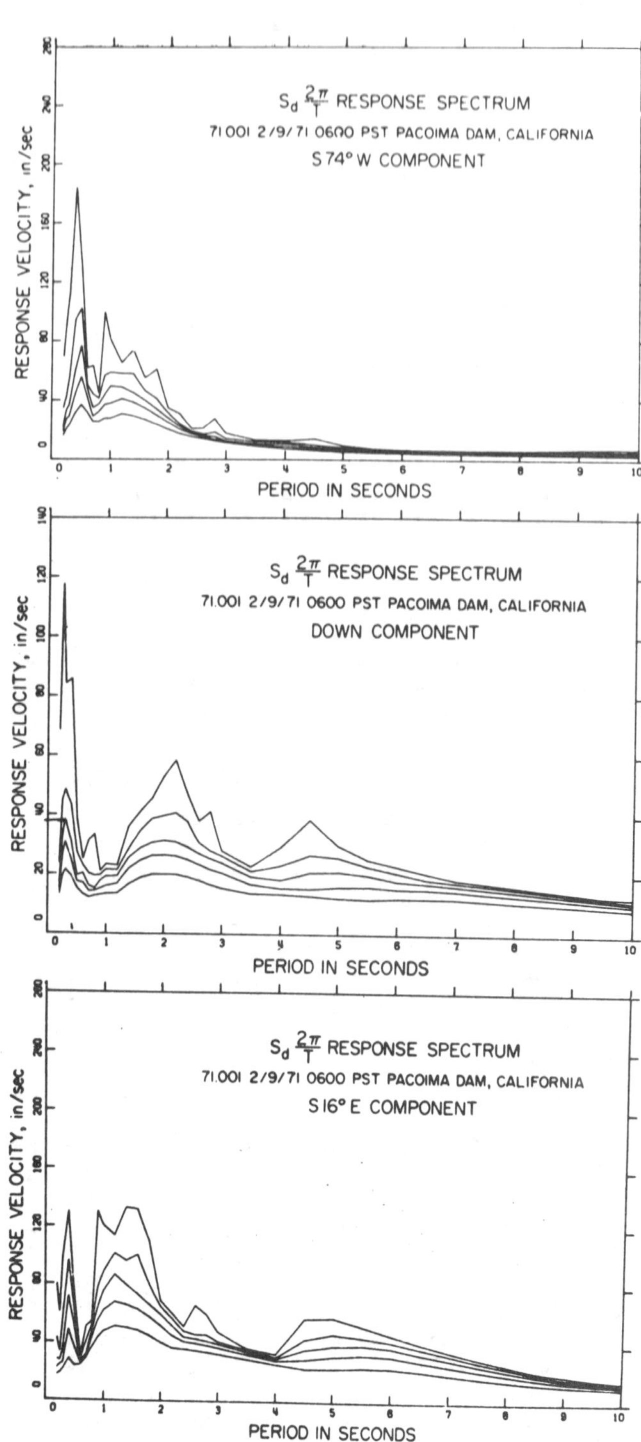

Fig. 12. Maximum velocity response spectra for the San Fernando earthquake of Feb. 9, 1971 (M:6.6)

100 m depth and located near recent faults. Quantitatively M_d (or M_{max}) $\sim f(\sigma)$. Deep hydrofracturing experimental data could be helpful in quantitative estimation of M_d. Where copious microearthquake activity is exhibited following impounding impending seismic status of the reservoir area could be assessed through periodic b-value estimations. In case of Koyna, Mula, Idukki, Kariba and Manicouagan 3 reservoirs, b-values could be used as premonitory index for prediction of larger seismic events. Thus intensive microearthquake survey in the reservoir area could be a powerful tool in the risk estimation and for prediction feasibility study of reservoir induced seismicity.

Table. Details of dams, reservoirs and level of RIS with magnitude of maximum induced earthquake

Sr. No.	Dam: Name, COUNTRY	Reservoir location (degrees)		Type	Reservoir (maximum): depth (in m)	length (in km)	capacity (in $10^6 m^3$)	M	Date d m y	Level of RIS
1	2	3		4	5	6	7	8	9	10
1.	Akosombo, GHANA	7.50N	0.25E	RF	109.0	400	165000	5.3	11.64	II
2.	Almendra, SPAIN	41.21N	6.16W	CA	185.0	32.5	2649	2	1.72	III
3.	Bajina Basta, YUGOSLAVIA	43.97N	19.37E	HCG	80.0	7.0	340	4.8	3.7.67	II
4.	Benmore, NEW ZEALAND	44.40S	170.23E	EF	96.0	65	2040	5.0	7.7.66 6.4.71	II
5.	Blowering, AUSTRALIA	35.50S	148.26E	EF	95.0	25	1628	3.5	6.1.73	II
6.	Cabin Creek, USA	39.62N	105.72W	RF	46.0	1.0	18	2	68	III
7.	Cajuru, BRAZIL	20.30S	44.70W	CG	20.7	18	192	4.7	23.1.72	II
8.	Camarills, SPAIN	38.36N	1.65W	CG	43.65	8.0	37	4.1	15.4.64	II
9.	Canelles, SPAIN	42.03N	0.65E	CA	132.0	21.0	678	4.7	9.1.62	II
10.	Clark Hill, USA	33.85N	82.38W	CG	54.0	53.0	3517	4.3	2.8.74	II
11.	Contra, SWITZERLAND	46.23N	8.83E	CA	190.0	4.8	86	3	10.65	III
12.	Coyote Valley, USA	39.23N	123.17W	E	22.0	5.5	151	5.2	6.6.62	II
13.	El Grado, SPAIN	42.38N	0.17E	CG	85	13	400	-	-	-
14.	Emosson, SWITZERLAND	46.09N	6.91E	CA	170	4.0	225	3	73-74	III
15.	Eucumbene, AUSTRALIA	36.08S	148.72E	EF	106	32	4761	5.0	18.5.59	II
16.	Fairfield, USA	34.34N	81.32W	E	49	-	-	2.8	10.78	III
17.	Ghirni, INDIA	(18.37N	76.83E)	EF	15.2	-	3	2	-	III
18.	Grancarevo, YUGOSLAVIA	42.75N	18.48E	CA	105	17	1280	3	68-70	III
19.	Grandval, FRANCE	44.97N	3.10E	CMA	78	18	292	V(MM)	5.8.63	II
20.	Hendrik Verwoerd, AFRICA	30.63S	25.78E	CDA	55	60	5000	2	71	III
21.	Hoover, USA	36.13N	114.43W	CAG	191	118	36703	5.0	10.3.40	II
21A	Idukki, INDIA	9.83N	76.97E	CA	160	30	1460	3.5	2.7.77 1.12.78	II
22.	Itezhitezhi, ZAMBIA	15.79S	26.07E	RF	62	45	5000	4.2	13.5.78	II
23.	Jocassee, USA	34.98N	82.94W	RF	107	18	1431	3.2	25.11.75	II
24.	Kamafusa, JAPAN	38.15N	140.50E	CG	42.3	-	45	3	-	III
25.	Kariba, ZAMBIA/RHODESIA	16.93S	27.93E	DK CA	122	260	175000	6.25	23.9.63	I
26.	Kastraki, GREECE	38.67N	21.70E	EF	91.2	-	1	4.6	-	II
27.	Keban, TURKEY	38.82N	39.33E	CG/RF	182	17	31000	3.5	74	II
28.	Kerr, USA	47.89N	114.11W	CA	54	40	1505	4.9	28.7.71	II
29.	Kinnersani, INDIA	17.68N	80.67E	-	61.75	-	-	5.3	13.4.69	II
30.	Koyna, INDIA	17.62N	73.76E	CG	100	33	2780	7.0	10.12.67	I
31.	Kremasta, GREECE	38.90N	21.53E	EF	120	28	4750	6.3	24.1.66	I
32.	Kurobe, JAPAN	36.53N	137.65E	CA	180	-	149	4.9	19.8.61	II
33.	La Cohilla, SPAIN	43.12N	4.54W	A	98	4	12	-	75	III
34.	La Fuensanta, SPAIN	38.38N	2.23W	G	74.7	11	235	-	73	III
35.	Mangalam, INDIA	10.63N	76.52E	EF	28.5	-	25	3	63	III
36.	Mangla, PAKISTAN	33.22N	73.68E	EF	104.4	35	7250	3.6	28.5.67	II
37.	Manicouagan 3, CANADA	50.11N	68.65W	EF	96	68	10423	4.1	23.10.75	II
38.	Marathon, GREECE	38.18N	23.90E	CG	60.3	5	41	5.75	20.7.38	II
39.	Mica, CANADA	52.07N	118.30W	RF	191	200	31300	4.1	5.1.74	II
40.	Monteynard, FRANCE	44.90N	5.70E	CA	125	9	275	4.9	25.4.63	II

1	2	3		4	5	6	7	8	9	10
41.	Mula, INDIA	19.37N	74.62E	EF	44	-	1017	1.0	72	III
42.	Nurek, USSR	(38.42N	69.27E)	EF	285.3	-	11000	4.5	27.11.72	II
43.	Oroville, USA	39.53N	121.43W	EF	204	30	4400	5.7	1.8.75	II
44.	Oued Fodda, ALGERIA	36.02N	1.60E	CG	83	6	225	3	5.33	III
45.	Palisades, USA	43.23N	111.12W	EF	67	26	1729	3.7	10.6.66	II
46.	Parambikulam, INDIA	10.38N	76.80E	G	66	8	504	3	63	III
47.	Piastra, ITALY	44.21N	7.21E	CG	83.7	2	13	4.4	7.4.66	II
48.	Pieve di Cadore, ITALY	46.45N	12.41E	CAG	98	8.6	69	V(MM)	13.1.60	II
49.	Porto Colombia, BRAZIL	20.12S	48.35W	EF/CG	50.35	30	1460	5.1	24.2.74	II
50.	Rocky Reach, USA	47.78N	120.17W	CG	53.1	56	802	-	-	-
51.	San Luis, USA	37.07N	121.13W	EF	104.4	9	2603	-	-	-
52.	Sanford, USA	35.63N	101.67W	EF	66.5	29	1736	-	-	-
53.	Schlegeis, AUSTRIA	47.07N	11.77E	CA	113	8	128	2	4.73	III
54.	Sefia Rud, IRAN	36.75N	49.37E	CG	80	35	1800	4.7	2.8.68	II
55.	Sharavathy, INDIA	14.10N	76.82E	E	38	-	-	2	-	III
56.	Shasta, USA	40.77N	122.30W	CAG	153	40	5615	3	-	III
57.	Sholayar, INDIA	10.31N	76.77E	CGE	59.4	8	154	2	-	III
58.	Talbingo, AUSTRALIA	35.72S	148.33E	ERF	142	18	935	3.5	6.1.73	II
59.	Ukai, INDIA	(21.25N	73.72E)	ERF/CG	63.55	-	8511	3	-	III
60.	Vajont, ITALY	46.27N	12.38E	CA	232	3.4	150	3	63	III
61.	Volta Grande, BRAZIL	20.14S	48.05W	EFCG	31.35	35	2300	5.1	24.2.74	II
62.	Vouglans, FRANCE	46.42N	5.68E	CA	112	6.5	605	4.5	21.6.71	II
63.	Warragamba, AUSTRALIA	33.97S	150.42E	G	104	52	2053	5.4	9.3.73	II
64.	(Xinfengjiang, CHINA) Hsinfengkiang, CHINA	23.78N	114.58E	G	80	40	13896	6.2	18.3.62	I

Bracketed locations are for centres of dams; other locations except those bracketed are for centres of reservoirs.

A : Arch; C : Concrete; D : Double
DK : Double curvature; E : Earth; EF : Earth fill
G : Gravity; H : Hollow; M : Multiple
RF : Rockfill; I : Intense seismicity ($M \geq 6$); II : Moderate seismicity ($3 < M \leq 5.9$)
III : Microseismicity ($M < 3$); MM : Modified Mercalli Scale Intensity

ACKNOWLEDGEMENT

The authors record with great appreciation the encouragement received from the Director, Central Water and Power Research Station, Pune, India during the preparation of the paper.

REFERENCES

1. PACKER D.R., CLUFF L.S., KNUEPFER P.L. and WITHERS R.L. Study of induced seismicity, Report, Woodward-Clyde Consultants, San Francisco, USA, 1979.
2. SIMPSON D.W. Seismicity changes associated with reservoir loading, Engineering Geology, Vol.10, 1976, 123-150.
3. GUHA S.K., GOSAVI P.D., VERMA M.M., AGARWAL S.P., PADALE J.G. and MARWADI S.C. Recent seismic disturbances in the Shivajisagar lake of the Koyna Hydroelectric Project Maharashtra, India, Central Water and Power Research Station, Khadakwasla, Poona (India), 1970, Report.
4. MORRISON P.W. Jr., STUMP B.W. and UHRHAMMER R. The Oroville earthquake sequence of August, 1975, Bulletin of the Seismological Society of America, Vol.66, 1976, 1065-1084.
5. LEBLANC G. and ANGLIN F. Induced seismicity at the Manic 3 reservoir, Quebec, Bulletin of the Seismological Society of America, Vol.68, 1978, 1469-1486.
6. MIAO-YUCH W., MAO-YUAN Y., YU-LIANG H., TZU-CHIANG L., YUN-TAI C., YEN C., and JUI F. Mechanism of the reservoir impounding earthquakes at Hsinfengkiang and a preliminary endeavour to discuss their cause, Engineering Geology, Vol.10, 1976, 331-351.
7. RIKITAKE T. Earthquake Prediction, Elsevier, 1976.
8. DRAKOPOULOS J. Characteristic parameters of fore- and aftershock sequences in the region of Greece, Dissertation, 1968, 1-29.

Session 4: Discussion

ENVIRONMENT AND RISK INCLUDING INDUCED SEISMICITY

Chonggang Shen (Session Chairman; Secretary General, China National Committee on Large Dams). Earthquake-resistant design of large dams must take into account the environment, especially the risk due to induced seismicity by impounding water of the reservoirs. But the problems are quite different for people living either around the reservoir or downstream of the dam. Because of the direct influence of induced seismicity and the secondary damage due to earthquake, proper attention must be given to the inhabitants. For people living downstream of the dam the risks of failure or of serious damage to the dam are more apparent. Damage to the Lower San Fernando Dam in the USA, after the earthquake in 1971, caused the temporary evacuation of 80 000 inhabitants from downstream due to the danger of failure of the embankment.

In Session 4 there are four papers concerning this topic: Papers 28, 29, 32 and 33. In other Sessions we can find some material dealing with this subject: Papers 2, 16 and 22. In recent years more attention has been paid to the research work on induced seismicity by impounding water of reservoirs. This is because more high dams and more large reservoirs are being operated. As some of the Authors mentioned, although the induced earthquake is not so severe (magnitude less than 4.5) and at present the percentage of occurrence is rather low (less than 7%), for the security of dams we must pay due consideration to this question and check their safety.

Every Author agreed that the mechanism of induced seismicity has not been solved yet. Due to the difficulty of researching the tectonic structures and the stresses of deep strata as well as the characteristics of rock mass, many factors are still unknown. Our investigations are usually based on the geophysical method, because deep drillings are generally too expensive. It is true that the tectonic aspects of the earth's surface may not coincide with those of deep strata. Therefore research on induced seismicity has a more or less speculative character. But I would like to stress the interest of Paper 32. The Authors attempt to explain and predict induced seismicity by the analysis of geological aspects. They also provide a good example. Perhaps this is one possible way to solve this complex problem.

To consider induced seismicity in the design of large dams it is very important to ascertain the real tectonic earthquake. Sometimes during the impounding of water in a reservoir, local stone falls or slides may also induce earthquake, especially in karstic regions; however, this type of earthquake usually exerts no serious influence on the dam. Sometimes natural earthquakes can also appear. So we have to make the analysis and differentiation cautiously. Both the depth and volume of reservoir water has an influence on the induced seismicity. But from experience obtained in China, the area of reservoir water surface must also be considered, because the water may be in contact with some tectonic strata and different rock mass which may exert a great influence. For example, in the Hsinfenjiang reservoir the first induced earthquake occurred when the reservoir water was only 20m deep. In the Danjiankao reservoir, from 1959 to November 1967 (i.e. before impounding water), there were some small earthquakes. From the beginning of impounding water to 1970 there was a series of 20 earthquakes, but not concentrated in one region; after November 1971, there were two concentrated regions of earthquake with magnitudes of 3.5 (April 1972). Another new example is Hunanzhen reservoir. This dam was completed, and the impounding of water began, in January 1979. On 17 June the induced seismicity was felt by the local people when water depth was 20m.

All the Authors share the same correct opinion that a set of earthquake recorders must be established before construction of the dam. In addition we must gain knowledge of the tectonic structure in the deep strata and investigate the inner stresses in the rock. From the comparison of the change of geophysics and stresses it may be possible in the future to speculate about the intensity of induced seismicity.

Of course, the mechanism of induced seismicity is a combination of different actions. My Chinese colleagues have emphasized the effect of water. Water filtered into the deep strata of rock, with the consequent influences such as changes of pore pressure, friction in the fault and the strength of the rock. But, to our regret, the research of change in the hydrogeological conditions is far from sufficient. We hope that detailed investigation will be made on this aspect. For the proper selection of the designed seismic acceleration for the dam Mr Londe, in the first Session, mentioned that for important reservoirs the risk can influence very seriously peoples' lives and property. It is preferable to consider additional acceleration in checking the stability and strength of dams. Usually this consideration will cause no, or only very little, appreciable increase in the volume or cost of a dam.

It may also be considered that induced seismicity always begins from low intensities; the water level is also rather low. If earthquake phenomena occurred, the water level of a reservoir could be lowered, allowing sufficient time for the necessary investigations and predictions of the maximum possible intensity, and for making the necessary strengthening works for the dam, as we have done in Hsinfenjiang Dam.

Some points which are worth discussing in this Session, and which are covered by the Authors are

(1) How to organize the monitoring systems to observe the behaviour of a dam during strong earthquake. How to develop simplified methods to compare with the natural conditions.

(2) What investigation and research work is necessary for the prediction of earthquake intensity caused by induced seismicity?

(3) What earthquake acceleration should be adopted for the sufficient safety of a dam, especially for the additional acceleration by induced seismicity?

(4) What is the mechanism of induced seismicity? How do we evaluate properly the role of reservoir water in induced earthquake?

Professor D.Resendiz (introducing Paper 30). During the first Session of this Conference, Professor Londe in his opening speech suggested the existence of a missing link in the development of seismic design methods for dams; namely data on the performance of actual structures. I fully agree with his obervation.

The first thing to consider when looking for information on the seismic performance of embankment dams is the definition of possible damage modes, as it determines the specific date that we should look for in the field. Of course, earthquake damage modes are many: longitudinal and transverse cracking, slumping, settlement, sliding, etc. But experience shows that they are not independent modes.

On shaking, the embankment distorts so that the slopes' profile becomes S-shaped, extension develops in the crest and in the bulging portions of the slopes, and it gives rise to longitudinal cracks, etc. Settlement occurs because of the spreading or slumping of the embankment, and because the dam's height is not the same for different cross sections, differential settlements ensue, giving rise to extension zones along the crest, which in turn result in transverse cracking. Furthermore, the sliding mode of damage can be viewed as the extreme case of the slumping mode, or in other terms, the difference between slumping and sliding is one of degree. Thus most of the earthquake damage modes are actually related to each other and can be traced to the shear distortion of the embankment, which therefore could be called the basic damage mode.

Professor Londe also pointed out that earthquake analysis of embankment dams is still speculative, and I agree. The scarcity of quantitative data on field performance determines that we have not been able to test many of the hypotheses that underlie the various methods of analysis that have been proposed.

In our Paper and elsewhere (Performance of El Infiernillo and La Villita Dams in Mexico, including the earthquake of March 14, 1979. Comision Federal de Electricidad, Mexico, 1980, 146 pp.) data are provided which might eventually help to build up the missing link referred to above. The data are used in the paper to calibrate some of the existing methods of analysis of the shear-distortion damage mode; yet it is important to remember that the main value of publishing field observations is that anybody can then analyse them to reach their own conclusions.

Professor T.A. Paskalov (Paper 31). I know that after completion of the El Infiernillo Dam the performance of full scale testing was retained; and I know that some vibration tests were performed on the second dam. Were the periods obtained during these tests compared with those which the dam shows, just elaborating the data from the records we have obtained?

Professor D. Resendiz. In the case of El Infiernillo Dam, vibration tests were in fact made, and resonance frequencies were obtained for some of the vibration modes of the embankment. However, comparison of those frequencies to features of the response spectrum corresponding to earthquake excitations is not straight forward. Analyses are being made of the various sets of dynamic response data available for each dam, and the vibration test results will be included.

Mr R.G.T. Lane (Sir Alexander Gibb and Partners). Regarding Professor Resendez' presentation, data on the behaviour of El Infiernillo Dam during construction and impounding has been given elsewhere (Behaviour of dams built in Mexico. Contribution to the 12th International Congress on Large Dams, Mexico, 1976). Under its own weight (considering the top 100m of the dam) the dam consolidated at 5% strain at the depth of 100m. That gives one measure of behaviour of material in the dam. By integration one finds that the average consolidation over 100m height is of the order of 250cm. Over this same 100m height, the average peak acceleration during the earthquake was 20% of g, and the duration of the earthquake was 66s. These figures are quoted from the paper. This vertical acceleration acts as a compactor of the material in the dam, and as a result of a quick mental calculation I find that, due entirely to the vertical acceleration during the earthquake, the top of the dam should have gone down about 12.5cm. It did. Is it really necessary to consider slip surfaces at all?

Professor D. Resendiz. Compaction due to vertical acceleration may contribute to permanent deformation of a rockfill dam during an earthquake. However, the shear stresses induced by the same earthquake give rise to dam distortion and thus to crest settlement even if clear-cut sliding surfaces do not develop. This was the point the paper was intended to make.

On the other hand, instrumentation of both El Infiernillo and La Villita dams is such that volume change of the embankment cannot be determined beyond doubt. (In fact, if the dams were to be instrumented today, we would do it differently for this very purpose). Thus a definite, separate quantification of volume change and shear distortion as sources of crest settlement has not been made as yet, although we are trying to accomplish it. Two ways will be followed: (1) fully analyzing the available information corresponding to past earthquakes; (2) performing more complete measurements during future earthquakes. In the mean time, the recorded behaviour is subject to more than one interpretation. Mr Lane's is one; ours is another, and there might be additional ones. Yet the behaviour records that we reported have a value of their own, as I pointed out in my presentation.

Mr E.T. Haws (introducing Paper 28). Table 1 of Paper 28 shows the environmental factors required for seismic design, and it synthesizes published international practice.

There is a common assumption that induced seismicity will not be greater than the local maximum credible natural earthquake. However, the inferred suggestion that induced seismicity is simply an acceleration in time of an event which would in any case occur must be treated with caution, particularly for events of moderate return period. It is conceivable that pore pressure changes could give an effect of lubrication on faults. A larger area of fault being activated could then lead to perhaps a higher induced, than natural, earthquake of moderate return period.

Table 1 also includes the possibility of major earth or rock slides in the reservoir area. Highly significant results have been quoted by Professor Seed of accelerations as high as 1.0g and 1.2g at high levels in the rock gorge at Pacoima. The Vaiont case was a severe lesson; clearly such very high accelerations as are quoted, acting on marginally stable reservoir banks, could lead to bad slips and possible overtopping problems. In such cases reservoir area seismicity requires close attention.

Other summary Tables presented in Paper 28 are based on very considerable literature research of 80 cases of dams damaged by earthquake. The complete tables are available from the Authors.

Study of these case histories showed comparatively few instances of significant damage to concrete dams. This is due only partly to the numerical predominance of earth dams. There may well be a sufficient factor of safety in the properties of normal dam concrete, related to transient dynamic stresses, for damage generally to be avoided. A contributor indicated some sophisticated arch dam analysis results, the end product being contours of concrete tensile strength which were required to resist transient stresses. Such analyses are a step in the direction of better understanding of tolerance levels required, but a warning should be sounded about searching for single properties of concrete and varying concrete mixes throughout a dam on the basis of that single property. Clearly this is neither desirable nor the intention.

Mr A.C.Allen (Allen, Gordon & Co., Consulting Engineers). I wish to relate design to the lower seismicity of the UK, and in particular to the traditional centres of seismicity of Comrie and Menstrie in central Scotland. In this room nearly 30 years ago I referred to Lednock Dam, a concrete buttress dam 40m high, 5 miles from Comrie, and probably the first dam in the UK to be designed for earthquake (Proc. Instn Civ. Engrs, Vol.1, No.3, May, 1952, p.290). The modest allowance was a horizontal acceleration of 7.5%g, incorporated statically, in structure and water load. This corresponded to an intensity of 8 on the Rossi-Forel scale then in use, and was adopted following a deterministic study of the seismicity records of the locality. Activity had occasionally been frequent, the most notable period being between 1839 and 1848 when a swarm of 320 shocks was experienced.

Currently there is a swarm of activity nearer to Menstrie, particularly around Glendevon, and it is fortunate that from about 12 years ago accurate records have been obtained on the Lownet instrumental system of the Global Seismology Unit of the Institute of Geological Sciences. In about ten years up to 1978, in an area of some 40km radius, there were no tremors of Richter magnitude 3 and above and only 7 of magnitude 2 and above. In the 18 months of the current swarm, in an area of 10km radius, on average over three have occurred every 6 months and over two in addition about every 2 months. Located focal depths are from 2 to 15km.

On the River Devon, Upper Glendevon Reservoir (1955) is immediately at the head of the Lower Glendevon Reservoir (1922) which is above Castlehill Reservoir (1978). The three Reservoirs are in cascade and there are built-up areas downstream. This is relevant to environment and risk. The centre of current activity is between the Upper (45m concrete gravity) and the Lower (35m earth) Glendevon Dams and some 4km to the north of them. Three low intensity epicentres are in a straight line between the two dams. Castlehill Dam, (40m concrete arch-gravity) is about 16km from the centre of activity. Slight structural damage in Glendevon, believed to be due to both historical and recent activity, is apparent locally in the vicinity of Lower Glendevon Dam only. It seems to be developing slowly while the low order of activity is continuing.

What is the significance of this position to the design of dams in the UK, as distinct from the maintenance and operation of them? My views are not markedly changed from what they were 30 years ago. A gatehouse that might have been constructed on the top of Lednock Dam then was sunk below road level. The associated intake structure was a shaft in rock and not a tower. In this, and similar cases, I would at present still favour a modest static allowance for earthquake, being not yet convinced of the appropriate dynamic analysis, but I am sure Professor Severn's methods of dynamic testing will help to establish a more relevant design method. In the case of Castlehill, a concrete arch gravity dam, and particularly with the known experience of Vaiont, to make no allowance for earthquake was a deliberate design decision. With the order of seismic activity concerned there does not appear to be any need to make special provisions in relation to the design of embankment dams, except to avoid features prone to damage. The 'slosh' mode of analysis demonstrated yesterday has relevance to understanding slow and delayed development of slight embankment movement that has been noted at Lower Glendevon Dam so far. From sensations experienced during minor earthquake I have no doubts regarding the significance of vertical accelerations.

Design is not only associated with future structures but it is necssary for improvement and remedial works at existing ones.

Dr P.L Willmore (Paper 24). I must confess to a slight uncertainty now as to the situation in central Scotland. I had thought, after a very brief conversation with Dr Burton of my staff who studied the situation, that there had been some rather substantial, rather conspicuous, very recent damage to the dam which was to be associated with the very small earthquakes which had taken place, the largest of them being of a magnitude of about 3.5 on the bodywave scale. The interesting point which we had made previously was that it had been an article of faith among the people who had talked about earthquakes in the Ochil Hills that the Ochil fault was the source of the earthquakes, but the Ochil fault fades away to the south under the coalfields, and the line of activity which has been active for the last eighteen months runs right through the reservoirs to which Mr Allen referred. That is all I know about it at the moment. I am sorry I have not got the actual plot of epicentres here but Dr. Burton is in the course of producing a paper which will describe the whole situation.

Turning now to my own Paper, I should like to emphasize the concept of perceptibility of earthquakes. Fig. 4 in the Paper shows that if you take the relationship between magnitude, intensity and distance and then compute the expectation of given levels of intensity as a function of magnitude, the expectation in respect of UK seismicity peaks at around magnitude 5 on the bodywave scale. For the eastern United States, the expectations peak in respect of a surface wave measure of mb at around magnitude 6, and in the Pacific they peak at around magnitude 7. You cannot, however, perform these calculations for linear magnitude frequency relations, because the curvature or truncation of the relation is the essential feature which brings down the perceptibility on the high magnitude side of the peak.

My second point is to emphasize the ease of acquiring and installing sophisticated networks. With modern portable equipment you can isolate vector components of earth motion and form cross-correlograms to recognize waveforms or transform the frequency content at will. Soon there will be no distinction between short-period and long-period instruments, or between sensitive and strong-motion types. Any 'good' instrument will produce a record covering the whole information content of the earth motion at the point of emplacement.

Finally, there have been a lot of comments on the difficulty of making adequate observations of actual deformations of structures in response to realistic inputs, and I should like to introduce the possibility of using medium range teleseisms (which are recorded several times a week in many parts of the world) for this purpose. Basically, you need three sorts of network, starting with a tripartite three-component enclosing the point of interest at a distance of several kilometres. From this you can extract bedrock motion at any point inside the triangle. Next you instrument the structure, so as to record the response to this known input. Thirdly, you can surround the stucture with a tight array of instruments which can, in principle, observe the radiation of energy from the structure back into its environment.

Dr B. Skipp (Paper 32). The problem which must often be faced in a region of apparent low seismicity is what effort and what methodology can really be justified in order to make certain statements, or to support the statements, regarding the potential for induced seismicity and the vibratory ground motion. It has been commented that review panels have become much more demanding over the years. This is characteristic not only of dams but also of other hazardous installations; and whereas two decades ago the opinion of a suitably qualified and learned geologist would be sufficient to satisfy both public opinion, represented by the journalists, and public opinion represented by politicians, this is no longer the case. Even the most elementary geological argument has to be formulated in a rather formal way.

Thus, on the matter of induced seismicity, one is faced with the problem of how to construct bricks without straw, because the procedures so far put forward are not scientifically substantiated. They owe much to those developed for nuclear power plants and in particular the approach which attempts to establish a maximum earthquake in a region has been used, often I think as a desperate measure to produce a figure where judgement would probably be better. Mr Haws (Paper 28) has pointed out correctly the dangers in this procedure. Examples where maximum magnitudes have been related to fault lengths, using globally derived relationships which have so much dispersion as to be almost useless, are not to be relied upon.

There is a more promising approach which involves the identification of what can be called current activity, not in a geological sense but as activity in the sense that there is some feature which is currently releasing low level seismic energy. We expect that one can estimate an appropriate earthquake from stress drop although this is a hope which is highly speculative. Then one has an approach which is nothing more or less than empiricism, which says that one can characterize sufficiently the geological and tectonic settings of particular sites so that with proper network, and with proper information from instrumentation over the years, enough information can be accumulated to set up some kind of pattern recognition procedure. This is probably the line to follow, because so far none of the other methods are credible.

The histograms shown in both our Paper and in other papers give some idea of what might be expected. A word of warning though; these histograms are not to be interpreted statistically. They are not proper samples, and I think they should be regarded with the same kind of caution that we regarded the data on peak accelerations in the early days of the collection of strong ground motion.

However, perhaps one should take a magnitude of 5, in or about the site, and look at the implications in terms of strong ground motion, characteristic of a particular strong motion which one thinks sensible for that region, but look more carefully at those situations where induced seismicity has been of an extreme variety to try to characterize the special features of such sites.

A speaker. It was asked how we persuade journalists on points at issue; but the important question is how we persuade the public. We must consider what we actually mean by induced seismicity, and whether this is the right term to use with the public, and what impressions are created by the term.

A more proper description than induced seismicity (which implies to the public that the reservoirs are the cause of the development of the earth tremors) would be the triggering of agents which are already present: these agents are present and are going to be activated at some time in the future whatever conditions prevail.

Mr D.A. Howells (Chairman, SECED). Dr Willmore (Paper 29) suggests that monitoring should start as early as possible: I support this strongly, and suggest that the moment any money is allocated for the first tentative feasibility study of a reservoir scheme, an adequate monitoring programme should be set up. I cannot give any advice about how to persuade the dam promoter to accept this. Reservoir schemes often go through a long study period and are often postponed for several years, but nonetheless are built in the end. If the monitoring is carried on through this doldrums period a convincingly long record may be built up.

This monitoring should be designed to serve geological purposes as well as purely seismological ones, and to clarify the structure at some depth around the reservoir. For this one needs good determinations of depth of focus and fault plane solutions. I imagine that for this purpose the instruments should not be clustered too closely around the dam site. Is this likely to add unreasonably to the cost of installation?

Dr P.L. Willmore. It does not add significantly to the cost once you have passed the threshold of using a telemeter system around your dam site. Then the range typically is determined by the size of the final reservoir you want to get out of that area.

Mr G. Rocke (Babtie Shaw and Morton, Glasgow). I should like to support Dr Willmore's plea for an agency to collect information. In the 1960s I was involved in the design and construction of two earth dams in Scotland both straddling the Highland Boundary fault, where there is seismic activity of a low order - of such a low order that it was decided that practically no damage could occur to the embankment dams. However, it was decided to put a provisional .05g pseudostatic horizontal force allowance into the structural element of the dams and the valve towers. That was in the early 1960s when earthquake effects were not attracting the attention they are today. My firm had taken advice at that time from the Seismology Unit in Edinburgh, and were assured that any disturbances would probably be of a small magnitude.

In the early 1970s I discussed with Dr Willmore the seismic history of the Kielder area in the north of England, in relation to the Kielder Water Scheme, and received from him an assessment of the cost of monitoring the reservoir and dam areas. Following the receipt of that advice it was decided not to monitor because of the low risk of seismic events to the embankment dam and also the high cost of monitoring.

When Kielder Dam was about 8m from its final crest level there was a seismic event, not connected with the Highland Boundary Fault but much closer to Kielder, on the border between England and Scotland. This event occurred in January 1980. The effects were felt between the Solway Firth and Glasgow, quite an extensive area. On reflection, the existence of some monitoring units at Kielder would have been instructive, and my firm are currently examining ways in which some simple instrumentation might yet be installed before impounding is significantly advanced in 1981.

A body is required within our profession to promote an organized approach to monitoring for its long term benefits to the nation. Kielder Dam will impound in 1981/82 and in about the same period so will Megget Dam in the south of Scotland. These two major works seem to offer a unique opportunity to embark on a swift course of action; perhaps ICOLD or a similar body might be prepared to take up this opportunity and explore the possibilities.

Closing address

MR R. G. T. LANE (Chairman of the Organizing Committee)
PROFESSOR H. BOLTON SEED (University of California, Berkeley)

Professor H. Bolton Seed. This has been an extremely interesting two days, and I shall try to put the problem of earthquake resistant design into some kind of perspective. It is true, as we heard at the outset of the programme, that relatively few dams have failed as a result of earthquake. It is important that the problem should not be blown out of proportion. Nevertheless, the few dams that have failed serve to show us that many others exist which are equally vulnerable. While future dams may be built to withstand earthquakes without difficulty, due to improved construction techniques, some of the dams that exist at the present time may not be able to meet the test if the need arises. It is to these dams, many built in another era, to which we should devote most of our attention. I suspect that with regard to these older structures the earthquake resistance design problem has for too long been treated as a rather trivial problem in design stages, and it is an enormous step forward that it should have been placed in its true perspective by the convening of congresses such as this and by the many excellent papers which have been presented during this two day meeting. I believe that as a result of this renewed interest in the earthquake resistant design problem we can look forward to an era of dam design which will provide both greater safety and economy than ever before. This Conference will contribute greatly to the mitigation of seismic hazards in dam design, and I am personally delighted to have had the opportunity to participate with you in this significant event.

Accordingly, on behalf of myself and all foreign delegates to this meeting, I would like to extend sincere thanks to the Institution of Civil Engineers, to the British Section of the International Commission on Large Dams, and to the Society for Earthquake and Civil Engineering Dynamics for providing us all with a most enjoyable, stimulating and thought-provoking programme over the last two days.

Mr R. G. T. Lane. This Conference has been successful in a number of ways, but particularly I would like to mention that we have gathered here together many disciplines. We have here, both as authors and as delegates, seimologists, geologists, geotechnical engineers, consulting civil engineers and academic civil engineers, and I think that getting so many people together, who so often think differently, is something of an achievement, and the more we do it the better.

I would also like to mention that although this is not an ICOLD meeting, the International Commission on Large Dams is very well represented. We have in fact not only the President of ICOLD, but we have the chairmen of four of their technical committees, that is, the Committees on Analysis and Design, Materials, Environment and Seismic Aspects. So here again we have been able to gather together people who have been engaged in this work actively for a long time.

It is quite evident from our discussions that in earthquake zones the engineer if faced with very special problems - and design must be considered in two stages. Firstly, the dam must be proportioned to fulfil its objectives and to fit into the valley. This requires a quick but effective way of accounting for earthquake hazard at the site. Secondly, a stress analysis must be carried out to verify the stability of the structure under all conditions of loading.

The design engineer has a responsibility which is extraordinarily great. If something happens, it is his responsibility, and to ignore putting into a design the very best and latest methods which are available is absolutely wrong.